Biopharmaceutical Informatics

Despite the phenomenal clinical success of antibody-based biopharmaceuticals in recent years, discovery and development of these novel biomedicines remains a costly, time-consuming, and risky endeavor with low probability of success. To bring better biomedicines to patients faster, we have come up with a strategic vision of Biopharmaceutical Informatics which calls for syncretic use of computation and experiment at all stages of biologic drug discovery and pre-clinical development cycles to improve probability of successful clinical outcomes. Biopharmaceutical Informatics also encourages industry and academic scientists supporting various aspects of biotherapeutic drug discovery and development cycles to learn from our collective experiences of successes and, more importantly, failures. The insights gained from such learnings shall help us improve the rate of successful translation of drug discoveries into drug products available to clinicians and patients, reduce costs, and increase the speed of biologic drug discovery and development. Hopefully, the efficiencies gained from implementing such insights shall make novel biomedicines more affordable for patients.

This unique volume describes ways to invent and commercialize biomedicines more efficiently:

- Calls for digital transformation of biopharmaceutical industry by appropriately collecting, curating, and making available discovery and pre-clinical development project data using FAIR principles.
- Describes applications of artificial intelligence and machine learning (AIML) in discovery of antibodies in silico (DAbI) starting with antigen design, constructing inherently developable antibody libraries, finding hits, identifying lead candidates, and optimizing them.
- Details applications of AIML, physics-based computational design methods, and other bioinformatics tools in fields such as developability assessments, formulation and excipient design, analytical and bioprocess development, and pharmacology.
- Presents pharmacokinetics/pharmacodynamics (PK/PD) and Quantitative Systems Pharmacology (QSP) models for biopharmaceuticals.
- Describes uses of AIML in bispecific and multi-specific formats.

Dr Sandeep Kumar has also edited a collection of articles dedicated to this topic which can be found in the Taylor and Francis journal *mAbs*.

Biopharmaceutical Informatics

Learning to Discover Developable Biotherapeutics

Edited by

Sandeep Kumar and Andrew E. Nixon

CRC Press
Taylor & Francis Group
Boca Raton London New York

CRC Press is an imprint of the
Taylor & Francis Group, an **informa** business

Designed cover image: Sandeep Kumar

First edition published 2025
by CRC Press
2385 NW Executive Center Drive, Suite 320, Boca Raton FL 33431

and by CRC Press
4 Park Square, Milton Park, Abingdon, Oxon, OX14 4RN

CRC Press is an imprint of Taylor & Francis Group, LLC

ISBN: 978-1-032-29167-3 (hbk)
ISBN: 978-1-032-29168-0 (pbk)
ISBN: 978-1-003-30031-1 (ebk)

DOI: 10.1201/9781003300311

Typeset in Times
by codeMantra

Dedications

To the cause of globally equitable access to medicines and healthcare

Sandeep Kumar dedicates this book to memories of his mother.

Andrew E. Nixon dedicates this book to his wife and children for their on-going support.

Contents

Foreword

The unparalleled versatility of monoclonal antibodies as therapeutics has inspired scientists for decades. As early as the 1980s, antibody engineers aimed to create novel versions of antibodies that would be more potent and engage alternate biological pathways compared to antibodies that are naturally produced in humans. Now, some 40 years later, this aim has been achieved, and the creation of antibody therapeutics is currently industrialized into a global enterprise. The commercial clinical pipeline has grown from a few dozen in the 1980s to over 1300 by 2023, driven by technological advances, global distribution of relevant knowledge, substantial financial investments, and, critically, the successful approval and marketing of these biotherapeutic macromolecules. Current protein engineering methods now allow the creation of molecules with a wide range of shapes and sizes. Antibody therapeutics may be monospecific or they may engage two or more different antigens or different epitopes on the same antigens. They may be conjugated to a variety of other biologically active components, such as small molecule cytotoxic agents or steroids, interleukins, or non-antibody protein-binding domains. These therapeutics may be composed of only an antibody fragment that has been stabilized through protein engineering (e.g., single-chain variable fragments) or they may be small antibody domains derived from non-human species (e.g., VHH). Importantly, over 200 antibody therapeutics have been granted marketing approvals or are the subject of marketing applications undergoing review in at least one country (https://www.antibodysociety.org/antibody-therapeutics-product-data/).

Despite the advances of the past decades, the discovery and development of antibody therapeutics remain inefficient, costly, and time-consuming endeavors with a relatively low approval success rate. The field of biopharmaceutical informatics, and in general the application of machine learning and artificial intelligence to antibody discovery, holds great promise in its ability to reduce inefficiencies in the discovery process, which now includes a substantial amount of experimental work. Typically, antibody discovery programs involve the generation of hundreds or thousands of molecules that need to be evaluated for numerous desired properties (e.g. specificity, affinity, developability, pharmacology), followed by iterations to create derivatives with improved properties. The improved efficiencies inherent in the ability to design, select, and further engineer molecules entirely in silico thus hold great appeal for the biopharmaceutical industry. If a discovery process that includes in silico work yields a higher percentage of fit-for-purpose molecules, then approval success rates may increase, which would yield substantial reductions in development costs. Currently, at least two-thirds of commercially sponsored antibody therapeutics that enter clinical studies, which is the most expensive stage of development, are terminated due to issues with safety, efficacy, or business reasons.

Considering the potential of the field to transform antibody discovery and development, the publication of *Biopharmaceutical Informatics: Learning to Discover*

Developable Biotherapeutics is timely. Editors Sandeep Kumar and Andrew Nixon, as well as the authors who contributed chapters, are renowned experts in the field. The book provides comprehensive coverage of the applications of in silico methods to numerous aspects of antibody therapeutics discovery, including identification of targets, antibody design strategies, structure-function relationships, and developability. *Biopharmaceutical Informatics: Learning to Discover Developable Biotherapeutics* will be a valuable resource for scientists involved in antibody therapeutics discovery, biopharmaceutical executives interested in developing more efficient and effective discovery processes, and all those who seek to advance the current state of the art.

Janice M. Reichert, Ph.D.
Director of Business Intelligence, The Antibody Society, Inc.; Editor-in-Chief, mAbs.

Preface

May everyone be happy. May everyone be healthy.
May everyone see. May there be no sorrow or misery.

A prayer by ancient sages of Rigveda.

Eons ago, our ancestors dreamed of a perfect world. They desired for everyone to be happy and healthy. Greek philosophers described being healthy and happy as *eudaimonia* and wished it for everyone. Unfortunately, this ancient endeavor remains largely unrealized, despite tremendous progress in human development over the course of many civilizations and millennia. The recent COVID-19 pandemic is an example of serious challenges we must overcome to assure *eudaimonia* for everyone. To achieve and remain in *eudaimonia* requires that we continue to invent novel medicines to speedily address unmet medical needs as they emerge. However, novel medicines, particularly biologic medicines, are very expensive to discover, test, and make in large amounts. Moreover, biologic drug discovery and development projects take too long and fail too often, with manufacturers often passing on the costs to the payers – nations, governments, insurance companies, pharmacies, hospitals, and eventually to the patients and their families. Empirical processes rooted in experimental trial and error along with our incomplete understanding of biomolecular structure-function and dynamics, the patient's genetic background, physiology, and disease history are among the leading root causes that underpin the failure of drug discovery and development projects.

But there is hope! Arrival of the digital age is enabling scientists to go beyond the cardinal principles of scientific progress, namely, generating hypotheses and inferring from observations. Computation has enabled us to explore new realms of scientific research via modeling, simulation, machine learning, and artificial intelligence. Digital transformation of human civilization started in the early 1970s and has revolutionized our lives since then. Many industries, such as banking, commerce, marketing, utilities, manufacturing, travel, shopping, entertainment and so on, have all benefited from digital transformation. Perhaps the biopharmaceutical industry can also benefit via greater availability of experimental data, modeling, simulation, machine learning, and artificial intelligence. This realization led us to the holistic vision of *Biopharmaceutical Informatics* a few years ago. *Biopharmaceutical Informatics* calls for synergistic use of experimentation and computation to reduce empiricism inherent to the discovery and development of biotherapeutics. Reducing the empiricism shall accelerate as well as improve the productivity of biotherapeutic drug discovery and development cycles along with reducing the costs associated with them. This book describes our first attempt to capture the promise of *Biopharmaceutical Informatics* and the excitement around it. We hope that our efforts shall inspire readers to explore this field further by launching their own investigations.

We are grateful to all the chapter authors who took time out of their busy schedules to educate us on their pathbreaking research efforts aimed at making the discovery and development of biotherapeutics more efficient. Although most of the book focuses on therapeutic antibodies, the scientific concepts can be extended to other classes of biopharmaceuticals as well. This book wouldn't have been feasible without the unwavering support and encouragement from the publisher, particularly Hilary Lafoe, Sukirti Singh, Varalika Kathuria and Karthik Orukaimani.

We are also grateful to our families for all the support over time and patience during the editing of this book. Modernity often reinvigorates antiquity as newer generations rekindle our ancient endeavors. We hope this book will inspire readers to develop new concepts and technologies that make the most advanced medicines accessible to everyone.

To everyone involved with this book, thank you.

Sandeep Kumar, Ph.D.

Andrew E. Nixon, Ph.D.

About the Editors

Dr. Sandeep Kumar is currently a Distinguished Fellow (Executive Director) at the department of Computational Science in Moderna Therapeutics, Cambridge, MA where he leads Molecular Design and Modeling team. Sandeep Kumar holds a Ph.D. in Computational Biophysics and has over 25 years of experience researching protein structure – Function relationships. Sandeep Kumar has so far contributed towards more than 100 research articles, reviews, book chapters, and has previously edited a book entitled "Developability of Biotherapeutics: Computational Approaches". Sandeep has been contributing towards discovery and development of numerous monoclonal antibodies, antibody drug conjugates, bispecific and multi-specific modalities, as well as vaccines. Based on the insights gained from these experiences, Sandeep has been advocating for Biopharmaceutical Informatics, a strategic vision dedicated to synergistic use of computation and experimentation towards a cost effective and more efficient discovery and development of Biotherapeutics. More recently, he is promoting the concept of DAbI (Discovery of Antibodies in silico) where he sees an opportunity for generative AI to not only accelerate biopharmaceutical drug design but also to expand the antigen space druggable by antibody-based biotherapeutics.

Dr. Andrew E. Nixon is currently the Senior Vice President & Global Head, Biotherapeutics Discovery at Boehringer Ingelheim Pharmaceuticals, Inc., Ridgefield, CT, USA. He earned his Ph.D. in Physical Biochemistry from the University of London for studies completed at the MRC's National Institute for Medical Research. He has over 20 years of experience in biologic drug discovery and has contributed to over 100 antibody discovery programs resulting in numerous clinical candidates and approved biologics, including TAKHZYRO, a fully human antibody inhibitor of plasma kallikrein.

Contributors

Rahmad Akbar
University of Oslo and Oslo University Hospital
Oslo, Norway

Alison Betts
Takeda Pharmaceuticals
Boston, Massachusetts

Adrian Carr
Large Molecules Research, Sanofi
Cambridge, Massachusetts

Charlotte M. Deane
Oxford Protein Informatics Group, Department of Statistics
University of Oxford
Oxford, United Kingdom

Venkata G. Dhara
Pfizer
Andover, Massachusetts

Paweł Dudzic
Natural Antibody
Szczecin, Poland

Andreas Evers
Antibody Discovery & Protein Engineering
Merck Healthcare KGaA
Darmstadt, Germany

Tonya Frolov
LabGenius Ltd
London, United Kingdom

Fernando Garces
BioMap
Palo Alto, California

Victor Greiff
University of Oslo and Oslo University Hospital
Oslo, Norway

Per Jr. Greisen
BioMap
Palo Alto, California

M. Michael Gromiha
Protein Bioinformatics Lab, Department of Biotechnology, Bhupat and Jyoti Mehta School of Biosciences
Indian Institute of Technology
Chennai, India
and
India and International Research Frontiers Initiative, School of Computing
Tokyo Institute of Technology
Yokohama, Japan

Tushar Jain
Adimab LLC
Lebanon, New Hampshire

Alexander Jung
Boehringer Ingelheim Pharma GmbH & Co. KG
Global Innovation and Alliance Management
Biberach, Germany

Venkata K. Kowthavarapu
University of Florida
Orlando, Florida

Eric Krauland
Adimab LLC
Lebanon, New Hampshire

Konrad Krawczyk
Natural Antibody
Szczecin, Poland

Sandeep Kumar
Biotherapeutics Discovery
Boehringer Ingelheim Inc.
Ridgefield, Connecticut

Daisuke Kuroda
Research Center of Drug and Vaccine Development
National Institute of Infectious Diseases
Tokyo, Japan

Shipra Malhotra
Takeda Oncology
Cambridge, Massachusetts

Hardik Mody
Genentech
San Francisco, California

Daniel A. Nissley
Oxford Protein Informatics Group, Department of Statistics
University of Oxford
Oxford, United Kingdom

Andrew E. Nixon
Biotherapeutics Discovery
Boehringer Ingelheim Pharmaceuticals, Inc.
Ridgefield, Connecticut

Ziwei Pang
BioMap
Beijing, China

Ponraj Prabakaran
Large Molecules Research
Sanofi
Cambridge, Massachusetts

R. Prabakaran
Protein Bioinformatics Lab, Department of Biotechnology
Bhupat and Jyoti Mehta School of Biosciences
Indian Institute of Technology
Chennai, India
and
India and Emory University
Atlanta, Georgia

Bianka Prinz
Adimab LLC
Lebanon, New Hampshire

Yu Qiu
Large Molecules Research
Sanofi
Cambridge, Massachusetts

Puneet Rawat
University of Oslo and Oslo University Hospital
Oslo, Norway
and
Protein Bioinformatics Lab, Department of Biotechnology
Bhupat and Jyoti Mehta School of Biosciences
Indian Institute of Technology
Chennai, India

Matthew I. J. Raybould
Oxford Protein Informatics Group, Department of Statistics
University of Oxford
Oxford, United Kingdom

Anahita Rouyan
Natural Antibody
Szczecin, Poland

Tadeusz Satława
Natural Antibody
Szczecin, Poland

Melody Shahsavarian
Large Molecules Research
Sanofi
Cambridge, Massachusetts

Divya Sharma
Protein Bioinformatics Lab, Department of Biotechnology
Bhupat and Jyoti Mehta School of Biosciences
Indian Institute of Technology
Chennai, India

Amrinder Singh
MORGI, Centre for Cancer Cell Reprogramming
Institute of Clinical Medicine
and
Department of Molecular Cell Biology
Institute of Cancer Research, Oslo University Hospital
University of Oslo
Oslo, Norway

Eva Smorodina
University of Oslo and Oslo University Hospital
Oslo, Norway

Vanita D. Sood
Fable Therapeutics
Boston, Massachusetts

Madhuresh Sumit
Genomic Medicine CMC
Sanofi
Great Boston, Massachusetts

Maximiliano Vásquez
GENEART Laboratory
Regensburg, Germany

Jack Wade
Technical University of Denmark
Kongens Lyngby, Denmark

Thomas Watkins
Large Molecules Research
Sanofi
Cambridge, Massachusetts

Maria Wendt
Large Molecules Research
Sanofi
Cambridge, Massachusetts

Trevor Wilkinson
AstraZeneca
Cambridge, United Kingdom

Wiktoria Wilman
Natural Antibody
Szczecin, Poland

Leonard Wossnig
University College
London, United Kingdom

Sonia Wróbel
Natural Antibody
Szczecin, Poland

Biopharmaceutical Informatics

1

An Introduction

Andrew E. Nixon and Sandeep Kumar

We have just lived through an incredible period in human history where our ability to deliver novel medicines to solve the global pandemic caused by coronavirus disease (COVID-19) changed long-held paradigms for discovery and development and opened the door for new therapeutic modalities. While the success of the COVID-19 treatments shows us what we as an industry can achieve when we focus our efforts and attention on a single problem, the need for safe and effective treatments and even better, cures, across the spectrum of diseases that we face remains a significant unmet challenge. With each therapeutic candidate that reaches patients, our understanding of the biology of disease is increased and with that understanding comes additional ideas for how we may approach cures. Our ability to leverage modern tools to shorten the time from insight to therapeutic candidate, and ultimately to cure, is the ambition that is encapsulated in this book.

Every year, scientists discover numerous novel antibody-based drug candidates to address a broad range of unmet medical needs. However, few will overcome the hurdles in their product and clinical development to win regulatory approval and make medicines available in the market.[1–7] By comparison with small molecules, it usually takes longer to discover and develop biotherapeutics with the associated greater consumption of both human and capital resources, although the probability of successfully reaching the market for a biologic is typically higher than for a small molecule drug.[1] There is clearly significant room for improvement in the overall probability of success from the Phase 1 clinical trials to launch, which in part will come from a deeper understanding of disease biology, a topic that is largely out of

DOI: 10.1201/9781003300311-1

scope for this book, and in part from the way that we approach drug discovery and development in the biopharmaceutical industry.

We would argue that syncretic use of computation and experimentation in biotherapeutic drug discovery and development, or biopharmaceutical informatics[8–10] (also see https://www.tandfonline.com/journals/kmab20/collections/biopharmaceutical-informatics for further reading), is inevitable for us to address the challenges we face in drug discovery and development, both currently and in the future. Biopharmaceutical informatics encourages laboratory automation and collection of experimental data under standardized conditions across all drug discovery and development projects. Access to such data can enable scientists to connect the intrinsic molecular properties of the drug candidates to their experimental behavior, thereby allowing for the design of antibody drug candidates with greater fit to the drug product development and clinical development platforms. This virtuous cycle between experimentation and computation offers the potential to reduce the time to iterate through design-build-test loops and to reduce the number of experiments needed to do so. The vision of biopharmaceutical informatics (see Figure 1.1) includes several efforts such as the de novo discovery of potential antibody binders to a given antigen/epitope using the computational means; obtaining a detailed understanding of antibody structure-function as well as developability property relationships both computationally and experimentally; intent as well as ability

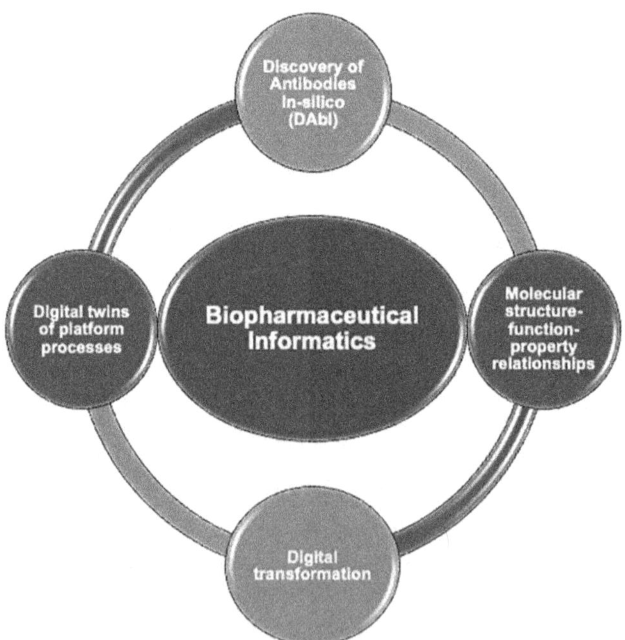

FIGURE 1.1 Strategic vision of biopharmaceutical informatics calls for syncretic use of computation and experimentation to bring affordable biotherapeutics to patients in a timely manner by helping reduce empiricism inherent to biotherapeutic drug discovery and development cycles and by accelerating drug discovery and development processes as well as improving the overall success rate of these projects.

to drive digital transformation by collecting experimental data using the standardized conditions, curating and sharing it within a biopharmaceutical organization using FAIR (findable, accessible, interoperable, and reusable) data management principles[11,12] so it can be used by the digital scientists to help improve drug discovery and development workflows as well as platforms. Many of these efforts are already ongoing within the biopharmaceutical industry and in the pharmaceutical sciences departments of the leading institutions. Ultimately, biopharmaceutical informatics strives to reduce empiricism inherent to the discovery and development of biotherapeutic drug products and help deliver them to patients rapidly by accelerating drug discovery and development processes as well as improving the overall success rate of such projects. Note that our vision of biopharmaceutical informatics does not call for the elimination of experimentation from biotherapeutic drug discovery and development cycles. On the contrary, the experiments are an integral part of biopharmaceutical informatics. However, it does call for reducing the empiricism and therefore wasteful effort inherent in the "trial and error" mindset. This directly translates into making the life of bench scientists easier because they would need to perform fewer experiments per project. Furthermore, biopharmaceutical informatics is about moving away from the "computation *versus* experimentation" debate to the "computation *and* experimentation" mindset in the interest of patients waiting for effective as well as affordable novel biotherapeutics.

The public availability of large well-curated data sets of antibody sequences[13–16] provides stepping stones for moving toward the in silico generation of antibody libraries of inherently developable molecules and toward antibody discovery using computational means. We are among the pioneers in computational developability assessments of biotherapeutics and have previously edited a book entitled *Developability of Biotherapeutics: Computational Approaches* in 2015. Our concept of "developability" is much broader than "manufacturability" and includes safety, efficacy, and pharmacology along with the mechanism of drug action and risks of target-mediated toxicity or other issues such as undesired immunogenicity. Therefore, a developable antibody-based biotherapeutic should possess good sequence–structural and physicochemical properties that assure desirable levels of target affinity and specificity, conformational and colloidal stabilities, high molecular integrity (low drug product heterogeneity due to physicochemical degradation such as deamidation, fragmentation, isomerization, oxidation, and aggregation), low off-target binding and poly-reactivity, flexible drug product presentations, clinical safety, efficacy, good pharmacology, and increased probability of winning approval from the regulatory agencies.

The ability to build and screen antibody sequence libraries in silico will require a revolution of the current antibody-based drug discovery paradigm. First, we need to be able to generate antibodies in silico with defined physicochemical properties, rather than searching for developable hits among the antibody libraries generated via animal immunizations, hybridomas, or display libraries.[17] With the accumulation of billions of antibody sequences and repertoires in the public domain, it has become feasible to generate novel antibody sequences via artificial intelligence and machine learning (AIML). Second, we need to pivot workflows and mindset from "function first, then developability" to "developability first, then function" by constructing large antibody sequence libraries with desirable developability attributes and screening them for antigen binding.

For example, developing an AIML-based platform for generating antibody sequence libraries whose sequence and physicochemical attributes resemble those of the clinical stage or even better the marketed antibodies and using them for drug discovery can be very useful.[18–23]

Greater use of computation can also potentially expand the target space druggable by the biotherapeutics, by including the targets that are difficult to express and purify in the laboratory. The ability to expand the druggable antigen space via de novo computational design for difficult targets also offers opportunities to upgrade the role of computation as an initiator of drug discovery projects rather than play an assistive role in supporting projects initiated and driven primarily via experimental trial and error.

Recently, we proposed a conceptual roadmap to the discovery of antibodies in silico (DAbI).[21] This roadmap, shown in **Figure 1.2**, identifies three areas for technological milestones before one can fully realize the potential of DAbI. These milestones are:

1. **In silico generation of antigen-agnostic/independent and antigen-specific antibody libraries.** We need the ability to generate human antibody sequence libraries in silico, both in antigen-specific and in antigen-independent manner.

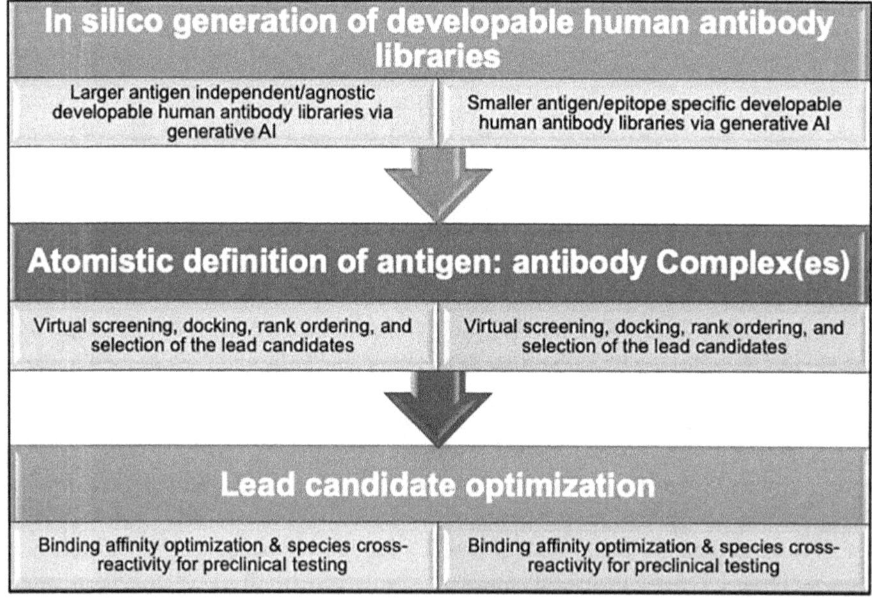

FIGURE 1.2 A conceptual workflow for the discovery of antibodies in silico (DAbI). Note that this roadmap is applicable to monospecific formats such as full-length antibodies, single domain antibodies, and single chain variable fragments. In case of multispecific antibodies, an additional milestone of computationally evaluating the compatibility of the physicochemical properties of the individual binders with one another and with the intended final molecule format will need to be developed.

The universe of antibody sequences is vast and much bigger than that can be explored using the display libraries or sequencing immune repertoires from immunized animals or human plasma samples. In silico generation of antibody sequences using quantum as well as classical computing can help us explore greater extents of this universe and build more diverse sequence libraries than feasible, experimentally. Moreover, unlike the classical antibody generation methods, the computationally generated antibody libraries can also be *a priori* endowed with desirable intrinsic physicochemical attributes important for their developability.

2. **Rapid virtual screening of in silico-generated antibody libraries against a given antigen or epitope.** This requires an ability to generate realistic models of antigen-antibody complexes in atomistic details. This should be seen in the context of rapidly screening large in silico-generated antibody libraries for potential binders for a given antigen or epitope. While docking methods have been traditionally used for predicting antigen-antibody complexes, the compute times required can become impractical if used as screening tools. Therefore, there is a need for adopting more rapid virtual screening tools for use with the antibodies. These abilities are needed for screening the in silico-generated antibodies for finding hits, assessing them, and identifying potential lead molecules.
3. **Reliable prediction of change in free energies of binding upon mutations.** Once we have selected a lead candidate, we need the ability to accurately predict and manipulate the affinity of the antibody toward the antigen to support its biological functional requirements. Furthermore, we should also be able to impart cross-reactivity to animal models for preclinical testing and validation of the lead molecule(s). This can be achieved by optimizing lead drug candidate(s) via single point or combinatorial mutations at the interfacial regions of the antigen-antibody complex(es) for appropriate affinity and species cross-reactivity.

These above-mentioned DAbI milestones do not account for novel antibody formats such as bispecific and multispecific biotherapeutics. In such cases, we need to augment DAbI with a fourth milestone. This milestone shall consist of our ability to combine all the different specific binders into the desired overall molecule design without losing on the developability as well as the potency of the individual binders.

Table 1.1 compares the DAbI with classical methods of biotherapeutic drug discovery. Most of the technologies required to enable DAbI already exist, at least in their nascent forms, and can be adapted rapidly. While DAbI is an attractive new approach to antibody discovery, there are still several practical advantages to the classical methods of antibody generation and their use in biotherapeutic discovery. Therefore, rather than calling for a complete switchover of the drug discovery to DAbI, it is currently more pragmatic to consider it as another arrow in the quiver of the drug hunters interested in discovering and developing novel biotherapeutics.

In this book, we offer potential technological solutions to address the challenges we face in biologic drug discovery and development. These solutions are rooted in the vision

TABLE 1.1 Comparison of classical antibody discovery and discovery of antibodies in silico(DAbI)

	CLASSICAL ANTIBODY DISCOVERY	*DAbI*
Track record	Highly successful track record with nearly 200 approved biotherapeutics all over the world since the 1990s.	Emerging technology. Not many examples yet.
Average timelines	1–2 years for discovery of hits that bind well-behaved targets; additional times up to a year for lead identification, humanization, affinity, and developability optimizations. 3+ years for finding initial hits for challenging targets.	No reliable estimates yet. But could vary from a few weeks to a couple of months for initial hits/ designs for experimental screening. Additional time may be required for computationally driven lead identification and optimization followed by experimental testing.
Resources and costs	Highly resource and cost intensive.	High initial investments to develop computer infrastructure and develop a digital platform, but resource and cost-effective in the longer term.
Developability	During classical discovery, the focus is on function. The hits and leads may encounter significant developability roadblocks during product and clinical development phases, thereby raising the costs of drug development.	One can pre-pay for developability while generating antigen-agnostic or antigen-specific antibody libraries and therefore lower the risk of late-stage failures.
Druggable antigen space	Suitable for antigens that express well in the lab under in vitro conditions because immunization experiments can require several milligrams of the antigen material. Otherwise, there will be significant hurdles.	Expanded druggable antigen space because no in vitro expression of antigens for immunization is required. Suitable for challenging targets such as GPCRs, claudins, intrinsically disordered proteins, and other low-solubility proteins
Modalities served	Very versatile, including Variable Heavy-chain domain of Heavy-chain-only antibodies (VHHs), domain antibodies, multispecifics, Antibody Drug Conjugates (ADCs), and so on.	Depending upon the availability of sequence–structural data, DAbI platform can be, in principle, used for all modalities. However, currently, scFvs, Fabs, and monospecific antibodies may be more amenable to DAbI.
Core dependencies and processes	Dependent on animals, hybridoma, or display technologies for antibody generation. Empirical process based on experiences gained from trial and error.	Does not require animals, hybridoma, or display libraries for antibody generation but may require experimental validation of the designs.

of biopharmaceutical informatics and address different aspects of antibody-based drug discovery and development cycles. The book begins by shedding light on the urgency of digital transformation in the biopharmaceutical industry. In the second chapter, Frolov et al. describe how digital transformation can fundamentally change the way biotherapeutics are discovered and developed and the opportunities afforded by making this change. The challenges associated with the digital transformation are also discussed in detail, along with calling for cultural change in drug discovery and development mindset.

The first step in a drug discovery project is to raise antibodies against a given antigen. This requires producing enough quantities of antigen (typically, a few to several milligrams) to enable animal immunizations, hybridoma cultures, or panning of display libraries and for functional validation of the antibodies raised against the targets. In case of an antigen that has good stability and solubility under in vitro conditions, then it is relatively easy to raise antibodies against it. An example of such an antigen is hen egg white lysozyme. Antibodies that bind it were instrumental in laying down the foundations of structural and computational Immunology in the 1990s and early 2000s.[24–30] However, many targets of medicinal interest such as G-Protein Coupled Receptors (GPCRs), claudins, and Wnts can be quite unstable as well as insoluble outside of their cellular environments and therefore raising antibodies against them poses numerous challenges. In the third chapter of this book, Wilkinson describes how computational tools can help optimize such proteins and improve chances of successful antibody generation campaigns against them.

The therapeutic concept and therefore targets of interest need to be validated at several stages as drug discovery and development projects progress. Analyses of the adaptive immune receptor repertoires can be a powerful tool for not only target validation but also for novel drug discovery. In the fourth chapter, Prabakaran Ponraj and team describe the utility of analyzing immune repertoires via computational means for both target validation and novel drug discovery.

Recently, AIML have emerged as useful tools for biological research and are finding innovative applications in the discovery and development of novel biotherapeutics. Biopharmaceutical informatics envisions early adoption of such tools to accelerate discovery and de novo design. In the fifth and sixth chapters, the pioneers in this field, Konrad Krawcyzk and Daisuke Kuroda, share their enthusiasm around AIML by describing the importance of self-consistent datasets toward facilitating generative modeling and opportunities for the de novo design of the therapeutic antibodies.

Methods rooted in computational biophysics, molecular modeling, and simulations have a long and rich history of helping both small molecule and biotherapeutic drug discovery. Notwithstanding the current enthusiasm around machine learning technologies, physics-based approaches remain highly relevant to biotherapeutic drug discovery, as reflected in Chapters 7 and 8, coming from the established research groups of leading bioinformaticians, Michael Gromiha and Charlotte Deane. These chapters remind us of the fundamental need of gaining further understanding of antibody structure–dynamics and function relationships to biotherapeutic drug discovery.

Biotherapeutic drug candidates need to be both functional and developable. Developability refers to the translation of a biotherapeutic drug candidate into an approved biotherapeutic drug product available in the hands of doctors to treat patients with unmet medical needs. In fact, developability considerations have emerged over the past decade from being an afterthought to being an integral part of the drug discovery process. This integration of developability into the molecule discovery process has been facilitated by a combination of several factors which include the emergence of computational approaches to developability, improved throughput of biophysical assays, and greater open communications between the drug discovery and development organizations within the biopharmaceutical industry. Chapters 9 and 10 describe these considerations and how computation can help design more developable biotherapeutics. These chapters come from teams that have not only pioneered developability but also championed it over the years. Their efforts have also helped differentiate biotherapeutic drug discovery and development cycles from those developed earlier for small-molecule drugs.

The discovery and development of successful biotherapeutics requires another ingredient. Approaches rooted in systems biology are being increasingly used to discover and validate novel drug targets as well as understand the drug mechanism of action and pharmacological properties of the biotherapeutic drug products. Chapter 11 by Madhuresh Sumit and team describes how systems biological approaches have been useful toward target discovery as well as drug development, while Betts et al. in Chapter 12 focus on the development of quantitative systems pharmacology models for biotherapeutics.

Lastly, the clinical success of monospecific monoclonal antibodies has spurred creativity of the drug discovery scientists to bind multiple targets simultaneously by designing bispecific and multispecific antibody-based formats as well as antibody–drug conjugates. While these formats have been attractive from the medicinal perspective, developing them as drug products has been challenging. In 2023, several bispecific and multispecific antibody-based biotherapeutics gained approval from regulatory agencies.[31,32] Each of these products is a testimony to the hard work and perseverance of the discovery and development scientists working in the biopharmaceutical industry over the years. However, AIML and computational design are finally beginning to make inroads into the development of such formats. There could not have been a better chapter to close this book than that written by Greisen et al. (Chapter 13) to describe these exciting new developments.

Biotherapeutic drug discovery and development is a multifaceted endeavor. In this book, we have tried to cover many topics and explored the opportunities available for the increased role of computation at various stages of biotherapeutic drug discovery and development. However, our focus has been primarily on molecule discovery and early-stage drug development. Therefore, this book by no means provides a comprehensive coverage of all topics in this field. Several topics that remain unserved in this book include understanding disease biology, target identification, therapeutic concept elaboration, and definitions of the research target and target product profiles. Opportunities for AIML, data science, and molecular simulations in formulation design and excipient selection, cell line development, bioreactor and other manufacturing processes, late-stage scale-up and process validation of drug product development as well as clinical trial design are also not covered in this book. There is a myriad of exciting opportunities in these aspects as well!

This book has focused on antibody-based biotherapeutics because of their success in the clinic over the recent decades. However, the overall landscape of biotherapeutics is continuously emerging with the arrival of several new modalities, including bioconjugates, peptides, oligonucleotides, and proteins, especially phage lysins, replacement enzymes, cell and gene therapies, mRNA-based vaccines, and many more. These developments bring exciting new opportunities and avenues for applications of biopharmaceutical informatics. As the drug modalities, development processes, and delivery methods diversify, there is a need for a paradigm shift in drug discovery and development cycles toward greater inclusion of computation and to move away from being primarily dependent on experimental trial and error.

REFERENCES

1. Smietana, K., Siatkowski, M. & Møller, M. Trends in clinical success rates. *Nat. Rev. Drug Discov.* **15**, 379–380 (2016).
2. Smietana, K., Ekstrom, L., Jeffery, B. & Møller, M. Improving R&D productivity. *Nat. Rev. Drug Discov.* **14**, 455–456 (2015).
3. Shnaydman, V. Industry drug development portfolio forecasting: productivity, risk, innovation, sustainability. *J. Commer. Biotechnol.* **25**, 2, 15–26 (2020).
4. Schuhmacher, A., Wilisch, L., Kuss, M., Kandelbauer, A., Hinder, M. & Gassmann, O.. R&D efficiency of leading pharmaceutical companies – A 20-year analysis. *Drug Discov. Today* **26**, 1784–1789 (2021).
5. Smietana, K., Quigley, D., Vyver, B. V. de & Møller, M. The fragmentation of biopharmaceutical innovation. *Nat. Rev. Drug Discov.* **19**, 17–18 (2020).
6. Schuhmacher, A., Hinder, M., Stein, A. von S. und, Hartl, D. & Gassmann, O. Analysis of pharma R&D productivity – a new perspective needed. *Drug Discov. Today* **28**, 103726 (2023).
7. Farid, S. S., Baron, M., Stamatis, C., Nie, W. & Coffman, J. Benchmarking biopharmaceutical process development and manufacturing cost contributions to R&D. *mAbs* **12**, 1754999 (2020).
8. Developability of biotherapeutics, computational approaches, edited by Satish Singh and Sandeep Kumar. (2015). Taylor and Francis, Boca Raton. doi:10.1201/b19023. https://www.taylorfrancis.com/books/mono/10.1201/b19023/developability-biotherapeutics-satish-kumar-singh-sandeep-kumar.
9. Deane, C. M. & Vásquez, M. Developability of biotherapeutics: computational approaches. Edited by Sandeep Kumar and Satish K. Singh. *mAbs* **9**, 12–14 (2017).
10. Kumar, S., Plotnikov, N. V., Rouse, J. C. & Singh, S. K. Biopharmaceutical Informatics: supporting biologic drug development via molecular modelling and informatics. *J. Pharm. Pharmacol.* **70**, 595–608 (2018).
11. Wise, J., de Barron, A. G., Splendiani, A., Balali-Mood, B., Vasant, D., Little, E., Mellino, G., Harrow, I., Smith, I., Taubert, J. & van Bochove, K. Implementation and relevance of FAIR data principles in biopharmaceutical R&D. *Drug Discov. Today* **24**, 933–938 (2019).
12. Wilkinson, M. D., Dumontier, M., Aalbersberg, I. J., Appleton, G., Axton, M., Baak, A., Blomberg, N., Boiten, J. W., da Silva Santos, L. B., Bourne, P. E. & Bouwman, J. The FAIR Guiding Principles for scientific data management and stewardship. *Sci. Data* **3**, 160018 (2016).
13. Ferdous, S. & Martin, A. C. R. AbDb: antibody structure database—a database of PDB-derived antibody structures. *Database,* 1–9 (2018), doi:10.1093/database/bay040.

14. Rubelt, F., Busse, C. E., Bukhari, S. A. C., Bürckert, J. P., Mariotti-Ferrandiz, E., Cowell, L. G., Watson, C. T., Marthandan, N., Faison, W. J., Hershberg, U. & Laserson, U. Adaptive immune receptor repertoire community recommendations for sharing immune-repertoire sequencing data. *Nat. Immunol.* **18**, 1274–1278 (2017).
15. Kovaltsuk, A., Leem, J., Kelm, S., Snowden, J., Deane, C. M. & Krawczyk, K Observed antibody space: a resource for data mining next-generation sequencing of antibody repertoires. *J. Immunol.* **201**, 2502–2509 (2018).
16. Olsen, T. H., Boyles, F. & Deane, C. M. Observed antibody space: a diverse database of cleaned, annotated, and translated unpaired and paired antibody sequences. *Protein Sci.: A Publ. Protein Soc.* **31**, 141–146 (2022).
17. Gray, A., Bradbury, A. R., Knappik, A., Plückthun, A., Borrebaeck, C. A. & Dübel, S.. Animal-free alternatives and the antibody iceberg. *Nat. Biotechnol.* **38**, 1234–1239 (2020).
18. Raybould, M. I., Marks, C., Krawczyk, K., Taddese, B., Nowak, J., Lewis, A. P., Bujotzek, A., Shi, J. and Deane, C. M.. Five computational developability guidelines for therapeutic antibody profiling. *Proc. Natl. Acad. Sci.* **116**, 4025–4030 (2019).
19. Licari, G., Martin, K. P., Crames, M., Mozdzierz, J., Marlow, M. S., Karow-Zwick, A. R., Kumar, S. and Bauer, J. Embedding dynamics in intrinsic physicochemical profiles of market-stage antibody-based biotherapeutics. *Mol. Pharm.* **20**, 1096–1111 (2022).
20. Ahmed, L., Gupta, P., Martin, K. P., Scheer, J. M., Nixon, A. E. & Kumar, S. Intrinsic physicochemical profile of marketed antibody-based biotherapeutics. *Proc. Natl. Acad. Sci.* **118**, e2020577118 (2021).
21. Bauer, J., Rajagopal, N., Gupta, P., Gupta, P., Nixon, A. E. & Kumar, S. How can we discover developable antibody-based biotherapeutics? *Front. Mol. Biosci.* **10**, 1221626 (2023).
22. Pejchal, R., Cooper, A. B., Brown, M. E., Vásquez, M. & Krauland, E. M. Profiling the biophysical developability properties of common IgG1 Fc effector silencing variants. *Antibodies* **12**, 54 (2023).
23. Martin, K. P., Grimaldi, C., Grempler, R., Hansel, S. & Kumar, S. Trends in industrialization of biotherapeutics: a survey of product characteristics of 89 antibody-based biotherapeutics. *mAbs* **15**, 2191301 (2023).
24. Silverton, E. W., Padlan, E. A., Davies, D. R., Smith-Gill, S. & Potter, M. Crystalline monoclonal antibody fabs complexed to hen egg white lysozyme. *J. Mol. Biol.* **180**, 761–765 (1984).
25. Xavier, K. A. & Willson, R. C. Association and dissociation kinetics of anti-hen egg lysozyme monoclonal antibodies HyHEL-5 and HyHEL-10. *Biophys. J.* **74**, 2036–2045 (1998).
26. Shick, K. A., Xavier, K. A., Rajpal, A., Smith-Gill, S. J. & Willson, R. C. Association of the anti-hen egg lysozyme antibody HyHEL-5 with avian species variant and mutant lysozymes. *Biochim. Biophys. Acta (BBA) – Protein Struct. Mol. Enzym.* **1340**, 205–214 (1997).
27. Pons, J., Rajpal, A. & Kirsch, J. F. Energetic analysis of an antigen/antibody interface: alanine scanning mutagenesis and double mutant cycles on the hyhel-10/lysozyme interaction. *Protein Sci.* **8**, 958–968 (1999).
28. Smith-Gill, S. J., Mainhart, C. R., Lavoie, T. B., Rudikoff, S. & Potter, M. VL-VH expression by monoclonal antibodies recognizing avian lysozyme. *J. Immunol.* **132**, 963–967 (1984).
29. Sinha, N. & Smith-Gill, S. J. Molecular dynamics simulation of a high-affinity antibody–protein complex: the binding site is a mosaic of locally flexible and preorganized rigid regions. *Cell Biochem. Biophys.* **43**, 253–274 (2005).
30. Sinha, N. & Smith-Gill, S. Electrostatics in protein binding and function. *Curr. Protein Pept. Sci.* **3**, 601–614 (2002).
31. Kinch, M. S., Kraft, Z. & Schwartz, T. 2023 in review: FDA approvals of new medicines. *Drug Discov. Today* **29**, 103966 (2024).
32. Crescioli, S., Kaplon, H., Chenoweth, A., Wang, L., Visweswaraiah, J. & Reichert, J. M. Antibodies to watch in 2024. *mAbs* **16**, 2297450 (2024).

Digital Transformation in the Biopharmaceutical Industry 2

Rebuilding the Way We Discover Complex Therapeutics

Tonya Frolov, Leonard Wossnig, and Alexander Jung

2.1 INTRODUCTION

The biopharmaceutical sector, driven by the possibility of curing diseases, has made tremendous progress over the last decades: the first biosynthetic human insulin in 1982; the discovery of clustered regularly interspaced short palindromic repeats (CRISPR) and the advent of synthetic biology; the Human Genome Project; the first approved monoclonal therapeutic antibody in 1996; the discovery of the Yamanaka factors in 2006; the explosion of omics; the approval of the first cell therapy in 2017; and the rapid growth of

DOI: 10.1201/9781003300311-2

biologics which in 2022 accounted for 40% of the 37 drugs approved by the Food and Drug Administration (FDA) (de la Torre & Albericio, 2023).

Correspondingly, the pharmaceutical sector has experienced continued and significant growth. Worldwide sales exceeded $1.13 trillion in 2017 and $1.48 trillion in 2022 (Mikulic, 2024). The total spending and global demand for medicines will increase to approximately $1.9 trillion by 2027 (IQVIA, 2023).

Yet scientific and technological gains have not increased the efficiency of research and development (R&D). As noted by Scannell and Bosley, inflation-adjusted costs per novel drug increased nearly 100-fold between 1950 and 2010 and drugs are more likely to fail in clinical development today than in the 1970s (Scannell & Bosley, 2016). Recent estimates for the mean cost of developing a new small-molecule drug range from $314 million to $2.8 billion (Wouters et al., 2020; Sarkar et al., 2023). Despite a lot of effort to improve success rates, the majority of clinical trials continue to fail and the probability of positive outcomes in phase I clinical trials remains low (~10%) (Takebe et al., 2018). Most importantly, despite tremendous progress, too many diseases remain incurable. Those that remain to cure are typically multi-factorial and increasingly complex.

The scale of operations required to develop drugs, diagnostics, and devices is unparalleled. The average time it takes to get a new drug approved is 13 years (Paul et al., 2010), characterised by clinical trials and regulatory approval which requires time. As an example, in 2019, Novartis alone was running 500 clinical trials across >70 countries, with over 80,000 patients participating. In parallel, scientists were designing and developing new therapeutics, products, and devices (Finelli & Narasimhan, 2020).

Digital transformation is the fundamental rewiring of how a company operates. It is the process of adoption and implementation of digital technology with the aim of being nimble, reducing costs, and improving efficiency and productivity (Hole et al., 2021). Digital transformation has already impacted every industry, although it is moving much more rapidly in some than in others. Notably, aerospace, automotive industry, and banking have been quick to adopt digital transformation that generates business value in measurable ways. Surprisingly, the biopharmaceutical industry has lagged behind this revolution. The last several years have seen a surge in digital transformation within this sector, bringing optimisations across all facets of healthcare, including research, drug discovery, clinical trials, hospital management, surgery, diagnostics, monitoring, and wearables.

The discovery and development of new therapeutics comprises target selection, lead discovery, lead optimisation, pre-clinical testing, drug product development, and clinical testing. This is one of the most investment-intensive, research-intensive, and strictly regulated sectors. The processes involve difficult, manually intensive, time-consuming tasks that are prone to human error. The complexity of human biology further renders drug discovery and development very risky.

In parallel, the number of products and amount of data under review by the regulatory authorities has been increasing significantly over the last 20 years, accompanied by a trend towards more comprehensive and cross-validated evaluation processes and regulations (ICH, 2002). The regulatory authorities, notably the US Food and Drug Administration (FDA), have also in recent years opted for improved drug data and knowledge life cycle management (ICH, 2019). As this is a critical part of the biopharmaceutical value chain, it may trigger a much more focussed data and knowledge transformation within the biopharmaceutical industry. In 2019, the FDA announced its Technology Modernization Action Plan (FDA, 2019). In 2021, it further announced the reorganisation of its information technology, data management, and cybersecurity functions, signalling continued commitment towards its technology and data modernisation efforts (FDA, 2021), the aim of which is to move closer to the speed of industry by improving timely delivery of services.

Beyond the need for companies to implement data transformation to keep up with regulatory requirements, it offers a significant opportunity to leverage cross-product and cross-project insights to avoid repeating mistakes and to improve productivity and success rates. Publicly available data generated from healthcare research and adjacent industries, such as from biobanks, biomarker studies, and large-population studies, is only growing. This results in increased potential for predictive modelling in disease understanding, prevention, and cure. Last but not least, it may already help to predict clinical success or failure in the very early stages of testing a hypothesis.

Here, we review the principles of successful digital transformation that can be applied *specifically* to the earlier steps of the drug development pipeline after target selection: lead discovery and lead optimisation. We discuss how automation, robotics, and the systemic application of machine learning (ML) and predictive analytics could generate novel insights and *better* molecules with lower non-mechanism toxicity, faster. The more complex the therapeutic and/or problem to solve for, the more we stand to gain from the potential of a digital transformation. We propose a framework for wider digital transformation, including a more strategic approach to experimentation, reduced empiricism, and reshaping company culture, using the discovery of complex therapeutic modalities, such as multi-specific antibodies, as a case study.

As with digital transformation across any sector, the goal is to facilitate faster decision-making and operational efficiency by generating novel data-driven insights, or in other words – better, faster, and more affordable. Within the biopharmaceutical sector, this includes testing and falsifying hypotheses as early as possible and with less experimental effort and cost. By not starting clinical studies that are unlikely to be successful, the biopharmaceutical sector would revolutionise the availability of resources needed to generate new, better, disease hypotheses, ultimately benefiting patients.

Although other sectors have witnessed a more rapid adoption of digital transformation, it is evident that the biopharmaceutical sector is capable of drug discovery 4.0. New analytical and technological developments, such as automation, targeted data acquisition, and data from external knowledge bases, can now be integrated to generate high-quality data for ML methods. The continuous acceleration of data generation, combined with unprecedented data storage availability and exponentially growing computing power are enabling data-driven decision-making. This, in turn, can redefine the iterative process of experimentation into one that is driven by insight.

2.2 CURRENT STATE OF DIGITALISATION IN PRE-CLINICAL R&D

Artificial intelligence (AI) has vitalised computer-aided drug design, enabled by the increasing availability of large and well-curated public and proprietary datasets, enhanced computing capabilities, and access to cloud storage (Sarkar et al., 2023). AI can be defined as the science and engineering of creating intelligent machines that have the ability to learn, reason, perceive, interact, and solve problems autonomously (Manning, 2020). Many specific methods underpin AI systems; deep neural networks are at the heart of the current AI renaissance but other ML approaches may form components of AI systems; in general, they all seek to make a prediction or estimation based on some input data via some learned mathematical function.

One application of AI and ML is through the use of trained prediction models from a known dataset, to make inferences about as yet unseen data. This is known as interpolation and extrapolation depending on whether we are working inside or outside of the known data distribution. This is particularly relevant for biology where our understanding of underlying molecular mechanisms is often limited, and for drug discovery where information is missing for novel therapeutics or novel biology. A well-known example is using ML methods and statistical modelling for predicting protein structures from amino acid sequences (Duch et al., 2007).

AI has already shown positive results in research and in speeding up clinical trials as a result of the growing abundance of pharmacological and patient data. A well-known application of AI is the assessment of molecular characteristics by *in silico* screening and the identification of molecules with desired properties (Klaus, 2023). For example, a model capable of predicting the binding strength of a molecule may be developed based on data from physically measuring binding affinities for a range of substrates (Duch et al., 2007). Other applications include but are not limited to, peptide synthesis, structure-based virtual screening (Thompson et al., 2022; Varkaris et al., 2024), toxicity prediction (Klambauer et al., 2023), and drug repositioning (Gupta et al., 2021).

Previously, there has been a tendency for high-throughput systems with low biological complexity, as is the case with large-scale screening for binding affinity. The challenge is that the data rarely translates to the desired clinical, functional response because existing proxy models are insufficient (Bender and Cortes-Ciriano, 2021a; Bender and Cortes-Ciriano, 2021b). But AI can now be deployed to solve unique or complex problems, where human intelligence is ill-equipped, or where traditional methods would simply take too long, or where earlier computational methods did not suffice, see e.g. (Lyu et al., 2024).

While it is indisputable that AI holds immense potential for accelerating drug discovery and development, the current focus needs to shift from building more models to acquiring 'good' high-quality data that will optimise the functionality of these models. There is little to no data available for novel therapeutic candidates or novel targets, and thus general models are unlikely to work for lead optimisation onwards. This is in contrast to large language models (LLMs), where large, averaged, datasets are readily available on the internet and suffice. For drug discovery, we need complex biological data with a higher predictive validity (Scannell et al., 2022), i.e. data from proxy models that are more predictive of patient outcomes, even if it is challenging to generate. Data needs to be both clinically relevant and of sufficient quality to train the models – i.e. machine learning-grade data. Key properties include data veracity, variety, volume, velocity, and value, or the five 'Vs' of drug discovery as we refer to them. For a more in-depth review of what we mean by high-quality data, refer to Wossnig (2023).

The integrity of data collection, storage, protocols, and standardisation is critical. This needs to be supported by a technology stack that maximises consistency and reproducibility. If satisfied, this will result in superior predictive models and meaningful insights.

2.3 CHALLENGES IN DIGITAL TRANSFORMATION OF PRE-CLINICAL R&D

The requirement for automation and standardisation of reproducible, scalable, and clinically relevant data is now largely accepted, and many of the hardware and software tools that facilitate the creation of automated experimental and data workflows are now either available or under development. Here, we discuss how companies can adopt these, and the challenges that need to be overcome.

2.3.1 Operational Challenges

The integration of different technologies comes with its challenges. Teams typically need to integrate software that ranges from the latest available technologies to those that may be a couple of decades old, and as such there is no standardised approach for achieving this. Reliance on web-based software only is not possible when instruments are connected to personal computers. A further challenge is that instrument processes may be complex, or adapted for a specific and unique lab purpose. The development of automation needs to be balanced with providing manual flexibility for scientists to adapt experimentation and workflows.

Furthermore, tools are not typically suitable for immediate use by a scientist and instead require the input from or execution by software engineers, which can limit broader adoption. Given the complexity of workflows and data, it is unlikely that scientists will acquire the skills of a software engineer, highlighting the ongoing requirement for a mixed team of experts. While some people view the scientist of the future as a hybrid of software developer and experimental biologist, we believe that the field will continue to require specialists for each area.

It is also difficult to maintain a consistent adoption of technologies across all teams, which is particularly challenging for larger organisations. Modern software designed for personal use is typically designed to be user-friendly (e.g. the iPhone), but this is not always the case in the biopharmaceutical sector. Teams from varied career backgrounds show a range of technical competence, making it difficult to integrate technologies that could be adopted by everyone. Scientists may desire to implement technologies they are most familiar with, rather than those that would objectively benefit the team at large. Training teams is imperative but requires time investment, which may be difficult given the pressure to deliver assets in biopharmaceutical organisations. Training is further challenged by employee turnover. One of the trends that are forseen for the near future would be, that the data generated by instruments and lab scientists feed directly into a holistic data standard aka knowledge graph, reducing the need for dedicated data acquisition software and all their logistcal disadvantages.

2.3.2 Cultural Challenges

It is widely accepted that in early drug development and research organisations, cultural resistance persists towards digital transformation. Perhaps a contradiction to the notion that scientists should thrive for innovation, or a result of scientific mistrust of the benefit that new technologies could provide, possibly owing to the lack of success stories in the long journey to bring a new therapeutic to the market.

There often appears to be a generational and diversity gap in current teams, and one conclusion that could be drawn is that there is a reluctance to adopt digital transformation triggered by external forces simply because it comes from elsewhere. However, given that most scientists have adopted the benefits of digital transformation in their personal lives, one could suppose that they would be equally willing to adopt digital transformation within their own field of expertise.

Here, we will see a huge shift in the teams within the biopharmaceutical industry, as the trend for agile and diverse product teams begins to take over the product development framework. Over time, this will lead to improved communication of scientists, engineers, and data scientists.

Until this transformation materialises on a larger scale, communicating the value of, and involving teams with, digital transformation is critical. For example, AI and generative AI tend to be vastly misunderstood in the sense that either expectation rise too high, or scepticism hinders the adoption. In order to become widely accepted, their properties, benefits, and use cases need to be explained during training, roadshows, and pilot projects across the R&D sector. Perhaps more fundamentally, organisations need to think differently about what constitutes a valuable data asset. Traditionally, data was merely a stepping stone to finding candidates whereas now data is in itself a highly valuable asset.

With respect to the digital transformation of datasets, the benefits are not always immediate. Efforts need to be made towards maturing these datasets and towards the FAIRification with an organisation that comprises 15 guiding principles outlined by Wilkinson et al. (2016) aimed at enhancing the Findability, Accessibility, Interoperability, and Reusability of data. A particular complexity is that experimental data is typically expensive to produce and still relatively small scale. In order to create large enough datasets for successful ML applications, datasets may need to be additive and aggregated over many experiments and iterations. This gives rise to the difficulty of asserting consistent data models over time and ensuring the compatibility of datasets. Other key considerations are that datasets should be representative, i.e. will want to include negative results as well as positive results, and will want to minimise bias such that models trained on the data have an accurate and complete understanding of the real-world processes (Bender and Cortes-Ciriano, 2021a; Bender and Cortes-Ciriano, 2021b).

Data scientists in biotech require a rich understanding not only of data and engineering but also of complex science. This in turn may encompass biology, chemistry, toxicology, etc. They need to be able to create the right datasets, process the data, and make it timely available in the right format to the right model, and hence anticipate what the scientists may want to infer from them (Benchling, 2023). Building a team with the necessary expertise across AI, data science, data engineering, software engineering, and science is critical. A strategy to enable this should be defined from the earliest stages of company creation. Additionally, a company-wide vision for the creation of the right technology stack is also important.

The leadership's challenge here is to link these activities directly to the company's business strategy and to the acceleration of bringing new therapeutics to patients. If done successfully, teams will feel they are contributing to a meaningful outcome. As more and more data roles and digitalisation roles are created within R&D organisations, this will lead to an organic adoption and integration of digital tools in the daily work of scientists.

2.3.3 Data Management, Data Analytics, and Integrity Concerns

The establishment of a standardised digital framework for experimental data collection that adheres to FAIR (Findable, Accessible, Interoperable, Reusable) principles, what we referred to as FAIRification earlier, is critical to enhance the ability to find, use, and reuse the data. But challenges to achieving this persist, including data integration, which requires managing and integrating large, diverse datasets; initial substantial investments are needed to build and integrate advanced technologies such as new data management systems (FTE time), and for clearing and organising existing data; incentivising support, for example from shareholders; and ensuring the accuracy and reproducibility of AI-driven results, particularly in absence of benchmarks.

Experimental data in the early stages of drug discovery and research is as diverse as the tasks and the purpose it is supposed to serve. There exist, however, critical inflection points throughout the value chain that result in the need for further data accumulation and consolidation.

A robust data strategy can facilitate the entire R&D pipeline, and must take into account three key attributes: data quality, data integrity, and data governance.

2.3.3.1 Data quality standards

While organisations need to ensure that data regulatory standards required for submissions are met (e.g. Standard for Exchange of Nonclinical Data (SEND), Identification of Medicinal Products (IDMP), etc.), the need for structuring data and the need for high-quality data extends far beyond this. The value of the data comes not only from insight-driven decision-making at critical inflection points but from its ability to generate new insights in earlier stages of R&D. This, however, requires that all data is collected to satisfy minimum quality criteria such as the capture of relevant metadata or common identifiers in order that it be usable.

2.3.3.2 Data integrity principles

ALCOA (Attributable, Legible, Contemporaneous, Original, and Accurate) and ALCOA plus principles, introduced by the FDA, are often applied as data integrity measures to data assets. Unfortunately, they impose high standards on data-generation and automation teams within R&D labs. However, if the data can be reused to generate high-value predictions going forward, then this apparent 'burden' turns into a competitive advantage. Furthermore, data integrity and organisation help statisticians, data scientists, and data engineers to work collaboratively and integrate the data into the prediction framework with less domain knowledge.

2.3.3.3 Data governance

Researchers may encounter situations where information or data requests are not directly denied but instead met with limitations based on perceived issues such as data quality, researcher expertise, or confidentiality concerns. This aligns with observations in individual data-sharing negotiations, where frameworks are established to address confidentiality, data protection regulations, and researcher qualifications. A well-established governance structure including roles such as data domain owners, and/or data stewards is needed to maintain a high level of trust in the data-sharing process. This fosters data transparency and data usage along the value chain for non-expert users to generate new insights.

The landscape of available data is hugely diverse in terms of data standards, maturity, and ontological data models. Building data assets around domain knowledge requires a huge initial effort to build the right taxonomies, ontologies, and data classification rules within the sector. However, it is worthwhile and necessary to invest here, mainly in the form of value-added pilot use cases and projects, in order to create the data platform to support AI and ML predictions in a sufficient way.

2.3.4 Barriers to Automation Adoption

Building out a high-throughput, automated platform requires a high level of upfront investment. But the appeal is that it can enable significant scientific discovery, particularly in the context of complex and high-impact problems that would otherwise take decades, or longer to solve.

Given the significant investment required, it is important to define the concrete problem that needs to be solved and to assess whether the development of the platform is warranted. Parameters to consider that increase the value of the investment include:

(i) the task is simple but highly repetitive and vulnerable to human error due to the sheer repetitiveness of the task; (ii) the workflow is highly complex and human interaction at multiple levels along the workflow would lead to an increased risk of error resulting in catastrophic failure; (iii) the return on investment is such that a significant saving in time and in human resource can be realised with the ability to run automated tasks either continuously or to completion with a high degree of precision and accuracy independent of human shift patterns; (iv) the task is either amenable to automation or adapted to run using automated processes, as automation is not always a like-for-like replacement for tasks designed for manual handling.

2.4 CORE INGREDIENTS FOR SUCCESSFUL DIGITAL TRANSFORMATION

Four elements influence digital transformation: (i) the steady acceleration of the rate at which new data is generated; (ii) cloud computing that offers universally available data storage and exponentially growing computing power; (iii) new ML methods/AI; and (iv) a shift in culture to build teams equipped to develop and deploy a data-driven approach (Finelli & Narasimhan, 2020). Here, we review developments enabling the rapid digital transformation of drug discovery.

2.4.1 New Data-Generation Methods and Laboratory Automation

In synthetic biology, considerable progress has been made towards achieving autonomous and scalable experimentation, largely due to the combination of automation and ML (Martin et al., 2023). While fully automating a laboratory is an ideal goal in some cases, it is not always the most practical or beneficial approach. In certain situations, simpler and more modular methods may be more effective (Stephenson et al., 2023). For antibody engineering, given the current technological advancements, a modular yet fully integrated platform is in our view most appropriate.

A key application of automation is in improving the Design-Build-Test-Learn (DBTL) cycle, commonly employed in drug discovery programmes. This cycle involves designing a series of candidates, constructing, and testing them, then using the gathered data to inform the design of the next series of candidates, often employing ML or statistical modelling.

Effective implementation of DBTL frameworks in laboratories hinges on several factors. Consistent, reliable data are vital for ML applications, necessitating rigorous tracking and monitoring of reagents, samples, protocols, and business rules as well as control compounds that are used in programmes. Automation in data handling and metadata management is crucial to reduce errors commonly associated with manual data entry or processing.

In individual labs, various factors can impede clear and reproducible experimental results. These include inconsistent sample tracking, varying lab conditions, manual process errors (including simple copy-paste errors of data), and non-uniform data collection and storage practices. Additionally, differences in commercial and proprietary instruments and protocols can introduce significant batch effects, leading to inconsistencies across different research groups and labs. Therefore, standardisation and automation across an organisation are essential.

The advent of advanced liquid handling technologies has revolutionised the movement of reagents and samples in automated experimental workflows, boosting reproducibility and throughput. When paired with data management tools that consolidate experimental parameters and data, these technologies improve sample tracking and error detection, aiding decision-making in drug discovery programmes. Technologies like next-generation sequencing (NGS) or new deep screening of sequence-sequence level interactions (Porebski et al., 2023), facilitate the generation of large datasets suitable for training ML models. These datasets can be used to evaluate system performance and shape future experimental designs.

However, further improvements are needed in hardware and software systems to enhance connectivity between automated steps and equipment. Additionally, standardising data infrastructures, exchange formats, protocol descriptions, ontologies, and data reporting will improve data interpretability and promote information sharing. In the following sections, we will examine the current status of automation systems pertinent to biologics engineering.

Automation technologies in synthetic biology exhibit a wide range in format and degree, encompassing various levels of human interaction (Figure 2.1). As automation

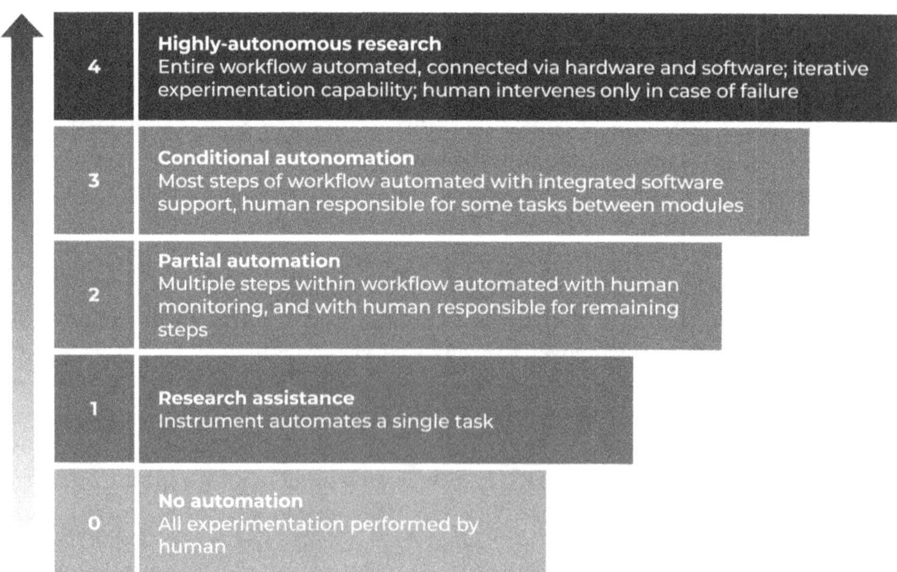

FIGURE 2.1 Levels of automation in synthetic biology laboratories achieved through hardware and software automation technologies. Adapted from Stephenson et al. (2023) and Martin et al. (2023).

tools and machinery become more sophisticated, higher levels of automation are attainable, consequently reducing human involvement. Common tools like pipettes are standard, alongside instruments automating single tasks (e.g. thermocyclers, centrifuges, plate readers). These instruments are integral to developing automated workflows, with the coordination of these diverse instruments and automated modules posing a significant challenge.

Beyond unit process automation equipment (level 1 automation), there is a plethora of tools for automating multi-step processes and entire experimental workflows. Liquid handling robots and microfluidic devices, for instance, have been designed in various formats to automate critical tasks and multiple steps in lengthy pipelines. Liquid handling robots enable more flexibility and throughput, whereas microfluidic devices enable precision and miniaturisation, allowing researchers to control liquids on a microscopic scale. Adaptable automation platforms are now being developed, notably by HighRes Biosolutions, offering highly versatile modular platforms with increased system flexibility, adaptive scheduling software, and improved protocol design. These technologies and an overview of their constraints and benefits are summarised in Figure 2.2. Combining robust physical automation with computational tools then enables closed-loop iterative experimentation, enhancing discovery speed and productivity.

Recent trends emphasise systems approaches that combine hardware and software automation, especially crucial in cell-based work for generating large, high-quality datasets. This is particularly evident in drug discovery's lead optimisation stage, where complex assay readouts are essential and processing of data is more complex. The DBTL framework becomes increasingly automated at level 3 and above. Physical automation, in tandem with software tools, can then enhance efficiency, foster reproducibility through standardisation, and minimise experimental error and variability (e.g. by using data normalisation). These improvements are essential for advancing the DBTL cycle. To effectively screen multiple designs concurrently and expedite DBTL cycles, comprehensive automation is indispensable at each stage of the DBTL process.

Design tools in synthetic biology have evolved to include functionalities for DNA part design and assembly, utilising established methods like Golden Gate assembly, Gibson assembly, SLIC, and digestion-ligation methods (Chao et al., 2015; Chao et al., 2017). Recent years have seen the integration of computational tools and ML, with innovations like generative design (Ingraham et al., 2023) and protein language models (Fenoy et al., 2022; Alley et al., 2019; Brandes et al., 2022; Rives et al., 2021; Wu et al., 2021; Choi, 2022; Dounas et al., 2024; You et al., 2021; Heinzinger et al., 2019; Lu et al., 2020) augmenting traditional computational biology approaches.

In the build phase, DNA assembly stands out as a particularly time-intensive and error-prone process, making it an ideal candidate for automation (Walsh et al., 2019; Storch et al., 2020; Linshiz et al., 2016; Kang et al., 2022). Transitioning manual assembly to high-throughput automated systems is complex, but using standardised parts and modular designs can facilitate the large-scale prototyping of combinatorial DNA libraries (Casini et al., 2015) as well as faster assembly and screening of combinatorial design spaces. Methods like Golden Gate assembly, which are executed as one-pot reactions with limited steps, such as liquid handling, incubation, and thermocycling, can be automated to significantly cut down on time and manual errors.

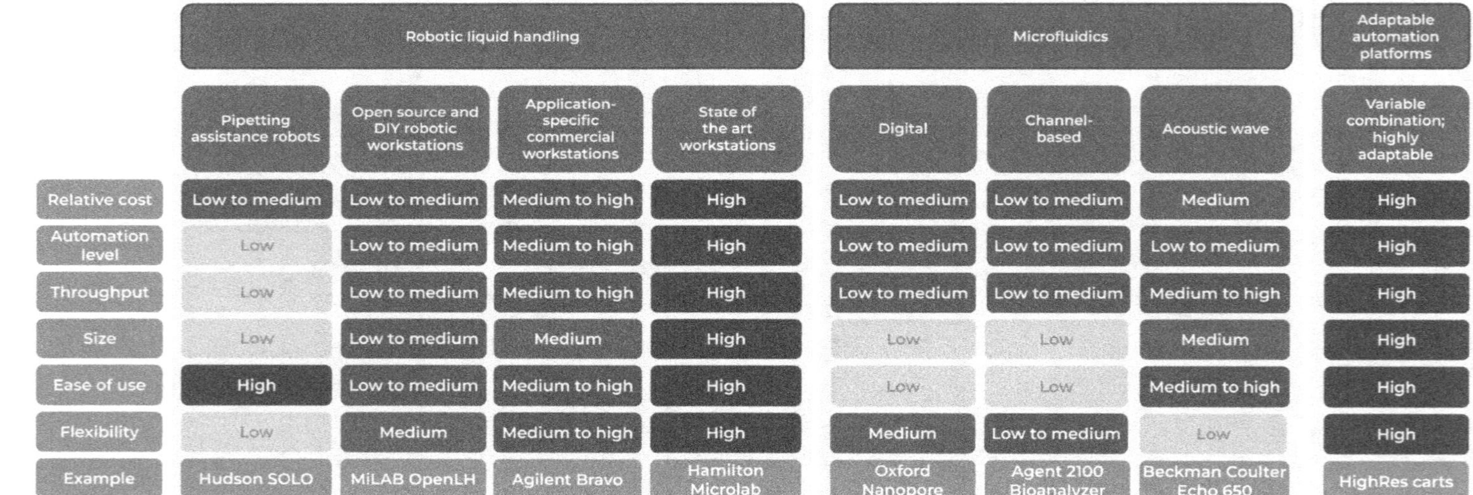

	Robotic liquid handling				Microfluidics			Adaptable automation platforms
	Pipetting assistance robots	Open source and DIY robotic workstations	Application-specific commercial workstations	State of the art workstations	Digital	Channel-based	Acoustic wave	Variable combination; highly adaptable
Relative cost	Low to medium	Low to medium	Medium to high	High	Low to medium	Low to medium	Medium	High
Automation level	Low	Low to medium	Medium to high	High	Low to medium	Low to medium	Low to medium	High
Throughput	Low	Low to medium	Medium to high	High	Low to medium	Low to medium	Medium to high	High
Size	Low	Low to medium	Medium	High	Low	Low	Medium	High
Ease of use	High	Low to medium	Medium to high	High	Low	Low	Medium to high	High
Flexibility	Low	Medium	Medium to high	High	Medium	Low to medium	Low	High
Example	Hudson SOLO	MiLAB OpenLH	Agilent Bravo	Hamilton Microlab	Oxford Nanopore	Agent 2100 Bioanalyzer	Beckman Coulter Echo 650	HighRes carts

FIGURE 2.2 Overview of automation technologies and characteristics that need to be considered when incorporating automation. Adapted from Stephenson et al. (2023).

Steps like cell transformation, transfection, or transduction are also challenging to automate due to specific experimental conditions and high labour intensity, but many successful implementations exist (Si et al., 2017).

The testing phase presents another automation challenge. Protocols vary widely, with some steps being more conducive to automation than others. Cell-based work, in particular, introduces unique challenges for automation, and current solutions often struggle to upscale or simply shift bottlenecks along the pipeline (Doulgkeroglou et al., 2020). Killing assays, for example, using the IncuCyte® by Sartorius, are notoriously hard to automate. Recent systems like the Tecan Freedom EVO or Tecan Cellerity allow for better integration of flasks (so-called RoboFlask) into the process and thereby simplify cell-based work. Automated analysis involving techniques like chromatography and mass spectrometry also presents distinct challenges. The complexity of the data and inefficient resolution renders it difficult for software to accurately make an assessment, such as correctly distinguishing between peaks and often requires human intervention (Cappadona et al., 2012). However, the integration of liquid handling technologies with component instruments that automate critical or labour-intensive steps, combined with advanced software, can significantly alleviate the workload on researchers and enhance experimental results.

A notable application of this approach is the modular DBTL platform developed by LabGenius, which focusses on the design, synthesis, and testing of multi-specific antibodies using cell-based assays. LabGenius has developed a flexible robotic liquid handling platform that can perform disease-relevant, functional, cell-based assays to produce machine learning-grade data at high throughput, minimising variability and noise and rendering it ideally suited for active learning. Computational models enable the design and learning steps, while the automated liquid handling system covers the automated build and test steps. By combining high-throughput experimentation with modelling, LabGenius can today evaluate the performance of thousands of antibody designs in disease-relevant cell-based assays in just a few weeks. This is possible through an active learning-based approach that generates data in an optimal way to train ML models (LabGenius, 2023).

While automation has made significant strides in fields like antibody engineering, manual intervention is often still necessary for monitoring and connecting different elements of the pipeline.

The connectivity among each step in the discovery process and integration of computing throughout makes it possible to capture a wealth of information: explicit representations of goals, hypotheses, experimental results, and findings, as well as metadata that can be critical to tracking why and how experiments were carried out in order to understand and repeat them or diagnose errors (e.g. environmental conditions, instrument settings, protocols, runtime logs). For this reason, many systems that are ideally capable of level 4 automation (e.g. King et al., 2009; Williams et al., 2015; Sparkes et al., 2010), often function as level 3 systems due to the regular need for human intervention in hardware and software issues. Nevertheless, they highlight the potential of combined human-robot efforts in accelerating research through strategic automation.

It is also important to recognise that fully autonomous systems often require a high degree of integration, which often coincides with less flexibility or modularity of the constituent parts. Such systems are not possible in many organisations due to the need for equipment to work across many different projects with a multitude of workflows.

In scientific research, pipetting assistant robots offer a basic level of automation, emulating manual pipettes with a small footprint and lower cost than more complex platforms. Their typical applications include reagent dispensing, serial dilutions, and basic sample preparations. While having lower throughput due to limited channels and deck size, they alleviate manual pipetting errors and save time. Manufacturers like Integra, Hudson Robotics, and Gilson offer various models, from dedicated microplate dispensers to versatile workstations that can integrate with other devices for expanded functions.

Some manufacturers provide pre-configured workstations for specific research areas such as NGS, polymerase chain reaction (PCR), and proteomics. These are larger and costlier, starting from $100 to 200k, but offer high throughput, flexibility, and reliability through advanced hardware and software. Companies like PerkinElmer, Agilent, and Beckman Coulter offer various models tailored to specific applications.

Contrastingly, microfluidic devices, primarily developed in-house by research labs, face a commercialisation gap, with notable exceptions in sequencing and analysis sample preparation. Commercial systems like the Agilent 2100 Bioanalyser and the Beckman Coulter Echo 650 series have succeeded due to their innovative approaches in reducing sample consumption, analysis time, and contamination. However, challenges remain in the commercial viability and reliability of microfluidic devices, as seen in the discontinuation of certain products. Efforts continue in commercialising microfluidics, with emerging low-cost or open-source platforms.

In synthetic biology, as automation and digitalisation advance, the need for standard data management and communication protocols is crucial for creating an integrated laboratory environment. While liquid handling robots and automated analytical units are widespread, they typically operate as standalone units with proprietary standards, command sets, device driver software, and data formats. The result of this is poor connectivity among devices, data silos, and difficulty in integrating multiple devices to form continuous, flexible automation pipelines. In automated laboratories, multiple devices are set up for steps, such as sample preparation and analysis, and connected to software systems that orchestrate the flow of samples through experimental pipelines. Piecing together diverse instruments needed for a given workflow without standardised hardware interfaces complicates the connection of equipment to a centralised control system. Variations in software among devices make integration very challenging, where even devices with similar functionality have vastly different command sets, behaviours, and capabilities for handling device states and errors. Furthermore, data collected from various instruments in a workflow are typically stored in proprietary formats that complicate visualisation, analysis, and sharing. Arguably, having a manufacturer-independent interface definition that applies to all devices and stores results in a common format would greatly simplify the process of setting up flexible and modular laboratory automation systems that can then be sustainably tailored to application and researcher-specific needs. There is hence a big need for standardisation of data formats across analytical devices and interfaces (communication standards) – both are needed for true end-to-end integration. Such standards (e.g. as introduced by the Standardisation in Laboratory Automation consortium or via the Analytical Information Markup

Language) would be able to facilitate easier data storage, visualisation, and sharing in a unified format and can enhance reproducibility, traceability, and remote monitoring of experiments. They can hence enable the automatic upload to global repositories via cloud computing, simplifying data management, enabling remote process monitoring, and making it possible to efficiently share experimental results with all stakeholders.

We note that large language models (LLMs) like Generative Pre-trained Transformer 3 (GPT-3) and ChatGPT are emerging as promising tools for laboratory automation. These models, trained on extensive datasets, show adaptability across various applications, notably in intuitive robot control via natural language inputs. However, it is too early to tell how they will actually impact the state of laboratory automation.

2.4.2 Cloud Computing

Cloud computing, a vital component in drug discovery, involves the on-demand delivery of configurable computing resources over a network. It relies on virtualisation technology, where a hypervisor software abstracts and shares hardware resources as virtual machines (VMs). This environment allows for rapid scaling of resources with minimal interaction with the service provider (Spjuth et al., 2021).

The growing reliance on ML and statistical modelling underscores the need for effective data management solutions that ensure rapid and efficient data access. The data collection phase in the ML life cycle involves various tasks, including conducting experiments to generate data, extracting existing data from databases, and selecting specific data for modelling. Cloud computing plays a crucial role in this context, offering scalable and on-demand infrastructure for storage, databases, and middleware without initial costs. Using cloud services for data hosting provides easy and quick global access to data, particularly when modelling is conducted within cloud environments (Dalpé and Joly, 2014). This accessibility not only streamlines the modelling process but also enhances collaborative efforts and simultaneous access.

Before the advent of cloud computing, drug discovery organisations requiring significant computational power were constrained to on-premise servers, entailing substantial overhead costs and maintenance responsibilities. This setup posed insurmountable challenges for smaller entities and formidable obstacles even for larger corporations, rendering it an impractical endeavour for most. The emergence of cloud computing, particularly through infrastructure or platform as a service, has revolutionised this landscape, democratising access to computing resources previously unattainable to biotechnology companies.

Cloud based platforms such as Microsoft's Azure or Amazon's AWS for example, have meanwhile entered majority of the corporate pharmaceutical world with a more or less complete set of governance, storage and computing tools on demand. In the absence of such solutions, companies like LabGenius would have required large amounts of additional human resources and may have deemed the endeavour financially unviable. Many small teams and companies find themselves in similar circumstances, where the cost of leveraging ML technologies would have otherwise been prohibitive, barring those organisations for which such technologies constitute the core business operations. Cloud computing has significantly lowered this barrier, enabling even the smallest of teams to engage in exploratory ML initiatives by leveraging appropriate infrastructure and platforms to assess the viability of integrating ML into their workflows.

From a data acquisition and analysis standpoint, cloud computing has also facilitated the adoption of distributed, asynchronous, simultaneous, and secure data handling practices. Notably, Laboratory Information Management Systems (LIMS) and Electronic Laboratory Notebooks (ELN) now benefit from automated backups, universal accessibility across connected devices, and instantaneous sharing and analysis capabilities across disparate locations and users. The cloud environment affords streamlined system management, allowing for immediate implementation of software modifications with unprecedented agility and efficiency. Increasingly, alongside traditional off-the-shelf LIMS, custom bespoke applications are being developed and integrated due to their low creation and deployment barriers. These tailored applications can be fine-tuned to meet the specific requirements of either a company as a whole or the distinct needs of individual scientists. This trend highlights a shift towards greater efficiency and adaptability in scientific research settings.

Cloud computing has three primary service levels: infrastructure as a service (IaaS), platform as a service (PaaS), and software as a service (SaaS), each providing an increasing level of abstraction from the hardware. There are four main cloud deployment models: public cloud (accessible to the general public), private cloud (deployed on-premise for a single organisation), community cloud (shared by multiple organisations with similar concerns), and hybrid cloud (a combination of any two models, often public and private, with a software layer for seamless application portability). OpenStack, a modular open-source software stack, is widely used for operating cloud environments in academia and by local cloud providers (www.openstack.org/).

Major public cloud providers offer extensive catalogues of virtual machine images (VMIs) with pre-installed software environments, enhancing tool distribution in drug discovery and improving reproducibility.

Containerisation offers a lightweight method for deploying VMs by packaging an application's code, runtime, dependencies, and settings into a portable and repeatable unit. It allows applications to run consistently across various platforms and is more resource-efficient than VMs. Kubernetes (K8s), developed by Google, is the leading container orchestration platform, enabling easy management of containerised applications. It's commonly used in private cloud platforms and major cloud providers for scalable access to computing resources.

While software as a service requires minimal IT skills, efficiently working with infrastructure as a service and platform as a service necessitates basic knowledge of Linux or programming languages like Python.

Private clouds and on-premises Kubernetes (K8s) clusters serve as valuable complements in this ecosystem (Emami et al., 2019). They provide scientists with the ability to swiftly repurpose existing hardware resources to meet the diverse application needs at different stages of the ML life cycle. This approach enables a more efficient and cost-effective way to manage computational resources, catering to the dynamic and sometimes unpredictable requirements of ML projects. The integration of private clouds and K8s clusters into the ML workflow thus represents a strategic approach to balancing resource demands while maintaining operational flexibility and control.

Using infrastructure as a service (IaaS) resources from cloud providers presents a viable alternative, offering access to extensive resources on a pay-per-minute basis, eliminating upfront and maintenance hardware costs. However, the costs for advanced cloud infrastructure, such as graphics processing units (GPUs) or large VMs with ample random-access memory (RAM), should not be underestimated. Teams with consistently high GPU demands are increasingly establishing private Kubernetes (K8s) infrastructures to support GPU-powered, containerised workflows.

2.4.3 Machine Learning and MLOps to Support the Machine Learning Life Cycle

The integration of machine learning (ML) in drug discovery involves comprehensive management of data and model life cycles. This process encompasses stages like data gathering, pre-processing, training, validating, and deploying models, along with predictions and inference. ML Operation (MLOps), a blend of practices and software, supports these stages, enhancing reproducibility and robustness in iterative drug research. Since machine learning in drug discovery is covered amply in other resources (e.g. Vijayan et al., 2022; Sarkar et al., 2023; Wossnig et al., 2024), we mainly focus on MLOps in the following section.

MLOps includes a range of functions primarily concerned with facilitating deployment, audit logging, and reproducibility of ML systems, vis à vis version control and automation for code, data, pipelines, and models. In drug discovery, one frequently encounters changes in the experimental data that suddenly changes model performance. Often, there is also a lack of metadata which hinders the diagnosis of errors and troubleshooting. MLOps, specifically model and data versioning, can enable the rapid diagnosis of errors, reverting to previous models, and generally better data-model-prediction lineage that ensures that the right method is applied to the right problem at the right time.

Cloud computing offers scalable, resilient computing resources crucial for MLOps and ML operations. Containerisation, coupled with platforms like Kubernetes and scientific workflows, enables sturdy, reproducible ML analysis (e.g. consistent use of environment and library versions), and higher speed in drug discovery. Additionally, emerging federated methods promise to facilitate collaborative ML across multiple organisations, leveraging both private and cloud-based infrastructure.

Most drug discovery projects, specifically in the lead optimisation stage follow the iterative DBTL process (Figure 2.3). ML is used to enhance decision-making, particularly in the Learn phase, using both project-specific and global models. MLOps emerges as a critical framework, integrating data engineering, science, and operations to manage the ML life cycle in production. It aims to streamline processes from data preparation to model deployment, ensuring up-to-date models are accessible. Cloud services support MLOps stages, with tools like Vertex AI, KubeFlow, STACKn, and H2O.ai aiding in public and private infrastructures. Continuous ML modelling requires the integration of data handling, pre-processing, quality control, and validation into a reproducible pipeline for optimal decision-making in drug discovery.

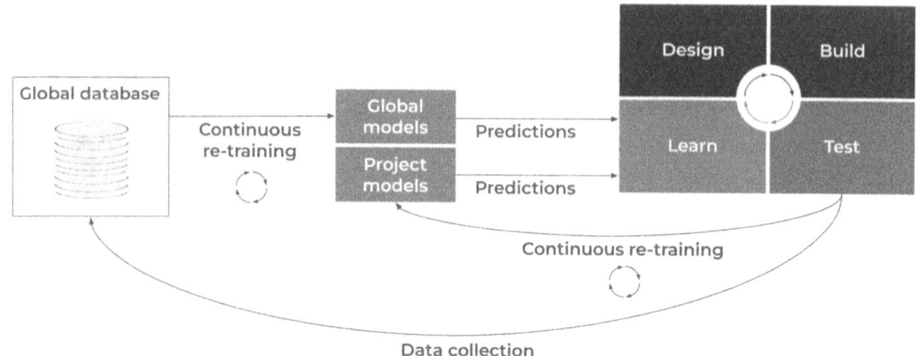

FIGURE 2.3 The use of ML models in the DBTL cycle in drug discovery. Within a specific project, models are trained using data from the Test phase and used to make predictions that support the Learn phase, such as running an assay to optimise a molecule for binding to a particular target. Global models can then be used for predictions, such as predicting binding to off-targets. Global models are typically trained using data collected from many projects or merged with publicly available data. By contrast, project models are developed for a particular objective within a drug discovery project. Project models are typically manually trained, and may need updating. Adapted from Spjuth et al. (2021).

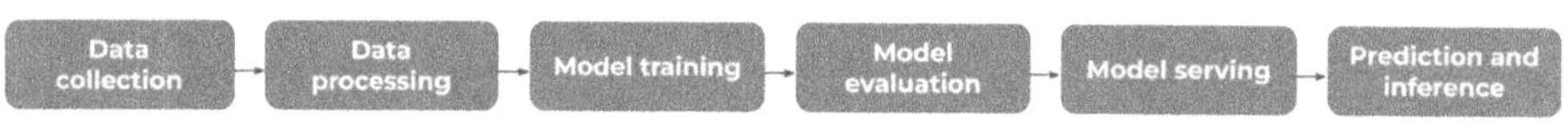

FIGURE 2.4 The machine learning life cycle. Diagram adapted from Spjuth et al. (2021).

The ML life cycle (Figure 2.4) in drug discovery involves multiple steps with specific infrastructure and software needs. Cloud IaaS reduces the need for on-premises computing infrastructure, offering scalable, on-demand resources. Data collection is crucial, requiring efficient data management solutions. Pre-processing involves handling artefacts and quality control, with cloud computing facilitating transparent, reproducible workflows. Model training and validation are resource-intensive, whereas cloud IaaS offers a cost-effective alternative to on-site infrastructure. Prominent tools like TensorFlow and PyTorch are readily available in cloud environments and have significantly simplified the process of training models on complex infrastructure such as GPUs – thereby reducing the barrier of entry. Kubernetes-based scientific workflows, exemplified by Recursion Pharma and BenevolentAI, orchestrate model development pipelines efficiently. For example, AlphaFold (Jumper et al., 2021) was quickly integrated by a large number of companies due to its availability in a container (c.f. the AlphaFold GitHub repository, https://www.github.com/google-deepmind/alphafold).

The initial step is data collection. MLOps should enable careful versioning of data (e.g. using Data Version Control (DVC), Pachyderm, or Neptune). This is essential to later on trace the lineage from a dataframe (a specific version of the data) to a model and the model predictions (results). Without proper data versioning and clear pointers to the most recent data version, prediction errors or delays can easily occur.

After selecting and assembling data in drug discovery, a crucial step involves pre-processing, which addresses issues like duplicate data, missing values, normalisation, augmentation, and quality control. Cloud computing significantly aids this process by simplifying the construction and execution of pipelines or workflows. This enhancement contributes to increased transparency, reproducibility, and robustness in data handling.

Many workflow systems in the cloud allow for declarative specifications of analysis pipelines, with capabilities to execute these workflows on both public and private clouds. Nextflow, a popular workflow engine in life science research, can directly execute workflows on infrastructure as a service (IaaS) resources and Kubernetes clusters. Additionally, Argo and Pachyderm are two Kubernetes-native workflow systems.

Notebooks deployed on cloud resources are commonly used for specifying pre-processing steps. This approach not only facilitates pre-processing execution but also supports visual interpretations, enhancing understanding and analysis. Implementing a complete workflow of all pre-processing steps in the cloud ensures that the pre-processing is reproducible, portable, and scalable. On the other hand, notebooks allow people to write lower-quality and potentially unsafe code. Here again, MLOps practices can be applied to mitigate risks and follow good standards.

Moreover, numerous platform as a service (PaaS) services focussed on big data pre-processing and analysis are available as public cloud services. An example is Databricks, which is based on Apache Spark, offering advanced capabilities for handling and analysing large datasets in cloud environments.

After dataset construction, the focus shifts to model development, including training, validation, optimisation, and model hyperparameter tuning. This phase can be resource-intensive and time-consuming. Cloud computing offers significant advantages here, providing flexibility and scalability than on-site workstations. These workstations, while powerful, are not scalable, require substantial upfront investment, and necessitate ongoing maintenance.

For AI modelling in drug discovery, software stacks like TensorFlow, PyTorch, and SciKit-Learn (the first two are primarily for deep learning) are prominent. These tools are readily available in cloud environments, with virtual machine images (VMIs) and container images facilitating rapid setup for single-node infrastructures. Kubernetes is widely utilised for scaling across multiple nodes, along with specialised AI frameworks for Kubernetes.

Scientific workflows, orchestrated using engines like SciPipe and AirFlow, are employed to manage complex model development processes, including tasks like nested cross-validations. These technologies and practices illustrate the evolving landscape of AI modelling in drug discovery, where cloud computing and containerisation play critical roles in enhancing efficiency and scalability.

Model serving and inference in the ML modelling life cycle involve making AI models accessible to end-users in a production-grade environment with appropriate governance. This step, often the most technically challenging, requires not only model validation (Smiatek et. al., 2021) and versioning but also critical features like privacy, access control, auditability, logging, monitoring, and a resilient infrastructure capable of recovering from failures.

Modern software engineering practices, particularly DevOps, which merges development (Dev) and operations (Ops), offer valuable insights for this stage. Cloud computing and infrastructure-as-code have revolutionised DevOps, providing a flexible, scalable environment suited for production-grade solutions. Kubernetes, popular in DevOps, supports these requirements, offering features essential for serving ML models, including resilience and scalability.

Public clouds typically provide services for serving and publishing ML models, and numerous cloud-native open-source frameworks have emerged, leveraging Kubernetes. These include TensorFlow Serving, Seldon, Kubeflow, and STACKn, or ones directly hosted by the cloud provider such as Google Kubernetes Engine (Google Cloud) or Amazon Elastic Kubernetes Service (AWS) which facilitate model management and serving.

An important aspect of model serving is making models available via an Application Programming Interface (API), allowing for integration with other software components and enabling cloud-based inference, essentially predictions as a service (SaaS). In drug discovery, platforms like OpenRiskNet and DeepCell Kiosk exemplify this approach. OpenRiskNet, built on OpenShift, a Kubernetes distribution, allows for publishing AI model services and adds layers of discoverability and interoperability. DeepCell Kiosk, designed for microscopy image analysis using TensorFlow, also utilises Kubernetes for scalable deployment and inference. These examples demonstrate the evolving integration of cloud computing and AI in drug discovery, highlighting the critical role of model serving and inference in the ML life cycle.

2.4.4 Company Culture That Drives a Data-Driven Approach

Senior management sponsorship and widespread cross-functional support of a data engineering culture, as opposed to just science, is integral to digital transformation and a data-driven approach (Saldanha, 2019). Take the example of MLOps, integrating data engineering, science, and operations to manage the ML life cycle in production. Teams need to work together with a shared understanding of the framework. However, challenges often remain with respect to a company-wide holistic adoption of digitisation. Reinhardt et al. (2020) note that knowledge of Industry 4.0 is concentrated at the more senior levels of organisations and highlights a disconnect based on seniority. To overcome this, data scientists should be hired early, and be given a senior position where they can shape the company-wide mindset from the get-go. Senior team members should articulate a compelling vision for becoming a data-driven organisation. As an example, GlaxoSmithKline has established the role of Senior Vice President Global Head of Artificial Intelligence and Machine Learning; Relay Therapeutics has established the roles of a Chief Data Officer and Senior Vice President, Artificial Intelligence; and Recursion the role of Chief Technology Officer – to name just a few. Teams should be trained and given the opportunity to learn from one another – for example, by encouraging biologists and data scientists to engage in conversation and to participate in seminars together.

2.5 CASE STUDIES OF SUCCESSFUL DIGITAL TRANSFORMATION

Case Study 1: ML-Driven Protein Engineering Platform for the Rapid Development of Complex and Uniquely Powerful Therapeutic Antibodies (LabGenius)

When progressing from target selection to candidate selection, it is important to build confidence in both the target and the candidate molecule's pharmacology and developability properties. Traditionally, the focus has been on building confidence in binding affinities against targets of interest, but binding optimisation campaigns often show limited correlation with the desired function (Wossnig et al., 2024). Focussing on function rather than affinity allows for the identification of unique and rarer functional biologics, particularly in the context of complex targets, and multi-specific or multi-valent antibodies with complex mechanisms of action such as T-cell engagers. LabGenius specialises in the systematic optimisation of therapeutic antibodies in disease-relevant, cell-based, functional assays. To do this, LabGenius has developed a modular platform, combining synthetic biology, robotic automation, and ML, to run DBTL cycles and uses a design-based ('combinatorial') active learning method called Multi-Objective Bayesian Optimisation (MOBO).

More complex molecules face greater manufacturability challenges. Developability properties are therefore also measured as part of the DBTL optimisation cycle to identify potent and selective molecules with a favourable manufacturability profile. Over successive DBTL cycles, active learning allows LabGenius to intelligently search more of the antibody design space. And by co-optimising for both function and developability, LabGenius selects molecules with a higher probability of being successful in the clinic.

This capability is underpinned by LabGenius' data-generation platform. Lab automation enables rapid and high-throughput standardised experimentation with increased consistency and precision. The data generation is supported by automated data capture and storage in a FAIR manner. Data processing capabilities are designed to make consistent datasets via normalisation based on consistent sets of both function and process controls. In this way, all data is made consistent before it is fed into the data analysis and ML models (Figure 2.5). Data analysis and ML (top layer) hence leverage a complex stack of processes that neatly integrate with each other. Errors or inconsistencies in the lower layers are minimised as errors in the individual layers can result in the accumulation and potentially amplification of errors in subsequent layers, making the data progressively less reliable and accurate. It is therefore essential to have a consistent, automated, and noise/error-free process from the bottom to top to enable good ML predictions and data analysis. Every platform project drives and directs continual refinement in every layer of the underlying tech stack. Any changes to each layer require a close collaboration between the science and the tech teams. Identifying which layer(s) have the biggest impact on data quality is crucial to maintaining the integrity of the stack, and should therefore be assessed continually.

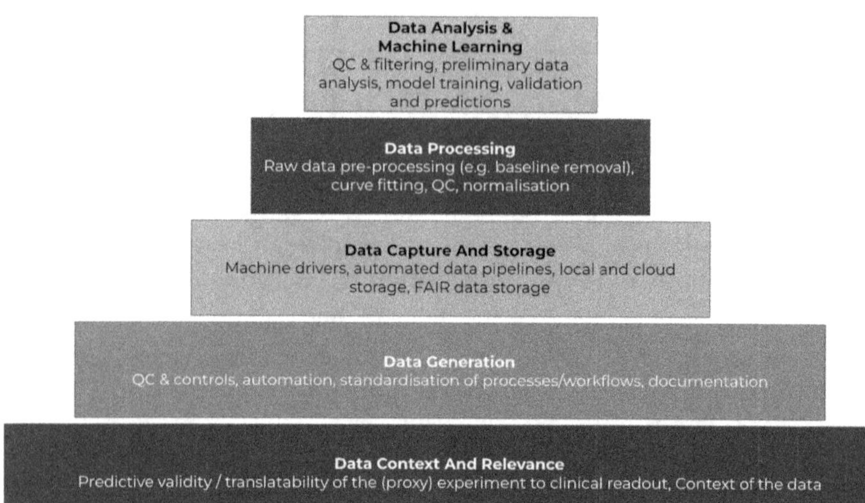

FIGURE 2.5 The complete tech stack for machine learning-driven drug discovery. The lower layers typically have the biggest impact on the actual outcome of a drug discovery programme. Without good foundations (i.e. predictive assays, data-generation, data capturing, and data pre-processing steps), the best analysis can only achieve so much. Data analysis and machine learning could be further broken down into data representation and machine learning models (Wossnig, 2023).

Case Study 2: Creating a Space for Biologicals Chemistry, Manufacturing, and Controls (CMC) Data (Boehringer Ingelheim)

Boehringer Ingelheim Dataland (Making a difference with data science, Boehringer Ingelheim) comprises the idea to provide a cloud platform to integrate data from the whole company in a data foundation layer for further (re-)use (Figure 2.6). The drivers behind the company-wide programme were to enable seamless data availability in one cloud system, connecting data from different domains, and organising data assets in a governed data catalogue. The creation of Dataland made Boehringer Ingelheim data findable, accessible, interoperable, and reusable, highlighting the company's commitment to data-driven thinking.

The first phase of Dataland was reserved for high-value use cases, such as for the clinical operations environment. The platform developed over time and became the standard repository for all data. In time, 'Biobase' began to transform the accessibility of data pertaining to the development and manufacture of biologics, facilitating future R&D pipelines and allowing the company to become more effective in developing breakthrough treatments.

Previously, all data from automated and manual lab processes were captured in LIMS, manufacturing execution systems, or similar systems. Metadata was mainly lined up in reports and spreadsheets. Below we summarise the characteristics of Dataland:

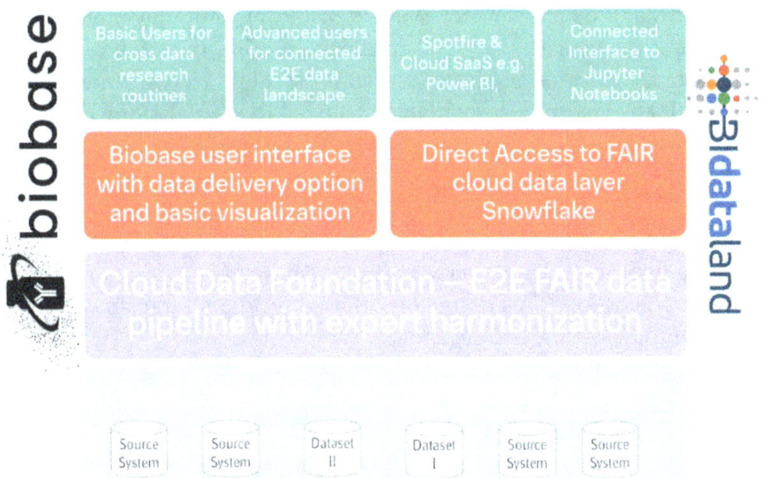

FIGURE 2.6 In the labs, the data is currently manually entered in the form of experimental results, attributes, or analytical values. The classical database structure of LIMS systems as well as manufacturing systems are not directly designed to extract the data for advanced analytics, statistics, ML models, and/or AI applications. The integration into a data layer that consists of harmonised and expert-curated data makes the data findable, accessible, and interoperable between projects, experiments, and scales. Users can tap into this data repository for both, direct project-specific visualisations and decisions, as well as for advanced applications and cross-project and scale comparisons. BIobase is also an advanced and curated data layer in the Boehringer Ingelheim Dataland, that can easily be requested by other applications, e.g. supply chain, etc., and be reused.

- *Goals:* Overcome the data gaps along the value chain
- *Guiding Principles:* (i) Introduce data standards. (ii) Avoid system point-to-point connections. While point-to-point integrations provide low latency when transferring data between systems, they can be extremely challenging to scale and maintain, requiring the user to instruct the workflow every step of the way (Brown & Gupta, 2023).
- *Value:* Data Availability, Interoperability, and Transparency
- *Setup:* Agile product team with Developers (IT) and Subject Matter Experts (Business), Product Owner from the Business

As one of the major learnings, we conclude that the transformation from individual lab solutions and spreadsheets to Biobase has taken a substantial time. Cultural resistance in the organisation was unexpectedly high. Counter intuitively, we expected most resistance from the fact that data has been made transparent and interoperable in the first place. But very few complaints have been made in that direction, while complaints about point-to-point workflows, e.g. in Spotfire, being replaced in favour of a more global data access platform have been frequent.

The breakthrough component for the use of the globally integrated data platform comes with an expensive strategy and roadmap exercise, in which requirements have been formulated, that could not be solved with the 'old' infrastructure anymore. Now, the full potential of a harmonised and integrated data access platform can be harvested, after roughly four years of development.

In conclusion, a technical team that covers the diversity of the task with developers and subject matter experts, and an agile environment, made it possible to develop the product to a technically mature stage within the data land environment of Boehringer Ingelheim. Its final implementation across many business processes required a strong push from the leadership. This ultimately allowed the digital transformation to make a large impact throughout the organisation.

2.6 CONCLUSION

While the pandemic delayed or suspended ~1,200 clinical trials (Xue et al., 2020), it also sped up digital innovation within the biopharmaceutical sector, which had traditionally been slow to incorporate new technologies. Fuelled by an urgency to respond, companies began prioritising investment in digital transformation and implemented changes to their operations. Healthcare was transformed by virtual appointments and digital test results. Amazon Web Services (AWS) launched their machine learning-powered COVID-19 Open Research Dataset (CORD-19) Search for researchers to rapidly access research information pertaining to Coronavirus disease 2019 (COVID-19) (Amazon, 2020). Pfizer used real-time predictive models of COVID-19 to target clinical trial site selection and optimisation; AI and ML were deployed to quickly quality-check and analyse clinical trial data; remote site monitoring enabled clinical trial sites to share source documentation with site monitors virtually for the first time (Pfizer, 2021). Ultimately, digital transformation provided an opportunity and an advantage for the rapid development of a vaccine or therapeutic.

Digital transformation extends far beyond the pandemic. Automation leveraged with AI drives significant scientific discovery, leveraging insight-driven decision-making, removing human bias, and helping to solve complex biological questions that are otherwise intractable using traditional methods. Importantly, it can also speed up the rate at which discoveries are made and can increase the success rate in later stages of the drug development. The impact of computational methods on drug discovery and development will only continue to grow. This presents vast opportunities for companies to expand the number of druggable targets, modalities, and treatable indications, and to design molecules with greater complexity. If adopted appropriately, the digital transformation from the earliest stages of drug discovery will significantly reduce failure rates, bring better therapeutics to the clinic, and incentivise the development of therapeutics for rare diseases. With that comes significant commercial opportunity and the ability to deliver global socio-economic impact.

REFERENCES

Alley, E.C., Khimulya, G., Biswas, S., AlQuraishi, M. and Church, G.M. (2019). Unified rational protein engineering with sequence-based deep representation learning. *Nature Methods* 16:1315–1322. https://doi.org/10.1038/s41592-019-0598-1

Amazon. (2020). AWS launches machine learning enabled search capabilities for COVID-19 dataset. [online] Available at: https://aws.amazon.com/blogs/publicsector/aws-launches-machine-learning-enabled-search-capabilities-covid-19-dataset/. [Accessed 25 Mar. 2024].

Benchling. (2023). Biotech is leaving AI opportunity on the table. Data and culture are the fix. [online] Available at: https://www.benchling.com/blog/data-and-culture-for-AI. [Accessed 25 Mar. 2024].

Bender, A. and Cortés-Ciriano, I. (2021a). Artificial intelligence in drug discovery: What is realistic, what are illusions? Part 1: Ways to make an impact, and why we are not there yet. *Drug Discovery Today* 26:511–524. doi: https://doi.org/10.1016/j.drudis.2020.12.009.

Bender, A. and Cortes-Ciriano, I. (2021b). Artificial intelligence in drug discovery: What is realistic, what are illusions? Part 2: A discussion of chemical and biological data used for AI in drug discovery. *Drug Discovery Today* 26:1040–1052. https://doi.org/10.1016/j.drudis.2020.11.037.

Boehringer Ingelheim. Making a difference with data science. [online] Available at: https://www.boehringer-ingelheim.com/about-us/corporate-profile/making-difference-data-science [Accessed 4 Apr. 2024].

Brandes, N., Ofer, D., Peleg, Y., Rappoport, N. and Linial, M. (2022). ProteinBERT: A universal deep-learning model of protein sequence and function. *Bioinformatics* 38:2102–2110. https://doi.org/10.1093/bioinformatics/btac020.

Brown, T. and Gupta, K. (2023). It's time to ditch point-to-point data integration | Blog | Fivetran. [online] Available at: https://www.fivetran.com/blog/its-time-to-ditch-point-to-point-data-integration [Accessed 4 Apr. 2024].

Cappadona, S., Baker, P.R., Cutillas, P.R., Heck, A.J. and van Breukelen, B. (2012 Sep). Current challenges in software solutions for mass spectrometry-based quantitative proteomics. *Amino Acids* 43(3):1087–1108. https://doi.org/10.1007/s00726-012-1289-8.

Casini, A., Storch, M., Baldwin, G.S. and Ellis, T. (2015). Bricks and blueprints: Methods and standards for DNA assembly. *Nature Reviews Molecular Cell Biology* 16(9):568–576. https://doi.org/10.1038/nrm4014.

Chao, R., Mishra, S., Si, T. and Zhao, H. (2017). Engineering biological systems using automated biofoundries. *Metabolic Engineering* 42:98–108. https://doi.org/10.1016/j.ymben.2017.06.003.

Chao, R., Yuan, Y. and Zhao, H. (2015). Recent advances in DNA assembly technologies. *FEMS Yeast Research* 15(1):1–9. https://doi.org/10.1111/1567-1364.12171.

Choi Y. (2022). Artificial intelligence for antibody reading comprehension: AntiBERTa. *Patterns (N Y)* 3(7):100535.

Dalpé, G. and Joly, Y. (2014). Opportunities and challenges provided by cloud repositories for bioinformatics-enabled drug discovery. *Drug Development Research* 75(6):393–401. https://doi.org/10.1002/ddr.21211.

de la Torre, B.G. and Albericio, F. (2023). The pharmaceutical industry in 2022:An analysis of FDA drug approvals from the perspective of molecules. *Molecules* 28:1038. https://doi.org/10.3390/molecules28031038.

Doulgkeroglou, M.N., Di Nubila, A., Niessing, B., König, N., Schmitt, R.H., Damen, J., Szilvassy, S.J., Chang, W., Csontos, L., Louis, S., Kugelmeier, P., Ronfard, V., Bayon, Y. and Zeugolis, D.I. (2020). Automation, monitoring, and standardization of cell product manufacturing. *Frontiers in Bioengineering and Biotechnology* 8:811. https://doi.org/10.3389/fbioe.2020.00811.

Dounas, A., Cotet, T-S. and Yermanos, A. (2024) Learning immune receptor representations with protein language models. *ArXiv*. https://doi.org/10.48550/arxiv.2402.03823.

Duch, W., Swaminathan, K. and Meller, J. (2007). Artificial intelligence approaches for rational drug design and discovery. *Current Pharmaceutical Design* 13(14):1497–1508. https://doi.org/10.2174/138161207780765954.

Emami Khoonsari, P., Moreno, P., Bergmann, S., Burman, J., Capuccini, M., Carone, M., Cascante, M., de Atauri, P., Foguet, C., Gonzalez-Beltran, A.N., Hankemeier, T., Haug, K., He, S., Herman, S., Johnson, D., Kale, N., Larsson, A., Neumann, S., Peters, K. and Pireddu, L. (2019). Interoperable and scalable data analysis with microservices: Applications in metabolomics. *Bioinformatics* 35(19):3752–3760. https://doi.org/10.1093/bioinformatics/btz160.

FDA (2019). FDA's technology modernization action plan. [online] Available at: https://www.fda.gov/about-fda/reports/fdas-technology-modernization-action-plan [Accessed 25 Mar. 2024].

FDA (2021). FDA Advances data, IT modernization efforts with new office of digital transformation. [online] Available at: https://www.fda.gov/news-events/press-announcements/fda-advances-data-it-modernization-efforts-new-office-digital-transformation [Accessed 25 Mar. 2024].

Fenoy, E., Edera, A.A. and Stegmayer, G. (2022). Transfer learning in proteins: Evaluating novel protein learned representations for bioinformatics tasks. *Brief Bioinform* 23(4):1–19. https://doi.org/10.1093/bib/bbac232

Finelli, L.A. and Narasimhan, V. (2020). Leading a digital transformation in the pharmaceutical industry: Reimagining the way we work in global drug development. *Clinical Pharmacology & Therapeutics* 108(4):756–761. https://doi.org/10.1002/cpt.1850.

Gupta, R., Srivastava, D., Sahu, M., Tiwari, S., Ambasta, R.K. and Kumar, P. (2021). Artificial intelligence to deep learning: Machine intelligence approach for drug discovery. *Molecular Diversity* 25(3):1–46. https://doi.org/10.1007/s11030-021-10217-3.

Heinzinger, M., Elnaggar, A., Wang, Y., Dallago, C., Nechaev, D., Matthes, F. and Rost, B. (2019). Modeling aspects of the language of life through transfer-learning protein sequences. *BMC Bioinformatics* 20(1):723–740. https://doi.org/10.1186/s12859-019-3220-8.

Hole, G., Hole, A.S. and McFalone-Shaw, I. (2021 Oct 6). Digitalization in pharmaceutical industry: What to focus on under the digital implementation process? *International Journal of Pharmaceutics: X* 33:100095. https://doi.org/10.1016/j.ijpx.2021.100095.

ICH. (2019). ICH harmonised guideline, technical and regulatory considerations for pharmaceutical product lifecycle management Q12. [online] Available at: https://database.ich.org/sites/default/files/Q12_Guideline_Step4_2019_1119.pdf [Accessed 25 Mar. 2024].

ICH. (2002). ICH harmonised tripartite guideline, the common technical document for the registration of pharmaceuticals for human use: Quality – M4q(R1). Quality overall summary of module 2, module 3: Quality. [online] Available at: https://database.ich.org/sites/default/files/M4Q_R1_Guideline.pdf [Accessed 25 Mar. 2024].

Ingraham, J.B., Baranov, M., Costello, Z., Barber, K.W., Wang, W., Ismail, A., Frappier, V., Lord, D.M., Ng-Thow-Hing, C., Van Vlack, E.R., Tie, S., Xue, V., Cowles, S.C., Leung, A., Rodrigues, J.V., Morales-Perez, C.L., Ayoub, A.M., Green, R., Puentes, K. and Oplinger, F. (2023). Illuminating protein space with a programmable generative model. *Nature* 623(7989):1070–1078. https://doi.org/10.1038/s41586-023-06728-8.

IQVIA (2023). Global market for medicines to rise to $1.9 trillion by 2027, says report from IQVIA Institute. [online] Available at: https://www.iqvia.com/newsroom/2023/01/global-market-for-medicines-to-rise-to-19-trillion-by-2027-says-report-from-iqvia-institute [Accessed 25 Mar. 2024].

Jumper, J., Evans, R., Pritzel, A., Green, T., Figurnov, M., Ronneberger, O., Tunyasuvunakool, K., Bates, R., Žídek, A., Potapenko, A., Bridgland, A., Meyer, C., Kohl, S.A.A., Ballard, A.J., Cowie, A., Romera-Paredes, B., Nikolov, S., Jain, R., Adler, J. and Back, T. (2021). Highly accurate protein structure prediction with AlphaFold. *Nature* 596(7873):583–589. https://doi.org/10.1038/s41586-021-03819-2.

Kang, D.H., Ko, S.C., Heo, Y.B., Lee, H.J. and Woo, H.M. (2022). RoboMoClo: A robotics-assisted modular cloning framework for multiple gene assembly in biofoundry. *ACS Synthetic Biology* 11(3):1336–1348. https://doi.org/10.1021/acssynbio.1c00628.

King, R.D., Rowland, J., Aubrey, W., Liakata, M., Markham, M., Soldatova, L.N., Whelan, K.E., Clare, A., Young, M., Sparkes, A., Oliver, S.G. and Pir, P. (2009). The robot scientist adam. *Computer* 42(8):46–54. https://doi.org/10.1109/MC.2009.270.

Klambauer, G., Clevert, D.A., Shah, I., Benfenati, E. and Tetko, I.V. (2023). Introduction to the special issue: AI meets toxicology. *Chemical Research in Toxicology* 36(8):1163–1167. https://doi.org/10.1021/acs.chemrestox.3c00217.

Klaus, B. (2023). AI models and drug discovery within pharmaceutical drug market. *Delaware Journal of Public Health* 9(4):52–53. https://doi.org/10.32481/djph.2023.11.009.

LabGenius. (2023). LabGenius debuts T-cell engager optimisation capability that yields molecules with >400-fold tumour killing selectivity versus clinical benchmark. [online] Available at: https://labgeni.us/news-1/pegsboston2023 [Accessed 25 Mar. 2024].

Linshiz, G., Jensen, E., Stawski, N., Bi, C., Elsbree, N., Jiao, H., Kim, J., Mathies, R., Keasling, J.D. and Hillson, N.J. (2016). End-to-end automated microfluidic platform for synthetic biology: From design to functional analysis. *Journal of Biological Engineering* 10(1):3–18. https://doi.org/10.1186/s13036-016-0024-5.

Lu, A.X., Zhang, H., Ghassemi, M. and Moses, A. (2020) Self-supervised contrastive learning of protein representations by mutual information maximization. BioRxiv. https://doi.org/10.1101/2020.09.04.283929.

Lyu, J., Kapolka, N., Gumpper, R., Alon, A., Wang, L., Jain, M.K., Barros-Álvarez, X., Sakamoto, K., Kim, Y., DiBerto, J., Kim, K., Tummino, T.A., Huang, S., Irwin, J.J., Tarkhanova, O.O., Moroz, Y., Skiniotis, G., Kruse, A.C., Shoichet, B.K. and Roth, B.L. (2024). AlphaFold2 structures template ligand discovery. *Science* 384:eadn6354.https://doi.org/10.1126/science.adn6354..

Manning, C. (2020). Artificial intelligence definitions. Stanford University. [online] Available at: https://hai.stanford.edu/sites/default/files/2020-09/AI-Definitions-HAI.pdf [Accessed 25 Mar. 2024].

Martin, H.G., Radivojevic, T., Zucker, J., Bouchard, K., Sustarich, J., Peisert, S., Arnold, D., Hillson, N., Babnigg, G., Marti, J.M., Mungall, C.J., Beckham, G.T., Waldburger, L., Carothers, J., Sundaram, S., Agarwal, D., Simmons, B.A., Backman, T., Banerjee, D. and Tanjore, D. (2023). Perspectives for self-driving labs in synthetic biology. *Current Opinion in Biotechnology* 79:102881. https://doi.org/10.1016/j.copbio.2022.102881.

Mikulic, M. (2024). Pharmaceutical market worldwide revenue 2001–2023. Statista. [online] Available at: https://www.statista.com/statistics/263102/pharmaceutical-market-worldwide-revenue-since-2001/ [Accessed 25 Mar. 2024].

Paul, S.M., Mytelka, D.S., Dunwiddie, C.T., Persinger, C.C., Munos, B.H., Lindborg, S.R. and Schacht, A.L. (2010). How to improve R&D productivity: The pharmaceutical industry's grand challenge. *Nature Reviews Drug Discovery* 9(3):203–214. https://doi.org/10.1038/nrd3078.

Pfizer (2021). Applying the latest digital technology to optimize COVID-19 vaccine efforts. [online] [Accessed 25 Mar. 2024].

Porebski, B.T., Balmforth, M., Browne, G., Riley, A., Jamali, K., Fürst, M.J.L.J., Velic, M., Buchanan, A., Minter, R., Vaughan, T. and Holliger, P. (2023). Rapid discovery of high-affinity antibodies via massively parallel sequencing, ribosome display and affinity screening. *Nature Biomedical Engineering* 8:214–232. https://doi.org/10.1038/s41551-023-01093-3.

Reinhardt, I.C., Oliveira, J.C. and Ring, D.T. (2020). Current perspectives on the development of industry 4.0 in the pharmaceutical sector. *Journal of Industrial Information Integration* 18:100131. https://doi.org/10.1016/j.jii.2020.100131

Rives, A., Meier, J., Sercu, T., Goyal, S., Lin, Z., Liu, J., Guo, D., Ott, M., Zitnick, C.L., Ma, J. and Fergus, R. (2021). Biological structure and function emerge from scaling unsupervised learning to 250 million protein sequences. *Proceedings of the National Academy of Sciences* 118(15):e2016239118. https://doi.org/10.1073/pnas.2016239118.

Saldanha, T. (2019). *Why Digital Transformations Fail: The Surprising Disciplines of How to Take Off and Stay Ahead.* Oakland, CA: Berrett-Koehler Publishers, A Bk Business Book, Cop.

Sarkar, C., Das, B., Rawat, V.S., Wahlang, J.B., Nongpiur, A., Tiewsoh, I., Lyngdoh, N.M., Das, D., Bidarolli, M. and Sony, H.T. (2023). Artificial intelligence and machine learning technology driven modern drug discovery and development. *International Journal of Molecular Sciences* 24(3):2026. https://doi.org/10.3390/ijms24032026.

Scannell, J.W. and Bosley, J. (2016). When quality beats quantity: Decision theory, drug discovery, and the reproducibility crisis. *PLoS One* 11(2):e0147215. https://doi.org/10.1371/journal.pone.0147215.

Scannell, J.W., Bosley, J., Hickman, J.A., Dawson, G.R., Truebel, H., Ferreira, G.S., Richards, D. and Treherne, J.M. (2022). Predictive validity in drug discovery: What it is, why it matters and how to improve it. *Nature Reviews Drug Discovery* 21:915–931. https://doi.org/10.1038/s41573-022-00552-x.

Si, T., Chao, R., Min, Y., Wu, Y., Ren, W. and Zhao, H. (2017). Automated multiplex genome-scale engineering in yeast. *Nature Communications*, 8(1):15187. https://doi.org/10.1038/ncomms15187.

Smiatek, J., Jung, A., Bluhmki, E., (2021) Validation Is Not Verification: Precise Terminology and Scientific Methods in Bioprocess Modeling, *Trends in Biotechnology*, Vol. 39, No. 11, 1117–1119. https://doi.org/10.1016/j.tibtech.2021.04.003

Sparkes, A., Aubrey, W., Byrne, E., Clare, A., Khan, M.N., Liakata, M., Markham, M., Rowland, J., Soldatova, L.N., Whelan, K.E., Young, M. and King, R.D. (2010). Towards robot scientists for autonomous scientific discovery. *Automated Experimentation* 2(1):1. https://doi.org/10.1186/1759-4499-2-1.

Spjuth, O., Frid, J. and Hellander, A. (2021). The machine learning life cycle and the cloud: Implications for drug discovery. *Expert Opinion on Drug Discovery* 16(9):1071–1079. https://doi.org/10.1080/17460441.2021.1932812.

Stephenson, A., Lastra, L.S., Nguyen, B.H., Chen, Y.-J., Nivala, J., Ceze, L. and Strauß, K. (2023). Physical laboratory automation in synthetic biology. *ACS Synthetic Biology* 12(11):3156–3169. https://doi.org/10.1021/acssynbio.3c00345.

Storch, M., Haines, M.C. and Baldwin, G.S. (2020). DNA-BOT: A low-cost, automated DNA assembly platform for synthetic biology. *Synthetic Biology* 5(1):ysaa010. https://doi.org/10.1093/synbio/ysaa010.

Takebe, T., Imai, R. and Ono, S. (2018). The current status of drug discovery and development as originated in United States academia: The influence of industrial and academic collaboration on drug discovery and development. *Clinical and Translational Science* 11(6):597–606. https://doi.org/10.1111/cts.12577.

Thompson, J., Patrick Walters, W., Feng, J.A., Pabon, N.A., Xu, H., Goldman, B.B., Moustakas, D., Schmidt, M. and York, F. (2022). Optimizing active learning for free energy calculations. *Artificial Intelligence in the Life Sciences* 2:100050. https://doi.org/10.1016/j.ailsci.2022.100050.

Varkaris, A., Pazolli, E, Gunaydin, H., Wang, Q., Pierce, L., Boezio, A.A., DiPietro, L., Frost, A., Giordanetto, F., Hamilton, E.P., Harris, K., Holliday, M., Hunter, T.L., Iskandar, A., Ji, Y., Larivée, A., LaRochelle, J.R., Lescarbeau, A., Llambi, F. and Lormil, B. (2024). Discovery and clinical proof-of-concept of RLY-2608, a first-in-class mutant-selective allosteric PI3Ka inhibitor that decouples anti-tumor activity from hyperinsulinemia. *Cancer Discovery* 14(2):240–257. https://doi.org/10.1158/2159-8290.cd-23-0944.

Vijayan, R.S.K., Kihlberg, J., Cross, J.B. and Poongavanam, V. (2022). Enhancing preclinical drug discovery with artificial intelligence. *Drug Discovery Today* 27(4):967–984. https://doi.org/10.1016/j.drudis.2021.11.023.

Walsh, D.I., Pavan, M., Ortiz, L., Wick, S., Bobrow, J., Guido, N.J., Leinicke, S., Fu, D., Pandit, S., Qin, L., Carr, P.A. and Densmore, D. (2019). Standardizing automated DNA assembly: Best practices, metrics, and protocols using robots. *SLAS Technology* 24(3):282–290. https://doi.org/10.1177/2472630318825335.

Wilkinson, M.D., Dumontier, M., Aalbersberg, Ij.J., Appleton, G., Axton, M., Baak, A., Blomberg, N., Boiten, J.-W., da Silva Santos, L.B., Bourne, P.E., Bouwman, J., Brookes, A.J., Clark, T., Crosas, M., Dillo, I., Dumon, O., Edmunds, S., Evelo, C.T., Finkers, R. and Gonzalez-Beltran, A. (2016). The FAIR guiding principles for scientific data management and stewardship. *Science Data* 3:160018. https://doi.org/10.1038/sdata.2016.18.

Williams, K., Bilsland, E., Sparkes, A., Aubrey, W., Young, M., Soldatova, L.N., De Grave, K., Ramon, J., de Clare, M., Sirawaraporn, W., Oliver, S.G. and King, R.D. (2015). Cheaper faster drug development validated by the repositioning of drugs against neglected tropical diseases. *Journal of the Royal Society Interface* 12(104):20141289. https://doi.org/10.1098/rsif.2014.1289.

Wossnig, L. (2023). The Right Data for Good Results: Introducing the 5 'V's of Drug Discovery Data. [online] Available at: https://medium.com/@leowossnig/the-right-data-for-good-results-introducing-the-5-vs-of-drug-discovery-data-331e29c683c5. [Accessed 25 Mar. 2024].

Wossnig, L., Furtmann, N., Buchanan, A., Kumar, S. and Greiff, V. (2024) Best practices for machine learning in antibody discovery and development. *Drug Discovery Today* 29(7):104025. https://doi.org/10.1016/j.drudis.2024.104025

Wouters, O.J., McKee, M. and Luyten, J. (2020). Estimated research and development investment needed to bring a new medicine to market, 2009–2018. *JAMA* 323(9):844. https://doi.org/10.1001/jama.2020.1166.

Wu, Z., Johnston, K.E., Arnold, F.H. and Yang, K.K. (2021). Protein sequence design with deep generative models. *Current Opinion in Chemical Biology* 65:18–27. https://doi.org/10.1016/j.cbpa.2021.04.004.

Xue, J.Z., Smietana, K., Poda, P., Webster, K., Yang, G. and Agrawal, G. (2020). Clinical trial recovery from COVID-19 disruption. *Nature Reviews Drug Discovery* 19:662–663. https://doi.org/10.1038/d41573-020-00150-9.

You, R., Yao, S., Mamitsuka, H. and Zhu, S. (2021). DeepGraphGO: Graph neural network for large-scale, multispecies protein function prediction. *Bioinformatics* 37:i262–i271. https://doi.org/10.1093/bioinformatics/btab270.

Computational Protein Design Strategies for Optimization of Antigen Generation to Drive Antibody Discovery

3

Trevor Wilkinson

3.1 INTRODUCTION

Monoclonal antibodies are established biotherapeutics approved for use in clinical practice. Antibody therapies are used in a wide range of therapeutic applications across disease areas, including oncology, immunomodulation, cardiovascular, metabolic, and infectious diseases. A recent review highlighted that 162 antibody therapeutics had been approved by at least one regulatory agency in different territories globally, underlining the growth of this therapeutic class [1]. This growth and success in the clinic are driven by their high target selectivity, generally good safety profiles, low toxicities, and favourable pharmacokinetics with slow clearance rates. The biologics field is also expanding to include a variety of next-generation formats, which include, but not limited to, antibody–drug conjugates,

 DOI: 10.1201/9781003300311-3

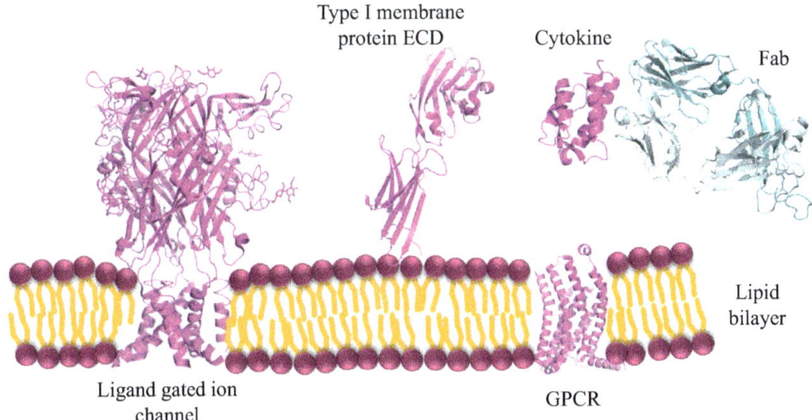

FIGURE 3.1 Structures illustrating diversity and complexity of protein targets of potential therapeutic interest. Each protein is represented in cartoon format generated in Pymol. The cytokine, a soluble secreted protein, is shown here bound to a Fab antibody fragment. Ligand-gated ion channel and GPCR are integral multi-pass membrane proteins. The type I membrane protein has a single-pass transmembrane region (not shown here) and an extracellular domain (ECD) extending beyond the lipid bilayer.

bispecific antibodies, radioimmunoconjugates, and single-domain antibodies, extending the potential clinical applications [2]. Antibody discovery is also able to isolate antibodies that bind a wide range of target proteins (Figure 3.1), which include soluble mediators (e.g. TNF-alpha, IL-17A), receptors (e.g. epidermal growth factor receptor), viral proteins (e.g. SARS-CoV-2 spike protein), transmembrane proteins (e.g. PD-1, PD-L1), and complex multi-spanning membrane proteins (e.g. CD20, CGRP receptor) (Figure 3.1) [3–8]. These target proteins, once identified as a valid therapeutic target, are used as antigens to drive the antibody discovery process to identify potential lead antibodies.

Antibody discovery uses either *in vivo* or *in vitro* strategies (Figure 3.2) to create a source of diverse candidate antibodies from which potential lead antibodies can be isolated [9]. *In vivo* strategies require the immunization of animal hosts with a target antigen to generate an immune response. This has typically involved the immunization of mice (wild type or humanized), followed by the fusion of immune cells with myeloma cells to generate hybridomas. Alternative hosts such as camelids or chickens may also be used [10]. In these cases, antibodies may be isolated by the generation of immune display libraries or B-cell screening [11,12]. *In vitro* strategies involve generation of display libraries (e.g. phage display), which can be used to select encoded antibodies that bind to a target antigen presented to the library [13–15].

To support and drive antibody discovery, an essential aspect is the production of the protein (antigen) targets that are used as immunogens or targets in the application of display technologies. This aspect of antibody discovery requires bioinformatic analysis of the target antigen at the sequence and structural levels, methods to produce high-quality target protein antigen, and formulations that enable antibody–drug discovery. This field has always relied on the application of databases and computational tools, but there is significant potential to enhance this area by the application of artificial intelligence (AI)

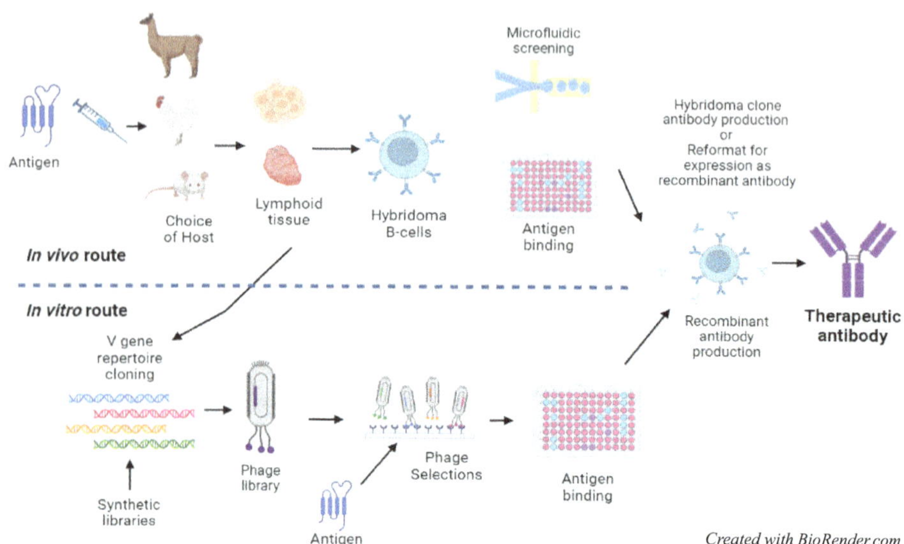

FIGURE 3.2 Strategies used to isolate therapeutic antibodies. Antibodies may be isolated by *in vivo* or *in vitro* methods. Immunizations are performed in a chosen host and lymphoid tissues isolated. For mice, hybridomas are made by fusing immune cells with myeloma followed by screening for antibodies binding to target antigen. Alternatively, B-cells can be screened directly using various microfluidic platforms. Phage display can be performed using phage libraries generated from immune cells derived from immunized hosts or via entirely synthetic libraries. The phage are exposed to target antigen (shown here immobilized on a surface) and binders identified. Antibody sequences can then be produced recombinantly by transient or stable cell expression followed by purification of the antibody chosen.

and machine learning (ML). In this article, methods for antigen design and production are discussed, and the potential of computational tools to enhance this area is explored.

3.2 TARGET PROTEIN (ANTIGEN) CONSIDERATIONS FOR ANTIBODY DISCOVERY

Protein targets that provide potential avenues for therapeutic antibody discovery may be identified by various routes. Approaches to this include deep understanding of the underlying biology of diseases and the proteins involved, genomics and proteomic methods to identify disease-relevant targets, and target-agnostic approaches such as phenotypic screening [16]. More recently, AI-driven approaches are being used to identify potential targets by mining and analysing data from various sources [17]. In the case of therapeutic antibody discovery, the target protein identified must be accessible to the antibody, which focuses attention on extracellular proteins secreted from cells or membrane proteins with accessible potential epitopes on the cell surface.

Once a target protein is selected, a key step in preparing a strategy for antibody discovery is the bioinformatic analysis of the target protein. This analysis is performed to define the optimal antigen design to facilitate antibody discovery according to the selected antibody lead identification technology. Proteins consist of one or several compact, autonomously folding substructures referred to as domains. Protein targets may also be complex in nature. For example, a receptor may function as a homo- or heterodimer or higher order complex. Once a drug discovery target is selected, several databases and computational tools are used to understand the structure–function relationships of proteins and their expression patterns. Thus, analysis is performed to understand the sequence and structure of the protein, prevalence of alternative forms such as splice variants, single-nucleotide polymorphisms (SNPs), post-translational modification, domain structure(s), and potential interaction partners (Figure 3.3). A wide range of databases and computational resources are available for this purpose: Protein features and sequences can be explored using UniProt [18] and Ensembl [19]; solved protein structures can be obtained from the Protein Data Bank (PDB) [20]; domains and boundaries can be predicted using tools based on homology, structure, and *ab initio* methods (reviewed by Wang et al. [21]) and more recent methods such as ResDom [22]. Protein haplotypes also need to be considered, as a single amino acid change in the target protein can impact the binding of an antibody, which would impact therapeutic efficacy. This can be evaluated using Haplosaurus, a tool available in Ensembl, which reflects real-world protein sequence variability and prevalence in populations [23]. A useful online resource describing and providing access to general tools in this area is the Protein Structural Bioinformatics Overview (PreStO) web tool [24]. In the case of membrane proteins, useful resources are the Orientations of Proteins in Membranes (OPM) database and the Positioning of Proteins in Membranes (PPM2.0 and PPM3.0) server [25,26]. This curated web resource provides information on the spatial positions of membrane proteins, topology, and membrane protein types, while PPM calculates the spatial position of proteins in membranes. This resource also provides protein images, structures, and visualization tools. There are also useful resources aimed at specific membrane

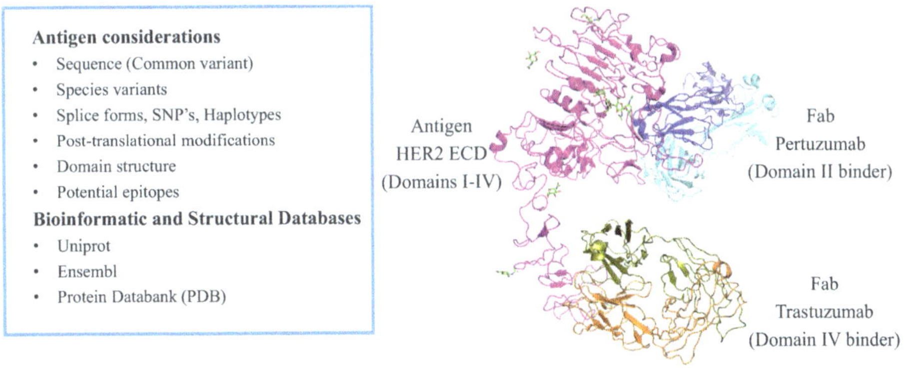

Antigen considerations

- Sequence (Common variant)
- Species variants
- Splice forms, SNP's, Haplotypes
- Post-translational modifications
- Domain structure
- Potential epitopes

Bioinformatic and Structural Databases

- Uniprot
- Ensembl
- Protein Databank (PDB)

Antigen
HER2 ECD
(Domains I-IV)

Fab
Pertuzumab
(Domain II binder)

Fab
Trastuzumab
(Domain IV binder)

FIGURE 3.3 Structural and sequence considerations for protein antigen generation. Cryo-EM structure of HER-2–trastuzumab–pertuzumab complex (PDB ID: 6OGE) demonstrates that proteins can be targeted by antibodies binding distinct epitopes. For each target antigen, key considerations that must be addressed to design an effective antigen are listed along with some key databases.

protein classes, including G protein-coupled receptors (GPCRs) (GPCRdb) and protein channels (ChannelsDB2.0) [27–29]. These databases are subject to regular revisions, with recent releases being linked to AlphaFold.

In addition to understanding the structure and sequence variants of the human target protein antigen, it is important to analyse sequence identity relationships to species variants of the target (e.g. mouse, non-human primate) and if there are any closely related human orthologues and paralogues. The reason for performing this analysis is to get a good understanding of aspects of the target which may impact a therapeutic antibody campaign and to understand and mitigate potential risks of off-target cross reactivity. It is important to understand if there are any closely related family members that could result in undesirable toxicities and should be included in screens to select antibodies that only bind the intended target proteins. An example of this was recently described for the membrane protein Claudin 6 (CLDN6), which is a potential oncotherapeutic target with high expression levels in solid tumours. CLDN6 is highly similar to other CLDN family members, with only three amino acid differences in extracellular loops to CLDN9, which is expressed in healthy tissue (Figure 3.4a) [30]. In identifying an antibody that targets CLDN6, it is necessary to avoid cross reactivity to CLDN9, which could drive toxicity. For therapeutic antibody discovery, the primary focus is on the human target – however, it is important to understand the relationship to species variants to understand if it is feasible to identify an antibody binding both humans and species variants (e.g.

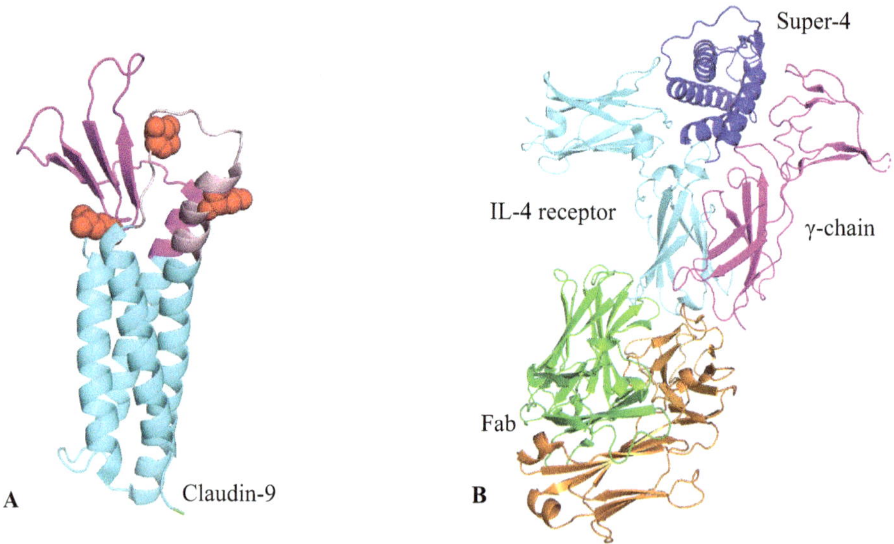

FIGURE 3.4 (a) Structure of Claudin-9 (PDB: 6OV2). The transmembrane helical region and intracellular region are shown as cyan cartoon, while the extracellular loops are shown in magenta. Red space-filled amino acids show the position of residues that differ between Claudin-9 and Claudin-6. An antibody binding selectively to Claudin-6 was isolated (see text), with the lack of Claudin-9 cross-reactivity being driven by a steric block from residue 156. (b). Structure of 'Stapler' antibody isolated by phage display against a stabilized ligand–receptor complex comprising IL-4, IL-4 receptor, and γ-chain. The antibody recognizes an interface formed by the juxtaposition of IL-4 receptor and the γ-chain (PDB: 3BPL).

mouse, non-human primate) to facilitate *in vivo* translational studies. This is also important in selecting the antibody discovery strategy – if a protein shares a very high (>95%) sequence homology with a mouse counterpart, an *in vivo* antibody discovery campaign may require the selection of a more divergent species host for immunizations (e.g. chicken) to avoid issues with tolerance. This is exemplified by the choice of immunisation strategy used for the target CLDN6, where the high (95%) sequence identity with the mouse homologue led to selection of chickens as the host for immunization to identify antibodies binding to CLDN6 [30]. Using this approach, high-affinity binders against CLDN6 were isolated, and antibodies were isolated, which showed minimal to no cross reactivity with CLDN9 and 22 other claudin family members.

3.3 ANTIGEN GENERATION STRATEGIES

The discovery of therapeutics based on antibodies and related molecules, such as single-domain antibodies (Vhh, nanobodies), requires the selection of an appropriate discovery platform. A diverse antibody binding panel can be derived from platforms such as hybridoma, single B-cell methods, or screening natural or synthetic antibody libraries via display technologies using phage, yeast, or mammalian display [13,31–33]. *In vivo* platforms requiring immunization are becoming increasingly diverse, with options for using laboratory mice, chickens, camelids, rabbits, and an expanding range of humanized/transgenic hosts such as the ATX-Gx™ mouse and OmniChicken® [7,34]. In all cases, a critical element is the production of the target protein antigen in sufficient quantity and quality to enable immunization, *in vitro* selection of antibodies binding to antigen, and functional activity assessment. This allows for the identification of lead antibodies that can be used as therapeutics or further engineered to enhance properties such as binding affinity and developability.

Protein antigens need to be produced and purified in a format suitable for antibody discovery via the chosen *in vivo* or *in vitro* platform. The simplest option for antigen generation is the use of synthetic peptides designed to represent the surface-accessible epitopes [35–38]. This approach directs the antibody response to a very specific epitope containing the peptide and has led to marketed biologics against a GPCR target, CCR4 [36]. It is more common for therapeutic antibody discovery to use intact protein antigen, as this will present both linear and conformational epitopes and is more physiologically relevant. Protein is typically produced recombinantly by designing expression plasmids or vectors that allow protein expression and purification from various hosts [39,40]. Selection of an appropriate expression host is somewhat empirical, with no single system being optimal. However, expression hosts that most closely resemble the native host are more likely to produce high-quality recombinant protein due to similar folding and trafficking machinery, cofactors, and post-translational modification pathways. The host selection will also depend on the type of protein being expressed. *Escherichia coli* remains very attractive as a host for the production of proteins, particularly if the target protein does not require post-translational modification [39]. Integral membrane proteins will typically require eukaryotic expression systems such as Human embryonic kidney (HEK) or Chinese hamster ovary (CHO) cells [41]. These expression hosts can be used as a source of recombinant target protein that can be extracted and purified (e.g. by detergents or polymers such as styrene maleic acid) or used to generate virus-like particles incorporating the target protein [42–44].

Designing appropriate constructs for expression requires an understanding of protein structure, the intended use of protein (e.g. immunogen and/or screening), and the quantity/quality required. In expression construct design, thought must be given to promoter strength, protein sequence and structure (full length and/or selected domain(s)), use of signal peptides (native or alternative), and fusion tags, which may be used for purification or to introduce a specific feature to facilitate screening (e.g. AviTag to allow biotinylation). Integral membrane proteins require special consideration depending on their type and structure. For example, type I membrane proteins have a single transmembrane region and may have a large extracellular domain(s), which can be expressed directly as a well-folded protein [45]. Multi-spanning membrane proteins (e.g. GPCRs, ion channels) are more challenging to produce, require detergent extraction prior to purification, and often need reconstitution into lipid membranes using formats such as nanodiscs to maintain function and stability [46–49]. Alternatively, the membrane protein can be expressed in a virus-like particle format, which avoids detergent extraction [50]. Finally, for membrane protein targets, it is always necessary to have a cell-based system expressing the protein in a native membrane lipid environment. This is to enable confirmation that any antibodies discovered retain binding to the native target protein. This could be a cell line naturally expressing the target protein or a recombinantly engineered cell line overexpressing the target protein of interest [51].

3.4 COMPUTATIONAL METHODS

Antibody discovery, whether following an immunization or display strategy, requires the production of a target antigen to enable both antibody selection/screening and functional characterization by binding or via biochemical or cell-based assays. This typically requires purified protein but may also use cells overexpressing target protein, Virus-like particles (VLPs), and immunization strategies employing priming with DNA or RNA encoding the target of interest, followed by a boost with an alternative source of antigen [10,12]. The production of recombinant target antigens can be very challenging, particularly for multi-pass membrane proteins such as G-protein-coupled receptors, ion channels, transporters, and tetraspanins [30,52–54]. Although this can be obviated using DNA/RNA as the immunogen, a source of target protein is essential for screening and functional validation of antibodies. These methods are currently mostly experimental and highly resource-intensive.

An emerging and accelerating trend in antibody discovery is the impact of improvements in computational and MI approaches, which hold the promise of revolutionizing biologics drug discovery. These developments have the potential to impact antigen design, antigen production, and antibody discovery/optimization. Developments in these areas are moving at different paces, with antigen design and antibody discovery/optimization being more advanced. It is somewhat challenging to separate these two areas, but several comprehensive reviews have explored the impact of computational methods and MI on antibody discovery and design [55–59]. In this chapter, the focus is on antigen design and computational methods that are impacting this field.

3.4.1 Impact of AI/ML on Target Protein Expression, Construct Design, and Protein Production

In designing target protein antigens for antibody discovery, a key element of a successful strategy is the generation of sufficient high-quality antigens. Experimentally, this typically requires expression of the protein antigen in a heterologous system, followed by purification of the protein and formulation. This has often required bespoke design of expression constructs and an empirical approach to expression leading to mixed or unpredictable outcomes. This can provide a bottleneck in recombinant antigen production. This is perhaps unsurprising given the complexity of the processes involved; parameters that need to be considered include choice of expression host, codon usage, mRNA synthesis, translation initiation, and protein solubility. In the case of secretory proteins, signal peptides are required for protein translocation and processing via the secretory pathway in cells. Just taking one of these considerations, mRNA abundance alone is not sufficient to explain protein expression abundance [60]. There is thus considerable interest in building models which may make sequence-to-expression models available that allow better prediction of expression outcomes. Recent progress in the areas of DNA synthesis, DNA sequencing, automation, and deep learning is now being explored to develop deep learning models that allow sequence-to-expression prediction [61].

This field is advancing, particularly in the area of *E. coli* protein expression reflecting the relative ease and low costs of performing expression experiments compared to other hosts. Computation-based methods are now emerging, which allow improvements in protein expression, stability, and function and provide tools to 'tune' protein expression based on more optimal protein designs, introduction of mutations conducive to protein expression, or synonymous codon changes to improve the performance of translation initiation sites [62–64]. Examples of these strategies and tools are ProteinMPNN [62], MPEPE (*m*utation *p*redictor for *e*nhanced *p*rotein *e*xpression) [63], and the TIsigner.com web service [65]. The TIsigner web service combines TIsigner (*t*ranslation *i*nitiation coding region de*signer*), SoDoPE (*s*o̲luble *do*main for *p*rotein *e*xpression), and Razor, a tool for signal peptide analysis. Codon optimization has also been explored using a deep learning approach called bidirectional long short-term memory conditional random field supported by experimental validation by comparison to commercially accessible algorithms from Genewiz and ThermoFisher [66]. Although these approaches have focused on bacterial expression, the application of deep learning methods to other hosts, such as mammalian systems, is to be expected, particularly given the increasing use of automation to generate sequence-to-expression datasets [67].

One area where we can expect further impact of ML and deep learning methodologies is for more complex targets such as membrane protein sequences, structures, and expression. The impact of MI in computational modelling of membrane proteins has been reviewed recently, suggesting a growing number of applications impacting membrane protein classification, topology identification, and interaction site detection [68,69]. Examples of practical applications of ML for multi-spanning membrane proteins are emerging. An example of this can be taken from the research area that enables structural studies of GPCRs. GPCRs play critical roles in cellular signalling and are major drug targets. However, their inherent instability in non-native environments (e.g. when extracted from the membrane using detergents) complicates their structural analysis and biophysical characterisation. A well established technique to improve GPCR stability is to introduce point mutations in transmembrane helices to enhance the thermostability of the GPCR [70]. This has proven useful in solving the structures of these inherently flexible proteins. Experimentally, this is a challenging and laborious process. Recently, a computational approach coupled with MI, trained on thermostability data of 1,231 mutants, has been used to predict thermostabilizing mutations in advance of experiments. Using the C5a receptor as an example, a blind prediction located 36% of thermostable mutants in the top 50 prioritized mutants versus 3% in the first 50 attempts using systematic alanine scanning [71]. In a further application of MI-guided engineering, Bedbrook et al. trained statistical models enabling the design of highly functional light-gated channelrhodopsins (ChR) [72]. Thirty designed ChR variants were made, which had improved properties and light sensitivity. Ongoing developments in this area hold promise for facilitating our ability to design and express membrane protein targets suitable for use as antigens. One note of caution here is that if mutated proteins are used as antigens, it is important to check their function and ensure that methods are in place to confirm that any antibodies identified bind to the native, wild-type protein.

3.4.2 Computational Protein Structure Prediction

Computational structure prediction of target protein antigens has long been part of therapeutic antibody discovery [73]. Homology modelling has been an important method for building 3D protein antigen structures based on using protein primary sequence, multiple sequence alignments, and knowledge gained from structural similarities to related proteins with solved structures. This involves a sequential process whereby sequence alignment is performed, structural templates are selected, protein backbones are built, and sidechains are added, followed by an optimization step [74,75]. This can be effective where the sequence identity between a target antigen and a protein homologue is at least 30%. This type of approach is very useful in providing structural information where a solved crystal structure is unavailable or too challenging to achieve experimentally. The information provided by the homology modelling can assist with recombinant antigen design by providing information on domain structure. This has a practical impact on issues such as where to place purification/epitope tags to facilitate protein production and use in screening and design of antigens, which may focus antibody generation on a desired domain/epitope.

An example of the use of antigen homology models for antibody design is demonstrated by engineering a functional antibody directed to IL-17A. In this study, IL-17A was modelled in its receptor-bound conformation using the experimentally solved structure of IL-17F in its receptor-bound conformation. These cytokines have a 61% sequence identity. Using this structure, a collection of sequence-unique antibodies was used to select a scaffold antibody based on *in silico* docking of each antibody to IL-17A using ZDOCK [76]. The selected antibody was used to design a focused yeast display library, which allowed isolation of a functional IL-17A binder that inhibited binding of IL-17A to its receptor with a nanomolar EC50. In another example, we have used homology modelling combined with protein–protein docking to guide affinity maturation of an antibody targeting murine CCL20 [77]. The crystal structure of murine CCL20 was available (PDB ID: 1HA6), and the structure of the antibody (AB1 Fv) was generated using SabPred [78]. Docking poses were obtained using RDOCK [79] and confirmed by a combination of knowledge gained from cross-reactivity profiles (AB1 does not bind human CCL20) and experimental alanine scanning. *In silico* affinity maturation was performed by adopting three protocols (Biovia Discovery Studio [80], Schrödinger Biologics Suite [81], and Rosetta [82]), and two single-point mutations were identified, which increased the affinity of AB1 for CCL20.

Recent advances in computational design are offering the prospect of a new vision for antibody discovery where prospects for *in silico* design of antibodies are a realistic aspiration [56,83]. One element of this is the reasonable expectation that accurately predicted structures of most proteins will be available [84]. Table 3.1 captures a selection of computational methods used to predict protein structures. In addition to computational methods, continuing advances in experimental technologies, particularly Cryogenic electron microscopy (Cryo-EM), are increasing the deposition of solved structures to the PDB, particularly for membrane proteins, where deposition has increased from 30–40 unique structures per year to 70–80 [85]. The deep neural network tools AlphaFold2 [86] and RoseTTAFold [87] have, and continue, to make a significant impact on protein structure availability, enabling high-accuracy prediction of protein structures from amino acid sequence. This opens the way to having an accurate structural model of any given target protein/antigen [86]. The availability of a structure for a protein antigen is important to assess protein stability and solubility computationally using tools such as CamSol [88,89]. The structure also allows the design of target

TABLE 3.1 Selected computational protein structure prediction tools

PREDICTION TOOL	*URL*	*REFERENCE*
AlphaFold	https://alphafold.ebi.ac.uk	[86]
I-TASSER	https://zhanggroup.org/I-TASSER/	[102]
Modeller	https://salilab.org/modeller/	[103]
Phyre2	http://www.sbg.bio.ic.ac.uk/phyre2	[104]
SWISS-MODEL	https://swissmodel.expasy.org/	[105]
Rosetta3	https://www.rosettacommons.org/software	[106]
C-QUARK	https://zhanggroup.org/QUARK/	[107]

antigens, which contain mutated residues to enhance protein stability or trap proteins in desirable conformations to focus immune responses to therapeutically important epitopes. This could, for example, involve the introduction of disulphide bonds to stabilize conformationally dynamic structures, as performed for the respiratory syncytial virus (RSV) fusion glycoprotein [90].

In considering the application of computational tools to improve the immunogenic potential of selected antigens, epitope prediction and the ability to target pre-selected epitopes are of key importance, whether in the field of therapeutic antibody generation or in vaccine development. In the case of B-cell epitope prediction, methods can be described as sequence-based or structure-based. Structure-based methods are generally considered more reliable, but their use is limited by the availability of solved antibody–antigen structures. Structural approaches are generally more powerful, as most epitopes are conformational. Computational epitope prediction methods include EpiPred [91], MabTope [92], and DiscoTope 3.0 [93]. This creates an interesting issue, however, as it is apparent that, in principle, any surface-accessible area on a protein antigen can be a potential epitope given the correct antibody binding partner [94]. This is observed in therapeutic antibodies where approved and clinically successful antibodies bind to the same target but via distinct epitopes (Figure 3.3). A further observation from antibody discovery campaigns is that screening procedures may select the tightest binders, which often target immunodominant epitopes and may miss other functionally valid epitopes [95]. The issue then becomes: How can we predict and select the most relevant functional epitope? In some cases (e.g. targeting a receptor–ligand interaction), we can focus attention on the protein–protein interaction surfaces of either the receptor or ligand as likely epitopes of interest. In other cases, multiple non-overlapping epitopes may be targeted and can only be resolved by functional testing and assays designed to select the required mechanism of action. In the following section, some case studies are presented of antigen design strategies to enhance antibody–drug isolation.

3.5 CASE STUDY EXAMPLES OF ANTIGEN DESIGN STRATEGIES TO DRIVE DRUG DISCOVERY AND IMMUNOGEN PERFORMANCE

Rational design of antigens is of considerable interest in driving therapeutic antibody discovery. This is also relevant in the design of immunogens for vaccination, particularly as we emerge from the COVID pandemic, which accelerated research for both vaccines and neutralizing monoclonal antibodies targeting SARS-CoV-2 [96]. There are thus opportunities for learning from the fields of antibody discovery and emerging vaccine strategies. This focuses attention on antigen design, which optimizes humoral response, prevents or reduces off-target antibody responses, and specifically targets preferred epitopes [97].

In generating antigens for therapeutic antibody campaigns, knowledge of the target protein structure, interacting partners, feasibility of production, and desired antibody effect (e.g. blocking antibody, antibody–drug conjugate) is key, and computational methods can enhance outcomes. This is illustrated in the following selected case studies.

In the first example, a strategy was designed to isolate antibodies that acted to 'staple' a molecular heterodimer by binding a composite epitope of two-receptor extracellular domains (Figure 3.4b). Interleukin-4 (IL-4) signals through either a type I heterodimeric receptor comprising IL-4Rα and the common γ-chain or a type II receptor composed of IL-4Rα and IL-13Rα1. A crystal structure of IL-4 in a ternary complex with the cytokine receptor extracellular domains showed that the two-receptor membrane proximal domains were in contact forming a neoepitope. An engineered variant of IL-4 was available (Super-4) with 3,700-fold higher affinity than IL-4, which stabilized the IL-4 receptor complex. With this information, an antigen preparation was made in which a receptor was tagged with a C-terminal biotin-acceptor peptide allowing biotinylation and expressed receptor subunits were complexed with Super-4. This biotinylated complex was used with a yeast scFv library to isolate a scFv, which bound specifically to the membrane–proximal receptor interface. This binding mode was confirmed by solving a structure of the antibody (reformatted as a Fab) with the ternary complex, which showed the antibody binding to the receptor interface [98].

In a second case study, the concept of *de novo* design of antibody Complementarity determining regions (CDR) loops was explored based on targeting epitopes for proteins in which a structure is available either from an experimentally solved structure or a computational model [95]. This approach required the exploitation of large structural databases to use a fragment-based procedure to design CDRs complementary to a selected target epitope. This approach was experimentally tested using single-domain antibodies based on their monomer structures, ease of production, and small size. To develop this approach, a database of CDR-like fragments and corresponding antigen-like regions was compiled from the non-redundant PDB to build the so-called AbAg database. Given a known structural target epitope, the database is searched to locate antigen-like regions similar to the epitope. To perform the search, the input epitope is fragmented into either linear or surface-patch fragments. The search method identifies CDR-like fragments that may interact with the target epitope. These fragments are evaluated for favourable interactions and ranked. Top-ranking designed CDR motifs are then grafted to an antibody scaffold that is tolerant to loop replacement. This method does require that the structure of the target antigen is known but is capable of working with both experimentally derived structures and models created by tools such as AlphaFold2 [95]. This technique was applied to three target antigens – human serum albumin, bovine trypsin, and SARS-CoV-2 spike protein receptor-binding domain (RBD). The six single-domain antibodies designed bound to the target proteins with affinities ranging from 120 to 1,800 nM, with two antibodies targeting the RBD exhibiting affinities of 130 and 210 nM.

A final case study example is drawn from the field of vaccine research targeting RSV, which is the leading cause of serious respiratory disease in infants. Although a prophylactic, humanized monoclonal antibody, palivizumab, which targets RSV F protein, is approved, vaccine design has lagged behind. RSV F protein has been the primary antigen target for vaccine development. This is a class I fusion glycoprotein and requires proteolytic cleavage for activation and formation of the mature trimer.

The RSV F protein is responsible for fusing the viral and host cell membranes during virus cell entry and exists in two antigenically unique forms – a pre-fusion form and a post-fusion form, which undergoes a dramatic conformational change. The pre-fusion form is the target of most neutralizing antibodies [99]. A key issue in targeting the pre-fusion form of RSV F is that the protein is metastable and can spontaneously change conformation when extracted from membranes with detergent or if subjected to various stresses (e.g. physical, chemical, or temperature stresses). To tackle this issue, stabilization of the pre-fusion protein is an attractive strategy. To achieve stabilization, pre-fusion stabilizing mutations were computationally designed using a combination of methods – introduction of cysteine residue pairs to form stabilizing disulphide bonds; introduction of non-polar amino acids to fill cavities in the pre-fusion protein that might otherwise allow for movement; and introduction of charged residue mutations to decrease ionic repulsion or increase ionic interaction between residues that are close in the pre-fusion state [90]. These residue mutations were computationally assessed using a combination of Schrodinger BioLuminate, Molecular Operating Environment [100], and Rosetta-based stability prediction protocols [101]. Using computational design, 398 hypothetical F domain constructs with combinations of disulphide, cavity filling, and charge mutations were expressed as ectodomain constructs, including a fibritin foldon trimerization domain at the C-terminus. In general, the disulphide mutations had the greater stabilizing effect on the pre-fusion conformation. Pre-fusion F constructs that exhibited greater stabilization were tested and elicited a 10-fold higher serum-neutralizing titre than a prototype vaccine F protein candidate. Introduction of the stabilizing mutations onto F-proteins of two major RSV subgroups, followed by immunization led to complete protection against RSV challenge.

These selected case studies illustrate how a combination of protein structure-informed design and computational techniques can enhance antibody generation in both a therapeutic antibody generation context and in vaccine immunogen design.

3.6 CONCLUSIONS: COMPUTATIONAL ANTIGEN DESIGN AND FUTURE DEVELOPMENTS

In this chapter, the topic of protein antigen and immunogen design has been explored, and the potential impact of computational tools in facilitating aspects of antigen generation from protein expression to protein structure has been examined. Advances in computing power, increasing capture of data from existing experimental approaches, and enhancements to the speed of determining protein structures, coupled with MI approaches, are leading to rapid breakthroughs and opportunities in biologics drug discovery. The clear, long-term ambition of structure-based and computational antigen and antibody design is the ability to design, entirely *in silico,* antibodies, which bind to a selected target. This breakthrough is imminent, although it seems certain to bring further challenges. Designing antibodies de novo presents several scientific and technical challenges due to the complexity of antibody structures and their interactions with antigens. Assuming we can computationally design an antibody which binds to a target antigen with high affinity there are several other challenges to consider. This includes, but is not limited to, issues such as the antibodies developability profile and its pharmacokinetic, pharmacodynamic and toxicity profile. We are also experiencing an era where novel formats are being explored such as bi- and multi-specifics, antibody–drug conjugates, and alternative scaffolds, each of which brings its own challenges. At the core of this challenge is the requirement to generate target antigen to drive biologics drug discovery and validate the performance of designed biologics. Given the diversity and complexity of potential protein targets, there remains much to be learned, and computational techniques must be central to advances in this area.

REFERENCES

1. Lyu, X., Zhao, Q. and Hui, J. et al., The global landscape of approved antibody therapies. *Antib Ther*, 2022. **5**(4): pp. 233–57.
2. Goulet, D.R. and W.M. Atkins, Considerations for the design of antibody-based therapeutics. *J Pharm Sci*, 2020. **109**(1): pp. 74–103.
3. Shi, L., Lehto, S.G., Zhu, D.X.D. et al., Pharmacologic characterization of AMG 334, a potent and selective human monoclonal antibody against the calcitonin gene-related peptide receptor. *J Pharmacol Exp Ther*, 2015. **356**(1): pp. 223–31.
4. Rouge, L., Chiang, N., Steffek, M. et al., Structure of CD20 in complex with the therapeutic monoclonal antibody rituximab. *Science*, 2020. **367**(6483): pp. 1224–30.
5. Scapin, G., Yang, X., Prosise, W.W. et al., Structure of full-length human anti-PD1 therapeutic IgG4 antibody pembrolizumab. *Nat Struct Mol Biol*, 2015. **22**(12): pp. 953–8.
6. Catley, M.C., Coote, J., Bari, M. et al., Monoclonal antibodies for the treatment of asthma. *Pharmacol Ther*, 2011. **132**(3): pp. 333–51.
7. Mullen, T.E., Abdullah, R., Boucher, J. et al., Accelerated antibody discovery targeting the SARS-CoV-2 spike protein for COVID-19 therapeutic potential. *Antib Ther*, 2021. **4**(3): pp. 185–96.
8. Jakobovits, A., Amado, R.G., Yang, X. et al., From XenoMouse technology to panitumumab, the first fully human antibody product from transgenic mice. *Nat Biotechnol*, 2007. **25**(10): pp. 1134–43.
9. Laustsen, A.H., Greiff, V., Karatt-Vellatt, A. et al., Animal immunization, in vitro display technologies, and machine learning for antibody discovery. *Trends Biotechnol*, 2021. **39**(12): pp. 1263–73.
10. Banik, S.S.R., Kushnir, N., Doranz, B.J. et al., Breaking barriers in antibody discovery: harnessing divergent species for accessing difficult and conserved drug targets. *MAbs*, 2023. **15**(1): pp. 2273018.
11. Pedrioli, A. and A. Oxenius, Single B cell technologies for monoclonal antibody discovery. *Trends Immunol*, 2021. **42**(12): pp. 1143–58.
12. Abdiche, Y.N., Harriman, R., Deng, X. et al., Assessing kinetic and epitopic diversity across orthogonal monoclonal antibody generation platforms. *MAbs*, 2016. **8**(2): pp. 264–77.
13. Alfaleh, M.A., Alsaab, H.O., Mahmoud, A.B. et al., Phage display derived monoclonal antibodies: from bench to bedside. *Front Immunol*, 2020. **11**: p. 1986.
14. Moraes, J.Z., Hamaguchi, B., Braggion, C. et al., Hybridoma technology: is it still useful? *Curr Res Immunol*, 2021. **2**: pp. 32–40.

15. Winters, A., McFadden, K., Bergen, J. et al., Rapid single B cell antibody discovery using nanopens and structured light. *MAbs*, 2019. **11**(6): pp. 1025–35.
16. Minter, R.R., A.M. Sandercock, and S.J. Rust, Phenotypic screening-the fast track to novel antibody discovery. *Drug Discov Today Technol*, 2017. **23**: pp. 83–90.
17. Pun, F.W., I.V. Ozerov, and A. Zhavoronkov, AI-powered therapeutic target discovery. *Trends Pharmacol Sci*, 2023. **44**(9): pp. 561–72.
18. UniProt, C., UniProt: the universal protein knowledgebase in 2023. *Nucleic Acids Res*, 2023. **51(D1)**: pp. D523–31.
19. Cunningham, F., Achuthan, P., Akanni, W. et al., Ensembl 2019. *Nucleic Acids Res*, 2019. **47(D1)**: pp. D745–51.
20. Goodsell, D.S., Zardecki, C., Di Costanzo, L. et al., RCSB protein data bank: enabling biomedical research and drug discovery. *Protein Sci*, 2020. **29**(1): pp. 52–65.
21. Wang, Y., Zhang, H., Zhong, H. et al., Protein domain identification methods and online resources. *Comput Struct Biotechnol J*, 2021. **19**: pp. 1145–53.
22. Wang, L., Zhong, H., Xue, Z. et al., Res-Dom: predicting protein domain boundary from sequence using deep residual network and Bi-LSTM. *Bioinform Adv*, 2022. **2**(1): p. vbac060.
23. Spooner, W., McLaren, W., Slidel, T. et al., Haplosaurus computes protein haplotypes for use in precision drug design. *Nat Commun*, 2018. **9**(1): p. 4128.
24. Paiva, V.A., Nagarajan, R., and Selvaraj, S Protein structural bioinformatics: an overview. *Comput Biol Med*, 2022. **147**: p. 105695.
25. Lomize, M.A., Pogozheva, I.D., Joo, H. et al., OPM database and PPM web server: resources for positioning of proteins in membranes. *Nucleic Acids Res*, 2012. **40(Database** issue): pp. D370–6.
26. Lomize, A.L., S.C. Todd, and I.D. Pogozheva, Spatial arrangement of proteins in planar and curved membranes by PPM 3.0. *Protein Sci*, 2022. **31**(1): pp. 209–20.
27. Spackova, A., Vávra, O., Raček, T. et al., ChannelsDB 2.0: a comprehensive database of protein tunnels and pores in AlphaFold era. *Nucleic Acids Res*, 2024. **52(D1)**: pp. D413–8.
28. Munk, C., Mutt, E., Isberg, V. et al., An online resource for GPCR structure determination and analysis. *Nat Methods*, 2019. **16**(2): pp. 151–62.
29. Pandy-Szekeres, G., Caroli, J., Mamyrbekov, A. et al., GPCRdb in 2023: state-specific structure models using AlphaFold2 and new ligand resources. *Nucleic Acids Res*, 2023. **51(D1)**: pp. D395–402.
30. Screnci, B., Stafford, L.J., Barnes, T. et al., Antibody specificity against highly conserved membrane protein Claudin 6 driven by single atomic contact point. *iScience*, 2022. **25**(12): p. 105665.
31. Bowley, D.R., Labrijn, A.F., Zwick, M.B. et al., Antigen selection from an HIV-1 immune antibody library displayed on yeast yields many novel antibodies compared to selection from the same library displayed on phage. *Protein Eng Des Sel*, 2007. **20**(2): pp. 81–90.
32. Robertson, N., Lopez-Anton, N., Gurjar, S.A. et al., Development of a novel mammalian display system for selection of antibodies against membrane proteins. *J Biol Chem*, 2020. **295**(52): pp. 18436–48.
33. Beerli, R.R. and C. Rader, Mining human antibody repertoires. *MAbs*, 2010. **2**(4): pp. 365–78.
34. Ching, K.H., Berg, K., Morales, J., et al., Expression of human lambda expands the repertoire of OmniChickens. *PLoS One*, 2020. **15**(1): p. e0228164.
35. Cox, J.H., Hussell, S., Søndergaard, H. et al., Antibody-mediated targeting of the Orai1 calcium channel inhibits T cell function. *PLoS One*, 2013. **8**(12): p. e82944.
36. Niwa, R., Shoji-Hosaka, E., Sakurada, M. et al., Defucosylated chimeric anti-CC chemokine receptor 4 IgG1 with enhanced antibody-dependent cellular cytotoxicity shows potent therapeutic activity to T-cell leukemia and lymphoma. *Cancer Res*, 2004. **64**(6): pp. 2127–33.

37. Sasaki, Y., Kosaka, H., Usami, K. et al., Establishment of a novel monoclonal antibody against LGR5. *Biochem Biophys Res Commun*, 2010. **394**(3): pp. 498–502.
38. Lee, J.H., Park, C.K., Chen, G. et al., A monoclonal antibody that targets a NaV1.7 channel voltage sensor for pain and itch relief. *Cell*, 2014. **157**(6): pp. 1393–404.
39. Kesidis, A., Depping, P., Lodé, A. et al., Expression of eukaryotic membrane proteins in eukaryotic and prokaryotic hosts. *Methods*, 2020. **180**: pp. 3–18.
40. Assenberg, R., Wan, P.T., Geisse, S. et al., Advances in recombinant protein expression for use in pharmaceutical research. *Curr Opin Struct Biol*, 2013. **23**(3): pp. 393–402.
41. Jain, N.K., Barkowski-Clark, S., Altman, R. et al., A high density CHO-S transient transfection system: comparison of ExpiCHO and Expi293. *Protein Expr Purif*, 2017. **134**: pp. 38–46.
42. Jamshad, M.,, Lin, Y.P., Knowles, T.J. et al., Surfactant-free purification of membrane proteins with intact native membrane environment. *Biochem. Soc. Trans.*, 2011. **39**: pp. 813–18.
43. Zeltins, A., Construction and characterization of virus-like particles: a review. *Mol Biotechnol*, 2013. **53**(1): pp. 92–107.
44. Schmidpeter, P.A.M., N. Sukomon, and C.M. Nimigean, Reconstitution of membrane proteins into platforms suitable for biophysical and structural analyses. *Methods Mol Biol*, 2020. **2127**: pp. 191–205.
45. Ravn, P., Madhurantakam, C., Kunze, S. et al., Structural and pharmacological characterization of novel potent and selective monoclonal antibody antagonists of glucose-dependent insulinotropic polypeptide receptor. *J Biol Chem*, 2013. **288**(27): pp. 19760–72.
46. Birch, J. and A. Quigley, The high-throughput production of membrane proteins. *Emerg Top Life Sci*, 2021. **5**(5): pp. 655–63.
47. Bayburt, T.H. and S.G. Sligar, Membrane protein assembly into Nanodiscs. FEBS Lett, 2009. **584**: pp. 1721–7.
48. Agharkar, A., Rzadkowolski, J., McBroom, M. et al., Detergent screening of the human voltage-gated proton channel using fluorescence-detection size-exclusion chromatography. *Protein Sci*, 2014. **23**(8): pp. 1136–47.
49. Wang, M., Wei, L., Xiang, H. et al., A megadiverse naive library derived from numerous camelids for efficient and rapid development of VHH antibodies. *Anal Biochem*, 2022. **657**: p. 114871.
50. Tucker, D.F., Sullivan, J.T., Mattia, K.A. et al., Isolation of state-dependent monoclonal antibodies against the 12-transmembrane domain glucose transporter 4 using virus-like particles. *Proc Natl Acad Sci U S A*, 2018. **115**(22): pp. E4990–9.
51. Elegheert, J., Behiels, E., Bishop, B. et al., Lentiviral transduction of mammalian cells for fast, scalable and high-level production of soluble and membrane proteins. *Nat Protoc*, 2018. **13**(12): pp. 2991–3017.
52. Gulezian, E.,, Crivello, C., Bednenko, J. et al., Membrane protein production and formulation for drug discovery. *Trends Pharmacol Sci*, 2021. **42**(8): pp. 657–74.
53. Hutchings, C.J., P. Colussi, and T.G. Clark, Ion channels as therapeutic antibody targets. *MAbs*, 2019. **11**(2): pp. 265–96.
54. Dodd, R., Schofield, D.J., Wilkinson, T. et al., Generating therapeutic monoclonal antibodies to complex multi-spanning membrane targets: overcoming the antigen challenge and enabling discovery strategies. *Methods*, 2020. **180**: pp. 111–26.
55. Guarra, F. and G. Colombo, Computational methods in immunology and vaccinology: design and development of antibodies and immunogens. *J Chem Theory Comput*, 2023. **19**(16): pp. 5315–33.
56. Bauer, J., Rajagopal, N., Gupta, P. et al., How can we discover developable antibody-based biotherapeutics? *Front Mol Biosci*, 2023. **10**: p. 1221626.
57. Schoeder, C.T., Schmitz, S., Adolf-Bryfogle, J. et al., Modeling immunity with Rosetta: methods for antibody and antigen design. *Biochemistry*, 2021. **60**(11): pp. 825–46.

58. Zhao, J., Nussinov, R., Wu, W.J. et al., In dilico methods in sntibody design. *Antibodies*, 2018. **7**(3): p. 22.
59. Bai, G., Sun, C., Guo, Z. et al., Accelerating antibody discovery and design with artificial intelligence: recent advances and prospects. *Semin Cancer Biol*, 2023. **95**: pp. 13–24.
60. Wang, D., Discrepancy between mRNA and protein abundance: insight from information retrieval process in computers. *Comput Biol Chem*, 2008. **32**(6): pp. 462–8.
61. Nikolados, E.M. and D.A. Oyarzun, Deep learning for optimization of protein expression. *Curr Opin Biotechnol*, 2023. **81**: pp. 102941.
62. Sumida, K.H., Núñez-Franco, R., Kalvet, I. et al., Improving protein expression, stability, and function with proteinMPNN. *J Am Chem Soc*, 2024. **146**(3): pp. 2054–61.
63. Ding, Z., Guan, F., Xu, G. et al., MPEPE, a predictive approach to improve protein expression in *E. coli* based on deep learning. *Comput Struct Biotechnol J*, 2022. **20**: pp. 1142–53.
64. Bhandari, B.K., Lim, C.S., Remus, D.M. et al., Analysis of 11,430 recombinant protein production experiments reveals that protein yield is tunable by synonymous codon changes of translation initiation sites. *PLoS Comput Biol*, 2021. **17**(10): p. e1009461.
65. Bhandari, B.K., C.S. Lim, and P.P. Gardner, TISIGNER.com: web services for improving recombinant protein production. *Nucleic Acids Res*, 2021. **49(W1)**: pp. W654–61.
66. Fu, H., Liang, Y., Zhong, X. et al., Codon optimization with deep learning to enhance protein expression. *Sci Rep*, 2020. **10**(1): p. 17617.
67. Gamage, N., Cheruvara, H., Harrison, P.J. et al., High-throughput production and optimization of membrane proteins after expression in mammalian cells. *Methods Mol Biol*, 2023. **2652**: pp. 79–118.
68. Sun, J., Kulandaisamy, A., Liu, J. et al., Machine learning in computational modelling of membrane protein sequences and structures: from methodologies to applications. *Comput Struct Biotechnol J*, 2023. **21**: pp. 1205–26.
69. Li, H., Sun, X., Cui, W. et al., Computational drug development for membrane protein targets. *Nat Biotechnol*, 2024. **42**(2): pp. 229–42.
70. Tehan, B.G. and J.A. Christopher, The use of conformationally thermostabilised GPCRs in drug discovery: application to fragment, structure and biophysical techniques. *Curr Opin Pharmacol*, 2016. **30**: pp. 8–13.
71. Muk, S.,, Ghosh, S., Achuthan, S. et al., Machine learning for prioritization of thermostabilizing mutations for G-protein coupled receptors. *Biophys J*, 2019. **117**(11): pp. 2228–39.
72. Bedbrook, C.N., Yang, K.K., Robinson, J.E. et al., Machine learning-guided channelrhodopsin engineering enables minimally invasive optogenetics. *Nat Methods*, 2019. **16**(11): pp. 1176–84.
73. Norman, R.A., Ambrosetti, F., Bonvin, A.M.J.J. et al., Computational approaches to therapeutic antibody design: established methods and emerging trends. *Brief Bioinform*, 2020. **21**(5): pp. 1549–67.
74. Haddad, Y., V. Adam, and Z. Heger, Ten quick tips for homology modeling of high-resolution protein 3D structures. *PLoS Comput Biol*, 2020. **16**(4): p. e1007449.
75. Hameduh, T., Haddad, Y., Adam, V. et al., Homology modeling in the time of collective and artificial intelligence. *Comput Struct Biotechnol J*, 2020. **18**: pp. 3494–506.
76. Nimrod, G., Fischman, S., Austin, M. et al., Computational design of epitope-specific functional antibodies. *Cell Rep*, 2018. **25**(8): pp. 2121–31 e5.
77. Cannon, D.A., Shan, L., Du, Q. et al., Experimentally guided computational antibody affinity maturation with de novo docking, modelling and rational design. *PLoS Comput Biol*, 2019. **15**(5): p. e1006980.
78. Dunbar, J., Krawczyk, K., Leem, J. et al., SAbPred: a structure-based antibody prediction server. *Nucleic Acids Res*, 2016. **44(W1)**: pp. W474–8.
79. Li, L., R. Chen, and Z. Weng, RDOCK: refinement of rigid-body protein docking predictions. *Proteins*, 2003. **53**(3): pp. 693–707.

80. Agarwal, D., Kumar, S., Ambatwar, R. et al., Lead identification through in silico studies: targeting acetylcholinesterase enzyme against Alzheimer's disease. *Cent Nerv Syst Agents Med Chem*, 2024. **24**(2): pp. 219–42.
81. Verma, D.K., Kapoor, S., Das, S. et al., Potential inhibitors of SARS-CoV-2 main protease (M(pro)) identified from the library of FDA-approved drugs using molecular docking studies. *Biomedicines*, 2022. **11**(1): p. 85.
82. Weitzner, B.D., Jeliazkov, J.R., Lyskov, S. et al., Modeling and docking of antibody structures with Rosetta. *Nat Protoc*, 2017. **12**(2): pp. 401–16.
83. AlQuraishi, M., Protein-structure prediction revolutionized. *Nature*, 2021. **596**(7873): pp. 487–8.
84. Tunyasuvunakool, K., Adler, J., Wu, Z. et al., Highly accurate protein structure prediction for the human proteome. *Nature*, 2021. **596**(7873): pp. 590–6.
85. Birch, J., Cheruvara, H., Gamage, N. et al., Changes in membrane protein structural biology. *Biology (Basel)*, 2020. **9**(11): p. 401.
86. Jumper, J., Evans, R., Pritzel, A. et al., Highly accurate protein structure prediction with AlphaFold. *Nature*, 2021. **596**(7873): p. 583–9.
87. Baek, M., DiMaio, F., Anishchenko, I. et al., Accurate prediction of protein structures and interactions using a three-track neural network. *Science*, 2021. **373**(6557): pp. 871–6.
88. Sormanni, P., F.A. Aprile, and M. Vendruscolo, The CamSol method of rational design of protein mutants with enhanced solubility. *J Mol Biol*, 2015. **427**(2): pp. 478–90.
89. Rosace, A.,, Bennett, A., Oeller, M. et al., Automated optimisation of solubility and conformational stability of antibodies and proteins. *Nat Commun*, 2023. **14**(1): p. 1937.
90. Che, Y., Gribenko, A.V., Song, X. et al., Rational design of a highly immunogenic prefusion-stabilized F glycoprotein antigen for a respiratory syncytial virus vaccine. *Sci Transl Med*, 2023. **15**(693): p. eade6422.
91. Krawczyk, K., Liu, X., Baker, T. et al., Improving B-cell epitope prediction and its application to global antibody-antigen docking. *Bioinformatics*, 2014. **30**(16): pp. 2288–94.
92. Bourquard, T., Musnier, A., Puard, V. et al., MAbTope: a method for improved epitope mapping. *J Immunol*, 2018. **201**(10): pp. 3096–105.
93. Hoie, M.H., Gade, F.S., Johansen, J.M. et al., DiscoTope-3.0: improved B-cell epitope prediction using inverse folding latent representations. *Front Immunol*, 2024. **15**: p. 1322712.
94. Sela-Culang, I., Y. Ofran, and B. Peters, Antibody specific epitope prediction-emergence of a new paradigm. *Curr Opin Virol*, 2015. **11**: pp. 98–102.
95. Aguilar Rangel, M., Bedwell, A., Costanzi, E. et al., Fragment-based computational design of antibodies targeting structured epitopes. *Sci Adv*, 2022. **8**(45): p. eabp9540.
96. Cervenak, J., R. Kurrle, and I. Kacskovics, Accelerating antibody discovery using transgenic animals overexpressing the neonatal Fc receptor as a result of augmented humoral immunity. *Immunol Rev*, 2015. **268**(1): pp. 269–87.
97. Caradonna, T.M. and A.G. Schmidt, Protein engineering strategies for rational immunogen design. *NPJ Vaccines*, 2021. **6**(1): p. 154.
98. Spangler, J.B., Moraga, I., Jude, K.M. et al., A strategy for the selection of monovalent antibodies that span protein dimer interfaces. *J Biol Chem*, 2019. **294**(38): pp. 13876–86.
99. Ngwuta, J.O., Chen, M., Modjarrad, K. et al., Prefusion F-specific antibodies determine the magnitude of RSV neutralizing activity in human sera. *Sci Transl Med*, 2015. **7**(309): p. 309ra162.
100. Roy, U. and L.A. Luck, Molecular modeling of estrogen receptor using molecular operating environment. *Biochem Mol Biol Educ*, 2007. **35**(4): pp. 238–43.
101. Kellogg, E.H., A. Leaver-Fay, and D. Baker, Role of conformational sampling in computing mutation-induced changes in protein structure and stability. *Proteins*, 2011. **79**(3): pp. 830–8.

102. Yang, J., Yan, R., Roy, A. et al., The I-TASSER suite: protein structure and function prediction. *Nat Methods*, 2015. **12**(1): pp. 7–8.
103. Webb, B. and A. Sali, Comparative protein structure modeling using MODELLER. *Curr Protoc Bioinform*, 2016. **54**: pp. 5.6.1–5.6.37.
104. Kelley, L.A., Mezulis, S., Yates, C.M. et al., The Phyre2 web portal for protein modeling, prediction and analysis. *Nat Protoc*, 2015. **10**(6): pp. 845–58.
105. Waterhouse, A., Bertoni, M., Bienert, S. et al., SWISS-MODEL: homology modelling of protein structures and complexes. *Nucleic Acids Res*, 2018. **46**(**W1**): pp. W296–303.
106. Leaver-Fay, A., Tyka, M., Lewis, S.M. et al., ROSETTA3: an object-oriented software suite for the simulation and design of macromolecules. *Methods Enzymol*, 2011. **487**: pp. 545–74.
107. Mortuza, S.M., Zheng, W., Zhang, C. et al., Improving fragment-based ab initio protein structure assembly using low-accuracy contact-map predictions. *Nat Commun*, 2021. **12**(1): p. 5011.

Bioinformatic Analyses of Antibody Repertoires and Their Roles in Modern Antibody Drug Discovery

4

Melody Shahsavarian, Thomas Watkins, Ponraj Prabakaran*, Adrian Carr, Maria Wendt, and Yu Qiu

4.1 INTRODUCTION

The field of antibody discovery has evolved dramatically with the integration of next-generation sequencing (NGS), high-throughput screening, and computational methods, reshaping the traditional paradigms of immunization and in vitro antibody discovery.[1–3] This has allowed for an expanded understanding of antibody diversity as well as the selection of highly specific antibodies with desirable therapeutic traits. Historically, antibody discovery relied heavily on animal immunization, a method with its own set of advantages, including the in vivo maturation of antibodies, which often results in high affinity

* author for correspondence

DOI: 10.1201/9781003300311-4

and specificity.[4] The process of antibody discovery has been significantly enhanced by the precision and scale afforded by NGS technologies, which facilitate a deeper exploration of the antibody repertoire.[5] Technological limitations and logistical complexities associated with animal use have driven the development of alternative methods. In vitro display technologies, such as phage, yeast, or mammalian display, represent a crucial shift toward non-animal-derived antibody discovery.[6] These methods not only minimize animal use but also allow for precise control over the selection process, targeting specificities that are challenging to achieve in vivo. Thus, in vitro display technologies have become foundational in the transition from animal models to more controlled experimental setups, where libraries of antibodies can be screened against a myriad of antigens with high throughput and specificity. Integration of NGS with these display platforms enables the rapid screening of vast combinatorial libraries that are directly cloned from a pool of B-cells or synthetically designed, thus capturing a broader diversity of antibody sequences.[7]

Deep sequencing of antibody repertoires with NGS in recent years has allowed an unprecedented quantitative understanding of immune repertoire dynamics applied to antibody discovery,[8] both from natural B-cells isolated either from human donors[9] or immunized animal models[10] and from synthetic antibody display libraries.[11] Paired sequencing strategies, more recently, can even provide information at the single-cell level on the natural pairing of antibody heavy and light chains, thus revealing essential information about the antigen specificity and function of the antibodies.[12–14] These technologies have had a significant impact on time, efficiency, and throughput of antibody discovery platforms.

Recently, machine learning (ML) algorithms have become instrumental in creating innovative approaches that impact antibody discovery and development.[15–26] By leveraging large datasets of antibody sequences generated from NGS, ML can help develop models that could predict antibody behavior, optimize binding affinities, and enhance the developability of therapeutic antibodies. This synergy between NGS and ML is setting new standards in the rapid identification and optimization of antibodies, pushing the boundaries of what can be achieved in therapeutic development.[27,28]

In this chapter, we will briefly overview the NGS technologies for antibody discovery. We will then explore NGS-enabled in vivo, in vitro, and in silico methods, representing the integration of NGS technology with traditional immunization, display libraries, and artificial intelligence/machine learning (AI/ML) for advanced antibody discovery. This integrated approach exemplifies how cutting-edge technologies are reshaping the field of antibody discovery, leading to the development of more rapid and effective antibody therapeutics.

4.2 NGS TECHNOLOGIES, TOOLS, AND DATA ANALYSIS FOR MODERN ANTIBODY DISCOVERY

NGS technologies have greatly transformed the field of antibody discovery, providing an unprecedented ability to explore antibody repertoires derived from immunizations and display libraries at the individual molecule level.[3,29,30] The shift from traditional methods, such as Sanger sequencing, which was useful for identifying dominant clones but provided only a glimpse into the antibody diversity landscape, is truly remarkable. NGS technologies for antibodies vary significantly in their applications and capabilities, as summarized in Table 4.1 and previously described.[31,32] For instance, Roche's 454 sequencing platform, although now discontinued, was once essential for its ability to produce long reads up to 700 bp, making it ideal for sequencing the entire variable (V) region of antibodies in a single run. On the other hand, the Illumina platform, which produces shorter reads of about 150 nucleotides, is particularly well-suited for

TABLE 4.1 NGS platforms for sequencing antibody repertoires from in vivo immunized animals and in vitro display libraries.

PLATFORM	*READ LENGTH*	*MAXIMUM READS PER RUN*	*RUN TIME*	*ERROR (%)*	*REMARKS*
Illumina MiSeq	2 X 300 bp	~ 25 million	4–55 hours	~0.1	Suitable for bulk sequencing of variable regions from heavy and light chains separately. Low error rate and high throughput.
Illumina HiSeq	2 X 150 bp	~ 1.5 billion	~48 hours	~0.1	Suitable for bulk sequencing of heavy and light chains separately. Low error rate and high throughput. Principle platform used for 10X Genomics sequencing, enabling barcoding of single cells for paired information such as CDR-H3 and CDR-L3 pairing.
454 GS FLX	~ 700 bp	~ 1 million	10–24 hours	~1	Offered longer read lengths suitable for scFv libraries with sequencing of heavy and light chains separately but discontinued technology with limited availability.
Ion Torrent	~ 400 bp	~80 million	2–7 hours	~1	Suitable for medium length reads, offering a balance of speed and detail.
PacBio	~ 20 kb	~4 million	10–30 hours	~11 (raw), <1 (CCS)	Long reads enable full-length sequencing of heavy/light chains in single reads suitable for scFv or Fab libraries.

The read lengths, number of reads per run, and run times can vary based on the specific model and configuration of the sequencing kit or chips used for each platform.

analyzing the heavy chain complementarity determining region 3 (HCDR3), a crucial segment in determining antibody specificity. Although the HCDR3 sequences are shorter, Illumina's high throughput facilitates a more extensive examination. Emerging platforms like Pacific Biosciences (PacBio) offer the potential to sequence even longer reads from single DNA or RNA molecules, potentially simplifying the complex library preparations required for scFv and Fab libraries. However, their application in antibody repertoire analysis may be constrained by the quality of the sequences they produce. Meanwhile, platforms like Ion Torrent strike a balance with their moderate read lengths and faster sequencing capabilities at a lower cost, although their error rate remains a concern. The continued development and refinement of these technologies, including error correction strategies such as the use of unique molecular identifiers (UMIs), is crucial as they evolve to meet the high standards required for precise and effective antibody discovery.

Once the raw data from NGS sequencing runs are available, bioinformatics tools are utilized to process and analyze the data, facilitating the antibody discovery process from understanding diversity to NGS-based screening and selection of functional clones. A selected list of bioinformatics software tools for analyzing antibody repertoire data generated from NGS is presented in Table 4.2. Initially, the process involves data cleaning, which includes removing low-quality sequences and sequencing adapters to ensure data integrity. Subsequent steps involve using tools such as IMGT/V-QUEST and IgBLAST to identify the variable (V), diversity (D), and joining (J) gene segments, and to define the framework regions and complementarity determining regions (CDRs) essential for antigen-binding specificity. The selection of these tools often depends on the volume of data and the user's expertise. For those less familiar with computational

TABLE 4.2 Selected bioinformatics software tools for NGS-generated antibody repertoire data analysis.

TOOLS	*DESCRIPTION*	*URL*
IgBLAST	V(D)J annotation, CDRs assignment, and SHM	https://www.ncbi.nlm.nih.gov/igblast/
IMGT/HighV-QUEST	A pioneering international information system in immunogenetics and immunoinformatics with tools available for V(D)J annotation, CDRs assignment, SHM, statistical analysis, and plots	https://www.imgt.org/
MiXCR	Raw sequences to clonotypes	https://github.com/milaboratory/mixcr/
PipeBio	User-friendly end-to-end workflows with visualization for antibody discovery, NGS analysis, and antibody engineering	https://pipebio.com/
ENPICOM	IGX Platform, data visualization and analysis, and AI-aided antibody development	https://enpicom.com/
Geneious Biologics	Powerful tools for antibody sequence annotation and analysis, statistical analysis, and visualization	https://www.geneious.com/biopharma/

IgBLAST is freely available for use, while the other tools listed require licensing and involve costs.

techniques, platforms like IMGT High-VQUEST offer user-friendly interfaces, while more sophisticated tools like MiXCR provide detailed immunological analyses for advanced users. Furthermore, clustering tools such as CD-hit[33] or UCLUST[34] are employed to group similar sequences, aiding in the identification of clonal expansions and enriching our understanding of immune response diversity. For handling larger datasets and facilitating collaborative research, cloud-based commercial software platforms like PipeBio, ENPICOM, and Geneious Biologics provide comprehensive, integrated solutions that streamline the workflow and enhance data analysis capabilities. These platforms combine powerful bioinformatics tools with user-friendly interfaces, significantly simplifying the management and analysis of complex data, thereby accelerating antibody research and development.

After performing antibody NGS data processing and analysis, several metrics and analytical techniques are employed to thoroughly examine the B-cell repertoire. Figure 4.1 illustrates various NGS-based antibody repertoire metrics, such as isotype analysis, V(D)J segment usage frequencies, CDR3 properties, somatic hypermutation (SHM) analysis, and clonal relationship and lineage analysis, which are depicted using trees or network graphs. Statistical analysis is also crucial to ensure sequencing depth is comparable between samples, estimate repertoire diversity, and assess convergence. By employing this comprehensive approach, researchers can gain deep insights into the antibody response, including the types of antibodies produced, namely isotypes, the genes utilized for antibody building blocks comprising V(D)J gene segments, the CDRs, the extent of antibody diversification through mutations, and the relationships between different antibody clones. This information is invaluable for understanding the immune system's response to pathogens and for developing novel antibody-based therapies.

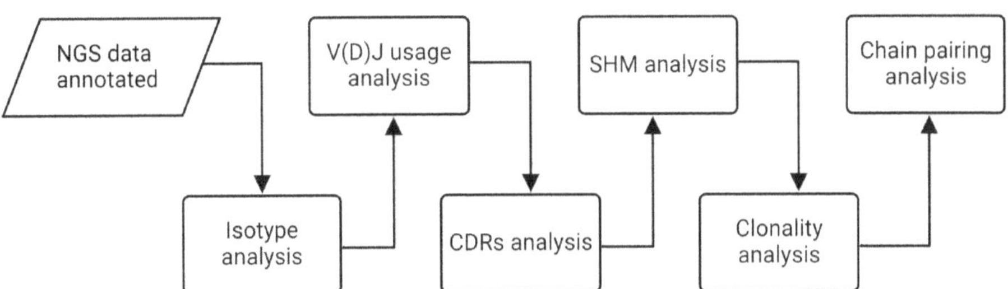

FIGURE 4.1 NGS antibody repertoire data analysis employs a multifaceted approach to define key repertoire features, serving as essential metrics for NGS-driven antibody discovery. This analysis includes isotype analysis to categorize sequences by constant regions influencing effector functions, and V(D)J usage analysis to assess gene segment frequencies, revealing specific germline contributions. Complementarity determining regions (CDRs) analysis, particularly focusing on CDR3, examines sequence length and amino acid properties crucial for antigen binding. Somatic hypermutation (SHM) analysis explores the diversity of antibody clones, a critical process for generating high-affinity antibodies. Clonal relationship and lineage analysis track antibody evolution, offering insights into immune response development. Lastly, with paired sequencing data, chain pairing analysis evaluates the pairing of heavy (VH) and light (VL) chain genes, enhancing our understanding of antibody specificity.

In the case of NGS-based antibody discovery using display libraries, the depth and breadth of analysis that can be performed on the outputs from selections of these libraries have dramatically increased with the advent of NGS.[11,35] Additionally, data on gene usage frequency, along with other immunogenetic features, such as specific gene lineages, CDR lengths, and amino acid patterns, can be analyzed to identify trends associated with different panning strategies.

4.3 NGS-ENABLED IN VIVO ANTIBODY DISCOVERY FROM IMMUNIZED ANIMALS

In vivo antibody discovery using a hybridoma-based approach was established several decades ago and has subsequently undergone several improvements.[36–38] This method involves immunizing animals with a specific antigen, followed by the laborious process of fusing B-cells from the animal's spleen with myeloma cells, creating immortalized hybridomas. The generated hybridomas continuously produce antibodies specific to the immunizing antigen. However, hybridoma technology suffers from several limitations. Firstly, it captures only a tiny fraction of the vast diversity of antibodies generated by the immune system. Secondly, the process is highly inefficient and time-consuming, requiring extensive screening of individual hybridoma clones to identify those producing high-affinity antibodies. The emergence of NGS technologies has revolutionized in vivo antibody discovery, offering a powerful and multifaceted approach.[39] This revolution extends far beyond traditional hybridoma technology by incorporating advanced technologies that work synergistically with NGS analysis. Particularly, human immunoglobulin transgenic mice offer a more human-like antibody repertoire for NGS sequencing. Furthermore, NGS combined with single B-cell antibody technologies, antigen-specific single B-cell sorting, and recombinant antibody cloning have become powerful approaches.[40,41] Additionally, natively paired immune libraries, constructed from B-cell antibody genes, preserve the crucial pairing of heavy and light chain genes for proper antibody function. NGS analysis of these libraries allows for the identification of a vast repertoire of diverse and functional antibodies. Figure 4.2 outlines the NGS-driven in vivo antibody discovery process, starting with B-cell isolation and proceeding through sequencing and comprehensive repertoire analysis. This includes examining V(D)J gene usage, SHM diversity, clonal expansion, and antibody lineage relationships. The method facilitates the identification of highly diverse, high-affinity antibodies that are not detectable by traditional in vivo hybridoma methods.

In one of the use-cases as adopted by researchers at Genentech, integrating NGS into hybridoma technology has significantly advanced the antibody discovery process.[42] This approach overcomes traditional limitations related to throughput and storage capacity by digitizing antibody variable domain sequences for high-throughput screening directly from hybridoma cells. The use of barcoded primers in a 96-well format allows for the simultaneous processing of multiple samples, enhancing sequencing output and efficiency substantially. Additionally, robust bioinformatics tools ensure precise

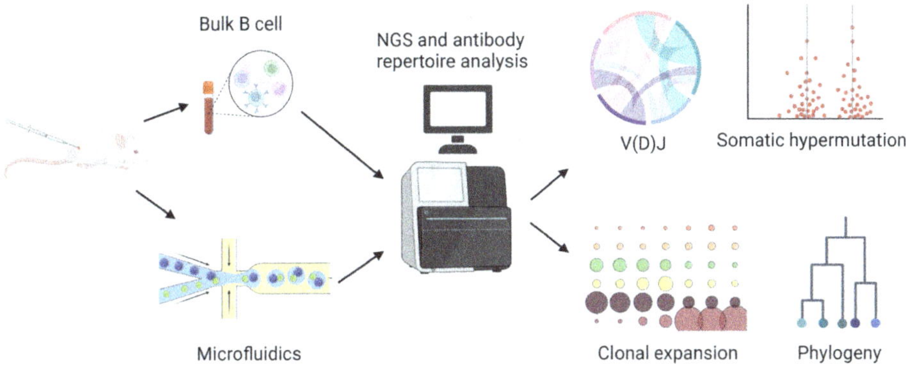

FIGURE 4.2 NGS-enabled in vivo antibody discovery process starts with isolating bulk or single B-cells from a model organism and utilizes NGS for in-depth repertoire analysis. This analysis investigates V(D)J gene usage, explores SHM diversity critical for high-affinity antibodies, examines clonal expansion of desirable B-cells, and elucidates antibody lineage relationships. This comprehensive NGS approach allows the discovery and characterization of high-affinity antibodies in vivo.

resolution and unambiguous identification of antibody sequences. This method dramatically reduces the need for physical storage by converting sequences into digital data, offering a scalable, cost-effective solution that aligns perfectly with the needs of modern biomedical research and drug development. Additionally, it facilitates a more dynamic, rapid, and expansive exploration of potential therapeutic antibodies, demonstrating a critical advancement in the field.

In another use-case study, George Georgiou and colleagues developed a groundbreaking method that bypasses the traditional high-throughput screening processes for isolating antigen-specific monoclonal antibodies (mAbs).[10] By leveraging NGS and bioinformatic analysis, this technique directly mines the antibody variable region (V)-gene repertoires from bone marrow plasma cells (BMPCs) of immunized mice. BMPCs, notable for producing most circulating antibodies but are unable to be immortalized, exhibit a highly polarized V-gene repertoire following immunization. The most abundant variable heavy (VH) and variable light (VL) genes are paired based on their relative frequencies, reconstructed through automated gene synthesis, and expressed in bacterial or mammalian systems as recombinant antibodies. This method significantly accelerates the progression from immunization to specific antibody production and yields antibodies with high antigen specificity and nanomolar affinities, underscoring its potential as a tool to rapidly respond to emerging infectious diseases and advancing both immunological research and therapeutic antibody development.

In another case study by GigaGen Inc., the authors leveraged NGS to significantly advance the discovery of therapeutic monoclonal antibodies by deeply sequencing yeast scFv libraries derived from immunized Trianni mice, both before and after fluorescence-activated B-cell sorting (FACS).[43] This approach allowed the authors to track the evolution and enrichment of antibody clones, showcasing the technology's ability to narrow down thousands of diverse clones to a few with potential therapeutic properties. The sequence analysis highlighted minimal overlap in antibody sequences enriched by different methods, pointing to the influence of antigen presentation on immune response specificity. Remarkably, sequences derived from soluble immunogen methods showed more similarity across FACS methods compared to those from cell/DNA methods, suggesting differences in immune response elicitation. This study exemplifies how integrating NGS with innovative immunization strategies can streamline the antibody discovery process, enhance the functional relevance of identified antibodies, and expedite the development of new therapeutics.

Overall, NGS-based in vivo antibody discovery offers a revolutionary approach, merging the strengths of animal immunization with advanced screening methods. This opens the door for exploring a vast universe of possibilities within the animal-derived antibody repertoire.

4.4 NGS-ENABLED IN VITRO ANTIBODY DISCOVERY FROM DISPLAY LIBRARIES

In vitro antibody library platforms such as phage and yeast display have become powerful tools for therapeutic antibody discovery, allowing the identification of fully human antibodies against virtually any desired target.[7,44–47] These platforms have several advantages as they provide: (i) large antibody sequence and structural diversities, typically up to 10^{11} unique clones, (ii) fast and easy selection methods able to apply specific enrichment pressures on the entire library, such as incubation with solid-surface immobilized antigen or FACS, and (iii) rapid identification of binder antibodies by providing a direct phenotype-genotype link. Traditional screening methods of in vitro display libraries and their enrichment outputs, however, can be time and resource intensive. These methods rely on the isolation of single colonies for various binding or functional assays and subsequent Sanger sequencing, thus limiting the number of tested antibodies to typically a few hundred. Additionally, most of the clones picked for characterization in this way represent antibodies present at the highest frequency in the enriched population. This inevitably means that clones picked at random have a large level of redundancy, with those high frequency clones being picked multiple times, reducing even further the number of tested unique antibodies. Furthermore, this dominance of high frequency clones does not represent only antibodies that have been enriched due to high-affinity antigen binding but also those that might have been enriched due to higher expression or display of the antibodies by the phage system. The presence of such clones at high frequency will often lead to the failure of the traditional screening methods in the detection of rarer but more desirable and functional antibodies.[48]

More recently, the combination of NGS and in vitro display technologies has largely removed the limitations of traditional screening methods and made the use of these platforms in therapeutic antibody discovery even more powerful.[32] NGS allows deep repertoire profiling and mining of antibody libraries and their enrichment outputs, providing a more holistic view of the population dynamics and the evolution of the antibody repertoires through consecutive enrichment rounds. As detailed in Figure 4.3,

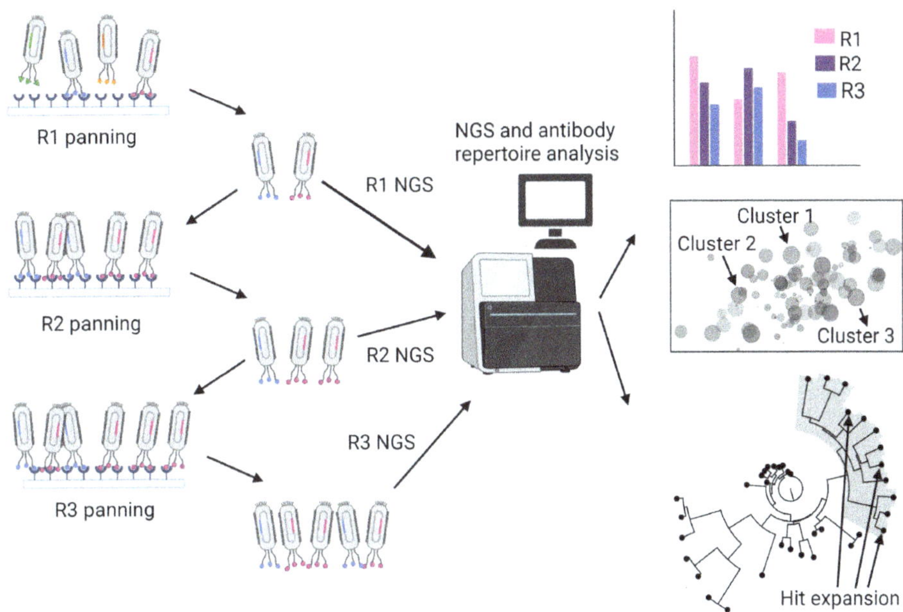

FIGURE 4.3 NGS-enabled in vitro antibody discovery process begins with the panning of phage-displayed libraries across multiple rounds, typically three rounds (R1, R2, R3), each targeting a specific antigen. Post-panning, phage particles from each round undergo NGS and antibody repertoire analysis. This analysis includes evaluating the frequency distribution of sequences from each round, clustering to identify related antibody sequences and constructing phylogenetic trees to track the evolution and expansion of antibody hits. This comprehensive approach facilitates the identification and optimization of high-affinity antibodies in vitro.

the NGS-enabled in vitro discovery process utilizes iterative panning of phage libraries against a target antigen, followed by NGS analysis, for example, frequency distribution, sequence clustering, and phylogenetic trees, to identify and optimize high-affinity antibody candidates. With NGS, not only the high frequency enriched clones but also those that are rarer with lower frequencies become visible, resulting in a more diverse panel of identified antibody hits.[11] Moreover, as almost all enriched clones are detected, larger clusters of enriched sequences (families of sequences closely related in sequence space) can be identified. Therefore, selection of representative clones from these clusters can be made in a more intelligent manner, addressing, for example, developability criteria of the selected clones early in the discovery phase.[49] Moreover, having access to these multi-member sequence clusters presents the possibility of "hit expansion." This is the strategy where variants of a known functional antibody sequence are identified in NGS data with the objective of improving some property of the parental antibody, that is, affinity, developability, cross-reactivity, etc.[50]

First studies of incorporating NGS in phage display antibody discovery, due to the limited capacity of the sequencing length of sequencing technologies existing at the time, focused only on HCDR3 or the heavy chain variable region (VH) sequences.[51,52]

The short-read sequencing approaches have the advantage of offering a very high sequencing depth and are very well-suited for the global analysis of library quality, for example, looking at the diversity of antibody chains separately. Several groups have also used these sequencing methods for NGS-based clone selection, using clone rescue methods to recover the full antibody sequence. However, these studies have been limited to libraries with variability introduced only in specific segments of the variable gene or to the selection of only a handful of clones for expression and characterization.[48] The Fischer lab, for example, used HCDR3 sequences with high enrichment ratio obtained from NGS to design oligos and PCR-amplify the full scFv sequence of six antibodies that were expressed and tested for binding to the target.[51] To perform clone selection at a larger scale and/or on fully randomized libraries, however, the sequence information of both VH and VL is needed. The Georgiou lab developed a yeast display method based on short-read sequencing but allowing the identification of full-length antibodies.[14] In this method, the two antibody chains are cloned in opposite orientations separated by a bidirectional promoter. Short-read sequencing can hence cover the combination of CDR3 sequences from both heavy and light chains (HCDR3-LCDR3). By also sequencing separately the full VH and VL genes, they can reconstruct bioinformatically the entire paired antibody sequence. More recently, new technologies have been developed which allow long-read sequencing and the coverage of the full paired sequence from an scFv or Fab library on a single read. PacBio sequencing, for example, has been successfully used for NGS-based repertoire analysis and identification of binders of phage display libraries.[35,53]

4.5 NGS-ENABLED IN SILICO ANTIBODY DISCOVERY VIA ARTIFICIAL INTELLIGENCE METHODS

Machine learning (ML) is a domain of artificial intelligence (AI) that focuses on 'learning' the relationship between input and output variables and can be used for both regression- and classification-based tasks. Deep learning (DL), a branch of ML science that has recently gained popularity, leverages interconnected neurons arranged into various layers to both transform and process input data. Despite the uptake of DL approaches in cancer screening and disease diagnosis,[54–56] their use in commercial antibody discovery pipelines is relatively recent. NGS platforms provide thousands of paired antibody sequences, which is now sufficient for training deep learning-based AI/ML models. Availability of large amounts of antibody sequence data via NGS has also resulted in the rapid growth of various databases, in the public domain, aimed at collecting and storing this information. For example, the Observed Antibody Space (OAS) currently contains over a billion antibody sequences collected from over 80 studies and includes both paired and unpaired sequences.[57] Other databases like SAbDab[58] and Thera-SAbDab[59] focus on collating structural data of public or therapeutic antibodies, respectively. It is the accessibility of well-annotated data alongside the almost exponential growth of new algorithms that has driven the accelerated development of antibody-specific ML tools.

Antibody 3D structure prediction is the domain which has benefited the most from the advent of ML, and methods of antibody structures have evolved rapidly. Much of this evolution is owed to the development of AlphaFold2,[60] which has revolutionized protein structure prediction. However, despite AlphaFold2's remarkable accuracy, the method still struggles with antibody structure prediction. AlphaFold2 requires a multiple alignment step using various homologues, which is problematic in the context of antibodies due to overall diversity of the HCDR3 region. This has led to the development of antibody-specific protein predictors such as ABodyBuilder2[61] and DeepAb,[62] both of which are DL approaches capable of accurately predicting the entire Fv modelling region,[61] whereas ABlooper specializes in CDR loop prediction by leveraging graph neural networks.[63] Protein language models have also shown success in predicting antibody structure and function by assigning probabilities to predicted amino acids or 'tokens.'[64] IgFold is an example of a structure prediction tool that leverages this type of approach and utilizes a model trained on embeddings generated from AntiBERTy, a natural antibody dataset consisting of over 500 million sequences.[65] Unlike ABlooper and DeepAb, however, IgFold is able to make use of template structures, which results in more accurate and faster predictions.[66] Lastly, EquiFold attempts to represent antibodies as geometrical structures, avoiding the need for either protein language model embeddings or multiple sequence alignments.[67] Obtaining accurate predicted structural representations is crucial for not only understanding the binding potential of candidate antibodies but also their biophysical properties.[68] Features such as solubility, hydrophobicity, shelf-life, viscosity, and immunogenicity have major roles in determining whether an antibody is suitable for large-scale manufacturing and commercialization.[69] Indeed, poor developability in post-Phase 1 antibodies can largely be attributed to a small number of these features[70]; therefore, determining developability concerns early in the development cycle is key. In-silico screening tools like FreeSASA,[71] PROPKA3,[72] and TAP[73] can leverage predicted structures to identify major developability concerns, thus addressing these problems much earlier in development when compared with classical approaches. Lastly, structure prediction can also help guide selection to a more diverse set of candidates. Tools such as SPACE2 leverage predicted structures to cluster antibodies based on their predicted epitope of binding,[74] thus allowing therapeutic candidates to target a wider range of epitopes.

The modular nature of the methods discussed above ushers in the concept of a more integrative approach to antibody discovery, where classic approaches are accompanied and enhanced using various AI/ML tools (Figure 4.4). For example, antigen-specific antibodies generated by the phage library could be supplemented by exploring the nearby sequence space using generative models. Generative models offer significant benefits to researchers, most notably a reduction in resources and time required for candidate selection, and removal of any downstream developability concerns.[75] ML models trained on binder/non-binder data can then subsequently be used to filter AI-generated sequences into those predicted to bind the same target antigen, with major developability issues identified and removed. Indeed, a combinatorial approach using a suite

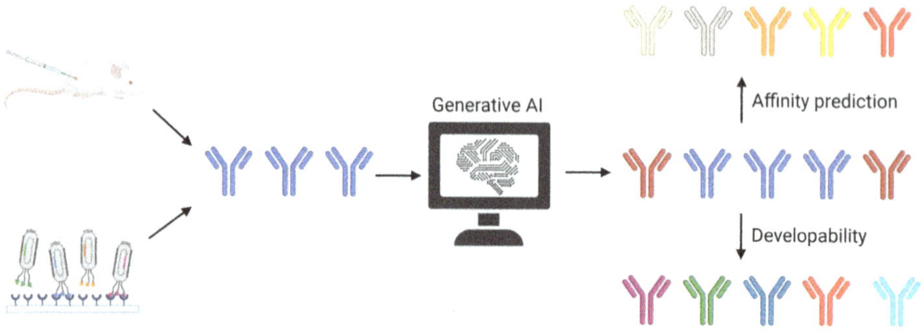

FIGURE 4.4 Illustration depicting how machine learning approaches can augment antibodies identified by traditional discovery workflows. The process begins with generating a list of candidate antibodies through conventional methods, such as mouse immunization or phage display (blue antibodies). These antibodies are then enhanced using generative AI tools to explore the nearby sequence space, generating additional variants (red antibodies). Affinity prediction is applied to the candidate list to identify antibodies with improved binding to the target antigen. Additionally, developability screening is performed to identify and mitigate potential developability issues in both the original and AI-generated antibody sequences, enhancing their suitability for therapeutic use.

of ML tools was recently demonstrated in the context of CTLA-4- and PD-1-specific antibodies.[76] Heavy and light chain CDR3 sequences were first used to train a convolutional neural network (CNN) capable of predicting binder or non-binder. In-silico mutagenesis was then carried out to determine antigen-binding motifs, which in turn allowed identification of sequences with a higher probability of binding. Lastly, generative adversarial networks (GANs) capable of predicting entire HCDR3 sequences were constructed, and the binding of the AI-generated sequences was queried using the previous CNN.[76] Combinatorial platforms using structural data are also emerging, with a recent example highlighting the power of DL in the context of reprogramming the binding capacity of a SARS-CoV-2-specific antibody.[77] In this example, an attention-based geometric neural network was trained using 3D protein structures and asked to learn the effect of single-point CDR3 mutations on binding affinity. This model was then applied to a single SARS-CoV-2-specific antibody to improve binding to the Delta variant of SARS-CoV-2, with the observed increase in binding affinity owed to the removal of various steric clashes between the antibody and Delta variant as determined by predicted 3D structures.[77]

Given the combinatorial potential and speed of development of AI/ML tools, a fully in silico platform capable of identifying a set of potent and diverse candidates is somewhat inevitable. Generative models can be used for candidate sequence generation, structure prediction and clustering models can be used for binding affinity and epitope prediction, and developability concerns can be removed from lead candidates. Despite this, in vitro experiments will always be required to fully validate candidate antibodies.[25] As such, at least for the immediate future, a synergistic approach that combines classic antibody development alongside AI/ML tools that aims to limit the amount of in vitro work required is the most feasible approach.

4.6 SUMMARY AND FUTURE DIRECTIONS

Over the past decade or so, NGS technologies have profoundly transformed the landscape of antibody discovery. These technologies provide a detailed insight into the extensive antibody repertoire present in both immunized animals and display libraries. This advancement has led to a revitalization of animal immunization-based methods through NGS-enabled in vivo antibody discovery, facilitating the identification of rare, diverse, and functionally significant antibodies. Simultaneously, NGS-enabled in vitro discovery using phage and yeast display libraries has been significantly enhanced, broadening the scope for detailed analysis and expanding the range of potential antibody candidates. Furthermore, in-silico antibody discovery leverages AI/ML to potentially generate entirely novel antibody sequences with desired properties and functions de novo, combined with recent breakthroughs in protein structure predictions. However, the need for and importance of experimental validation remains paramount. Future directions involve a seamless integration of NGS with traditional methods to capture a wider antibody repertoire and the use of AI for clone selection for the discovery of powerful therapeutic antibodies. Additionally, NGS-derived antibody sequence data will be mined to enhance our understanding of B-cell biology and repertoire diversity, guiding the selection of novel antibody candidates, previously unattainable through traditional methods. With advanced sequencing techniques and bioinformatics tools on the horizon, the future promises AI-driven creation of comprehensive antibody libraries and de novo antibody design, all informed by data-driven insights derived from NGS-based antibody repertoires. This transformative approach sets the stage for the development of next-generation antibody therapeutics, potentially revolutionizing the field of medicine.

ACKNOWLEDGMENTS

We would like to thank Ankit Mahendra from Large Molecules Research for critically reading the manuscript and Ally Hatton for providing legal guidance. We thank the editor, Sandeep Kumar, for his valuable comments and suggestions.

REFERENCES

1. Naso MF, Lu J, Panavas T. Deep sequencing approaches to antibody discovery. *Curr Drug Discov Technol* 2014; 11:85–95. doi: 10.2174/15701638113106660040.
2. Glanville J, D'Angelo S, Khan TA, Reddy ST, Naranjo L, Ferrara F, Bradbury AR. Deep sequencing in library selection projects: what insight does it bring? *Curr Opin Struct Biol* 2015; 33:146–60. doi: 10.1016/j.sbi.2015.09.001.

3. Prabakaran P, Glanville J, Ippolito GC. Editorial: next-generation sequencing of human antibody repertoires for exploring B-cell landscape, antibody discovery and vaccine development. *Front Immunol* 2020; 11:1344. doi: 10.3389/fimmu.2020.01344.
4. Kohler G, Milstein C. Continuous cultures of fused cells secreting antibody of predefined specificity. *Nature* 1975; 256:495–7. doi: 10.1038/256495a0.
5. Wine Y, Boutz DR, Lavinder JJ, Miklos AE, Hughes RA, Hoi KH, Jung ST, Horton AP, Murrin EM, Ellington AD, et al. Molecular deconvolution of the monoclonal antibodies that comprise the polyclonal serum response. *Proc Natl Acad Sci U S A* 2013; 110:2993–8. doi: 10.1073/pnas.1213737110.
6. McCafferty J, Griffiths AD, Winter G, Chiswell DJ. Phage antibodies: filamentous phage displaying antibody variable domains. *Nature* 1990; 348:552–4. doi: 10.1038/348552a0.
7. Alfaleh MA, Alsaab HO, Mahmoud AB, Alkayyal AA, Jones ML, Mahler SM, Hashem AM. Phage display derived monoclonal antibodies: from bench to bedside. *Front Immunol* 2020; 11:1986. doi: 10.3389/fimmu.2020.01986.
8. Glanville J, Zhai W, Berka J, Telman D, Huerta G, Mehta GR, Ni I, Mei L, Sundar PD, Day GM, et al. Precise determination of the diversity of a combinatorial antibody library gives insight into the human immunoglobulin repertoire. *Proc Natl Acad Sci U S A* 2009; 106:20216–21. doi: 10.1073/pnas.0909775106.
9. Zhu J, Ofek G, Yang Y, Zhang B, Louder MK, Lu G, McKee K, Pancera M, Skinner J, Zhang Z, et al. Mining the antibodyome for HIV-1-neutralizing antibodies with next-generation sequencing and phylogenetic pairing of heavy/light chains. *Proc Natl Acad Sci U S A* 2013; 110:6470–5. doi: 10.1073/pnas.1219320110.
10. Reddy ST, Ge X, Miklos AE, Hughes RA, Kang SH, Hoi KH, Chrysostomou C, Hunicke-Smith SP, Iverson BL, Tucker PW, et al. Monoclonal antibodies isolated without screening by analyzing the variable-gene repertoire of plasma cells. *Nat Biotechnol* 2010; 28:965–9. doi: 10.1038/nbt.1673.
11. Yang W, Yoon A, Lee S, Kim S, Han J, Chung J. Next-generation sequencing enables the discovery of more diverse positive clones from a phage-displayed antibody library. *Exp Mol Med* 2017; 49:e308. doi: 10.1038/emm.2017.22.
12. DeKosky BJ, Kojima T, Rodin A, Charab W, Ippolito GC, Ellington AD, Georgiou G. In-depth determination and analysis of the human paired heavy- and light-chain antibody repertoire. *Nat Med* 2015; 21:86–91. doi: 10.1038/nm.3743.
13. Rajan S, Kierny MR, Mercer A, Wu J, Tovchigrechko A, Wu H, Dall Acqua WF, Xiao X, Chowdhury PS. Recombinant human B cell repertoires enable screening for rare, specific, and natively paired antibodies. *Commun Biol* 2018; 1:5. doi: 10.1038/s42003-017-0006-2.
14. Wang B, DeKosky BJ, Timm MR, Lee J, Normandin E, Misasi J, Kong R, McDaniel JR, Delidakis G, Leigh KE, et al. Functional interrogation and mining of natively paired human V(H):V(L) antibody repertoires. *Nat Biotechnol* 2018; 36:152–5. doi: 10.1038/nbt.4052.
15. Amimeur T, Shaver JM, Ketchem RR, Taylor JA, Clark RH, Smith J, Van Citters D, Siska CC, Smidt P, Sprague M, et al. Designing feature-controlled humanoid antibody discovery libraries using generative adversarial networks. bioRxiv 2020:2020.04.12.024844. doi: 10.1101/2020.04.12.024844.
16. Akbar R, Bashour H, Rawat P, Robert PA, Smorodina E, Cotet TS, Flem-Karlsen K, Frank R, Mehta BB, Vu MH, et al. Progress and challenges for the machine learning-based design of fit-for-purpose monoclonal antibodies. *MAbs* 2022; 14:2008790. doi: 10.1080/19420862.2021.2008790.
17. Kim J, McFee M, Fang Q, Abdin O, Kim PM. Computational and artificial intelligence-based methods for antibody development. *Trends Pharmacol Sci* 2023; 44:175–89. doi: 10.1016/j.tips.2022.12.005.
18. Cheng J, Liang T, Xie XQ, Feng Z, Meng L. A new era of antibody discovery: an in-depth review of AI-driven approaches. *Drug Discov Today* 2024; 29:103984. doi: 10.1016/j.drudis.2024.103984.

19. Makowski EK, Wang T, Zupancic JM, Huang J, Wu L, Schardt JS, De Groot AS, Elkins SL, Martin WD, Tessier PM. Optimization of therapeutic antibodies for reduced self-association and non-specific binding via interpretable machine learning. *Nat Biomed Eng* 2024; 8:45–56. doi: 10.1038/s41551-023-01074-6.
20. Elias S, Wrzodek C, Deane CM, Tissot AC, Klostermann S, Ros F. Prediction of polyspecificity from antibody sequence data by machine learning. *Front Bioinform* 2023; 3:1286883. doi: 10.3389/fbinf.2023.1286883.
21. Li L, Gupta E, Spaeth J, Shing L, Jaimes R, Engelhart E, Lopez R, Caceres RS, Bepler T, Walsh ME. Machine learning optimization of candidate antibody yields highly diverse sub-nanomolar affinity antibody libraries. *Nat Commun* 2023; 14:3454. doi: 10.1038/s41467-023-39022-2.
22. Graves J, Byerly J, Priego E, Makkapati N, Parish SV, Medellin B, Berrondo M. A review of deep learning methods for antibodies. *Antibodies (Basel)* 2020; 9:1–22. doi: 10.3390/antib9020012.
23. Bauer J, Rajagopal N, Gupta P, Gupta P, Nixon AE, Kumar S. How can we discover developable antibody-based biotherapeutics? *Front Mol Biosci* 2023; 10:1221626. doi: 10.3389/fmolb.2023.1221626.
24. Wilman W, Wrobel S, Bielska W, Deszynski P, Dudzic P, Jaszczyszyn I, Kaniewski J, Mlokosiewicz J, Rouyan A, Satlawa T, et al. Machine-designed biotherapeutics: opportunities, feasibility and advantages of deep learning in computational antibody discovery. *Brief Bioinform* 2022; 23:1–20. doi: 10.1093/bib/bbac267.
25. Shanehsazzadeh A, McPartlon M, Kasun G, Steiger AK, Sutton JM, Yassine E, McCloskey C, Haile R, Shuai R, Alverio J, et al. Unlocking de novo antibody design with generative artificial intelligence. bioRxiv 2024: 2023.01.08.523187. doi: 10.1101/2023.01.08.523187.
26. Bennett NR, Watson JL, Ragotte RJ, Borst AJ, See DL, Weidle C, Biswas R, Shrock EL, Leung PJY, Huang B, et al. Atomically accurate de novo design of single-domain antibodies. bioRxiv 2024: 2024.03.14.585103. doi: 10.1101/2024.03.14.585103.
27. Arras P, Yoo HB, Pekar L, Clarke T, Friedrich L, Schroter C, Schanz J, Tonillo J, Siegmund V, Doerner A, et al. AI/ML combined with next-generation sequencing of VHH immune repertoires enables the rapid identification of de novo humanized and sequence-optimized single domain antibodies: a prospective case study. *Front Mol Biosci* 2023; 10:1249247. doi: 10.3389/fmolb.2023.1249247.
28. Saka K, Kakuzaki T, Metsugi S, Kashiwagi D, Yoshida K, Wada M, Tsunoda H, Teramoto R. Antibody design using LSTM based deep generative model from phage display library for affinity maturation. *Sci Rep* 2021; 11:5852. doi: 10.1038/s41598-021-85274-7.
29. Dimitrov DS. Therapeutic antibodies, vaccines and antibodyomes. *MAbs* 2010; 2:347–56. doi: 10.4161/mabs.2.3.11779.
30. Georgiou G, Ippolito GC, Beausang J, Busse CE, Wardemann H, Quake SR. The promise and challenge of high-throughput sequencing of the antibody repertoire. *Nat Biotechnol* 2014; 32:158–68. doi: 10.1038/nbt.2782.
31. Chaudhary N, Wesemann DR. Analyzing immunoglobulin repertoires. *Front Immunol* 2018; 9:462. doi: 10.3389/fimmu.2018.00462.
32. Rouet R, Jackson KJL, Langley DB, Christ D. Next-generation sequencing of antibody display repertoires. *Front Immunol* 2018; 9:118. doi: 10.3389/fimmu.2018.00118.
33. Fu L, Niu B, Zhu Z, Wu S, Li W. CD-HIT: accelerated for clustering the next-generation sequencing data. *Bioinformatics* 2012; 28:3150–2. doi: 10.1093/bioinformatics/bts565.
34. Edgar RC. Search and clustering orders of magnitude faster than BLAST. *Bioinformatics* 2010; 26:2460–1. doi: 10.1093/bioinformatics/btq461.
35. Nannini F, Senicar L, Parekh F, Kong KJ, Kinna A, Bughda R, Sillibourne J, Hu X, Ma B, Bai Y, et al. Combining phage display with SMRTbell next-generation sequencing for the rapid discovery of functional scFv fragments. *MAbs* 2021; 13:1864084. doi: 10.1080/19420862.2020.1864084.

36. Wu M, Ke Q, Bi J, Li X, Huang S, Liu Z, Ge L. Substantially improved electrofusion efficiency of hybridoma cells: based on the combination of nanosecond and microsecond pulses. *Bioengineering (Basel)* 2022; 9:1–14. doi: 10.3390/bioengineering9090450.
37. Listek M, Honow A, Gossen M, Hanack K. A novel selection strategy for antibody producing hybridoma cells based on a new transgenic fusion cell line. *Sci Rep* 2020; 10:1664. doi: 10.1038/s41598-020-58571-w.
38. Zhang W, Li R, Jia F, Hu Z, Li Q, Wei Z. A microfluidic chip for screening high-producing hybridomas at single cell level. *Lab Chip* 2020; 20:4043–51. doi: 10.1039/d0lc00847h.
39. Prabakaran P, Rao SP, Wendt M. Animal immunization merges with innovative technologies: a new paradigm shift in antibody discovery. *MAbs* 2021; 13:1924347. doi: 10.1080/19420862.2021.1924347.
40. Tiller T. Single B cell antibody technologies. *N Biotechnol* 2011; 28:453–7. doi: 10.1016/j.nbt.2011.03.014.
41. Pedrioli A, Oxenius A. Single B cell technologies for monoclonal antibody discovery. *Trends Immunol* 2021; 42:1143–58. doi: 10.1016/j.it.2021.10.008.
42. Chen Y, Kim SH, Shang Y, Guillory J, Stinson J, Zhang Q, Hotzel I, Hoi KH. Barcoded sequencing workflow for high throughput digitization of hybridoma antibody variable domain sequences. *J Immunol Methods* 2018; 455:88–94. doi: 10.1016/j.jim.2018.01.004.
43. Medina-Cucurella AV, Mizrahi RA, Asensio MA, Edgar RC, Leong J, Leong R, Lim YW, Nelson A, Niedecken AR, Simons JF, et al. Preferential identification of agonistic OX40 antibodies by using cell lysate to pan natively paired, humanized mouse-derived yeast surface display libraries. *Antibodies (Basel)* 2019; 8:1–15. doi: 10.3390/antib8010017.
44. Bradbury AR, Sidhu S, Dubel S, McCafferty J. Beyond natural antibodies: the power of in vitro display technologies. *Nat Biotechnol* 2011; 29:245–54. doi: 10.1038/nbt.1791.
45. Frenzel A, Schirrmann T, Hust M. Phage display-derived human antibodies in clinical development and therapy. *MAbs* 2016; 8:1177–94. doi: 10.1080/19420862.2016.1212149.
46. Chao G, Lau WL, Hackel BJ, Sazinsky SL, Lippow SM, Wittrup KD. Isolating and engineering human antibodies using yeast surface display. *Nat Protoc* 2006; 1:755–68. doi: 10.1038/nprot.2006.94.
47. Huang A, Jin W, Fahad AS, Mussman BK, Nicchia GP, Madan B, de Souza MO, Griffin JD, Bennett JL, Frigeri A, et al. Strategies to screen anti-AQP4 antibodies from yeast surface display libraries. *Antibodies (Basel)* 2022; 11:1–17. doi: 10.3390/antib11020039.
48. Ferrara F, Teixeira AA, Naranjo L, Erasmus MF, D'Angelo S, Bradbury ARM. Exploiting next-generation sequencing in antibody selections – a simple PCR method to recover binders. *MAbs* 2020; 12:1701792. doi: 10.1080/19420862.2019.1701792.
49. Azevedo Reis Teixeira A, Erasmus MF, D'Angelo S, Naranjo L, Ferrara F, Leal-Lopes C, Durrant O, Galmiche C, Morelli A, Scott-Tucker A, et al. Drug-like antibodies with high affinity, diversity and developability directly from next-generation antibody libraries. *MAbs* 2021; 13:1980942. doi: 10.1080/19420862.2021.1980942.
50. Miyazaki N, Kiyose N, Akazawa Y, Takashima M, Hagihara Y, Inoue N, Matsuda T, Ogawa R, Inoue S, Ito Y. Isolation and characterization of antigen-specific alpaca (Lama pacos) VHH antibodies by biopanning followed by high-throughput sequencing. *J Biochem* 2015; 158:205–15. doi: 10.1093/jb/mvv038.
51. Ravn U, Gueneau F, Baerlocher L, Osteras M, Desmurs M, Malinge P, Magistrelli G, Farinelli L, Kosco-Vilbois MH, Fischer N. By-passing in vitro screening–next generation sequencing technologies applied to antibody display and in silico candidate selection. *Nucleic Acids Res* 2010; 38:e193. doi: 10.1093/nar/gkq789.
52. Zhang H, Torkamani A, Jones TM, Ruiz DI, Pons J, Lerner RA. Phenotype-information-phenotype cycle for deconvolution of combinatorial antibody libraries selected against complex systems. *Proc Natl Acad Sci U S A* 2011; 108:13456–61. doi: 10.1073/pnas.1111218108.

53. Han SY, Antoine A, Howard D, Chang B, Chang WS, Slein M, Deikus G, Kossida S, Duroux P, Lefranc MP, et al. Coupling of single molecule, long read sequencing with IMGT/HighV-QUEST analysis expedites identification of SIV gp140-specific antibodies from scFv phage display libraries. *Front Immunol* 2018; 9:329. doi: 10.3389/fimmu.2018.00329.
54. Hunter B, Hindocha S, Lee RW. The role of artificial intelligence in early cancer diagnosis. *Cancers (Basel)* 2022; 14:1–20. doi: 10.3390/cancers14061524.
55. Ting DSW, Pasquale LR, Peng L, Campbell JP, Lee AY, Raman R, Tan GSW, Schmetterer L, Keane PA, Wong TY. Artificial intelligence and deep learning in ophthalmology. *Br J Ophthalmol* 2019; 103:167–75. doi: 10.1136/bjophthalmol-2018–313173.
56. Kumar Y, Koul A, Singla R, Ijaz MF. Artificial intelligence in disease diagnosis: a systematic literature review, synthesizing framework and future research agenda. *J Ambient Intell Humaniz Comput* 2023; 14:8459–86. doi: 10.1007/s12652-021-03612-z.
57. Olsen TH, Boyles F, Deane CM. Observed antibody space: a diverse database of cleaned, annotated, and translated unpaired and paired antibody sequences. *Protein Sci* 2022; 31:141–6. doi: 10.1002/pro.4205.
58. Dunbar J, Krawczyk K, Leem J, Baker T, Fuchs A, Georges G, Shi J, Deane CM. SAbDab: the structural antibody database. *Nucleic Acids Res* 2014; 42:D1140–6. doi: 10.1093/nar/gkt1043.
59. Raybould MIJ, Marks C, Lewis AP, Shi J, Bujotzek A, Taddese B, Deane CM. Thera-SAbDab: the therapeutic structural antibody database. *Nucleic Acids Res* 2020; 48:D383–8. doi: 10.1093/nar/gkz827.
60. Jumper J, Evans R, Pritzel A, Green T, Figurnov M, Ronneberger O, Tunyasuvunakool K, Bates R, Zidek A, Potapenko A, et al. Highly accurate protein structure prediction with AlphaFold. *Nature* 2021; 596:583–9. doi: 10.1038/s41586-021-03819-2.
61. Abanades B, Wong WK, Boyles F, Georges G, Bujotzek A, Deane CM. ImmuneBuilder: deep-learning models for predicting the structures of immune proteins. *Commun Biol* 2023; 6:575. doi: 10.1038/s42003-023-04927-7.
62. Ruffolo JA, Sulam J, Gray JJ. Antibody structure prediction using interpretable deep learning. *Patterns* (*N* Y) 2022; 3:100406. doi: 10.1016/j.patter.2021.100406.
63. Abanades B, Georges G, Bujotzek A, Deane CM. ABlooper: fast accurate antibody CDR loop structure prediction with accuracy estimation. *Bioinformatics* 2022; 38:1877–80. doi: 10.1093/bioinformatics/btac016.
64. Bepler T, Berger B. Learning the protein language: evolution, structure, and function. *Cell Syst* 2021; 12:654–69 e3. doi: 10.1016/j.cels.2021.05.017.
65. Ruffolo JA, Gray JJ, Sulam J. Deciphering antibody affinity maturation with language models and weakly supervised learning. arXiv 2021. doi: 10.48550/arXiv.2112.07782.
66. Ruffolo JA, Chu LS, Mahajan SP, Gray JJ. Fast, accurate antibody structure prediction from deep learning on massive set of natural antibodies. *Nat Commun* 2023; 14:2389. doi: 10.1038/s41467-023-38063-x.
67. Lee JH, Yadollahpour P, Watkins A, Frey NC, Leaver-Fay A, Ra S, Cho K, Gligorijevic V, Regev A, Bonneau R. EquiFold: protein structure prediction with a novel coarse-grained structure representation. *bioRxiv* 2022; https://doi.org/10.1101/2022.10.07.511322.
68. Jain T, Boland T, Vasquez M. Identifying developability risks for clinical progression of antibodies using high-throughput in vitro and in silico approaches. *MAbs* 2023; 15:2200540. doi: 10.1080/19420862.2023.2200540.
69. Bailly M, Mieczkowski C, Juan V, Metwally E, Tomazela D, Baker J, Uchida M, Kofman E, Raoufi F, Motlagh S, et al. Predicting antibody developability profiles through early stage discovery screening. *MAbs* 2020; 12:1743053. doi: 10.1080/19420862.2020.1743053.
70. Raybould MIJ, Deane CM. The therapeutic antibody profiler for computational developability assessment. *Methods Mol Biol* 2022; 2313:115–25. doi: 10.1007/978-1-0716–1450–1_5.

71. Mitternacht S. Free SASA: An open source C library for solvent accessible surface area calculations. *F1000Res* 2016; 5:189. doi: 10.12688/f1000research.7931.1.
72. Olsson MH, Sondergaard CR, Rostkowski M, Jensen JH. PROPKA3: consistent treatment of internal and surface residues in empirical pKa predictions. *J Chem Theory Comput* 2011; 7:525–37. doi: 10.1021/ct100578z.
73. Raybould MIJ, Marks C, Krawczyk K, Taddese B, Nowak J, Lewis AP, Bujotzek A, Shi J, Deane CM. Five computational developability guidelines for therapeutic antibody profiling. *Proc Natl Acad Sci U S A* 2019; 116:4025–30. doi: 10.1073/pnas.1810576116.
74. Spoendlin FC, Abanades B, Raybould MIJ, Wong WK, Georges G, Deane CM. Improved computational epitope profiling using structural models identifies a broader diversity of antibodies that bind to the same epitope. *Front Mol Biosci* 2023; 10:1237621. doi: 10.3389/fmolb.2023.1237621.
75. Ingraham JB, Baranov M, Costello Z, Barber KW, Wang W, Ismail A, Frappier V, Lord DM, Ng-Thow-Hing C, Van Vlack ER, et al. Illuminating protein space with a programmable generative model. *Nature* 2023; 623:1070–8. doi: 10.1038/s41586-023-06728-8.
76. Lim YW, Adler AS, Johnson DS. Predicting antibody binders and generating synthetic antibodies using deep learning. *MAbs* 2022; 14:2069075. doi: 10.1080/19420862.2022.2069075.
77. Shan S, Luo S, Yang Z, Hong J, Su Y, Ding F, Fu L, Li C, Chen P, Ma J, et al. Deep learning guided optimization of human antibody against SARS-CoV-2 variants with broad neutralization. *Proc Natl Acad Sci U S A* 2022; 119:e2122954119. doi: 10.1073/pnas.2122954119.

Applications of Artificial Intelligence and Machine Learning toward Antibody Discovery and Development

5

Anahita Rouyan, Paweł Dudzic, Wiktoria Wilman, Tadeusz Satława, Sonia Wróbel, and Konrad Krawczyk

5.1 INTRODUCTION

Antibodies are proteins produced by the immune system to recognize and, in some cases, eliminate antigens or self-antigens found in healthy tissues in autoimmune diseases (Kindt and Goldsby 2007; Kelly-Scumpia et al. 2011); (Peng et al. 2014). Antibody-based biotherapeutics and monoclonal antibodies (mAbs) are a unique class of biologics that have transformed the pharmaceutical industry since their introduction in the mid-1980s. The global market for mAb-based drugs is currently estimated to value 152.5 billion US dollars and is projected to grow rapidly in the next decade, reaching 300 billion US dollars by 2025 (Kaplon and Reichert 2021; Lu et al. 2020). More than

DOI: 10.1201/9781003300311-5

550 antibodies are in early stage commercial clinical pipelines (Kaplon et al. 2020) for treating conditions such as cancer, cardiovascular disease, inflammation, neurological disorders, autoimmune, and infectious diseases. Research into the use of mAbs for the neutralization of viral agents, such as human immunodeficiency virus (HIV), influenza, and severe acute respiratory syndrome Coronavirus 2 (SARS-CoV-2), is ongoing (Wang et al. 2020a; Liu et al. 2020; Laustsen et al., 2022; Mannar et al. 2022).

The early discovery and design of antibody-based biotherapeutics typically take more than three years, with anti-SARS-CoV-2 antibodies excluded (Torjesen, n.d.; Laustsen et al. 2021; Carter and Lazar, n.d.). One of the reasons behind that is that mAb development pipelines utilize a combination of large screening libraries and experimental heuristics (Fischman and Ofran 2018). Most of the mAb-based therapeutics available on the market were developed using expensive and time-consuming techniques such as phage display or animal immunization platforms (Ferrara et al. 2022; Laustsen et al. 2021). With the increasing maturity and integration of computational methods within pharmaceutical company pipelines, therapeutic antibody development may become faster and more cost-effective, making immunotherapy more affordable to patients and expanding its applicability to more diseases (Greiff, Yaari, and Cowell 2020; Narayanan et al. 2021; Wilman et al. 2022).

Computational antibody design already employs established bioinformatic methods such as homology modeling, protein-protein docking, and protein interface prediction (X. Liu et al. 2017; Baran et al. 2017; Raybould et al. 2019). Pharmaceutically focused computational methods support the assessment of antibody immunogenicity and biophysical properties (Abhinandan and Martin 2007; Obrezanova et al. 2015). However, structure-based antibody design has been curbed by the lack of accurate antibody and antigen structures – until the emergence of machine learning (ML)-based methods capable of utilizing large data volumes. Examples of that are breakthroughs in computational structure prediction (Jumper et al. 2021; Tunyasuvunakool et al. 2021). These advances go hand in hand with the rapidly increasing volumes of available antibody sequences (Olsen et al. 2022a; Kovaltsuk et al. 2018). Structural, sequence, and experimental developability data are deposited in the public domain, providing a solid foundation for the expansion and improvement of data-driven methods (Dunbar et al. 2014; Kovaltsuk et al. 2018; Vita et al. 2010; Sirin et al. 2016; Koenig et al. 2017; Jain et al. 2017). Prominent examples are next-generation sequencing (NGS) of B-cell receptor (antibody) repertoires that offer a snapshot of millions of antibody sequences sampled from the theoretically possible 10^{16}–10^{18} antibody sequences found in the human repertoire (Miho et al. 2018; Glanville et al. 2009; Briney et al. 2019). Large volumes of data combined with the advances in method development indicate that the computational antibody analysis methods have now reached the level of maturity that allows for broader therapeutic development applications.

5.2 DATABASES

Data availability is one of the primary drivers accelerating the science of antibody discovery since it provides the basic resource to train computational models. Public antibody databases are crucial resources for discovery in the biopharmaceutical industry, serving as baselines to provide benchmarks and train proof-of-concept models (Khetan et al. 2022). Public antibody databases range from low-throughput structural data (Dunbar et al. 2014) to high-throughput, more error-prone NGS datasets (Kovaltsuk et al. 2018). Distinct data sources feed specific modeling methods (e.g., structural modeling) – however, synergies also exist among the usage of different datasets (Kovaltsuk et al. 2017). Applications of datasets to different modeling tasks involve recognizing the benefits and limitations of different resources (e.g., considering that few structures for the task of antibody-antigen recognition are available).

Structural and assay data are arguably the most expensive and low throughput, resulting in few precious data points. Sequence data are widely available owing to the increasing use of NGS technologies. One of the chief issues when tapping into such data is the heterogeneity of depositions. For instance, even though structural data can be almost uniquely mapped to a single repository (for instance, Protein Data Bank (PDB)), the affinity assay data associated with them are not deposited systematically. Such assay data need to be extracted directly from scientific papers. Linking antibody sequence information to literature is challenging as well. Despite the available resources such as GenBank, protein sequences do not always find their way from supplementary materials to such repositories. Therefore, one needs to curate data from different sources to get the full information picture.

The benefit of curating different antibody databases is the added information from linking heterogeneous records. For instance, current NGS analysis processes typically use sequence data, rarely enriching it with structures (Kovaltsuk et al. 2017; Krawczyk et al. 2018). As another application example, comparing the naturally occurring and therapeutically produced molecules that can provide insight into similarities and differences among sources (Krawczyk et al. 2019). The Integrated Nanobody Database for Immunoinformatics (Deszyński et al. 2021) containing nanobody data from patents, NCBI GenBank, PDB, and scientific literature, is an example of curating heterogeneous data types for the added benefit of a single access point to nanobody data in the public domain. In many guises, the effort to curate and harmonize such databases is to make such data readily available for training computational models.

5.2.1 Databases in Machine Learning Approaches

Computational methods and ML, in particular, are becoming increasingly used to facilitate the therapeutic antibody discovery (Wilman et al. 2022). ML algorithms require well-curated datasets to achieve the acceptable levels of accuracy (Akbar et al. 2022). In an ideal scenario, one would use plentiful data points combining the sequence, structure, biophysical, and biological activity of these molecules. In reality, however, researchers only have subsets of different features (e.g., sequence without binding information). Nonetheless, ML algorithms already use the resources at hand, even when trained on such incomplete data representations.

A good example of ML application in the antibody space is the use of sequences in the Observed Antibody Space (OAS) database to train generative adversarial networks (GANs) aiming to propose mAbs with specific developmental parameters

(Amimeur et al. 2020). OAS data were also employed to streamline structure prediction in DeepAb (Ruffolo et al. 2022). The plentiful NGS data in OAS have also found applications in training transformers (Leem et al. 2022; Bachas et al. 2022) as well as predicting immunogenicity (Marks et al. 2021; Prihoda et al. 2022). Such applications were made possible by combining curated dataset availability and an understanding of their benefits and limitations for few-shot learning, such as support from samples of different sizes, availability of datasets focused on one or multiple indication/s or population/s, or variation in wet-laboratory protocols.

5.2.2 Database Types

There are primarily four types of antibody databases available: structure databases (e.g., SABDab (Dunbar et al. 2014)), sequence databases (e.g., OAS (Kovaltsuk et al. 2018; Olsen et al. 2022a)), databases that hold experimental assay data (e.g., affinity or immunogenicity (Vita et al. 2015)) or focus on a specific topic (e.g., Coronavirus antibody database (Raybould et al. 2021), or Immune Epitope Database (IEDB) (Vita et al. 2015)). Examples of such databases are collated in Table 5.1.

5.2.2.1 Structure databases

For structural data, the PDB is the most important global repository of 3D protein structural information. Some databases derive antibody-specific information from the PDB, such as canonical classes (PyIgClassify (Adolf-Bryfogle et al. 2015; Kelow et al. 2022)), antibody-antigen interaction data (PCLICK), or their entire structures (Ferdous and Martin 2018; Dunbar et al. 2014). Up to 6,600 of the approximately 200,000 structures deposited in the PDB so far have been confirmed to contain at least one antibody chain. In databases such as SAbDab or ABDB, users can specify a query antibody sequence (SAbDab) or use advanced features such as complementarity determining region (CDR) canonical classes (Abysis) to retrieve specific structures (Norman et al. 2020). Structural databases are typically used for structure prediction (Leem et al. 2016; Ruffolo et al. 2022) and antibody-antigen binding prediction (Krawczyk et al. 2013, 2014; Schneider et al. 2021).

5.2.2.2 Sequence databases

The primary source of antibody germline genes is the International ImMunoGeneTics Information System (IMGT), though there exist alternatives such as OGRDB (Lees et al. 2020). The majority of other sequence databases provide recombined antibodies. These databases range from those storing lower throughput data (e.g., GenBank (Clark et al. 2016)) to the NGS databases storing high-throughput information (e.g., iReceptor (Corrie et al. 2018), OAS (Kovaltsuk et al. 2018; Olsen et al. 2022a), cAb-Rep (Guo et al. 2019)). The lower throughput databases have the advantage of linking antibody sequences with experimental data (e.g., binding) in an unsystematic way. The NGS databases typically do not have any means of linking a particular sequence to either antigen specificity or other biophysical properties of an antibody. However, due to their large volume, they are typically used for pre-training language models (Olsen et al. 2022b; Leem et al. 2022).

TABLE 5.1 Examples of different antibody database types

DATABASE NAME	*DATABASE TYPE*	*DATABASE CONTENT DESCRIPTION*	*ACCESS LINK*	*REFERENCES*
AbMiner	Experimental specific	Matches commercially available antibodies to their respective genomic identifiers	https://discover.nci.nih.gov/abminer/	(Major et al. 2006)
AntiJen	Experimental specific	Quantitative binding data for peptides	http://www.ddg-pharmfac.net/antijen/AntiJen/antijenhomepage.htm	(Toseland et al. 2005)
IEDB	Experimental specific	Experimental data on humans, non-human primates, and other animal species antibody and T-cell epitopes	https://www.iedb.org/	(Vita et al. 2015)
IgPdb	Experimental specific	Inferred allelic variants for immunoglobulins	http://cgi.cse.unsw.edu.au/~ihmmune/	NA
PCLICK	Experimental specific	Antibody-antigen structures from a dataset of antibody-antigen complexes using the CLICK method	http://mspc.bii.a-star.edu.sg/minhn/cluster_pclick.html	(Nguyen et al. 2017)
SKEMPI 2.0	Experimental specific	Kinetics and energetics data upon mutation, for protein-protein interactions including antibody-antigen complexes	https://life.bsc.es/pid/skempi2/	(Jankauskaite et al. 2019)
ABCD	Sequence	Manually curated depository of sequenced antibodies	https://web.expasy.org/abcd/	(Lima et al. 2020)
cAb-Rep	Sequence	Curated human B-cell immunoglobulin sequence repertoires	https://cab-rep.c2b2.columbia.edu/	(Guo et al. 2019)
EMBLIg	Sequence	Antibody sequences automatically extracted from EMBL-ENA	http://www.abybank.org/emblig/	NA
Integrated Nanobody Database for Immunoinformatics (INDI)	Sequence	Structure data and sequence information of nanobodies created using an integrated curation approach from several sources	http://research.naturalantibody.com/nanobodies	(Deszyński et al. 2021)

(*Continued*)

TABLE 5.1 (*Continued*) Examples of different antibody database types

DATABASE NAME	*DATABASE TYPE*	*DATABASE CONTENT DESCRIPTION*	*ACCESS LINK*	*REFERENCES*
iReceptor	Sequence	Antibody/B-cell and T-cell receptor repertoire data from multiple independent repositories	https://gateway.ireceptor.org/	(Corrie et al. 2018)
OAS	Sequence	Annotated immune repertoires	http://opig.stats.ox.ac.uk/webapps/oas/	(Olsen, Boyles et al. 2022a; Kovaltsuk et al. 2018)
Patented Antibody Database	Sequence	Sequence information found in patent documents	https://www.naturalantibody.com/pad	(Krawczyk, Buchanan, and Marcatili 2021)
AAAAA	Sequence/Structure	Antibody receptor sequences and structures	https://plueckthun.bioc.uzh.ch/antibody/	(Honegger and Plückthun 2001)
abYsis	Sequence/Structure	Antibody sequence and structure data	http://www.abysis.org/abysis/	(Swindells et al. 2017)
IMGT	Sequence/Structure	Immunoglobulins, T-cell receptors, MHC, and in the immunoglobulin and MH superfamilies, and related proteins of the immune system of vertebrates and invertebrates	http://www.imgt.org/	(Lefranc et al. 1999)
Tabs – Therapeutic Antibody Database	Sequence/Structure/Experimental specific	Clinical trials, patents, papers, news, and regulatory agencies	https://tabs.craic.com/	NA
AB-Bind	Specific experimental	Contain experimental results for wild-type and mutant antibodies and antigens	https://github.com/sarahsirin/AB-Bind-Database	(Sirin et al. 2016)

(*Continued*)

TABLE 5.1 (*Continued*) Examples of different antibody database types

DATABASE NAME	*DATABASE TYPE*	*DATABASE CONTENT DESCRIPTION*	*ACCESS LINK*	*REFERENCES*
AbDb	Structure	Extracted from the PDB with standard numbering schemes applied and redundancy information	http://www.abybank.org/abdb/	(Ferdous and Martin 2018)
SAbDab	Structure	Annotated antibody and nanobody structures available in the PDB	http://opig.stats.ox.ac.uk/webapps/sabdab	(Dunbar et al. 2014)
SACS	Structure	Summary of antibody crystal structures in the PDB	http://www.abybank.org/sacs/	(Ferdous and Martin 2018)
Bcipep	Target specific	B-cell epitopes	https://webs.iiitd.edu.in/raghava/bcipep/info.html	(Saha et al. 2005)
CoV-AbDab	Target specific	Published or patented binding antibodies and nanobodies to coronaviruses	http://opig.stats.ox.ac.uk/webapps/covabdab	(Raybould et al. 2021)
Non-redundant Nanobody database	Target specific	Non-redundant structures of nanobodies	https://www.sciencedirect.com/science/article/pii/S2352340919301052	(Zavrtanik and Hadži 2019)
sdAb-DB	Target specific	Single-domain antibody repository	http://www.sdab-db.ca/	(Wilton et al. 2018)
Thera-SAbDab	Target specific	WHO-recognized antibody-related therapeutics	http://opig.stats.ox.ac.uk/webapps/newsabdab/therasabdab/	(Raybould et al. 2020)

For each antibody database type, multiple sources are presented above with a short description and access link.

5.2.2.3 *Databases with domain-specific information*

Sequence and structural data can be further enriched with antibody-specific experimental information. Data on epitopes targeted by antibodies can be easily downloaded from the IEDB (Vita et al. 2015), which now links such data to epitope-specific antibody sequences. One of the key pieces of information characterizing antibody-epitope interactions is the binding affinity. There are other more specialized resources such as Ab-Bind (Sirin et al. 2016), which contains data on 1101 mutations in 32 antibody complexes, and SKEMPI (Jankauskaite et al. 2019), which curates binding energy data for available structures but is not limited to antibody information only (Norman et al. 2020). There also exist resources that are specific to an antibody target (e.g., CoV-ABDAB) (Raybould et al. 2021), application (e.g., therapeutics) (Raybould et al. 2020), or format (e.g., nanobodies) (Deszyński et al. 2021).

5.2.2.4 *Mixed databases*

Some sources attempt to consolidate sequence and structure data. One example is abYsis, a web-based antibody research system that contains both types of information (Swindells et al. 2017). Another example is the IMGT, a sequence, genomic, and structural knowledge resource dedicated to immunoglobulins, T-cell receptors, MHC, the immunoglobulin and MH superfamilies, and related proteins of the vertebrate and invertebrate immune systems (Lefranc et al. 1999). A single source of information simplifies and shortens the time to arrive at a data point and streamlines data curation for computational modeling (Deszyński et al. 2021).

5.3 APPLICATIONS OF MACHINE LEARNING IN ANTIBODY DISCOVERY AND DEVELOPMENT

Recent developments in computational antibody engineering enabled by ML and its subset, deep learning, have led to improvements in the previous state-of-the-art methods such as structure prediction, binding prediction, and developability prediction. This section explores how deep learning has transformed the solutions to the previous antibody problems.

5.3.1 Structure Prediction with Deep Learning

Antibody structure prediction has a broad range of applications in antibody engineering since the molecular shape of the paratope determines antibody-antigen recognition (Kovaltsuk et al. 2017). Defining the crystal structure of proteins poses many technical challenges, generating significant interest in methods for predicting the 3D coordinates from sequences (Kryshtafovych et al. 2021). There are only about 6,500 crystal structures of antibodies and their fragments (Fabs, Fcs, etc.), some of them in complex with their antigen (Dunbar et al. 2014). This number recapitulates the natural structural diversity surprisingly well (Krawczyk et al. 2018), though inferring the structure of CDR-H3 from its sequence remains a challenge.

The need to represent millions of possible antibody structures was traditionally addressed using homology modeling and energy-based methods (Wilman et al. 2022). Such methods started ceding a way to deep learning methods, partly because of the solution to the general protein folding demonstrated by AlphaFold2 (Jumper et al. 2021; Kryshtafovych et al. 2021). Such protein structure prediction methods use data from coevolutionary signals, while this type of information is unavailable for antibodies. Paratope prediction and CDR-H3, in particular, are particularly elusive (Jones et al. 2015; Marks et al. 2011; Regep et al. 2017). The original work of Chothia and colleagues (Chothia and Lesk 1987) and much later work (Adolf-Bryfogle et al. 2015; North, Lehmann, and Dunbrack 2011; Kelow et al. 2022) demonstrated that CDRs have specific clusters of conformations. Incorporating such classical papers into the development of ML methods may potentially improve the accuracy of antibody structure prediction, though in some cases, ML models implicitly infer such properties (Ruffolo et al. 2022).

Antibody modeling methods build on deep learning architectures, devising solutions specific to this class of molecules. In what follows, we highlight four of the available recent methods that tackle the problem of antibody structure prediction (all methods are given in Table 5.2).

TABLE 5.2 Structural prediction models

METHOD	*METHOD DESCRIPTION*	*REFERENCES*
Generative Modeling for Protein Structures	Generative model predicting completions of corrupted protein structures	(Anand and Huang 2018)
DeepH3	Model predicting the inter-residue distances and orientations into a discrete set of bins, which is then used to score poses generated by RosettaAntibody	(Ruffolo et al. 2020)
DeepAb	Deep residual convolutional network predicting bins of distances and orientation angles, with a Rosetta-based protocol that realizes the structure of the predicted distance and angle constraints	(Ruffolo et al. 2022)
IG-VAE	Variational autoencoder generating the 3D coordinates of immunoglobulins	(Eguchi et al. 2022)
DeepSCAb	Model predicting inter-residue geometries as well as side-chain dihedrals of the antibody variable fragment	(Akpinaroglu et al. 2022)
Ablooper	Model predicting the structure of CDR loops and provides a confidence estimate for each of its predictions	(Abanades et al. 2022a)
Nanonet	Model given a sequence directly produces the 3D coordinates of the Cα atoms of the entire VH domain	(Cohen et al. 2022)

(*Continued*)

TABLE 5.2 (*Continued*) Structural prediction models

METHOD	*METHOD DESCRIPTION*	*REFERENCES*
AbodyBuilder2	Deep learning model behind ABodyBuilder2 is an antibody-specific version of the structure module in AlphaFold-Multimer with several modifications	(Abanades et al. 2022b)
EquiFold	Model relies on geometrical structure, uses a novel coarse-grained representation of protein structures that do not require multiple sequence alignments or protein language model embeddings	(Lee et al. 2022)
T-FoldAb	Model generating simultaneous prediction of backbone and side-chain conformations	(Wu et al. 2022)
xTrimoABFold	Model predicts antibody structure from antibody sequence based on a pre-trained antibody language model and homologous templates	(Anonymous 2022)

A short description and reference are presented above for each structural prediction method.

5.3.1.1 *DeepH3*

Based on RaptorX (Källberg et al. 2012), DeepH3 is a deep residual network used for de novo CDRH3 prediction. When given a one-hot encoding, DeepH3 predicts the inter-residue distances and orientations into a discrete set of bins, which is then used to score poses generated by RosettaAntibody (Weitzner et al. 2014). The training dataset included records from SAbDab (Dunbar et al. 2014), with 99% sequence identity and thresholds and 3.0 Å resolution. The benchmark dataset used was based on 49 variable fragment (Fv) structures chosen from the PyIgClassify database (Adolf-Bryfogle et al. 2015) with quality in mind, including CDRH3 loop of lengths between 9 and 20 residues. The method achieves accuracies in the 2 Å region rather than the 4 Å in trRosetta, a solution designed for general protein structure prediction (Du et al. 2021). DeepH3 revealed that the distributions of orientation angles find better use as discriminators than distance distributions alone, which were used in many previous methods for general protein structure prediction (Jumper et al. 2021; Marks et al. 2011; Källberg et al. 2012).

5.3.1.2 *ABlooper*

ABlooper proposed an alternative architecture for CDR prediction as Equivariant Graph Neural Networks (Abanades et al. 2022a). The solution encodes input data from SAbDab into 41-dimensional vectors with amino acid type, atom type, and loop to which the residue belongs. Moreover, it gives sinusoidal positional embeddings to each residue to describe how close it is located to the anchors. SAbDab data were used to train five E(G) NNs, each containing four layers. The prediction confidence was the agreement of these five networks on the generated coordinates. The benefit of its method is its speed, as it does not rely on other structure generation algorithms and is capable of rapid generation of coordinates, given a framework structure from homology modeling.

5.3.1.3 DeepAb

Predicting CDR-H3 loops is a challenging task driven by the broader goal of reconstructing the entire antibody variable region. Built on top of the DeepH3 method, DeepAb was designed to predict the whole variable regions. The method includes two main stages: (1) a deep residual convolutional network that predicts bins of distances and orientation angles and (2) a Rosetta-based protocol that realizes the structure of the predicted distance and angle constraints. Apart from the structural information given as input, the network includes a bidirectional-long-short term memory (LSTM) network trained on a set of 118,386 paired heavy and light chain sequences from OAS, with the goal of teaching the network the general features of antibody sequences (Kovaltsuk et al. 2018). This feature extraction enabled implicitly capturing certain structural properties – for example, recreating the PyIgClassify annotations (Adolf-Bryfogle et al. 2015). The network also tracks residues contributing to the coordinate/angle prediction of each other using an attention mechanism. The model primarily attends to residues surrounding each loop analyzed, with the difference that CDRH3 predictions generate results from a broader set of dependencies across heavy and light chains.

5.3.1.4 NanoNet

Methods such as DeepAb show the potential of ML techniques in antibody structural predictions. However, they may be hindered by slower methods required for generating coordinates. NanoNet addressed this issue (Cohen et al. 2022). The residual convolutional network was originally designed to predict structures of single-chain nanobodies but could also be used for mAb chain structure prediction; it generates predicted alpha carbon coordinates as output when fed a one-hot-encoded variable region sequence. It aligns the input structures with developing a single frame of reference for the generated predictions. Since the structure realization is carried out within a single network, NanoNet can produce thousands of structures in a few seconds.

The methods discussed above have enabled a dramatic improvement in the accuracy of antibody structure prediction, achieving CDRH3-level accuracy in the region of 2–3 Å rather than 3–4 Å previously (Almagro et al. 2014). These predictions are approaching sufficient levels that allow them to be reliably used as substitutes for crystal structures, providing models at the speed and scale (Krawczyk et al. 2018) required to address the antibody-binding problem for the virtual screening of drug candidates.

5.3.2 Antibody-Binding Prediction by Deep Learning (Paratope Prediction)

Three-dimensional structures of both antibodies and antigens play a key role in determining antibody-antigen interactions. Computational methods such as homology modeling, docking, and interface prediction can be employed during the lead identification and optimization phases to generate 3D models of antibodies and predict or identify the key residues that are part of antigen binding.

The development of novel binders was typically limited to animal immunizations of phage-display methods. The methods for analyzing antibody-antigen interactions were previously categorized into paratope, epitope, or docking, as reviewed previously (Norman et al. 2020). Parapred (Liberis et al. 2018) pioneered deep learning usage in antibody-antigen interface prediction (specifically, paratopes). However, the current combination of NGS and ML approaches opens the door to novel applications going beyond these three categories.

5.3.2.1 Paratope prediction

Solutions to paratope prediction were first proposed by proABC (Olimpieri et al. 2013), antibody i-Patch (Krawczyk et al. 2013), and PARATOME (Kunik et al. 2012). These methods relied on structural information and were based on statistical scores. Recently, the paratope prediction problem was addressed by deep learning methods that, surprisingly, do not even consider the structure.

Liberis and colleagues (Liberis et al. 2018) presented Parapred, a sequence-based probabilistic ML algorithm for paratope prediction. The method improved on the current state-of-the-art methodology and only required a stretch of the amino acid sequence corresponding to a hypervariable region as an input. No information about the antigen was required. Liberis and colleagues showed that such predictions could be used to improve both the speed and the accuracy of a rigid docking algorithm. Since then, the proposed improvements to paratope prediction employed fine-tuning a language model via AntiBerta (Leem et al. 2022) and a structure as input by Paragraph (Chinery et al. 2022).

Within paratope prediction, one needs to keep in mind that most methods do not perform the predictions with respect to the antigen, which does not consider the possible polyspecificity of different paratopes. Nevertheless, despite such radical simplification, such methods have proven utility in their therapeutic applications – for example, in the orthogonal identification of antigen-specific antibodies (Richardson et al. 2021).

5.3.2.2 Epitope prediction

The prediction of the paratope and epitope – the binding sites in antibody-antigen interactions – is essential for vaccine and synthetic antibody development. Del Vecchio and colleagues (Del Vecchio et al. 2021) argued that asymmetric treatment was required for paratope and epitope predictors. They proposed neural message-passing architectures that aim toward specific paratope and epitope prediction aspects, achieving significant improvements on both tasks. Another interesting development in this space is ScanNet (Tubiana et al. 2022), an end-to-end, interpretable geometric deep learning model capable of learning features directly from 3D structures. The model builds representations of atoms and amino acids using the spatio-chemical arrangement of their neighbors. Tubiana and colleagues trained the model for detecting protein-protein and protein-antibody-binding sites, demonstrating its accuracy.

Though representing state of the art, the methods do not achieve quality comparable to experimental epitope mapping experiments. Notably, deep learning methods that use more plentiful data (Tubiana et al. 2022) do not achieve better performance by order of magnitude compared to non-ML methods developed close to a decade ago (Krawczyk et al. 2014). Therefore, specifically in the field of epitope prediction, many issues still need to be addressed to reliably replace experimental epitope mapping methods.

5.3.2.3 Docking

A key application of predicting antibody-antigen interactions is allowing researchers to leverage the large volume of NGS data to mine for novel binders (virtual screening). Virtual screening initiatives in the area of small molecules often come together with large-scale docking (Maia et al. 2020). The increase in deep learning model usage applies to scoring different docking poses for protein-protein functions in general (Geng et al. 2019; Renaud et al. 2021; Wang et al. 2020b). Due to the different interface statistics (Krawczyk et al. 2013), antibody-antigen docking is a subproblem of protein docking in general (Ambrosetti et al. 2020; Kilambi and Gray 2017).

Within the remit of antibody-antigen docking, one focuses on generating poses by general-purpose docker such as ZDOCK (Chen et al. 2003) by constraining the paratope and proceeding to rescoring poses in an antibody-specific fashion (Sircar and Gray 2010; Krawczyk et al. 2014). Deep learning for antibodies employed docking to address virtual screening through rescoring ZDOCK poses in an antibody-specific way (Chen et al. 2003). The network in use was a deep convolutional network that classified docking poses into a bucket, showing an interval of the fraction of native reconstructed contacts. This method further demonstrated how such predictions could be employed in a special case of discriminating SARS-COV-2 binders via a mock virtual screening exercise.

5.3.2.4 Binding prediction from sequence

Advances in microfluidics technologies and NGS contributed a vast amount of antibody repertoire sequencing data, opening the door to new opportunities for deep learning-based applications. Combining experimental data with ML methods shows the possibility of predicting binding affinity (Bachas et al. 2022) and specificity (Zhang et al. 2022).

This is illustrated by the example of a study by Lim and colleagues (Lim et al. 2022). They started by encoding antibody light- and heavy-chain CDR3s into antibody images. Next, they built and trained a convolutional neural network model to classify binders and non-binders. To improve model interpretability, they performed in silico mutagenesis to identify CDR3 residues playing an important role in binder classification. The team built generative deep learning models using GAN models to produce synthetic antibodies against PD-1 and CTLA-4. The models generated variable-length CDR3 sequences resembling real-world sequences, showing that deep learning methods can be used to mine and learn patterns in antibody sequences, uncovering insights about antibody engineering, optimization, and discovery.

Generating such target-specific predictors requires significant data generation by deep mutational scans. This might seem counterintuitive to the purpose of computational methods that should not require lengthy and costly experimental runs. However, in such cases, training the model on several thousand experimentally generated sequences opens the door toward generation/enumeration and subsequent binding prediction of millions of sequences. Therefore, such approaches open the doors toward practical computational scanning for binders in the entire antibody space rather than sampling small subsets of it, as is the case in most experimental approaches.

5.3.2.5 Simulated Ab–Ag binding data

Even though experimentally generated data on binding becomes more high-throughput, it still falls short of covering all the possible antibody-antigen interaction pairs. To overcome this hurdle, researchers tested this concept in silico using simulated Ab–Ag binding data (Akbar et al. 2022). This allowed them to also perform a theoretical study of a general "binding objective function" and address the problem of "learnability" of the Ab–Ag recognition. Akbar and colleagues trained the LSTM-recurrent neural network (RNN) on CDR-H3 that were apriori computationally associated with developability data (Robert et al. 2021). The simulations included a wide range of physiological antibody-binding parameters and enabled the 3D structure computation of synthetic antibody-antigen binding. The network generated sequences that exceeded the developability parameters of sequences on which the model was trained. RNN-LSTM models were trained separately on binders as positive data and non-binders as negative data, suggesting that the design of native-like CDR-H3 sequences is possible without negative examples. Also, accuracy is likely to be further improved with more training examples. This fact could potentially reduce the cost of generating training datasets, given that the HER2 generation rate of the RNN was remarkably high despite only being trained on positive data. In addition, by delivering binders against this target for training, the network could generate specific binders against HER2. This demonstrated that antibody properties are, in principle, trainable on multiple modalities – binding and developability. The ability of in silico generative frameworks to generate antigen-specific and developable immune receptor sequences on a large scale and on demand is a critical feature for the rapid synthesis of antibodies.

5.3.3 Developability

Creating an antibody binder that exhibits therapeutic action is only part of the drug development requirements. Such a molecule must meet various biophysical, manufacturability, safety, efficacy, and pharmacology features termed "developability" (Jain et al. 2017). There are multiple self-association, polyspecificity, solubility, and immunogenicity assays, among others, that need to be performed to derisk the molecule and let it into clinical trials. To reduce the experimental burden, researchers attempted to solve this issue using statistical models (Raybould et al. 2019) or non-deep learning approaches – for instance, random forests (Obrezanova et al. 2015). Here, we focus on two aspects, singling out deimmunization or immunogenicity prediction and generalistic models for developability prediction.

5.3.3.1 Immunogenicity

Mice are often used as a source of therapeutic antibodies upon being challenged with a target antigen. The significant drawback of this approach is that animal antibodies can elicit unwanted immune responses in humans. To reduce such risks, researchers need to engineer antibodies to resemble human antibody molecules without losing activity as part of the humanization process (Kim and Hong 2012). Computational methods quantifying the nativeness of human sequences have long played a significant role in this process (Table 5.3).

Initially, computational humanization was tackled by frequency-based methods that quantified the similarity between animal and human sequences – for example, T20 (Gao et al. 2013) or humanness scores (Abhinandan and Martin 2007). These approaches were based on a small number of sequences (in thousands), offering a limited ability to determine the correlation between different residues. The availability of NGS data increased the antibody sequence samples from thousands to millions, leading to the creation of improved positional frequencies (Schmitz et al. 2020; Sheng et al. 2017; Kovaltsuk et al. 2018).

Positional profiles may have been enriched with NGS data but do not reflect positional interdependencies. To quantify positional correlations, researchers developed a multivariate Gaussian (MG) statistical score based on the OAS data (Clavero-Álvarez et al. 2018), later expanded to the LSTM model by Wollacott (Wollacott et al. 2019). Both models concentrated on predicting what constitutes the human sequence and brought to light the required element of correlations between positions. The authors of MG compared their score to therapeutic sequence immunogenicity. It resulted in a weak correlation ($r^2=0.18$), suggesting that sequence identities might not encode the immunogenicity information alone. This finding is consistent with the industry experience around immunogenicity origins. Immunogenicity toward biotherapeutic drugs can be observed in clinical trials through the generation of anti-drug antibodies (ADAs) by patients who receive immunotherapy. Immunogenicity emerges in patients due to multiple factors related to drug product quality (formulation, presence of aggregates in the product, or aggregation of the product in vivo during administration), patient's specific disease history, their genetic background, and the humanness of the antibody sequence (Kumar et al. 2011; Fathallah et al. 2015; Singh 2011).

A much larger scale attempt at employing NGS data for deimmunization was proposed by Hu-mab (Marks et al. 2021). The method is based on a random forest model trained to distinguish human and non-human sequences of a particular V gene type from ones from other species. Hu-mab correctly differentiated human and other animal sequences in both validation and test sets, noting slightly worse performance on the light chain, which may have been caused by the greater volume of negative training data available for variable heavy chain (VH) than variable light chain (VL) models. Another reason could be the smaller variability of light chains regarding isotypes and CDRs. The previous LSTM model (Wollacott et al. 2019) could not discriminate between human and other animal sequences, possibly because LSTM models were only trained on sequences originating in a single species (human).

TABLE 5.3 Humanization methods

HUMANIZATION METHOD	*METHOD DESCRIPTION*	*REFERENCES*
T20	Based on thousands of sequences, offers a limited ability to determine the correlation between a selected set of residues	(Gao et al. 2013)
Humanness scores	Based on thousands of sequences, one of the pioneering methods to quantify nativeness of antibody sequences	(Abhinandan and Martin 2007)
Improved positional frequencies	Based on millions of sequences, quantify nativeness of antibody sequences by determining the correlation between different residues	(Schmitz et al. 2020; Sheng et al. 2017; Kovaltsuk et al. 2018)
Multivariate Gaussian (MG) statistical score	Based on the OAS data, statistical score to quantify the nativeness of human sequence	(Clavero-Álvarez et al. 2018)
Multivariate Gaussian (MG) statistical score with the LSTM model	Based on the OAS data, predicts what constitutes the human sequence with usage of LSTM model	(Wollacott et al. 2019)
Hu-mab	Based on a random forest model trained to distinguish human and non-human sequences of a particular V gene type from ones from other species	(Marks et al. 2021)
BioPhi (Sapiens and OASis)	Uses language models for capturing the diversity of natural human antibody repertoires and humanizes a sequence	(Prihoda et al. 2022)
AbBERT	Transformer-based language model trained on up to 20 million unpaired heavy/light chain sequences from the OAS database	(Vashchenko et al. 2022)
Llamanade	For capturing unique features of nanobodies, uses large-scale analysis of nanobodies and IgG	(Sang et al. 2022)
IgReconstruct	Based on single nucleotide frequencies, a germline gene rearrangement tailored to the nucleotide frequency observations made in the repertoire is generated to estimate the similarity of a target Ab amino acid sequence to a given repertoire	(Schmitz et al. 2020)
AbDiver	Positional frequencies from OAS	(Młokosiewicz et al. 2022)

A short description and reference are presented above for each humanization method.

Another computational method that aims to accelerate the process of humanization is BioPhi, an antibody design interface that uses automated methods for capturing the diversity of natural human antibody repertoires (Prihoda et al. 2022). BioPhi offers two data-driven approaches – humanization (Sapiens) and humanness evaluation methods (OASis). Sapiens is a deep learning humanization approach based on masked language modeling (MLM). The method is trained with human variable region antibody sequences from the OAS. Its goal is to recognize and repair masked or mutated positions in unaligned amino acid sequences. OASis, on the other hand, is a humanness metric based on peptide search in the OAS, which evaluates the humanness of a sequence by dividing it into overlapping 9-mer peptides (inspired by human string content (Lazar et al. 2007)). Next, the metric compares them against the OAS database to predict their universality across the human population. Based on the in silico humanization benchmark of 177 antibodies, this solution offers mutational choices that are similar to those achieved by experimental humanization methods. The primary advantage of BioPhi is its attention layer and granularity, allowing users to examine the residue dependencies and mutational effects on the score.

5.3.3.2 Prediction of biophysical properties: Aggregation, viscosity, solubility, and thermostability

An mAb-based new drug development relies on the developability properties of these molecules, such as low aggregation propensity and low viscosity (Xu et al. 2019; Jarasch et al. 2015; Makowski et al. 2021). Given the limited number of molecules with the available sequence and biophysical property data, assessing the stability profiles of antibodies at high concentrations is challenging during early stage discovery. Predictive tools based on ML may help researchers to evaluate the developability of high-concentration antibody formulation early in the discovery process. Researchers have already applied computational tools to identify drug-like antibodies characterized by favorable stability (Makowski et al. 2021).

Viscosity prediction received attention as well. For example, Sharma and colleagues (Sharma et al. 2014) found viscosity to be highly correlated with Fv net charge and charge symmetry. At the same time, viscosity was determined to be weakly correlated with hydrophobicity. The team proposed a linear equation based on these parameters to calculate the viscosity at 180 mg/ml (pH 5.5 and 200 mM arginine-HCl). Another viscosity predictive tool is the spatial charge map, calculated using a molecular dynamics (MD) simulation that explains the exposed surface-negative charge distribution on the Fv region (Agrawal et al. 2016). Tomer and colleagues (Tomar et al. 2017) developed an equation for predicting the concentration-dependent viscosity curves, which used charges on the heavy and light chain variable regions, the hinge region, as well as the hydrophobic surface area of a full-length antibody. A summary of a comparison of the viscosity prediction tools described above can be found in Kuroda et al. (Kuroda and Tsumoto 2020). Lai and colleagues (Lai et al. 2022) recently applied ML to this problem area. The team developed a model based on 27 mAbs to predict antibody viscosity at

150 mg/ml, based on the decision tree classification method including two features: net charge and high viscosity index (HVI). To predict viscosity at different concentrations, the team developed a coarse-grained model combined with hydrodynamic calculations and HVI-derived parameters.

Regarding aggregation, researchers can leverage several in silico models for predicting the solubility/protein aggregation rates: developability index (Lauer et al. 2012), Solubis (De Baets et al. 2015), and Camsol (Sormanni et al. 2015). Other models help to identify aggregation-prone regions – for example, spatial aggregation propensity (Chennamsetty et al. 2009), Aggrescan 3D (Kuriata et al. 2019), and ANuPP (Prabakaran et al. 2021). ML has also been applied to predict protein aggregation kinetics (regression) based on sequence features (Tartaglia et al. 2005; Rawat et al. 2021b) and antibody amyloidogenesis (classification) (Liaw et al. 2013; David et al. 2010; Rawat et al. 2021a; Garofalo et al. 2021). The latter has limited application in therapeutic development but is greatly relevant for human diseases (Roberts 2014). Another ML-based model proposed by Lai and colleagues (Lai et al. 2021) was trained on 21 mAbs to predict therapeutic antibody aggregation rates at 150 mg/ml with the help of structural-based features extracted from MD simulations.

Computational approaches and solubility prediction tools open the door to fast high-throughput screening without any material requirements (Feng et al. 2022). Current approaches use molecular descriptors extracted from protein sequence – sequence-based predictors (Hebditch et al. 2017; Sormanni et al. 2017) – or from structures – structure-based predictors (Sormanni et al. 2015). The former often fail to consider tertiary structure information, which differentiates poorly soluble residues that drive protein folding from the residues exposed to the solvent, potentially eliciting aggregation. Structure-based tools, on the other hand, can be applied only when researchers have a reliable 3D model, limiting the throughput and applications to a massive volume of early stage mAb candidates. Some computational methods only output a binary classification (soluble/insoluble) instead of a numerical value (Hebditch et al. 2017; Smialowski et al. 2012). Feng and colleagues (Feng et al. 2022) presented a strategy called solPredict that uses the embeddings from pre-trained protein language model to predict the apparent solubility of mAbs in histidine (pH 6.0) buffer. The team used a dataset of 220 diverse, in-house mAbs for model training and hyperparameter tuning through fivefold cross-validation. The solPredict was found to achieve a high correlation with experimental solubility through an independent test set of 40 mAbs, delivering a good performance for both IgG1 and IgG4 subclasses. Hashemi and colleagues (Hashemi et al. 2022) modeled a soluble production of the anti-EpCAM extracellular domain single-chain variable fragment (scFv). The solution was then optimized as a function of four numerical factors: post-induction temperature, post-induction time, cell density of induction time, and inducer concentration, as well as one categorical variable using artificial neural network (ANN) and response surface methodology (RSM). The predicted value obtained by ANN for the response (106.1 mg/L) approached the experimental result closer than the one obtained by RSM (97.9 mg/L).

The application of computational tools to predict stability-enhancing mutations has increased in popularity over the past decade. Researchers use the state-of-the-art approach called protein consensus design, which leverages phylogenetic information from multiple sequence alignments (MSAs) to output the most frequent 34 amino acids for a residue position. However, such residues are characterized by improved thermostability in only 50% of the cases, with poorer performance noted for antibody sequences that have highly conserved framework regions. ML approaches employed large datasets such as ProTherm (Kumar et al. 2006), which predicts thermostability by collating mutant 40 effects on protein stability from mesophilic and thermophilic sequences. However, the publicly available thermostability data exclude mAb and scFv sequences, placing a limit on their generalization. It is still a challenge for researchers to predict thermally enhancing scFv sequences or develop thermostable mutations within approved scFv candidates. New computational approaches will need to be designed to combine sequence information to predict biophysical attributes (thermostability) and highly accurate, generalizability-enabling design. Harmalkar and colleagues (Harmalkar et al. 2022) addressed this by equipping unsupervised and supervised learning approaches over a thermostability prediction task that has been tuned specifically for generated scFv sequences.

Altogether, the prediction of biophysical properties of therapeutic antibodies is challenging since most of the studies cited perform the predictions within a specific experimental framework. For instance, "aggregation" can be measured by a variety of assays (e.g., hydrophobic interaction chromatography (HIC)) that, though correlating with one another, do not need to give the same reading for the same molecule. This makes generalizability extremely difficult and rendering benchmarking different methods developed with different datasets almost impossible.

5.3.3.3 Deep learning applications

Current efforts in developability prediction using deep learning methods are directed toward generating novel sequences computationally and selecting for good manufacturability properties. This can be done by generating the molecules and filtering them using general-purpose developability predictors such as therapeutic antibody profiler (TAP) (Mason et al. 2021). Alternatively, a GAN trained on large-scale natural data can be fine-tuned on sequences with specific developability properties (Amimeur et al. 2020). All in all, there is an interplay between how to efficiently sample the antibody sequence/structure space and how to predict the developability properties well.

Deep learning models aim to learn higher order dependencies missed by position frequency analysis, reducing the chance of generating non-functional proteins. However, current limitations of deep learning approaches focus only on CDR regions or heavy chains and a lack of experimental validation of predicted properties (Khetan et al. 2022). For instance, 74% antigen binding was achieved in a mouse library designed by a variational autoencoder that generated novel CDR-H regions (Friedensohn et al. 2020). However, such an approach ignores non-CDR region contributions to the paratope, and the diversity of sequences in this library is unknown.

Other generative approaches, such as GANs, can also be trained on natural human antibody repertoires and biased via transfer learning to generate sequences predicted to have the desired biophysical properties (Amimeur et al. 2020). However, researchers need more insights to better understand how such properties impact the overall developability of an antibody. Experimental validation of the predicted properties is also required.

It has already been conducted for an enzymatically active protein library (Repecka et al., 2021) and a nanobody library created using a generative deep neural network-powered autoregressive model trained on a native llama repertoire (Shin et al. 2021).

Previous studies have demonstrated the use of libraries of mammalian antibodies for selecting antibodies with optimal biophysical properties, reduced polyreactivity, and immunogenicity (Dyson et al. 2020). Dyson and colleagues described the use of a nuclease-directed integration system to generate antibodies with differing biophysical properties only based on their display level on mammalian cells. Other researchers have demonstrated the use of ML-guided directed evolution on combinatorial sequences (Wu et al. 2019). Recently, Chen and colleagues formulated ML pipelines to predict the developability of a library of 2,400 antibodies using sequence alone (Chen et al. 2020). Such advances in bioinformatics and in silico methods allowed for the efficient development of commercially viable antibodies. Antibody library variants of an original candidate are created to exhibit better developability than the parent molecule.

Generative ML has been proposed as a means to design antigen-specific mAbs, but efforts to confirm this hypothesis have been hampered by the cost and time required to test large numbers of antibody sequences. To address this challenge, Akbar and colleagues (Akbar et al. 2022) developed a lattice-based antibody-antigen binding simulation framework incorporating physiological parameters. They found that a deep generative model trained only on antibody sequence data can be used to design conformational epitope-specific antibodies matching or exceeding the training dataset in affinity and diversity, showing that lower sequence diversity is necessary for high-accuracy generative antibody modeling.

5.4 ANTIBODY GENERATION AND DESIGN BY LANGUAGE MODELS

Designing novel antibodies without resorting to experimental methods has long been the end goal of computational antibody engineering. The earliest attempts indicated the possibility of performing exhaustive in silico mutagenesis combined with binding energy estimation on a cocrystal structure (Lippow et al. 2007). Designing the antibodies without the cocrystal structure has been performed by variations on sampling the available structures (Li et al. 2014; Adolf-Bryfogle et al. 2018). One of the primary challenges in this area relates to creating an antibody sequence/structure that is plausible in a biological sense. Language models, such as Bidirectional encoder representations from transformers (BERT), have been applied to push antibody manifolds into representation space, sampling which facilitates novel antibody discovery.

Here, we describe the importance of antibody representation in terms of data interpretability and downstream model performance. We show how such representations can be enriched with domain knowledge by both manual feature engineering processes and representation learning techniques. Finally, we present how language models are used to produce antibody embedding and improve the antibody discovery process.

5.4.1 Antibody Representations

Data representation is a vital aspect of all ML projects as raw unprocessed data often cannot be used directly by the model. It is important that feature engineering becomes an intrinsic step of the ML development pipeline, as good representation significantly boosts downstream model performance and stability or reduces the input dimensionality. In this process, input data are modified and augmented with additional information and domain knowledge relevant to the problem being solved.

Several approaches to representing proteins can be used in parallel, each capturing a different perspective. For example, one can represent an antibody as a sequence of amino acids. Then, with the presence of structural data, it is possible to represent protein as a structure – for instance, by extracting voxelated atomic maps. Despite being very easy to interpret, such an approach has a major drawback: it is sensitive to rotations. That is why graphs are another popular option for representing protein structural data.

The function and the structure of a protein derive from the interaction between all of its residues. Such a residue interaction network can be abstracted to a graph where amino acids constitute nodes and relationships between them are the edges. One can use several definitions to construct such a graph: for example, using Cα atoms as nodes with connecting edge if Cα atoms are within 7 Å (Chakrabarty and Parekh 2016; Pittala and Bailey-Kellogg 2020), analogously for Cβ atoms with distance thresholds between 7 Å and 10 Å (Chakrabarty and Parekh 2016; Pittala and Bailey-Kellogg 2020), or using residue center of mass as a node with edge if any atoms are closer than 5 Å. Another approach introduces edges if the non-covalent interaction strength between interacting residues is above some predefined threshold (Brinda and Vishveshwara 2005). There also exist methods using different graph levels (atom-atom interactions followed by residue-residue) at the same time (Tubiana et al. 2022).

Another challenge is posed by representing molecular surfaces of a protein for ML. Gainza and colleagues (Gainza et al. 2020) applied geometric deep learning to learn surface fingerprints that are important for biological interactions and tested their model on three distinct problems: classification of ligand binding pockets, protein binding site prediction, and protein-protein complex search based on interaction fingerprints.

In the context of enriching representations, one can augment biological data on residue, sequence, and structure layers in the feature engineering process. The easiest residue representation is likely one-hot encoding where each residue is represented as a 20-dimensional unit vector with 1 in place of represented amino acid. This does not add any domain knowledge to the residue representation but enables the model to treat all residues equally, as all amino acid representations differ only by rotation, and all pairwise distances are the same.

Evolutionary relationships can be encapsulated into data. In this case, one can represent each residue as a 20- (or 21-) dimensional vector, where each value is taken from the BLOSUM substitution matrix (Lim et al., 2022). A common practice is to include the physicochemical properties of residues by adding Vectors of Hydrophobic, Steric, and Electronic properties to the representation (Liberis et al. 2018).

One can embed residues' positional information on the sequence level into sequence representation. To compare protein sequences in a meaningful way, one needs to align them so that the proteins from the same family will have corresponding evolutionary-dependent residues in the same positions. Upon sequence alignment, inputs have a fixed size, which licenses the use of simple models that can discover patterns and residue relationships from absolute input positions, which may be important for model interpretability.

Detlefsen and colleagues (Detlefsen et al. 2022) showed that alignment of the input sequences improved the model's ability to separate sequences from one protein family (b-lactamase) into different phyla. It remains unclear if it is possible to generalize this capability across multiple protein families. For antibodies, sequence alignment is achieved by numbering methods, which also provide the functional position context (e.g., CDR, framework) for each residue. At the structural level, one can augment input data with shape annotations for each residue. For example, ASA or paratope labels.

Most ML models natively employ vector data format of constant dimensionality. To utilize those models, one needs to transform variable-in-length biological sequences to point in fixed-size dimensional space so that similar input sequences, or functionally similar proteins, reside close to each other in the vector space.

Such transformation is called embedding, and to define it, one needs a similarity measure and the transformation function that transforms input to a point in the embedded space. Proteins are sequences of amino acids with variable lengths that fold into a three-dimensional shape. Therefore, as an example of a similarity measure, one can use Levenshtein distance between sequences or some shape difference measures.

5.4.2 Representation Learning

As opposed to the manual process of feature engineering, which involves manually adding domain knowledge to the representation, it is possible to apply representation learning techniques to render meaningful interpretable data representations. The basic representation learning approach is PCA, where the representation features are the output of linear transformations of input data. Representation learning is often used in combination with transfer learning, especially when there is little data available. That is because the model can learn to recognize meaningful biological features by making use of vastly available NGS data.

The biological knowledge embedded into the vector and the meaningfulness of such learned representations can be then assessed by the performance of downstream model predictions.

Detlefsen and colleagues (Detlefsen et al. 2022) demonstrated that low-dimensional sequence representation space topology shows good correspondence between the ancestral lineage of the proteins that were rendered on latent space. The team also revealed that the learned representation space is not Euclidean. Hence, arithmetic operations on vector representations are not a good approach to interpolate points in the latent space as one might get unpredictable results. They equipped learned representations with a Riemannian metric, so measuring the distance in the data space was possible. This demonstrated that the usage of geodesic distances enables more meaningful navigation over the latent space.

5.4.3 Language Models

Antibody diversity is estimated to be 10^{18} unique molecules (Briney et al. 2019). Therefore, with plenty of publicly available NGS datasets, efficient sequence embedding that captures its biological properties holds the promise of the possibility to navigate through the whole immunoglobulin sequence space, providing new insights into the biology of antibodies. Such navigation could be potentially used in the protein engineering process to interpolate between known sequences in latent space to yield new proteins functionally similar or to guide mutations to design antibodies with desired characteristics.

An emerging trend in antibody analysis is the use of natural language processing (NLP) methods to develop antibody embeddings by making analogies between amino acid k-mers, whole antibody sequences, antibody repertoires, and words, sentences, and documents. In NLP, the most widely used family of models to produce word embeddings is word2vec (Mikolov et al. 2013). Two architectures are used for representation learning:

- Skip-Gram, where given an input word, the neural network tries to predict its context (preceding and following words).
- Continuous Bag of Words architecture, where given context (neighbor words), the model is tasked with predicting the target word.

Because of the training process, output vectors embed contextual information and are capable of capturing both syntactic and semantic relationships, so distinct words that are related by the context are adjacent to each other in the vector space. Such context-aware embedding can be called distributed representation. Below, we highlight some of the language models that are used to analyze antibodies, with a compilation in Table 5.4.

5.4.3.1 ProtVec

ProtVec (Asgari and Mofrad 2015) is a skip-gram model trained on 546,790 sequences from Swiss-Prot to produce 100-dimensional protein representations. The study showed that the resulting embeddings were able to embed a range of meaningful chemical and physical properties, e.g., by showing they were able to classify protein families with 93% accuracy.

Inyoung and colleagues (Kim et al 2021) utilized the ProtVec model to represent the whole BCR repertoire as a 100-dimensional vector, by selecting the 100 most frequent CDR3 sequences from the BCR repertoire and summing their respective embeddings. Using this approach, they were able to effectively distinguish between repertoires of healthy subjects and COVID-19 patients. Furthermore, they analyzed how repertoire representation changes over the course of the disease and showed that it is possible to track disease progression using this approach.

TABLE 5.4 Language models applied to proteins

METHOD	*DESCRIPTION*	*REFERENCES*
ProtVec	Word2Vec-based model trained on Swiss-Prot protein data, produces 100-dimensional embeddings	(Asgari and Mofrad 2015)
Immune2vec	Word2Vec model trained on BCR data, shown to produce meaningful embeddings for 3-mers, sequences, and repertoires	(Ostrovsky-Berman et al. 2021)
ProtBERT	Transformer-based model, used to produce protein embeddings	(Elnaggar et al., n.d.) 2020
AbLang	Transformer-based model for restoring missing residues in antibody sequence data	(Olsen et al. 2022b)
AntiBERTa	Sequence representation utilizing attention-based language modeling for downstream tasks, used in paratope prediction, achieving SoTa results	(Leem et al. 2022)
AntiBERTy	BERT model used to analyze affinity maturation process, used in MIL setup, and shown to effectively classify bags of sequences containing potential binders	(Ruffolo et al. 2021)
AbBERT	ProtBERT-based model fine-tuned on antibodies. Scores that it produced for sequences correlated with immunogenicity, protein expression in cells, and stability metrics	(Vashchenko et al. 2022)
BioPhi (Sapiens)	Transformer model for antibody humanization	(Prihoda et al. 2022)

A short description and reference are presented above for each method.

5.4.3.2 Immune2vec

Immune2vec (Ostrovsky-Berman et al. 2021) used word2vec modeling to explore BCR sequences. As opposed to Inyoung and colleagues who employed the pre-trained ProtVec neural network, Ostrovsky-Berman et al. have trained the embedding model on their own. Using the analogies between biological sequences and natural language mentioned beforehand, they first applied the Immune2Vec model to all possible combinations of 8000 3-mers. The output embeddings were then projected into a 2D plane using t-distributed stochastic neighbor embedding (t-SNE) and colored according to biophysical properties. This showed that the resulting two-dimensional space can be divided into clusters with homogenous biophysical properties. On the sequence level, they trimmed CDR3 sequences to remove fragments coded by the V gene. Then, they used their model to produce embeddings and tried to assign one of six V genes to the resulting vectors. This demonstrated that there is an association between trimmed CDR3 and V gene, even though the trimmed CDR3 sequence is not coded by it. Finally, at the repertoire level, they produced embeddings for chronically infected individuals with hepatitis C virus (HCV) and spontaneous clearers and attempted to classify the repertoires, achieving roughly 90% accuracy for BCR data and 70% for T-cell receptor (TCR) data.

5.4.3.3 *Transformer architectures*

Recently, transformer architectures have been gaining popularity in the representation learning area as they achieved state-of-the-art results on a wide range of NLP tasks (Wolf et al. 2020). Using self-attention mechanisms, transformers are able to learn relationships between words in sentences and, therefore, produce distributed representations. Such neural networks are often trained with the MLM approach in which one masks or changes part of the input and the model learns to predict the altered part.

5.4.3.4 *ProtBERT*

To this time, there have been several adaptations of transformer architectures to antibody data. For example, Elnaggar and colleagues (Elnaggar et al., n.d.) showed that embeddings obtained from transformers captured relevant biological information as small-size models trained solely on those representations were able to compete with bigger architectures on various tasks such as protein classification. Their neural networks reached comparable performance to methods utilizing MSA, which indicates that transformer-produced embeddings could be a good starting point for various downstream tasks.

5.4.3.5 *AbLang*

This transformer architecture was also successfully applied in the field of antibody representation by Olsen and colleagues (Olsen et al. 2022b) who trained the AbLang model on antibody sequences from OAS. The resulting architecture consists of two parts: AbRep and AbHead, which produce representations for sequences and predict the probability of each amino acid on all positions, respectively. Pre-training was based on the RoBERTa approach (Liu et al. 2019). The team showed the biological information encoded into the vectors by drawing 10,000 naïve and 10,000 memory B-cell sequence representations (Ghraichy et al. 2021) using t-SNE and compared the results to evolutionary scale modeling 1b (ESM-1b) embeddings (Rives et al. 2021). Both models could separate antibody sequences by their V gene families, but AbLang yielded better separation of naïve and memory B cells. The resulting transformer model was capable of restoring missing residues in immunoglobulin sequences, obtaining similar or better results than using IMGT germlines, but without the knowledge of the germlines.

5.4.3.6 *AntiBERTa*

As proposed by Leem and colleagues (Leem et al. 2022), AntiBERTa (Antibody-specific Bidirectional Encoder Representation from Transformers) is another example of transformer architecture. The model was pre-trained using the RoBERTa approach on 57 million human BCR sequences from 61 studies available in OAS. The team selected random 1000 BCR heavy-chain sequences (Ghraichy et al. 2021) from naïve and memory B-cell sequences and showed that on the top of mutational load and V gene used, embeddings carry information about B-cell type – they were able to partition naïve and memory B-cell sequences in the representation space. This result was compared to embeddings prepared by ProtBERT (Elnaggar et al. 2020), where the separation was not as clear, which indicates that compared to the general protein transformer model, AntiBERTa representations encapsulate more antibody-specific information. Leem and colleagues noticed that high self-attention scores presented residue pairs of contacts indicating that the model is capable of understanding the structural information. So, they applied it to the paratope prediction problem, where the model classified each residue from the input sequence. This was achieved by adding a classification head on top of the already existing 12 layers. The prediction results were compared to Parapred (Liberis et al. 2018) and ProABC (Olimpieri et al. 2013), which demonstrated SoTa results in paratope prediction. The team used the model to produce embeddings of known therapeutic antibodies, showing that it was possible to determine their origin (human, murine, humanized, chimeric) as well as, to a certain degree, the correlation with immunogenicity response scores – ADA. This demonstrates that embeddings learned biologically relevant information, and the learned representations correspond to B-cell origin, immunogenicity, and structure.

5.4.3.7 AntiBERTy

AntiBERTy, another model based on the BERT architecture, was proposed by Ruffolo and colleagues (Ruffolo et al. 2021). It was trained on 558 million sequences from OAS with MLM objective. The team analyzed repertoires from donors with HIV-1 neutralizing VRC01 antibodies. For each sample, they created a k bearers neighbour (kNN) graph using model embeddings and visualized it in two-dimensions using Uniform Manifold Approximation and Projection (UMAP). Using these plots, they observed trajectories from germline sequences and mutated derivatives corresponding to sequence changes in the affinity maturation process. With repertoire data, individual sequences are not labeled. Hence – relying on clonal expansion – the team produced noisy labels, and frequently observed sequences were assumed to be binders. Next, they applied multiple instance learning (MIL) to predict whether the sets of sequences contain binding antibodies. They created single-instance bags of sequences from known VRC01 antibodies, confirming that the model produces the correct positive predictions. Finally, they annotated each antibody structure with attention, and in most cases (7 out of 10 sequences), attention-pointed binding residues.

5.4.3.8 AbBERT

Another transformer architecture called AbBERT trained on 20 million heavy and light sequences from OAS was published by Vashchenko and colleagues (Vashchenko et al. 2022). The model family is based on ProtBERT, but fine-tuned on antibodies. Both heavy and light sequences were used to train the models, during which the team annotated functional regions of input sequences by inserting additional annotation tokens, before and after all CDRs. First, the authors showed that the model is capable of predicting CDR regions. The team introduces the term "humanness" that is used to measure the similarity between input immunoglobulin sequence and antibodies sampled from people. This score was used to evaluate 600 known therapeutic sequences that have passed various clinical trials and showed that antibodies with a low assigned metric tend to be immunogenic. Then, the model was applied for in silico antibody optimization in which the baseline anti-SARS-CoV-1 antibody sequence was modified so that the optimized immunoglobulin was able to bind to another target – SARS-CoV-2. Model with AbBERT embeddings on input was used to solve the optimization problem. Finally, they performed in vitro experiments which demonstrated that poorly scored sequences were weakly expressed in the cells. They have also observed a correlation between the model scores and the protein stability metrics calculated using Free Energy Perturbation.

5.4.3.9 BioPhi (Sapiens module)

Sapiens is one of the two BioPhi models aiming at antibody humanization. Similarly, to other SoTa models, it is a transformer-based model trained toward the MLM goal. Two separate models for light and heavy chains have been created, each having 568,857 parameters and being based on the RoBERTa model. The training dataset consisted of human-only and unaligned sequences from OAS: 20 million heavy and 19 million light sequences. Analysis of the averaged attention matrix showed high importance between CDR loops, which are close structurally but apart in sequence – thus proving that the model is able to recognize long-range interactions.

Antibody humanization works by leveraging the fact that only human mAbs were used for training. Input variable region sequence is processed by the model, giving probabilities for all 20 amino acids for all positions. The most probable residues for frameworks are selected, keeping unchanged CDRs from input, which minimizes the risk of affecting binding properties but making antibodies more similar to human ones. Such a procedure has been performed on 177 antibodies (25 with known parental sequence and 152 humanized mAbs with presumed original sequence), obtaining results comparable to human experts.

5.5 CONCLUSIONS AND FUTURE PERSPECTIVES IN AI FOR ANTIBODY DISCOVERY

Over the past 40 years, antibodies have firmly established their role as the most important group of biologics. Up until now, the development of currently 100 approved antibody therapeutics relied on a "discovery" process driven by experimental laboratory-based methods. Thanks to advances in high-throughput experimental data generation as well as progress in computational model development, it is possible to shift the paradigm from antibody discovery toward "design."

For designing a novel biologic computationally, one requires two elements. First, one needs to generate biologically or physically plausible sequences and structures. Second, one requires objective functions to gauge whether the molecule has the properties expected of it. The sampling of novel molecules has been greatly facilitated by generative modeling such as variational auto encoders (VAEs), GANs, and language models that learn the representation of antibodies from large-scale NGS data. The performance of predicting objective antibody features such as antibody-antigen binding or developability still needs to be addressed. Nevertheless, researchers have started to combine the two features, generating molecules that are either biased or filtered for those with better biophysical features.

As such, computational methods are now capable of producing naturally viable starting points that are free from statistically obvious liabilities. Achieving the goal of fully computational antibody design – as opposed to "discovery" – still requires improving the prediction of the molecule's therapeutic features, chiefly binding and developability. On the binding front, one could hope for a modeling revolution, on par with structure prediction as the two problems bear many parallels. However, on the developability front, hoping for such progress is fanciful, mostly due to the lack of data.

Developability is an umbrella term uniting multiple biological assays. Even though many of these are regularly performed at organizations developing biologics, such a plethora of data was scarcely envisaged for training models. Therefore, data are oftentimes not comparable between different runs, projects, and teams since they were generated with a specific therapeutic challenge rather than to develop a generalistic developability prediction method. For this reason, a new paradigm emerges called "prediction-first," where data are generated specifically with model training in mind. Over the short term, they might not contribute to any therapeutic projects, but rather act as a long-term investment into the development of a foundation for a broadly applicable computational model.

All in all, shifting from discovery to design and from project-driven data generation toward prediction-first requires a sizable shift within the organizations responsible for biologics development. Understandably, it is a large diversion of resources from well-proved experimental methods to the development of innovative methods that still need to be validated. Nevertheless, with the ongoing progress in the development of computational models for antibodies, such a shift has become far more realistic.

REFERENCES

Abanades, Brennan, Guy Georges, Alexander Bujotzek, and Charlotte M. Deane. 2022a. "ABlooper: Fast Accurate Antibody CDR Loop Structure Prediction with Accuracy Estimation." *Bioinformatics,* January. https://doi.org/10.1093/bioinformatics/btac016.

Abanades, Brennan, Wing Ki Wong, Fergus Boyles, Guy Georges, Alexander Bujotzek, and Charlotte M. Deane. 2022b. "ImmuneBuilder: Deep-Learning Models for Predicting the Structures of Immune Proteins." *Commun Biol.* 6 (1):575. doi: 10.1038/s42003-023-04927-7.

Abhinandan, K. R., and Andrew C. R. Martin. 2007. "Analyzing the 'Degree of Humanness' of Antibody Sequences." *Journal of Molecular Biology* 369 (3): 852–62.

Adolf-Bryfogle, Jared, Oleks Kalyuzhniy, Michael Kubitz, Brian D. Weitzner, Xiaozhen Hu, Yumiko Adachi, William R. Schief, and Roland L. Dunbrack Jr. 2018. "RosettaAntibodyDesign (RAbD): A General Framework for Computational Antibody Design." *PLoS Computational Biology* 14 (4): e1006112.

Adolf-Bryfogle, Jared, Qifang Xu, Benjamin North, Andreas Lehmann, and Roland L. Dunbrack Jr. 2015. "PyIgClassify: A Database of Antibody CDR Structural Classifications." *Nucleic Acids Research* 43 (Database issue): D432–38.

Agrawal, Neeraj J., Bernhard Helk, Sandeep Kumar, Neil Mody, Hasige A. Sathish, Hardeep S. Samra, Patrick M. Buck, Li, and Bernhardt L. Trout. 2016. "Computational Tool for the Early Screening of Monoclonal Antibodies for Their Viscosities." *mAbs* 8 (1): 43–48.

Akbar, Rahmad, Philippe A. Robert, Cédric R. Weber, Michael Widrich, Robert Frank, Milena Pavlović, Lonneke Scheffer, et al. 2022. "In Silico Proof of Principle of Machine Learning-Based Antibody Design at Unconstrained Scale." *mAbs* 14 (1): 2031482.

Akpinaroglu, Deniz, Jeffrey A. Ruffolo, Sai Pooja Mahajan, and Jeffrey J. Gray. 2022. "Simultaneous Prediction of Antibody Backbone and Side-Chain Conformations with Deep Learning." *PloS One* 17 (6): e0258173.

Almagro, Juan C., Alexey Teplyakov, Jinquan Luo, Raymond W. Sweet, Sreekumar Kodangattil, Francisco Hernandez-Guzman, and Gary L. Gilliland. 2014. "Second Antibody Modeling Assessment (AMA-II)." *Proteins* 82 (8): 1553–62.

Ambrosetti, Francesco, Brian Jiménez-García, Jorge Roel-Touris, and Alexandre M. J. J. Bonvin. 2020. "Modeling Antibody-Antigen Complexes by Information-Driven Docking." *Structure* 28 (1): 119–29.e2.

Amimeur, Tileli, Jeremy M. Shaver, Randal R. Ketchem, J. Alex Taylor, Rutilio H. Clark, Josh Smith, Danielle Van Citters, et al. 2020. "Designing Feature-Controlled Humanoid Antibody Discovery Libraries Using Generative Adversarial Networks." *bioRxiv*. https://doi.org/10.1101/2020.04.12.024844.

Anand, Namrata, and Possu Huang. 2018. "Generative Modeling for Protein Structures." https://papers.nips.cc/paper_files/paper/2018/hash/afa299a4d1d8c52e75dd8a24c-3ce534f-Abstract.html.

Anonymous. 2022. "xTrimoABFold: Improving Antibody Structure Prediction without Multiple Sequence Alignments." https://arxiv.org/abs/2212.00735.

Asgari, Ehsaneddin, and Mohammad R. K. Mofrad. 2015. "Continuous Distributed Representation of Biological Sequences for Deep Proteomics and Genomics." *PloS One* 10 (11): e0141287.

Bachas, Sharrol, Goran Rakocevic, David Spencer, Anand V. Sastry, Robel Haile, John M. Sutton, George Kasun, et al. 2022. "Antibody Optimization Enabled by Artificial Intelligence Predictions of Binding Affinity and Naturalness." *bioRxiv*. https://doi.org/10.1101/2022.08.16.504181.

Baran, Dror, M. Gabriele Pszolla, Gideon D. Lapidoth, Christoffer Norn, Orly Dym, Tamar Unger, Shira Albeck, Michael D. Tyka, and Sarel J. Fleishman. 2017. "Principles for Computational Design of Binding Antibodies." *Proceedings of the National Academy of Sciences of the United States of America* 114 (41): 10900–905.

Brinda, K. V., and Saraswathi Vishveshwara. 2005. "A Network Representation of Protein Structures: Implications for Protein Stability." *Biophysical Journal* 89 (6): 4159–70.

Briney, Bryan, Anne Inderbitzin, Collin Joyce, and Dennis R. Burton. 2019. "Commonality despite Exceptional Diversity in the Baseline Human Antibody Repertoire." *Nature* 566 (7744): 393–97.

Carter, and Lazar. n.d. "Next Generation Antibody Drugs: Pursuit of The 'high-Hanging Fruit'." *Nature Reviews. Drug Discovery*. https://www.nature.com/articles/nrd.2017.227.

Chakrabarty, Broto, and Nita Parekh. 2016. "NAPS: Network Analysis of Protein Structures." *Nucleic Acids Research* 44 (W1): W375–82.

Chen, Rong, Li Li, and Zhiping Weng. 2003. "ZDOCK: An Initial-Stage Protein-Docking Algorithm." *Proteins* 52 (1): 80–87.

Chen, Xingyao, Thomas Dougherty, Chan Hong, Rachel Schibler, Yi Cong Zhao, Reza Sadeghi, Naim Matasci, Yi-Chieh Wu, and Ian Kerman. 2020. "Predicting Antibody Developability from Sequence Using Machine Learning." *bioRxiv*. https://doi.org/10.1101/2020.06.18.159798.

Chennamsetty, Naresh, Vladimir Voynov, Veysel Kayser, Bernhard Helk, and Bernhardt L. Trout. 2009. "Design of Therapeutic Proteins with Enhanced Stability." *Proceedings of the National Academy of Sciences of the United States of America* 106 (29): 11937–42.

Chinery, Lewis, Newton Wahome, Iain Moal, and Charlotte M. Deane. 2022. "Paragraph – Antibody Paratope Prediction Using Graph Neural Networks with Minimal Feature Vectors." *Bioinformatics.* 39 (1):btac732. doi: 10.1093/bioinformatics/btac732. PMID: 36370083.

Chothia, C., and A. M. Lesk. 1987. "Canonical Structures for the Hypervariable Regions of Immunoglobulins." *Journal of Molecular Biology* 196 (4): 901–17.

Clark, Karen, Ilene Karsch-Mizrachi, David J. Lipman, James Ostell, and Eric W. Sayers. 2016. "GenBank." *Nucleic Acids Research* 44 (D1): D67–72.

Clavero-Álvarez, Alejandro, Tomas Di Mambro, Sergio Perez-Gaviro, Mauro Magnani, and Pierpaolo Bruscolini. 2018. "Humanization of Antibodies Using a Statistical Inference Approach." *Scientific Reports* 8 (1): 14820.

Cohen, Tomer, Matan Halfon, and Dina Schneidman-Duhovny. 2022. "NanoNet: Rapid and Accurate End-to-End Nanobody Modeling by Deep Learning." *Frontiers in Immunology* 13 (August): 958584.

Corrie, Brian D., Nishanth Marthandan, Bojan Zimonja, Jerome Jaglale, Yang Zhou, Emily Barr, Nicole Knoetze, et al. 2018. "iReceptor: A Platform for Querying and Analyzing antibody/B-Cell and T-Cell Receptor Repertoire Data across Federated Repositories." *Immunological Reviews* 284 (1): 24–41.

David, Maria Pamela C., Gisela P. Concepcion, and Eduardo A. Padlan. 2010. "Using Simple Artificial Intelligence Methods for Predicting Amyloidogenesis in Antibodies." *BMC Bioinformatics* 11 (February): 79.

De Baets, Greet, Joost Van Durme, Rob van der Kant, Joost Schymkowitz, and Frederic Rousseau. 2015. "Solubis: Optimize Your Protein." *Bioinformatics* 31 (15): 2580–82.

Del Vecchio, Alice, Andreea Deac, Pietro Liò, and Petar Veličković. 2021. "Neural Message Passing for Joint Paratope-Epitope Prediction." *arXiv [q-bio.QM]*. arXiv. https://arxiv.org/abs/2106.00757.

Deszyński, Piotr, Jakub Młokosiewicz, Adam Volanakis, Igor Jaszczyszyn, Natalie Castellana, Stefano Bonissone, Rajkumar Ganesan, and Konrad Krawczyk. 2021. "INDI—integrated Nanobody Database for Immunoinformatics." *Nucleic Acids Research* 50 (D1): D1273–81.

Detlefsen, Nicki Skafte, Søren Hauberg, and Wouter Boomsma. 2022. "Learning Meaningful Representations of Protein Sequences." *Nature Communications* 13 (1): 1914.

Du, Zongyang, Hong Su, Wenkai Wang, Lisha Ye, Hong Wei, Zhenling Peng, Ivan Anishchenko, David Baker, and Jianyi Yang. 2021. "The trRosetta Server for Fast and Accurate Protein Structure Prediction." *Nature Protocols* 16 (12): 5634–51.

Dunbar, James, Konrad Krawczyk, Jinwoo Leem, Terry Baker, Angelika Fuchs, Guy Georges, Jiye Shi, and Charlotte M. Deane. 2014. "SAbDab: The Structural Antibody Database." *Nucleic Acids Research* 42 (Database issue): D1140–46.

Dyson, Michael R., Edward Masters, Deividas Pazeraitis, Rajika L. Perera, Johanna L. Syrjanen, Sachin Surade, Nels Thorsteinson, et al. 2020. "Beyond Affinity: Selection of Antibody Variants with Optimal Biophysical Properties and Reduced Immunogenicity from Mammalian Display Libraries." *mAbs* 12 (1): 1829335.

Eguchi, Raphael R., Christian A. Choe, and Po-Ssu Huang. 2022. "Ig-VAE: Generative Modeling of Protein Structure by Direct 3D Coordinate Generation." *PLoS Computational Biology* 18 (6): e1010271.

Elnaggar, Ahmed, Michael Heinzinger, Christian Dallago, Ghalia Rehawi, Yu Wang, Llion Jones, Tom Gibbs, et al. n.d. "ProtTrans: Towards Cracking the Language of Life's Code Through Self-Supervised Learning." https://ieeexplore.ieee.org/document/9477085/.

Elnaggar, Ahmed, Michael Heinzinger, Christian Dallago, Ghalia Rihawi, Yu Wang, Llion Jones, Tom Gibbs, et al. 2020. "ProtTrans: Towards Cracking the Language of Life's Code through Self-Supervised Deep Learning and High Performance Computing." *arXiv [cs.LG]*. arXiv. https://arxiv.org/abs/2007.06225.

Fathallah, Anas M., Manting Chiang, Anshul Mishra, Sandeep Kumar, Li Xue, C. Russell Middaugh, and Sathy V. Balu-Iyer. 2015. "The Effect of Small Oligomeric Protein Aggregates on the Immunogenicity of Intravenous and Subcutaneous Administered Antibodies." *Journal of Pharmaceutical Sciences* 104 (11): 3691–3702.

Feng, Jiangyan, Min Jiang, James Shih, and Qing Chai. 2022. "Antibody Apparent Solubility Prediction from Sequence by Transfer Learning." *iScience* 25 (10): 105173.

Ferdous, Saba, and Andrew C. R. Martin. 2018. "AbDb: Antibody Structure Database-a Database of PDB-Derived Antibody Structures." *Database: The Journal of Biological Databases and Curation* 2018 (January). https://www.ncbi.nlm.nih.gov/pmc/articles/PMC5925428/.

Ferrara, Fortunato, M. Frank Erasmus, Sara D'Angelo, Camila Leal-Lopes, André A. Teixeira, Alok Choudhary, William Honnen, et al. 2022. "A Pandemic-Enabled Comparison of Discovery Platforms Demonstrates a Naïve Antibody Library Can Match the Best Immune-Sourced Antibodies." *Nature Communications* 13 (1): 462.

Fischman, Sharon, and Yanay Ofran. 2018. "Computational Design of Antibodies." *Current Opinion in Structural Biology* 51 (August): 156–62.

Friedensohn, Simon, Daniel Neumeier, Tarik A. Khan, Lucia Csepregi, Cristina Parola, Arthur R. Gorter de Vries, Lena Erlach, Derek M. Mason, and Sai T. Reddy. 2020. "Convergent Selection in Antibody Repertoires Is Revealed by Deep Learning." *bioRxiv*. https://www.biorxiv.org/content/10.1101/2020.02.25.965673v1.

Gainza, P., F. Sverrisson, F. Monti, E. Rodolà, D. Boscaini, M. M. Bronstein, and B. E. Correia. 2020. "Deciphering Interaction Fingerprints from Protein Molecular Surfaces Using Geometric Deep Learning." *Nature Methods*. https://www.nature.com/articles/s41592-019-0666-6.

Gao, Sean H., Kexin Huang, Hua Tu, and Adam S. Adler. 2013. "Monoclonal Antibody Humanness Score and Its Applications." *BMC Biotechnology* 13 (July): 55.

Garofalo, Maura, Luca Piccoli, Margherita Romeo, Maria Monica Barzago, Sara Ravasio, Mathilde Foglierini, Milos Matkovic, et al. 2021. "Machine Learning Analyses of Antibody Somatic Mutations Predict Immunoglobulin Light Chain Toxicity." *Nature Communications* 12 (1): 3532.

Geng, Cunliang, Yong Jung, Nicolas Renaud, Vasant Honavar, Alexandre M. J. J. Bonvin, and Li C. Xue. 2019. "iScore: A Novel Graph Kernel-Based Function for Scoring Protein–protein Docking Models." *Bioinformatics* 36 (1): 112–21.

Ghraichy, Marie, Valentin von Niederhäusern, Aleksandr Kovaltsuk, Jacob D. Galson, Charlotte M. Deane, and Johannes Trück. 2021. "Different B Cell Subpopulations Show Distinct Patterns in Their IgH Repertoire Metrics." *eLife* 10:e73111. doi: 10.7554/eLife.73111. PMID: 34661527; PMCID: PMC8560093.

Glanville, Jacob, Wenwu Zhai, Jan Berka, Dilduz Telman, Gabriella Huerta, Gautam R. Mehta, Irene Ni, et al. 2009. "Precise Determination of the Diversity of a Combinatorial Antibody Library Gives Insight into the Human Immunoglobulin Repertoire." *Proceedings of the National Academy of Sciences of the United States of America* 106 (48): 20216–21.

Greiff, Victor, Gur Yaari, and Lindsay G. Cowell. 2020. "Mining Adaptive Immune Receptor Repertoires for Biological and Clinical Information Using Machine Learning." *Current Opinion in Systems Biology* 24 (December): 109–19.

Guo, Yicheng, Kevin Chen, Peter D. Kwong, Lawrence Shapiro, and Zizhang Sheng. 2019. "cAb-Rep: A Database of Curated Antibody Repertoires for Exploring Antibody Diversity and Predicting Antibody Prevalence." *Frontiers in Immunology* 10 (October): 2365.

Harmalkar, Ameya, Roshan Rao, Jonas Honer, Wibke Deisting, Jonas Anlahr, Anja Hoenig, Julia Czwikla, et al. 2022. "Towards Generalizable Prediction of Antibody Thermostability Using Machine Learning on Sequence and Structure Features." *bioRxiv*. https://pubmed.ncbi.nlm.nih.gov/36683173/.

Hashemi, Atieh, Majid Basafa, and Aidin Behravan. 2022. "Machine Learning Modeling for Solubility Prediction of Recombinant Antibody Fragment in Four Different E. Coli Strains." *Scientific Reports* 12 (1): 5463.

Hebditch, Max, M. Alejandro Carballo-Amador, Spyros Charonis, Robin Curtis, and Jim Warwicker. 2017. "Protein–Sol: A Web Tool for Predicting Protein Solubility from Sequence." *Bioinformatics* 33 (19): 3098–3100.

Honegger, A., and A. Plückthun. 2001. "Yet Another Numbering Scheme for Immunoglobulin Variable Domains: An Automatic Modeling and Analysis Tool." *Journal of Molecular Biology* 309 (3): 657–70.

Kim, Inyoung, Sang Yoon Byun, Sangyeup Kim, Sangyoon Choi, Jinsung Noh, Junho Chung, Byung Gee Kim. 2021. "Computational analysis of B cell receptor repertoires in COVID-19 patients using deep embedded representations of protein sequences." *bioRxiv.* https://www.biorxiv.org/content/10.1101/2021.08.02.454701v3.full

Jain, Tushar, Tingwan Sun, Stéphanie Durand, Amy Hall, Nga Rewa Houston, Juergen H. Nett, Beth Sharkey, et al. 2017. "Biophysical Properties of the Clinical-Stage Antibody Landscape." *Proceedings of the National Academy of Sciences of the United States of America* 114 (5): 944–49.

Jankauskaite, Justina, Brian Jiménez-García, Justas Dapkunas, Juan Fernández-Recio, and Iain H. Moal. 2019. "SKEMPI 2.0: An Updated Benchmark of Changes in Protein-Protein Binding Energy, Kinetics and Thermodynamics upon Mutation." *Bioinformatics* 35 (3): 462–69.

Jarasch, Alexander, Hans Koll, Joerg T. Regula, Martin Bader, Apollon Papadimitriou, and Hubert Kettenberger. 2015. "Developability Assessment during the Selection of Novel Therapeutic Antibodies." *Journal of Pharmaceutical Sciences* 104 (6): 1885–98.

Jones, David T., Tanya Singh, Tomasz Kosciolek, and Stuart Tetchner. 2015. "MetaPSICOV: Combining Coevolution Methods for Accurate Prediction of Contacts and Long Range Hydrogen Bonding in Proteins." *Bioinformatics* 31 (7): 999–1006.

Jumper, John, Richard Evans, Alexander Pritzel, Tim Green, Michael Figurnov, Olaf Ronneberger, Kathryn Tunyasuvunakool, et al. 2021. "Highly Accurate Protein Structure Prediction with AlphaFold." *Nature* 596 (7873): 583–89.

Källberg, Morten, Haipeng Wang, Sheng Wang, Jian Peng, Zhiyong Wang, Hui Lu, and Jinbo Xu. 2012. "Template-Based Protein Structure Modeling Using the RaptorX Web Server." *Nature Protocols* 7 (8): 1511–22.

Kaplon, Hélène, Mrinalini Muralidharan, Zita Schneider, and Janice M. Reichert. 2020. "Antibodies to Watch in 2020." *mAbs* 12 (1): 1703531.

Kaplon, Hélène, and Janice M. Reichert. 2021. "Antibodies to Watch in 2021." *mAbs* 13 (1): 1860476.

Kelly-Scumpia, Kindra M., Philip O. Scumpia, Jason S. Weinstein, Matthew J. Delano, Alex G. Cuenca, Dina C. Nacionales, James L. Wynn, et al. 2011. "B Cells Enhance Early Innate Immune Responses during Bacterial Sepsis." *The Journal of Experimental Medicine* 208 (8): 1673–82.

Kelow, Simon, Bulat Faezov, Qifang Xu, Mitchell Parker, Jared Adolf-Bryfogle, and Roland L. Dunbrack. 2022. "A Penultimate Classification of Canonical Antibody CDR Conformations." *bioRxiv*. https://doi.org/10.1101/2022.10.12.511988.

Khetan, Rahul, Robin Curtis, Charlotte M. Deane, Johannes Thorling Hadsund, Uddipan Kar, Konrad Krawczyk, Daisuke Kuroda, et al. 2022. "Current Advances in Biopharmaceutical Informatics: Guidelines, Impact and Challenges in the Computational Developability Assessment of Antibody Therapeutics." *mAbs* 14 (1): 2020082.

Kilambi, Krishna Praneeth, and Jeffrey J. Gray. 2017. "Structure-Based Cross-Docking Analysis of Antibody–Antigen Interactions." *Scientific Reports* 7 (1): 1–15.

Kim, Jin Hong, and Hyo Jeong Hong. 2012. "Humanization by CDR Grafting and Specificity-Determining Residue Grafting." *Methods in Molecular Biology* 907: 237–45.

Kindt, T. J., and R. A. Goldsby. 2007. "*Osborne BA Kuby Immunology*." WH Freeman and Co., NY.

Koenig, Patrick, Chingwei V. Lee, Benjamin T. Walters, Vasantharajan Janakiraman, Jeremy Stinson, Thomas W. Patapoff, and Germaine Fuh. 2017. "Mutational Landscape of Antibody Variable Domains Reveals a Switch Modulating the Interdomain Conformational Dynamics and Antigen Binding." *Proceedings of the National Academy of Sciences of the United States of America* 114 (4): E486–95.

Kovaltsuk, Aleksandr, Konrad Krawczyk, Jacob D. Galson, Dominic F. Kelly, Charlotte M. Deane, and Johannes Trück. 2017. "How B-Cell Receptor Repertoire Sequencing Can Be Enriched with Structural Antibody Data." *Frontiers in Immunology* 8 (December): 1753.

Kovaltsuk, Aleksandr, Jinwoo Leem, Sebastian Kelm, James Snowden, Charlotte M. Deane, and Konrad Krawczyk. 2018. "Observed Antibody Space: A Resource for Data Mining Next-Generation Sequencing of Antibody Repertoires." *Journal of Immunology* 201 (8): 2502–9.

Krawczyk, Konrad, Terry Baker, Jiye Shi, and Charlotte M. Deane. 2013. "Antibody I-Patch Prediction of the Antibody Binding Site Improves Rigid Local Antibody–antigen Docking." *Protein Engineering, Design & Selection: PEDS* 26 (10): 621–29.

Krawczyk, Konrad, Andrew Buchanan, and Paolo Marcatili. 2021. "Data Mining Patented Antibody Sequences." *mAbs* 13 (1): 1892366.

Krawczyk, Konrad, Sebastian Kelm, Aleksandr Kovaltsuk, Jacob D. Galson, Dominic Kelly, Johannes Trück, Cristian Regep, et al. 2018. "Structurally Mapping Antibody Repertoires." *Frontiers in Immunology*. https://doi.org/10.3389/fimmu.2018.01698.

Krawczyk, Konrad, Xiaofeng Liu, Terry Baker, Jiye Shi, and Charlotte M. Deane. 2014. "Improving B-Cell Epitope Prediction and Its Application to Global Antibody-Antigen Docking." *Bioinformatics* 30 (16): 2288–94.

Krawczyk, Konrad, Matthew I. J. Raybould, Aleksandr Kovaltsuk, and Charlotte M. Deane. 2019. "Looking for Therapeutic Antibodies in next-Generation Sequencing Repositories." *mAbs* 11 (7): 1197–1205.

Kryshtafovych, Andriy, Torsten Schwede, Maya Topf, Krzysztof Fidelis, and John Moult. 2021. "Critical Assessment of Methods of Protein Structure Prediction (CASP)-Round XIV." *Proteins* 89 (12): 1607–17.

Kumar, M. D. Shaji, K. Abdulla Bava, M. Michael Gromiha, Ponraj Prabakaran, Koji Kitajima, Hatsuho Uedaira, and Akinori Sarai. 2006. "ProTherm and ProNIT: Thermodynamic Databases for Proteins and Protein-Nucleic Acid Interactions." *Nucleic Acids Research* 34 (Database issue): D204–6.

Kumar, Sandeep, Satish K. Singh, Xiaoling Wang, Bonita Rup, and Davinder Gill. 2011. "Coupling of Aggregation and Immunogenicity in Biotherapeutics: T- and B-Cell Immune Epitopes May Contain Aggregation-Prone Regions." *Pharmaceutical Research* 28 (5): 949–61.

Kunik, Vered, Shaul Ashkenazi, and Yanay Ofran. 2012. "Paratome: An Online Tool for Systematic Identification of Antigen-Binding Regions in Antibodies Based on Sequence or Structure." *Nucleic Acids Research* 40 (Web Server issue): W521–24.

Kuriata, Aleksander, Valentin Iglesias, Jordi Pujols, Mateusz Kurcinski, Sebastian Kmiecik, and Salvador Ventura. 2019. "Aggrescan3D (A3D) 2.0: Prediction and Engineering of Protein Solubility." *Nucleic Acids Research*. https://doi.org/10.1093/nar/gkz321.

Kuroda, Daisuke, and Kouhei Tsumoto. 2020. "Engineering Stability, Viscosity, and Immunogenicity of Antibodies by Computational Design." *Journal of Pharmaceutical Sciences* 109 (5): 1631–51.

Lai, Pin-Kuang, Amendra Fernando, Theresa K. Cloutier, Jonathan S. Kingsbury, Yatin Gokarn, Kevin T. Halloran, Cesar Calero-Rubio, and Bernhardt L. Trout. 2021. "Machine Learning Feature Selection for Predicting High Concentration Therapeutic Antibody Aggregation." *Journal of Pharmaceutical Sciences* 110 (4): 1583–91.

Lai, Pin-Kuang, Austin Gallegos, Neil Mody, Hasige A. Sathish, and Bernhardt L. Trout. 2022. "Machine Learning Prediction of Antibody Aggregation and Viscosity for High Concentration Formulation Development of Protein Therapeutics." *mAbs* 14 (1): 2026208.

Lauer, Timothy M., Neeraj J. Agrawal, Naresh Chennamsetty, Kamal Egodage, Bernhard Helk, and Bernhardt L. Trout. 2012. "Developability Index: A Rapid in Silico Tool for the Screening of Antibody Aggregation Propensity." *Journal of Pharmaceutical Sciences* 101 (1): 102–15.

Laustsen, Andreas H., Markus-Frederik Bohn, and Anne Ljungars. 2022. "The Challenges with Developing Therapeutic Monoclonal Antibodies for Pandemic Application." *Expert Opinion on Drug Discovery* 17 (1): 5–8.

Laustsen, Andreas H., Victor Greiff, Aneesh Karatt-Vellatt, Serge Muyldermans, and Timothy P. Jenkins. 2021. "Animal Immunization, in Vitro Display Technologies, and Machine Learning for Antibody Discovery." *Trends in Biotechnology* 39 (12): 1263–73.

Lazar, Greg A., John R. Desjarlais, Jonathan Jacinto, Sher Karki, and Philip W. Hammond. 2007. "A Molecular Immunology Approach to Antibody Humanization and Functional Optimization." *Molecular Immunology* 44 (8): 1986–98.

Lee, Jae Hyeon, Payman Yadollahpour, Andrew Watkins, Nathan C. Frey, Andrew Leaver-Fay, Stephen Ra, Kyunghyun Cho, Vladimir Gligorijevic, Aviv Regev, and Richard Bonneau. 2022. "EquiFold: Protein Structure Prediction with a Novel Coarse-Grained Structure Representation." *bioRxiv*. https://doi.org/10.1101/2022.10.07.511322.

Leem, Jinwoo, James Dunbar, Guy Georges, Jiye Shi, and Charlotte M. Deane. 2016. "ABodyBuilder: Automated Antibody Structure Prediction with Data–driven Accuracy Estimation." *mAbs* 8 (7): 1259–68.

Leem, Jinwoo, Laura S. Mitchell, James H. R. Farmery, Justin Barton, and Jacob D. Galson. 2022. "Deciphering the Language of Antibodies Using Self-Supervised Learning." *Patterns of Prejudice*, May, 18;3(7):100513. doi: 10.1016/j.patter.2022.100513. PMID: 35845836; PMCID: PMC9278498.

Lees, William, Christian E. Busse, Martin Corcoran, Mats Ohlin, Cathrine Scheepers, Frederick A. Matsen, Gur Yaari, et al. 2020. "OGRDB: A Reference Database of Inferred Immune Receptor Genes." *Nucleic Acids Research* 48 (D1): D964–70.

Lefranc, M. P., V. Giudicelli, C. Ginestoux, J. Bodmer, W. Müller, R. Bontrop, M. Lemaitre, A. Malik, V. Barbié, and D. Chaume. 1999. "IMGT, the International ImMunoGeneTics Database." *Nucleic Acids Research* 27 (1): 209–12.

Li, Tong, Robert J. Pantazes, and Costas D. Maranas. 2014. "OptMAVEn--a New Framework for the de Novo Design of Antibody Variable Region Models Targeting Specific Antigen Epitopes." *PloS One* 9 (8): e105954.

Liaw, Chyn, Chun-Wei Tung, and Shinn-Ying Ho. 2013. "Prediction and Analysis of Antibody Amyloidogenesis from Sequences." *PloS One* 8 (1): e53235.

Liberis, Edgar, Petar Velickovic, Pietro Sormanni, Michele Vendruscolo, and Pietro Liò. 2018. "Parapred: Antibody Paratope Prediction Using Convolutional and Recurrent Neural Networks." *Bioinformatics* 34 (17): 2944–50.

Lim, Yoong Wearn, Adam S. Adler, and David S. Johnson. 2022. "Predicting Antibody Binders and Generating Synthetic Antibodies Using Deep Learning." *mAbs* 14 (1): 2069075.

Lima, Wanessa C., Elisabeth Gasteiger, Paolo Marcatili, Paula Duek, Amos Bairoch, and Pierre Cosson. 2020. "The ABCD Database: A Repository for Chemically Defined Antibodies." *Nucleic Acids Research* 48 (D1): D261–64.

Lippow, Shaun M., K. Dane Wittrup, and Bruce Tidor. 2007. "Computational Design of Antibody-Affinity Improvement beyond in Vivo Maturation." *Nature Biotechnology* 25 (10): 1171–76.

Liu, Cynthia, Qiongqiong Zhou, Yingzhu Li, Linda V. Garner, Steve P. Watkins, Linda J. Carter, Jeffrey Smoot, et al. 2020. "Research and Development on Therapeutic Agents and Vaccines for COVID-19 and Related Human Coronavirus Diseases." *ACS Central Science* 6 (3): 315–31.

Liu, Xiaofeng, Richard D. Taylor, Laura Griffin, Shu-Fen Coker, Ralph Adams, Tom Ceska, Jiye Shi, Alastair D. G. Lawson, and Terry Baker. 2017. "Computational Design of an Epitope-Specific Keap1 Binding Antibody Using Hotspot Residues Grafting and CDR Loop Swapping." *Scientific Reports* 7 (January): 41306.

Liu, Yinhan, Myle Ott, Naman Goyal, Jingfei Du, Mandar Joshi, Danqi Chen, Omer Levy, Mike Lewis, Luke Zettlemoyer, and Veselin Stoyanov. "RoBERTa: A Robustly Optimized BERT Pretraining Approach." *arXiv*. https://doi.org/10.48550/arXiv.1907.11692

Lu, Ruei-Min, Yu-Chyi Hwang, I- Ju Liu, Chi-Chiu Lee, Han-Zen Tsai, Hsin-Jung Li, and Han-Chung Wu. 2020. "Development of Therapeutic Antibodies for the Treatment of Diseases." *Journal of Biomedical Science* 27 (1): 1.

Maia, Eduardo Habib Bechelane, Letícia Cristina Assis, Tiago Alves de Oliveira, Alisson Marques da Silva, and Alex Gutterres Taranto. 2020. "Structure-Based Virtual Screening: From Classical to Artificial Intelligence." *Frontiers in Chemistry* 8 (April): 343.

Major, Sylvia M., Satoshi Nishizuka, Daisaku Morita, Rick Rowland, Margot Sunshine, Uma Shankavaram, Frank Washburn, et al. 2006. "AbMiner: A Bioinformatic Resource on Available Monoclonal Antibodies and Corresponding Gene Identifiers for Genomic, Proteomic, and Immunologic Studies." *BMC Bioinformatics* 7 (April): 192.

Makowski, Emily K., Lina Wu, Priyanka Gupta, and Peter M. Tessier. 2021. "Discovery-Stage Identification of Drug-like Antibodies Using Emerging Experimental and Computational Methods." *mAbs* 13 (1): 1895540.

Mannar, Dhiraj, James W. Saville, Zehua Sun, Xing Zhu, Michelle M. Marti, Shanti S. Srivastava, Alison M. Berezuk, et al. 2022. "SARS-CoV-2 Variants of Concern: Spike Protein Mutational Analysis and Epitope for Broad Neutralization." *Nature Communications* 13 (1): 4696.

Marks, Claire, Alissa M. Hummer, Mark Chin, and Charlotte M. Deane. 2021. "Humanization of Antibodies Using a Machine Learning Approach on Large-Scale Repertoire Data." *Bioinformatics,* June. https://doi.org/10.1093/bioinformatics/btab434.

Marks, Debora S., Lucy J. Colwell, Robert Sheridan, Thomas A. Hopf, Andrea Pagnani, Riccardo Zecchina, and Chris Sander. 2011. "Protein 3D Structure Computed from Evolutionary Sequence Variation." *PloS One* 6 (12): e28766.

Mason, Derek M., Simon Friedensohn, Cédric R. Weber, Christian Jordi, Bastian Wagner, Simon M. Meng, Roy A. Ehling, et al. 2021. "Optimization of Therapeutic Antibodies by Predicting Antigen Specificity from Antibody Sequence via Deep Learning." *Nature Biomedical Engineering* 5 (6): 600–612.

Miho, Enkelejda, Alexander Yermanos, Cédric R. Weber, Christoph T. Berger, Sai T. Reddy, and Victor Greiff. 2018. "Computational Strategies for Dissecting the High-Dimensional Complexity of Adaptive Immune Repertoires." *Frontiers in Immunology* 9 (February): 224.

Mikolov, Tomas, Ilya Sutskever, Kai Chen, Greg Corrado, and Jeffrey Dean. 2013. "Distributed Representations of Words and Phrases and Their Compositionality." *arXiv [cs.CL]*. arXiv. https://proceedings.neurips.cc/paper/2013/hash/9aa42b31882ec039965f3c4923ce901b-Abstract.html.

Młokosiewicz, Jakub, Piotr Deszyński, Wiktoria Wilman, Igor Jaszczyszyn, Rajkumar Ganesan, Aleksandr Kovaltsuk, Jinwoo Leem, Jacob D. Galson, and Konrad Krawczyk. 2022. "AbDiver: A Tool to Explore the Natural Antibody Landscape to Aid Therapeutic Design." *Bioinformatics* 38 (9): 2628–30.

Narayanan, Harini, Fabian Dingfelder, Alessandro Butté, Nikolai Lorenzen, Michael Sokolov, and Paolo Arosio. 2021. "Machine Learning for Biologics: Opportunities for Protein Engineering, Developability, and Formulation." *Trends in Pharmacological Sciences* 42 (3): 151–65.

Nguyen, Minh N., Mohan R. Pradhan, Chandra Verma, and Pingyu Zhong. 2017. "The Interfacial Character of Antibody Paratopes: Analysis of Antibody-Antigen Structures." *Bioinformatics* 33 (19): 2971–76.

Norman, Richard A., Francesco Ambrosetti, Alexandre M. J. J. Bonvin, Lucy J. Colwell, Sebastian Kelm, Sandeep Kumar, and Konrad Krawczyk. 2020. "Computational Approaches to Therapeutic Antibody Design: Established Methods and Emerging Trends." *Briefings in Bioinformatics* 21 (5): 1549–67.

North, Benjamin, Andreas Lehmann, and Roland L. Dunbrack Jr. 2011. "A New Clustering of Antibody CDR Loop Conformations." *Journal of Molecular Biology* 406 (2): 228–56.

Obrezanova, Olga, Andreas Arnell, Ramón Gómez de la Cuesta, Maud E. Berthelot, Thomas R. A. Gallagher, Jesús Zurdo, and Yvette Stallwood. 2015. "Aggregation Risk Prediction for Antibodies and Its Application to Biotherapeutic Development." *mAbs* 7 (2): 352–63.

Olimpieri, Pier Paolo, Anna Chailyan, Anna Tramontano, and Paolo Marcatili. 2013. "Prediction of Site-Specific Interactions in Antibody-Antigen Complexes: The proABC Method and Server." *Bioinformatics* 29 (18): 2285–91.

Olsen, Tobias H., Fergus Boyles, and Charlotte M. Deane. 2022a. "Observed Antibody Space: A Diverse Database of Cleaned, Annotated, and Translated Unpaired and Paired Antibody Sequences." *Protein Science: A Publication of the Protein Society* 31 (1): 141–46.

Olsen, Tobias H., Iain H. Moal, and Charlotte M. Deane. 2022b. "AbLang: An Antibody Language Model for Completing Antibody Sequences." *bioRxiv*. https://doi.org/10.1101/2022.01.20.477061.

Ostrovsky-Berman, Miri, Boaz Frankel, Pazit Polak, and Gur Yaari. 2021. "Immune2vec: Embedding B/T Cell Receptor Sequences in RN Using Natural Language Processing." *Frontiers in Immunology* 12. https://doi.org/10.3389/fimmu.2021.680687.

Peng, Hung-Pin, Kuo Hao Lee, Jhih-Wei Jian, and An-Suei Yang. 2014. "Origins of Specificity and Affinity in Antibody–protein Interactions." *Proceedings of the National Academy of Sciences* 111 (26): E2656–65.

Pittala, Srivamshi, and Chris Bailey-Kellogg. 2020. "Learning Context-Aware Structural Representations to Predict Antigen and Antibody Binding Interfaces." *Bioinformatics* 36 (13): 3996–4003.

Prabakaran, R., Puneet Rawat, Sandeep Kumar, and M. Michael Gromiha. 2021. "ANuPP: A Versatile Tool to Predict Aggregation Nucleating Regions in Peptides and Proteins." *Journal of Molecular Biology* 433 (11): 166707.

Prihoda, David, Jad Maamary, Andrew Waight, Veronica Juan, Laurence Fayadat-Dilman, Daniel Svozil, and Danny A. Bitton. 2022. "BioPhi: A Platform for Antibody Design, Humanization, and Humanness Evaluation Based on Natural Antibody Repertoires and Deep Learning." *mAbs* 14 (1): 2020203.

Rawat, Puneet, R. Prabakaran, Sandeep Kumar, and M. Michael Gromiha. 2021a. "Exploring the Sequence Features Determining Amyloidosis in Human Antibody Light Chains." *Scientific Reports* 11 (1): 13785.

Rawat, Puneet, R. Prabakaran, Sandeep Kumar, and M. Michael Gromiha. 2021b. "AbsoluRATE: An in-Silico Method to Predict the Aggregation Kinetics of Native Proteins." *Biochimica et Biophysica Acta: Proteins and Proteomics* 1869 (9): 140682.

Raybould, Matthew I. J., Aleksandr Kovaltsuk, Claire Marks, and Charlotte M. Deane. 2021. "CoV-AbDab: The Coronavirus Antibody Database." *Bioinformatics* 37 (5): 734–35.

Raybould, Matthew I. J., Claire Marks, Konrad Krawczyk, Bruck Taddese, Jaroslaw Nowak, Alan P. Lewis, Alexander Bujotzek, Jiye Shi, and Charlotte M. Deane. 2019. "Five Computational Developability Guidelines for Therapeutic Antibody Profiling." *Proceedings of the National Academy of Sciences of the United States of America* 116 (10): 4025–30.

Raybould, Matthew I. J., Claire Marks, Alan P. Lewis, Jiye Shi, Alexander Bujotzek, Bruck Taddese, and Charlotte M. Deane. 2020. "Thera-SAbDab: The Therapeutic Structural Antibody Database." *Nucleic Acids Research* 48 (D1): D383–88.

Regep, Cristian, Guy Georges, Jiye Shi, Bojana Popovic, and Charlotte M. Deane. 2017. "The H3 Loop of Antibodies Shows Unique Structural Characteristics." *Proteins* 85 (7): 1311–18.

Renaud, Nicolas, Cunliang Geng, Sonja Georgievska, Francesco Ambrosetti, Lars Ridder, Dario F. Marzella, Manon F. Réau, Alexandre M. J. J. Bonvin, and Li C. Xue. 2021. "DeepRank: A Deep Learning Framework for Data Mining 3D Protein-Protein Interfaces." *Nature Communications* 12 (1): 7068.

Repecka, D., V. Jauniskis, and L. Karpus, et al. 2021. "Expanding Functional Protein Sequence Spaces Using Generative Adversarial Networks." *Nature Machine Intelligence* 3: 324–333. https://www.nature.com/articles/s42256-021-00310-5.

Richardson, Eve, Jacob D. Galson, Paul Kellam, Dominic F. Kelly, Sarah E. Smith, Anne Palser, Simon Watson, and Charlotte M. Deane. 2021. "A Computational Method for Immune Repertoire Mining That Identifies Novel Binders from Different Clonotypes, Demonstrated by Identifying Anti-Pertussis Toxoid Antibodies." *mAbs* 13 (1): 1869406.

Rives, Alexander, Joshua Meier, Tom Sercu, Siddharth Goyal, Zeming Lin, Jason Liu, Demi Guo, et al. 2021. "Biological Structure and Function Emerge from Scaling Unsupervised Learning to 250 Million Protein Sequences." *Proceedings of the National Academy of Sciences of the United States of America* 118 (15). https://doi.org/10.1073/pnas.2016239118.

Robert, Philippe A., Rahmad Akbar, Robert Frank, Milena Pavlović, Michael Widrich, Igor Snapkov, Andrei Slabodkin, et al. 2021. "Unconstrained Generation of Synthetic Antibody-Antigen Structures to Guide Machine Learning Methodology for Real-World Antibody Specificity Prediction." *bioRxiv*. bioRxiv. https://doi.org/10.1101/2021.07.06.451258.

Roberts, Christopher J. 2014. "Therapeutic Protein Aggregation: Mechanisms, Design, and Control." *Trends in Biotechnology* 32 (7): 372–80.

Ruffolo, Jeffrey A., Jeffrey J. Gray, and Jeremias Sulam. 2021. "Deciphering Antibody Affinity Maturation with Language Models and Weakly Supervised Learning." *arXiv [q-bio.BM]*. arXiv. https://arxiv.org/abs/2112.07782.

Ruffolo, Jeffrey A., Carlos Guerra, Sai Pooja Mahajan, Jeremias Sulam, and Jeffrey J. Gray. 2020. "Geometric Potentials from Deep Learning Improve Prediction of CDR H3 Loop Structures." *Bioinformatics* 36 (Suppl_1): i268–75.

Ruffolo, Jeffrey A., Jeremias Sulam, and Jeffrey J. Gray. 2022. "Antibody Structure Prediction Using Interpretable Deep Learning." *Patterns* (New York, N.Y.) 3 (2): 100406.

Saha, Sudipto, Manoj Bhasin, and Gajendra P. S. Raghava. 2005. "Bcipep: A Database of B-Cell Epitopes." *BMC Genomics* 6 (May): 79.

Sang, Zhe, Yufei Xiang, Ivet Bahar, and Yi Shi. 2022. "Llamanade: An Open-Source Computational Pipeline for Robust Nanobody Humanization." *Structure* 30 (3): 418–29.e3.

Schmitz, Samuel, Cinque Soto, James E. Crowe Jr, and Jens Meiler. 2020. "Human-Likeness of Antibody Biologics Determined by Back-Translation and Comparison with Large Antibody Variable Gene Repertoires." *mAbs* 12 (1): 1758291.

Schneider, Constantin, Andrew Buchanan, Bruck Taddese, and Charlotte M. Deane. 2021. "DLAB-Deep Learning Methods for Structure-Based Virtual Screening of Antibodies." *Bioinformatics,* September. https://doi.org/10.1093/bioinformatics/btab660.

Sharma, Vikas K., Thomas W. Patapoff, Bruce Kabakoff, Satyan Pai, Eric Hilario, Boyan Zhang, Charlene Li, et al. 2014. "In Silico Selection of Therapeutic Antibodies for Development: Viscosity, Clearance, and Chemical Stability." *Proceedings of the National Academy of Sciences of the United States of America* 111 (52): 18601–6.

Sheng, Zizhang, Chaim A. Schramm, Rui Kong, NISC Comparative Sequencing Program, James C. Mullikin, John R. Mascola, Peter D. Kwong, and Lawrence Shapiro. 2017. "Gene-Specific Substitution Profiles Describe the Types and Frequencies of Amino Acid Changes during Antibody Somatic Hypermutation." *Frontiers in Immunology* 8 (May): 537.

Shin, Jung-Eun, Adam J. Riesselman, Aaron W. Kollasch, Conor McMahon, Elana Simon, Chris Sander, Aashish Manglik, Andrew C. Kruse, and Debora S. Marks. 2021. "Protein Design and Variant Prediction Using Autoregressive Generative Models." *Nature Communications* 12 (1): 2403.

Singh, Satish Kumar. 2011. "Impact of Product-Related Factors on Immunogenicity of Biotherapeutics." *Journal of Pharmaceutical Sciences* 100 (2): 354–87.

Sircar, Aroop, and Jeffrey J. Gray. 2010. "SnugDock: Paratope Structural Optimization during Antibody-Antigen Docking Compensates for Errors in Antibody Homology Models." *PLoS Computational Biology* 6 (1): e1000644.

Sirin, Sarah, James R. Apgar, Eric M. Bennett, and Amy E. Keating. 2016. "AB-Bind: Antibody Binding Mutational Database for Computational Affinity Predictions." *Protein Science: A Publication of the Protein Society* 25 (2): 393–409.

Smialowski, Pawel, Gero Doose, Phillipp Torkler, Stefanie Kaufmann, and Dmitrij Frishman. 2012. "PROSO II--a New Method for Protein Solubility Prediction." *The FEBS Journal* 279 (12): 2192–2200.

Sormanni, Pietro, Francesco A. Aprile, and Michele Vendruscolo. 2015. "The CamSol Method of Rational Design of Protein Mutants with Enhanced Solubility." *Journal of Molecular Biology* 427 (2): 478–90.

Sormanni, Pietro, Damiano Piovesan, Gabriella T. Heller, Massimiliano Bonomi, Predrag Kukic, Carlo Camilloni, Monika Fuxreiter, et al. 2017. "Simultaneous Quantification of Protein Order and Disorder." *Nature Chemical Biology* 13 (4): 339–42.

Swindells, Mark B., Craig T. Porter, Matthew Couch, Jacob Hurst, K. R. Abhinandan, Jens H. Nielsen, Gary Macindoe, James Hetherington, and Andrew C. R. Martin. 2017. "abYsis: Integrated Antibody Sequence and Structure-Management, Analysis, and Prediction." *Journal of Molecular Biology* 429 (3): 356–64.

Tartaglia, Gian Gaetano, Andrea Cavalli, Riccardo Pellarin, and Amedeo Caflisch. 2005. "Prediction of Aggregation Rate and Aggregation-Prone Segments in Polypeptide Sequences." *Protein Science: A Publication of the Protein Society* 14 (10): 2723–34.

Tomar, Dheeraj S., Li, Matthew P. Broulidakis, Nicholas G. Luksha, Christopher T. Burns, Satish K. Singh, and Sandeep Kumar. 2017. "In-Silico Prediction of Concentration-Dependent Viscosity Curves for Monoclonal Antibody Solutions." *mAbs* 9 (3): 476–89.

Torjesen, Ingrid. n.d. "Drug Development: The Journey of a Medicine from Lab to Shelf." *The Pharmaceutical Journal*, doi: 10.1211/PJ.2015.20068196

Toseland, Christopher P., Debra J. Clayton, Helen McSparron, Shelley L. Hemsley, Martin J. Blythe, Kelly Paine, Irini A. Doytchinova, Pingping Guan, Channa K. Hattotuwagama, and Darren R. Flower. 2005. "AntiJen: A Quantitative Immunology Database Integrating Functional, Thermodynamic, Kinetic, Biophysical, and Cellular Data." *Immunome Research* 1 (1): 4.

Tubiana, Jérôme, Dina Schneidman-Duhovny, and Haim J. Wolfson. 2022. "ScanNet: An Interpretable Geometric Deep Learning Model for Structure-Based Protein Binding Site Prediction." *Nature Methods* 19 (6): 730–39.

Tunyasuvunakool, Kathryn, Jonas Adler, Zachary Wu, Tim Green, Michal Zielinski, Augustin Žídek, Alex Bridgland, et al. 2021. "Highly Accurate Protein Structure Prediction for the Human Proteome." *Nature* 596 (7873): 590–96.

Vashchenko, Denis, Sam Nguyen, Andre Goncalves, Felipe Leno da Silva, Brenden Petersen, Thomas Desautels, and Daniel Faissol. 2022. "AbBERT: Learning Antibody Humanness via Masked Language Modeling." *bioRxiv*. https://doi.org/10.1101/2022.08.02.502236.

Vita, Randi, James A. Overton, Jason A. Greenbaum, Julia Ponomarenko, Jason D. Clark, Jason R. Cantrell, Daniel K. Wheeler, et al. 2015. "The Immune Epitope Database (IEDB) 3.0." *Nucleic Acids Research* 43 (Database issue): D405–12.

Vita, Randi, Laura Zarebski, Jason A. Greenbaum, Hussein Emami, Ilka Hoof, Nima Salimi, Rohini Damle, Alessandro Sette, and Bjoern Peters. 2010. "The Immune Epitope Database 2.0." *Nucleic Acids Research* 38 (Database issue): D854–62.

Wang, Chunyan, Wentao Li, Dubravka Drabek, Nisreen M. A. Okba, Rien van Haperen, Albert D. M. E. Osterhaus, Frank J. M. van Kuppeveld, Bart L. Haagmans, Frank Grosveld, and Berend-Jan Bosch. 2020a. "A Human Monoclonal Antibody Blocking SARS-CoV-2 Infection." *Nature Communications* 11 (1): 2251.

Wang, Xiao, Genki Terashi, Charles W. Christoffer, Mengmeng Zhu, and Daisuke Kihara. 2020b. "Protein Docking Model Evaluation by 3D Deep Convolutional Neural Networks." *Bioinformatics* 36 (7): 2113–18.

Weitzner, Brian D., Daisuke Kuroda, Nicholas Marze, Jianqing Xu, and Jeffrey J. Gray. 2014. "Blind Prediction Performance of RosettaAntibody 3.0: Grafting, Relaxation, Kinematic Loop Modeling, and Full CDR Optimization." *Proteins* 82 (8): 1611–23.

Wilman, Wiktoria, Sonia Wróbel, Weronika Bielska, Piotr Deszynski, Paweł Dudzic, Igor Jaszczyszyn, Jędrzej Kaniewski, et al. 2022. "Machine-Designed Biotherapeutics: Opportunities, Feasibility and Advantages of Deep Learning in Computational Antibody Discovery." *Briefings in Bioinformatics* 23 (4). https://doi.org/10.1093/bib/bbac267.

Wilton, Emily E., Michael P. Opyr, Senthilkumar Kailasam, Ronja F. Kothe, and Hans-Joachim Wieden. 2018. "sdAb-DB: The Single Domain Antibody Database." *ACS Synthetic Biology* 7 (11): 2480–84.

Wolf, T., L. Debut, V. Sanh, and J. Chaumond. 2020. "Transformers: State-of-the-Art Natural Language Processing." Proceedings of the *2020 Conference on Empirical Methods in Natural Language Processing: System Demonstrations*. https://aclanthology.org/2020.emnlp-demos.6/?ref=https://codemonkey.link.

Wollacott, Andrew M., Chonghua Xue, Qiuyuan Qin, June Hua, Tanggis Bohnuud, Karthik Viswanathan, and Vijaya B. Kolachalama. 2019. "Quantifying the Nativeness of Antibody Sequences Using Long Short-Term Memory Networks." *Protein Engineering, Design & Selection: PEDS* 32 (7): 347–54.

Wu, Jiaxiang, Fandi Wu, Biaobin Jiang, Wei Liu, and Peilin Zhao. 2022. "tFold-Ab: Fast and Accurate Antibody Structure Prediction without Sequence Homologs." *bioRxiv*. https://doi.org/10.1101/2022.11.10.515918.

Wu, Zachary, S. B. Jennifer Kan, Russell D. Lewis, Bruce J. Wittmann, and Frances H. Arnold. 2019. "Machine Learning-Assisted Directed Protein Evolution with Combinatorial Libraries." *Proceedings of the National Academy of Sciences of the United States of America* 116 (18): 8852–58.

Xu, Yingda, Dongdong Wang, Bruce Mason, Tony Rossomando, Ning Li, Dingjiang Liu, Jason K. Cheung, et al. 2019. "Structure, Heterogeneity and Developability Assessment of Therapeutic Antibodies." *mAbs* 11 (2): 239–64.

Zavrtanik, Uroš, and San Hadži. 2019. "A Non-redundant Data Set of Nanobody-Antigen Crystal Structures." *Data in Brief* 24 (June): 103754.

Zhang, Jie, Yishan Du, Pengfei Zhou, Jinru Ding, Shuai Xia, Qian Wang, Feiyang Chen, et al. 2022. "Predicting Unseen Antibodies' Neutralizability via Adaptive Graph Neural Networks." *Nature Machine Intelligence*, November, 1–13.

From Deep Generative Models to Structure-Based Simulations

6

Computational Approaches for Antibody Design

Daisuke Kuroda

6.1 INTRODUCTION

In the rapidly evolving field of biotherapeutics, the intersection of computational science and protein engineering has revolutionized the approach to drug discovery and development. Central to this transformation is the art of antibody design, a complex and critical component of modern therapeutic strategies.[1] Antibodies, with their unique specificity and versatility, have emerged as a leading modality in the treatment of various diseases, ranging from cancers to autoimmune disorders and infectious diseases. The process of designing these therapeutic antibodies, however, presents a multifaceted challenge, encompassing aspects of protein engineering, bioinformatics, and developability.[2]

The advent of machine learning (ML) and generative models, including large language models (LLMs), has opened new frontiers in deciphering the complex language of proteins, particularly in understanding and predicting the vast diversity of the antibody repertoire.[3–5] These computational tools, coupled with advanced molecular simulations,

DOI: 10.1201/9781003300311-6

enable scientists to explore the structural and functional nuances of antibodies. By leveraging these technologies, the field of antibody design has transitioned from a largely empirical endeavor, relying on experimental trial and error, to one that is increasingly predictive and rational.

Developability, a key consideration in antibody engineering, involves assessing the suitability of antibody candidates for therapeutic use, focusing on their manufacturability, stability, and efficacy.[6] This process has been greatly enhanced by bioinformatics and computational science, which provide insights into the molecular characteristics that govern the behavior of antibodies in biological systems. Through computational approaches, the scope of protein engineering has expanded, allowing for the design of antibodies with improved properties from their amino acid sequences (Figure 6.1). The properties amenable to computational enhancement encompass physicochemical attributes, such as binding affinity and stability, as well as biological and pharmacological aspects, including immunogenicity.

The integration of ML into antibody design is particularly noteworthy. Through the analysis of vast datasets, including sequences and structures of existing antibodies, enabled by high-throughput sequencing of immune repertoires, ML algorithms can predict the antigen-binding affinity and specificity of novel antibody candidates.[7] This capability is pivotal in accelerating the drug discovery process, reducing the time and cost associated with the development of new biotherapeutics.

Among the various algorithms in ML, deep learning (DL) has garnered significant attention due to its unparalleled success in mimicking human-like decision-making and

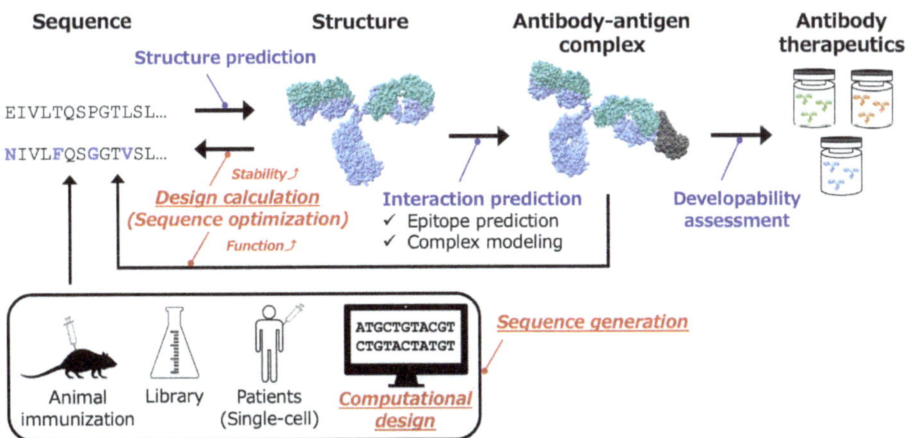

FIGURE 6.1 The computational workflow in antibody drug discovery: Seed antibodies are generated through various methods, including animal immunization, synthetic libraries, single-cell analysis, and computational design. The sequences of these seed antibodies can be further optimized using computational design calculations based on either sequences or structures, yielding lead antibody sequences. Structures of the antibody alone, as well as antibody-antigen complexes, can be predicted using computational methods. Additionally, developability assessments of lead antibodies can be conducted *in silico*. Topics discussed in this chapter are underlined and italicized in red for emphasis.

learning patterns. This has been especially evident since the early 2010s with breakthroughs in image and speech recognition.[8,9] Deep learning employs deep neural networks, which comprise several core architectures, such as convolutional neural networks (CNNs),[10] recurrent neural networks (RNNs) or long short-term memory (LSTM),[11] generative adversarial networks (GANs),[12] variational autoencoders (VAEs)[13], and transformer[14] models (Figure 6.2). Each architecture is distinguished by its specific strengths, weaknesses, and applications. These models autonomously learn features directly from data, thereby eliminating the need for manual feature extraction. The training process, which optimizes weights through backpropagation and gradient descent, refines learning over time. Although DL models reduce the necessity for manual feature engineering by learning complex data patterns, data pre-processing is still crucial in the bioinformatics workflow. This ensures data quality and compatibility, which are essential for the success of DL applications. This aspect becomes particularly critical in antibody design, where the sizes, amino acid compositions, and structural diversity of the functional sites, namely the complementarity-determining regions (CDRs), vary across antibodies. Such variability complicates the direct comparison of features among antibodies.

Molecular simulations, another vital component of computational antibody engineering, often rely on structural information of antibodies and their antigens.[15,16] These simulations play a crucial role in visualizing the dynamic interactions between antibodies

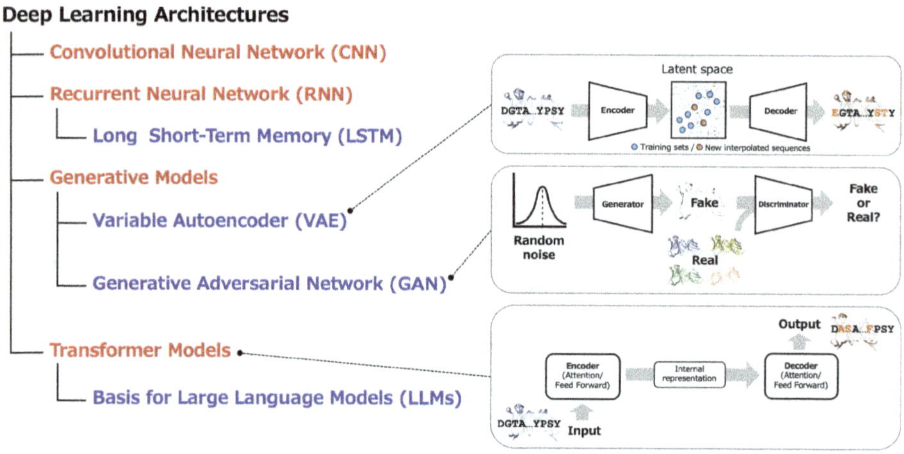

FIGURE 6.2 Classification of deep learning architectures: CNN: Designed to process data with a grid-like topology, using convolutional layers to efficiently extract and learn spatial hierarchies in data, particularly useful in image analysis. **RNN**: Designed to handle sequential data, where the output from previous steps is fed back into the network to inform responses at later steps, making them ideal for tasks like language modeling and time-series analysis. **LSTM**: An advanced type of RNN that is capable of learning long-term dependencies in sequential data. **VAE**: Designed to encode data into a latent space and then reconstruct it, facilitating data generation by sampling from the learned distribution in the latent space. **GAN**: Consists of two competing networks: a generator that creates data and a discriminator that evaluates it, working together to produce highly realistic data samples. **LLM**: Designed to understand, generate, and manipulate natural language, often built on architectures like transformers.

and their targets, offering valuable insights into their mechanisms of action and potential off-target effects. This informs the design of more effective and safer antibodies.

The synergy between ML, molecular simulations, and immune repertoire analysis is revolutionizing the field of antibody design and protein engineering.[17,18] Numerous studies in the literature have utilized these technologies for a variety of antibody-related prediction tasks, including the prediction of antibody structures,[19–24] antibody-antigen complexes,[25–30] their biophysical properties,[31–40] and the identification of lead candidates from large antibody libraries.[41–45] However, this chapter specifically focuses on the use of these methodologies for *de novo* generation and optimization of antibody sequences. Therefore, the objective of this chapter is to explore these technological advancements, with a particular emphasis on their impact on antibody sequence design for next-generation biotherapeutics and their influence on the future of antibody drug discovery.

6.2 ANTIBODY GENERATION THROUGH DEEP GENERATIVE MODELS

Generating antibody sequences is a critical task in developing therapeutic antibodies and diagnostic tools and conducting basic immunology research (Figure 6.1). Traditional methods like animal immunization and synthetic libraries offer distinct approaches to acquiring antibodies with the desired specificities and affinities. Each method has its strengths and limitations, with the choice often dictated by project-specific factors such as the nature of antigens, the intended antibody use, ethical considerations, and available resources.

The transition from these conventional experimental methods to incorporating deep generative models into antibody design signifies a paradigm shift in biotherapeutic development. Originally conceived for processing human languages, LLMs are now ingeniously repurposed to unravel the complex language of proteins.[46] This adaptation not only highlights the versatility of LLMs but also underscores the parallels in pattern recognition and sequence analysis between linguistics and molecular biology. Leveraging their core capabilities in sequence, context, and pattern recognition, LLMs offer a fresh approach to interpreting protein sequences, analogous to parsing sentences in a natural language. This innovative convergence of computational linguistics and molecular biology equips us with potent tools to advance protein engineering and antibody design. This section reviews the considerations involved in computationally generating antibodies *de novo* through LLMs and other DL-based techniques.

6.2.1 B-Cell Repertoires in the Era of Artificial Intelligence

ML methods fundamentally rely on data to uncover the intricate patterns of life. This is equally true for LLMs focused on generating antibody sequences, which benefit from the rich data provided by B-cell repertoires. These repertoires, now more accessible due to breakthroughs in high-throughput sequencing technologies, have revolutionized fields such as immunology, vaccine development, and therapeutic antibody discovery.[3] Enhancements in sequencing and computational analysis, coupled with a deeper understanding of the immune system, have made these advances possible. The decreasing cost of sequencing and increased throughput capacity have made large-scale genomic projects and the sequencing of vast cohorts and diverse species more practical, achieving previously unimaginable depth.

Integrating B-cell repertoire sequencing with other "omics" data, like proteomics and transcriptomics, offers a comprehensive view of the immune response.[47,48] Such a holistic approach can elucidate the evolution of repertoires in response to disease progression, vaccination, or therapeutic interventions. Signature changes in the B-cell repertoire, linked to various diseases, including autoimmune disorders, infectious diseases, and cancers, have been identified.[49–53] These signatures have potential as biomarkers for diagnosis and prognosis. Sequencing efforts pre- and post-vaccination have shed light on vaccine-induced immunity, informing the design of more effective vaccines.[54–56] Moreover, high-throughput sequencing has been instrumental in identifying antibodies with therapeutic potential against targets like severe acute respiratory syndrome coronavirus 2 (SARS-CoV-2).[57]

The development of specialized bioinformatics pipelines and software tools has enhanced the accuracy of sequence assembly, clonotype identification, and lineage tracing.[58] These tools are designed to manage the vast datasets produced, enabling detailed analyses of B-cell repertoires. By analyzing intricate patterns in repertoire data, computational algorithms guide both vaccine design and therapeutic antibody design.[4,59] Notable successes include employing ML, trained on high-throughput sequencing data, to detect changes in B-cell repertoire patterns in patients with dengue infection[60] and relapsing-remitting multiple sclerosis.[61] These studies have demonstrated the capability of ML algorithms to capture the nuances of B-cell repertoires, a task that would be challenging for humans without the aid of ML.

In the realm of artificial intelligence, representation learning, a key technique in deep learning, automates the discovery of data representations for feature detection or classification, bypassing the need for manual feature engineering. This approach allows models to identify complex patterns, enhancing performance in tasks such as classification and prediction. In protein modeling and design, representation learning is critical for encoding protein sequences, structures, or functional features for computational tasks. The significance of representation stems from its ability to translate the biological essence of a protein into a format that computational models, particularly those rooted in deep learning, can efficiently interpret and learn from.[62] The choice of representation—ranging from direct sequences and structures to more abstract feature-based representations and embeddings—significantly influences model performance in predicting protein structure and function and designing novel proteins (Figure 6.3).

Effective representation captures essential protein features relevant to biological functions, facilitating accurate predictions and the design of proteins with new properties. It is particularly vital in generative models aimed at creating novel protein sequences with specific functions. Here, representation learning navigates the protein

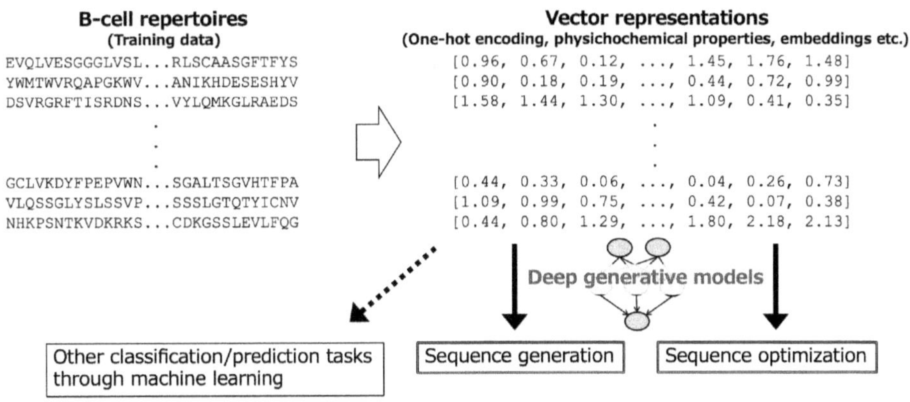

FIGURE 6.3 B-cell repertoires, feature representations, and computational tasks: Amino acid sequences are initially converted into representations or features that computational models can process. These representations are then used to address a range of protein science challenges, including the classification of existing sequences, the generation of new sequences, the optimization of existing sequences, and the prediction of structure from sequences.

space, guiding the design of proteins with intended functionalities. B-cell repertoires, with their rich sequence-function relationships, have been invaluable sources for ML-based computational antibody design. To explore the potential effects of the choice of representations on model performance, Wang et al. evaluated multiple embedding models for predicting sequence properties and receptor specificity of B-cell receptors or antibodies.[63] These models included the antibody-specific embedding model (Immune2vec[64]), transformer-based protein language models (ESM2,[65] ProtT5,[66] and AntiBERTy[67]), and traditional amino acid encoding methods (physicochemical encoding and amino acid frequency). Despite the varying architectures of these models, most embeddings effectively captured the sequence properties and specificity. Notably, the antibody-specific embeddings slightly outperformed general protein language models in predicting specificity. Moreover, the inclusion of full-length heavy chains and paired light chain sequences enhanced the predictive performance across all embedding types, indicating that functional signatures of antibodies are encoded within both the heavy and light chains.

However, despite these technological strides, challenges persist, including the need for standardized repertoire analysis methods, management of data complexity, and interpretation of the functional implications of sequence diversity. Future efforts may focus on enhancing single-cell sequencing technologies for more comprehensive repertoire capture, integrating functional assays for biological activity correlation, and developing advanced computational models for antigen specificity and antibody efficacy prediction. The accumulated B-cell repertoire data offers a valuable foundation for generative models and LLMs engaged in antibody sequence generation and optimization, signaling a promising frontier at the confluence of immunology and artificial intelligence.

6.2.2 Deep Generative Models and Large Language Models in Protein Science

More specifically, generative models and LLMs emerge as foundational elements within the domains of ML and artificial intelligence. Generative models, algorithms crafted to generate new data samples that mirror the distribution of a specific dataset, play a crucial role in protein design. These models enable the creation of novel protein sequences or structures with desired attributes such as improved stability, binding affinity, or enzymatic activity. Meanwhile, LLMs, particularly those built on the transformer architecture, have advanced to decode and comprehend the "language" of proteins, a complex sequence of amino acids. The transformer architecture employs attention mechanisms that allow simultaneous analysis of various sequence parts, enabling the model to assess the importance of each segment within the context of the entire sequence. This capacity is crucial for understanding protein structures and functions, as it considers both local and global interactions within the protein.

By harnessing extensive datasets of known protein sequences, including those from B-cell repertoires, transformer-based LLMs can predict the three-dimensional (3D) structures,[68–70] functions,[71,72] and mutational effects[73,74] of proteins from amino acid sequences. This comprehensive understanding and predictive capability underline the transformative impact of transformer architecture in decoding the complex language of proteins, offering a profound tool for advancements in protein engineering and enhancing our understanding of biological processes. However, while LLMs excel in capturing patterns within sequences, proteins exhibit intricate 3D interactions that extend beyond linear sequence relationships. The exploration of LLMs in protein science signifies a convergence with computational biology, particularly within natural language processing. There is a notable parallel between the linear structure of human language and the sequential nature of proteins, yet the 3D structures of proteins introduce complexities beyond those linear relationships. While LLMs can predict protein property and structure from sequences, the dynamic and complex nature of proteins requires consideration of their full 3D interactions, which can significantly influence their function and behavior.

Ultimately, the integration of LLMs into protein modeling and design promises a new frontier in understanding protein sequences, although the complexity of protein interactions necessitates ongoing refinement of these computational approaches.

6.2.3 Antibody Sequence Generation through Deep Generative Models

Generative models and LLMs have been employed in antibody generation (Table 6.1). An early example is the humanoid antibody library developed by Amimeur et al., which used GAN architecture to process 400,000 sequences from human antibody repertoires.[75] All the antibody repertoire sequences were computationally modeled for structural alignment. This alignment allowed for the direct comparison of residues at

TABLE 6.1 Machine learning applications for antibody sequence generation

ANTIBODY SEQUENCE GENERATION	*DEEP LEARNING ARCHITECTURE*	*TRAINING SETS FOR SEQUENCE GENERATIONS*	*REGIONS GENERATED IN THE LITERATURE*	*EXPERIMENTAL VALIDATIONS OF COMPUTATIONALLY GENERATED ANTIBODIES*
Antibody-GAN (Amimeur et al., 2020)[75]	GAN	OAS;[79] 400,000 F_V sequences	F_V	DSF (thermal stability), SINS (self-interaction), PEG solubility, SEC (purity)
SeqDesign (Shin et al., 2021)[77]	CNN	Naïve llama repertoire (~1.2 million)[76]	CDR-H3	Yeast display library (antigen binding)
AB-Gen (Xu et al., 2023)[78]	Transformer	OAS; 75,204,905 sequences of CDR-H3	CDR-H3	N/A
AbLang (Olsen et al., 2022)[80]	Transformer	OAS; 14,126,724 V_H, 187,068 V_L	Missing/ ambiguous residues	N/A
ProGen2-OAS (Nijkamp et al., 2023)[83]	Transformer	OAS; ~554 million F_V sequences	F_V	N/A
IgLM (Shuai et al., 2023)[82]	Transformer	OAS;[79] ~558 million F_V sequences	F_V	N/A
Ig-VAE (Eguchi et al., 2022)[86]	VAE	PDB;[87,88] 10,768 Ig domains	3D coordinates of V_H domain	N/A

identical structural positions across the dataset, simplifying the relationships the GAN needed to capture for generating new sequences. Experimental validation of the computationally generated antibodies involved expressing two examples in Chinese hamster ovary (CHO) cells, which exhibited the expected biophysical characteristics. This validation demonstrated the ability of the computer-generated library to explore meaningful antibody sequence space effectively.

Utilizing B-cell repertoires from llamas,[76] Shin et al. applied autoregressive models with CNN architecture to create a computational nanobody library.[77] Despite its smaller size compared to existing synthetic libraries, it contained functional sequences with micromolar-binding affinities, offering the potential for further maturation through additional experiments or computations.

AB-Gen employs a generative pre-trained transformer as the policy network of a reinforcement learning agent for antibody library design.[78] This transformer explores the vast space of CDR-H3 sequences, training on over 75 million CDR-H3 sequences from the Observed Antibody Space (OAS) database.[79] Through this training, the generative transformer of AB-Gen learns the sequence landscape, enabling the generation of novel sequences with property distributions akin to those in the training dataset. This method allows AB-Gen to create antibody libraries meeting multi-property constraints, such as binding affinity to major histocompatibility complex (MHC) class II, clearance, and viscosity, offering a robust resource for antibody discovery and development.

The challenge of missing or ambiguous residues due to sequencing errors or technological limitations is also addressed by generative models such as AbLang, featuring a transformer-based architecture.[80] Trained on B-cell repertoires,[79] this model predicts amino acid likelihoods, aiding in the restoration and generation of sequences with enhanced properties. Comparing AbLang with the general protein language model ESM-1b[81] demonstrated its superior capacity to capture antibody-specific information.

IgLM, another antibody-specific language model with a transformer-based architecture, generates variable-length CDR libraries, intending for easier lead sequence identification.[82] Trained on a massive dataset of 558 million antibody sequences,[79] IgLM enables tailored sequence creation based on species and chain type (e.g., heavy, kappa, and lambda). The model validation, which involved evaluating the sequence diversity and computational developability profiles of generated libraries, suggested its efficacy in antibody design.

ProGen2,[83] a more general protein language model, was also employed in generating antibody sequences. The models under ProGen2 were enhanced through the training of a suite ranging from 151 million to 6.4 billion parameters on diverse datasets, cumulatively encompassing 1 billion protein sequences from genomic, metagenomic, and B-cell repertoire databases. ProGen2 demonstrated its efficacy in the generation of antibody sequences by utilizing a 764 million parameter model, ProGen2-OAS, tailored specifically for natural antibodies from the B-cell repertoire database.[79] This particular model was instrumental in producing antibody sequences, highlighting its capacity to generate diverse and potentially therapeutically relevant sequences suitable for antibody library construction. This capability was evaluated through the distribution of CDR lengths and computational analysis of predicted aggregation propensity[84] and solubility.[85]

The antibody-specific generative models mentioned above do not generate 3D coordinates of antibodies. Consequently, while these models might create antibody libraries with potential utility, they lack a rational basis, leaving us unable to comprehend the efficacy of the generated antibodies. This dilemma mirrors the challenges encountered with traditional synthetic antibody libraries. Predicting structures could shed light on the physics underlying sequence generation, enhancing our understanding. For the exploration of antibody libraries targeting specific antigens, 3D structures prove even more indispensable. To address this gap, Eguchi et al. employed VAEs to generate 3D coordinates of antibodies, focusing on direct structure generation rather than sequences.[86] The specialized loss functions used in Ig-VAE for distance matrix and torsion angle inference, alongside Rosetta-based sequence design calculations, ensure structural accuracy. This approach facilitates the development of detailed epitope-tailored antibody libraries.

6.3 ANTIBODY OPTIMIZATION THROUGH STRUCTURE-BASED SIMULATIONS AND MACHINE LEARNING

In advancing antibody design, structure-based simulations and ML techniques have each emerged as pivotal strategies, though often employed independently. Structure-based simulations examine the complex molecular architecture of antibodies, utilizing molecular dynamics (MD) and Monte Carlo (MC) simulations to explore the vast conformational landscape of these proteins. Such simulations are crucial for understanding the dynamic behavior of antibodies, their interactions with antigens, and their stability under various physiological conditions.

Alongside these computational efforts, classical ML has transformed the field in its own right. By analyzing extensive datasets of protein structures and sequences, ML algorithms uncover patterns and relationships that might not be immediately observable. Techniques ranging from regression models to clustering and advanced ML algorithms like deep learning are utilized to predict antibody behavior, enhance binding affinities, and improve stability. For ML-based approaches, the training sets from which predictive models learn are key factors. While several data resources are public,[89–91] some standardization and normalization are often necessary to integrate data from various sources, and caution should be exercised when comparing experimental observables from different studies. To address this, simulations can generate training datasets for building ML-based predictive models, as seen in the Absolut! framework,[18] which enables parameter-based unconstrained generation of synthetic data relating to antibody-antigen binding. Although simulations themselves still require improvement, high-throughput simulated functional data, otherwise inaccessible, may be acceptable. Thus, although simulations and ML have each propelled antibody design forward, integrating these approaches represents a promising frontier yet to be fully explored.

This section first outlines two critical concepts in computational protein modeling and design—sampling and scoring—as they relate to structure-based simulations, and discusses their implications for ML-based protein design. Subsequently, it explores how these structure-based simulations and ML-based approaches have been independently applied in successful examples of computer-aided antibody design. It highlights the individual strengths of structure-based simulations and ML-based methods, as well as the potential synergies that could emerge from their integration.

6.3.1 Sampling and Scoring

In computational protein modeling, "sampling" refers to exploring the vast conformational space of proteins to identify potential structures they may adopt. Given the immense number of possible protein folds, sampling is crucial for finding the most relevant conformations for subsequent analysis. Two primary methods used in sampling are MD and MC simulations. On the one hand, MD simulations explore the atomic motions within a protein over time, leveraging principles of classical mechanics. This approach provides insights into the dynamic nature of proteins, including how they move and change shape. On the other hand, MC simulations engage in the random exploration of protein conformations, often utilizing the Metropolis algorithm to predict protein folding and interactions. The stochastic nature of MC simulations facilitates the examination of a wide array of conformations, including those that might not be captured by more deterministic methodologies. In both types of simulations, protein dynamics often depend on initial structures, and they can still become trapped in local minima of protein energy landscapes. Therefore, to sample a wider array of conformations through MD or MC simulations, generalized ensemble algorithms,[92] such as multi-canonical or replica exchange techniques, are often necessary to escape from these local minima.

Once a variety of protein conformations has been sampled, the challenge shifts to evaluating and ranking these conformations through "scoring". Scoring utilizes scoring functions or potentials to assess the physical plausibility and stability of each sampled conformation. There are two main types of scoring functions: (1) physics-based force fields, which are derived from fundamental physical principles to calculate protein conformation energies based on atomic interactions, playing a crucial role in MD simulations; and (2) statistical potentials, which, in contrast to physics-based approaches, are derived from the statistical analysis of known protein structures to estimate the likelihood of specific atomic interactions and conformations in proteins. Protein design calculations often rely on predicting the change in free energy upon mutations (i.e., $\Delta\Delta G = \Delta G_{\text{Mut}} - \Delta G_{\text{WT}}$), with each ΔG value calculated using scoring functions. Although statistical potentials heavily depend on experimental data, physics-based force fields are also partly parametrized using experimental observations, including structures of model compounds and results from spectroscopy experiments.[93,94]

The interplay between sampling and scoring is pivotal in computational protein modeling and design. While sampling produces a broad spectrum of potential protein conformations, identifying the most naturally occurring or functionally relevant ones without an efficient scoring system is challenging. Scoring functions assess and rank each conformation based on calculated stability or occurrence likelihood, enabling the refinement of the vast conformational space to the most plausible and significant structures for further exploration or application.

As discussed in the previous section, ML-based generative models possess the unique ability to directly generate protein sequences, thereby seamlessly integrating the processes of sampling and scoring. A prime example of such a model tailored for computational protein optimization is ProteinMPNN.[95] This model utilizes geometric features of proteins, trained on a message-passing neural network (MPNN), to innovate in the realm of sequence generation. When provided with a specific pair—an amino acid sequence alongside its structural counterpart—ProteinMPNN is proficient at generating modified sequences better suited to the existing structure. While Dauparas et al. demonstrated the success of ProteinMPNN in various computational protein design tasks,[95] the method relies on certain assumptions about the rigidity or prior knowledge of protein backbone conformations for sequence generation. Therefore, while it may be possible to improve the stability of antibodies due to the conserved nature of their framework regions through ProteinMPNN, enhancing their function poses a greater challenge. This difficulty arises from the structural diversity of CDR-H3, which plays a critical role in antigen recognition and is the most variable and functionally significant of the CDRs.

6.3.2 Machine Learning in the Context of Protein Sequence Design

In recent decades, ML algorithms have emerged as pivotal tools in analyzing extensive datasets of antibody sequences and structures. These algorithms identify patterns and relationships among sequences, structures, and experimental data from physicochemical measurements.[32,89,91] As a result, they enable accurate predictions of crucial physicochemical properties such as stability, viscosity, immunogenicity, and antigen-binding affinity.[33] A wide array of ML-based methods has been developed, pushing the boundaries of predictive protein engineering. Typically, these models use sequences or structures as inputs and provide outputs that predict whether a sequence or structure is stable, has low or high viscosity, and is immunogenic or non-immunogenic, either qualitatively or quantitatively.

However, while ML-based methods excel in predicting the properties of given amino acid sequences, they do not inherently design these sequences. In essence, traditional ML approaches have been more aligned with the scoring aspects of protein modeling and design rather than with direct sequence generation. Protein engineering aims to enhance wild-type sequences to improve protein characteristics. Generating a pool of random sequences and filtering them through prediction algorithms to find enhanced sequences is both laborious and inefficient. Considering the vastness of sequence space—for instance, the F_V region of antibodies, which consists of about 200 amino acids, resulting in an impractical number of possible sequences (approximately 10^{260})—this method proves to be unfeasible.

To address these challenges, more efficient methodologies that combine property prediction with sequence sampling and scoring have been developed. Known as computational design calculations, these innovative methods seamlessly merge the prediction of protein properties with the creative process of designing amino acid sequences. This synergy significantly enhances the efficiency of antibody engineering and the development of protein therapeutics, representing a notable advancement in the field.

6.3.3 Computational Strategy to Improve Stability of Antibodies

In the initial stages of biotherapeutic development, "stability" emerges as one of the primary concerns. An antibody that is challenging to isolate and express or is structurally fragile will not be suitable for long-term storage, rendering it impractical, regardless of its superior functionality. Protein expression level is sometimes referred to as an indicator of protein stability and could serve as an initial metric to assess antibody stability.

However, the terms "expression level" and "stability" are not always interchangeable. The relationship between protein expression level and stability remains an area of ongoing research.

Within the rigorous framework of protein engineering, protein stability can be categorized into three main types: thermodynamic stability, colloidal stability, and chemical stability. Significant computational efforts have been dedicated to improving thermal and colloidal stability and, more recently, the chemical stability of proteins. This initially involved structure-based simulations and later incorporated ML-based methods. Since the functional sites of antibodies, or CDRs, are exposed and localized on the surface, design efforts often target the framework regions across the F_V region to enhance stability without compromising functionality.

Advancements in computational power and prediction algorithms have now made it feasible to explore a point mutation-based approach. This method involves assessing stability changes (ΔΔG) through exhaustive replacement with 19 amino acids (i.e., all natural amino acids other than the wild type) at each position, from the N-terminal region to the C-terminal region, along an antibody F_V sequence. Despite the rise of the big data era and ML-based approaches, structure-based simulations continue to play a pivotal role in antibody design.[16,96–99] Using protein design software like Rosetta[100] or FoldX,[101] it is possible to sample all potential single mutations along an F_V region in several hours or less, even with a standard laptop computer. Therefore, in the realm of computational stability engineering, the scoring step emerges as a crucial challenge.

Thermodynamic stability of proteins is often experimentally assessed using differential scanning calorimetry (DSC) and differential scanning fluorimetry (DSF), which measure the melting temperature (T_m). A single mutation can increase the T_m of an antibody by up to ~8°C, particularly when the target site is a part of a cavity and deeply buried within the structure.[102] However, cavity-filling stabilizing mutations could alter the structural dynamics of antibodies, potentially leading to diminished functionality. Thus, to significantly enhance antibody thermal stability without compromising function, multi-point mutations are often necessary. However, the number of potential mutations grows exponentially when sampling two or more amino acids simultaneously, and addressing long-range interactions—where a mutation at one site may influence another at a distant site—remains a challenge. Protein design simulations typically treat the protein backbone as fixed or permit only minor adjustments, failing to account for allosteric-like effects that might emerge from distant mutation sites. MD simulations, supplemented by mathematical graph theory analysis, are frequently used to investigate the allosteric effects on protein structures.[103] Nevertheless, these approaches are still in the exploratory phase, and there is no universally accepted method for computationally analyzing the allosteric effects of mutations. Therefore, combinations of mutations are often applied by integrating the outcomes of computationally identified single-point mutations.

Recent studies exemplify the efficacy of computational point mutation scanning in stability engineering. Tomimoto et al. leveraged the dStability score from MOE molecular modeling software to assess the impact of single-point mutations across an entire domain on the thermal stability of a V_HH antibody, yielding a mutation that enhanced stability by over 5°C, as verified by DSF.[104] In an approach that uses a more rational strategy for selecting design positions, Okazaki et al. improved the stability of a scFv antibody by targeting its flexible regions.[105] Identified through triplicate 50-ns MD simulations, four mutations were strategically chosen to promote hydrogen bond formation between these flexible sites and surrounding residues. The feasibility of forming such hydrogen bonds was then reassessed through additional MD simulations, wherein the sidechain conformations of these mutations underwent evaluation via a continuous rotamer search, guided by the CHARMM36 force field. Consequently, one mutant displayed a thermal stability (T_m) increase of 5°C over the wild type.

Stability often poses a challenge in the process of humanizing antibodies, where there is a concerted effort to generate sequences that are more human-like than those derived from other animals.[106] To address this challenge, Tennenhouse et al. introduced CUMAb, a simulation-based methodology facilitating the systematic grafting of animal CDRs onto human frameworks.[107] Utilizing Rosetta for selecting frameworks based on energy and structural integrity, CUMAb-designed humanized antibodies matched or surpassed the affinities of their animal counterparts, with notable stability enhancements observed.

Furthermore, by leveraging immune repertoire sequencing data, ML-based methods have emerged as viable alternatives for stabilizing antibodies in the context of antibody humanization. Prihoda et al. developed a deep learning-based framework called BioPhi to be used for the antibody humanization process.[108] BioPhi optimizes the humanization process for therapeutic antibodies by offering an automated humanization workflow that incorporates novel methods based on deep learning and natural antibody repertoires. Specifically, BioPhi introduces a humanization method called Sapiens, which is trained on the OAS[79] using masked language modeling.[109] This deep learning approach allows BioPhi to generate humanized antibody sequences at scale while achieving results comparable to those of human experts.

The integration of structure-based simulations and ML-based methods represents an innovative and growing trend in the field of protein engineering. Lai et al. tackled the prediction of antibody aggregation and viscosity by integrating molecular dynamics simulations with ML models, yielding insights critical for therapeutic protein development.[110] Their study demonstrated that features derived from MD simulations could effectively inform ML models to predict aggregation rates and viscosity classifications. Notably, the k-nearest neighbors (KNN) regression model showed remarkable accuracy in predicting antibody aggregation rates, utilizing spatial positive charge maps and solvent-accessible surface areas of hydrophobic residues. This model achieved a correlation coefficient (r) of 0.89 with experimental aggregation rate measurements. Meanwhile, a logistic regression model excelled in classifying antibodies by viscosity, drawing on spatial charge distribution features. In a related effort, Harmalkar et al. developed a CNN-based classifier, trained on Rosetta's energetic features, illustrating the potential of integrating structure-based simulations with ML to identify thermostable residues with a 90% success rate.[111] These examples, from Lai et al.[110] and Harmalkar et al.,[111] highlight the synergistic potential of structure-based simulations and ML in refining antibody sequences, offering new pathways for antibody engineering.

6.3.4 Computational Strategy to Improve Functionality of Antibodies through Structure-Based Simulations

Affinity maturation, a critical process in the adaptive immune response, refers to the evolutionary mechanism by which B cells produce antibodies with increasingly higher affinity for their specific antigens. This process occurs through somatic hypermutation (SHM), a targeted mutation mechanism affecting the variable regions of antibody genes within the germinal centers of lymph nodes. During an immune response, B cells that recognize an antigen undergo rapid proliferation and mutation. This results in an antibody repertoire that is increasingly specific and effective against the target antigen, enhancing our ability to fight infections and respond to vaccines. Affinity maturation is fundamental to the development of long-lasting immunity and a key consideration in the design of therapeutic antibodies. Similarly, but more rapidly, antibodies can be evolved in computers. Several such approaches have been proposed to improve binding affinities and/or other biophysical properties of antibodies. Many of such successful examples were based on molecular simulations (Table 6.2).

In cases where structures of antibody-antigen complexes are unavailable, modeling such complex structures becomes a crucial starting point for functional design. Despite steady improvements in protein-protein docking algorithms,[112] accurately predicting the structures of antibody-antigen complexes remains a formidable challenge. There are a few instances where homology modeling followed by docking simulations has been employed, leading to computational affinity maturation based on these modeled complex structures.[113,114] However, these approaches often necessitate iterative and labor-intensive processes, such as experimental mutagenesis studies and advanced MD techniques, such as replica exchange MD and umbrella sampling, to refine predictions. While the exploration of conformational space in docking simulations is somewhat constrained due to the nature of antibodies (which recognize antigens primarily via CDRs), probing the conformational space on the antigen side of the antibody-antigen interface or identifying epitopes continues to be highly complex.

The recent emergence of AlphaFold2 (AF2) has significantly impacted the fields of protein science in various ways.[115] Originally developed for predicting the structures of monomers from their amino acid sequences, it is now becoming feasible to utilize AF2 for modeling protein-protein complexes.[116] Although the initial version of AF2 was not specifically designed for predicting structures of antibody-antigen complexes, given the independent evolutionary paths of antigens and antibodies, a recent update to AF2 has reportedly achieved success in this area, although details are yet to be disclosed at the time of writing this chapter.[117] Given the growing repository of structural information on antibody-antigen complexes[118] and the availability of B-cell repertoire data with functional insights,[119] integrating machine learning tools like AF2 with simulation-based approaches appears to be a promising direction for accurately modeling antibody-antigen complexes.[26,28,120]

Due to the difficulty of predicting antibody-antigen complex structures, early efforts in computational antibody design focused on enhancing species cross-reactivity[121] and

TABLE 6.2 Structure-based optimization of antibody function through simulations and machine learning

INITIAL STRUCTURES	*RESOLUTION (Å)*	*TARGET ANTIGEN*	K_D^{WT}	K_D^{DESIGN}	K_D^{WT}/K_D^{DESIGN}	*DESIGNED MUTATIONS (CHOTHIA NUMBERING)*	*METHODS*[A]	*REF.*
1MHP	2.80	Integrin	7 nM	850 pM	10	**L:** S28Q, N53E **H:** T50V, K64E	Sim	Clark et al. (2006)[123]
1MLC	2.50	Lysozyme	4.4 nM	30 pM	140	**L:** N92A **H:** T28D, E35S, S56V, T57D, G95D	Sim	Lippow et al. (2007)[122]
N/A	N/A	Gastrin	6 μM	13 nM	454	**L:** H91F, Q92F, R94P, V96A **H:** I100L, S102V	Sim (library design)	Barderas et al. (2008)[126]
3BN9	2.17	Protease	4.8 nM	340 pM	14	**H:** T98R	Sim	Farady et al. (2009)[121]
2BDN	2.53	MCP-1	0.8 nM	0.17 nM	4.6	**L:** N31R	Sim	Kiyoshi et al. (2014)[124]
3U4E	2.19	HIV-1 gp120	N/A	N/A	4.2[b]	**H:** $N100_fY$	Sim	Willis et al. (2015)[125]
N/A	N/A	α-synuclein	3.22 μM	0.0664 μM	48.5	**H:** N76D	Sim	Mahajan et al. (2018)[113]
N/A	N/A	CCL20	9.1 nM	2.3 nM	4.0	**L:** $G30_dY$ **H:** S27R	Sim	Cannon et al. (2019)[114]
2FJG	2.80	VEGF	270 pM	60 pM	4.5	**L:** A43P, Q89L, T94D, P95T **H:** Y58H, Y91F	Sim	Warszawski et al. (2019)[148]

(*Continued*)

TABLE 6.2 (*Continued*) Structure-based optimization of antibody function through simulations and machine learning

INITIAL STRUCTURES	*RESOLUTION (Å)*	*TARGET ANTIGEN*	K_D^{WT}	K_D^{DESIGN}	K_D^{WT}/K_D^{DESIGN}	*DESIGNED MUTATIONS (CHOTHIA NUMBERING)*	*METHODS*[a]	*REF.*
1ZV5	2.00	Lysozyme	7.6 nM	0.65 nM	11.7	**H:** $A100_cY$	Sim (library design)	Akiba et al. (2020)[149]
3BDY	2.60	VEGF	100 nM	8 nM	13	**Replacement of CDR-H3 (93-102):** From SRWGGDGFYAMDY to ARGGAVAGTGVYYFDY	Sim	Corbeil et al. (2021)[146]
2DD8	2.30	SARS-CoV-2 (ancestral)	N/A	177 nM	-	**L:** S30R, W91Y, S93L, S94F **H:** S30G, S31N, Y32W, T33S, T52D, I53S, L54R, I56E, N58T, D95E, V97E, M98Q	Sim	Jeong et al. (2022)[140]
2GHW	2.30	SARS-CoV-2	N/A	0.47 nM	-	**L:** R24K, Q27E, S28D, V29I, R30G, S31T, N32A, Y49F, A51G, S52A, T53I, R54L, T56P, Q89H, Q90N, R91V, S92Y, N93S **H:** A23K, S31N, N56Y	Sim	Hernandez et al. (2023)[144]
8HES	2.20	SARS-CoV-2 (XBB.1.5)	N/A	56.78 ng/mL[c]	-	**L:** Y49M, D50N, R53V, R54L, S56D **H:** V2W, T28A, Y30W, S31M, Y32F, K52A, E54Q, R94V, D95G, Y98W, N99I, I100W, $T100_bI$, $Y100_dF$, $H100_fF$, S101I	Sim	Moriyama et al. (2023)[138]
5VJ6	11.5 (EM)	HIV-1 gp120	0.047 µg/mL[c]	0.017 µg/mL[c]	2.8	**H:** $Y100_kD$	Sim	Holt et al. (2023)[136]
7X7O	3.75	SARS-CoV-2 (BA.1)	N/A	0.93 nM[c]	-	**H:** T28R, N56D	Sim	Ozawa et al. (2024)[137]
7FJC	2.96	SARS-CoV-2 (ancestral)	22.9 nM	0.82 nM	27.9	**H:** N56L, R99M, L100F	ML (deep learning)	Shan et al. (2022)[150]

[a] Mutations were identified through structure-based simulations (Sim) or machine learning (ML).
[b] Geometric mean of fold improvement in EC_{50} for nine viruses compared to the wild type.
[c] IC_{50} values.

improving binding affinity[122–124] or neutralizing potency[125] toward antigens, guided by crystal structures of antibody-antigen complexes or, in one case, by a homology model of an antibody alone.[126] These studies selected mutations based solely on energetic considerations, with design calculations concentrating on a single state of the antibody-antigen complex. Additionally, affinity maturation in our bodies encompasses somatic mutations not only within the antigen-binding site but also across framework regions that typically do not make physical contact with antigens. Current computational methods for assessing the effects of mutations primarily focus on the physical contacts between mutated residues and their immediate surroundings. Consequently, the targets for computational functional design are predominantly confined to the CDRs. This approach, however, overlooks the potential for long-range effects, where mutations at one site may influence another site distantly.[127,128]

In contrast to designs that focus on a single structure or a state, multistate design (MSD) aims to minimize the free energy across multiple protein conformations simultaneously.[129] MSD seeks to capture the dynamic nature of proteins, aiming to engineer proteins capable of adopting multiple, predefined structures or states. This advanced approach employs computational algorithms to navigate the energy landscapes of these different protein states, predicting amino acid sequences that ensure compatibility with each state. Consequently, designed proteins can seamlessly transition between states, mirroring the versatility of natural proteins. Recent developments have seen the introduction of several software packages enabling MSD. Sevy et al. applied this concept by developing a multistate design algorithm named RECON within the Rosetta framework.[130] Unlike traditional methods that apply a single sequence across all states, RECON allows for unique sequence adaptations for each state during the design process. This innovation was applied to the design of anti-influenza antibodies.[131] The authors generated and experimentally evaluated redesigned variants of an anti-influenza antibody targeting the hemagglutinin (HA) protein. These variants indeed displayed improved breadth and affinity against a range of influenza H1 subtypes, aligning with the predictions made by the computational multistate design.

In addition to the energetics of antigen recognition, complementarity, encompassing hydrophobic, electrostatic, and shape compatibility, plays a crucial role in antibody-antigen interactions. Historically, shape complementarity at antibody-antigen interfaces was thought to be inferior to that of other protein-protein interfaces, attributed to the independent evolution of antigen interfaces from antibodies.[132] However, later studies corrected this misconception, revealing that poor shape complementarity assessments were due to the low quality of crystal structures used in initial analyses.[133,134] It has been established that shape complementarity at antibody-antigen interfaces is comparable to that of other protein-protein interfaces.[134] Furthermore, research on electrostatic complementarity at the antibody-antigen interface has been conducted, exploring the impact of multiple charged residues that form positive-negative pairs at the interface.[135] This study involved experimental modulation of these charged residue pairs to examine their effects on antibody stability and binding affinity. The findings suggest that charged residues play a crucial role not only in binding affinity but also in the thermodynamic stability and secondary structures of antibodies. Another independent studies also highlighted the importance of electrostatic complementarity at antibody-antigen interfaces; mutations to improve electrostatic complementarity indeed enhanced binding affinity and neutralizing potency of antibodies.[113,124] However, several studies have shown that while mutations computationally predicted to introduce electrostatic interactions did enhance the potency and breadth of antibody neutralization, as confirmed by experimental validations, cryo-electron microscopy (cryo-EM) structural analyses revealed that the predicted electrostatic interactions failed to form, due to minor shifts in the conformations of the CDR backbones.[136,137] These studies highlight both the crucial role and the challenges associated with modulating electrostatic complementarity at antibody-antigen interfaces.

Therefore, enhancing complementarity at antibody-antigen interfaces has become a primary strategy in the functional design of antibodies. For instance, Moriyama et al. undertook the computational design of an anti-SARS-CoV-2 neutralizing antibody to broaden its specificity.[138] The original wild-type antibody had shown neutralizing effectiveness against Omicron subvariants like BA.5 and BQ.1.1 but was less effective against the XBB subvariants. To address this, a putative model structure of the antibody-XBB.1.5 RBD complex was computationally prepared, starting from a crystal structure of the antibody bound to an ancestral RBD. For computational redesign, residues on CDRs within a 5 Å distance from the antigen (XBB.1.5 RBD) were selected. After introducing multiple mutations on the antibody interface with the FastDesign algorithm in Rosetta,[139] resulting in approximately 2000 design constructs, these constructs were evaluated based on metrics such as interaction energy (< -35 kcal/mol), shape complementarity (Sc > 0.65), and the number of hydrogen bonds at the interfaces (> 4). An optimized antibody demonstrated enhanced neutralizing potency against various variants of concern of SARS-CoV-2, including XBB and XBB.1.5. It is also worth noting that traditional energy-based selection might not have identified the functionally optimized antibody reported by Moriyama et al., as it ranked beyond the 100th position in terms of interaction energy alone. Starting from an anti-SARS-CoV neutralizing antibody, another study utilized the FastDesign algorithm to generate new specificity toward SARS-CoV-2.[140] This computational design effort explicitly incorporated water molecules and ensured high shape complementarity (Sc > 0.7) at the antibody-antigen interface. However, the initial design yielded only a weak binding affinity (177 nM) toward the wild-type SARS-CoV-2 RBD. Subsequent structural determination of a design variant through X-ray crystallography, coupled with CDR3 extensions via library-based random mutagenesis experiments, was crucial to producing a final design variant. This designed antibody was capable of neutralizing early SARS-CoV-2 variants, including the Alpha and Delta variants.

These studies highlight the critical role of shape complementarity and hydrogen bonds in antibody-antigen interactions, emphasizing the necessity of considering multiple metrics for a comprehensive evaluation of design candidates. Moreover, the aforementioned study by Jeong et al.[140] further emphasized the importance of CDR lengths in computational affinity maturation, reinforcing the notion that both structural nuances and sequence characteristics are pivotal in optimizing antibody-antigen binding. Given that accurately predicting loop conformations remains a formidable challenge in computational protein science,[141] extending CDR conformations to arbitrary lengths during design calculations is not straightforward yet. Achieving this would necessitate both sub-angstrom accuracy in loop prediction and flexible backbone design calculations, areas still in need of substantial improvement. Within this framework, RosettaAntibodyDesign (RAbD) emerges as one of the pioneering computational protocols that allow for the modification of CDR lengths and backbone conformations.[142] It does so by replacing parent wild-type CDRs with existing counterparts from the PDB while concurrently designing the amino acid sequences. Since all CDRs, except for H3, can be categorized into canonical structures,[143] extending non-H3 CDRs based on existing structures may suffice for computational affinity maturation. However, the crucial role of CDR-H3 in antigen recognition and its high diversity mean that merely replacing existing CDR-H3 structures in the PDB is insufficient. The RAbD protocol was successfully applied to redesign an antibody targeting SARS-CoV, leading to the generation of a designed antibody capable of binding to various SARS-CoV-2 variants.[144] To overcome the limitation in CDR H3 design, Corbeil et al. introduced an antibody design protocol that broadens both the sequence and the structural diversity by leveraging human germline sequences of CDR-H3 from the IMGT database,[145] alongside pre-generated loop structure conformations.[146] This innovative approach was applied to an antibody targeting human VEGF, resulting in several design variants with varying lengths of CDR-H3 that exhibited enhanced binding affinities compared to the parental antibody. This breakthrough highlights the potential of integrating detailed CDR length considerations and advanced computational techniques to optimize antibody-antigen recognitions.

Balancing high functionality with stability in proteins often presents a trade-off, where enhancing one attribute may compromise the other.[147] Drawing inspiration from deep mutational scanning experiments that identified mutations improving both stability and binding affinity of antibodies, Warszawski et al. developed an automated web server.[148] This platform combines simulated ΔΔG upon mutations with evolutionary conservation scores derived from multiple sequence alignments of antibodies. It enables the design of antibodies with both higher stability and better binding affinity compared to their wild-type counterparts by pinpointing mutations at the V_L/V_H interfaces. These residues at the V_L/V_H interfaces are also critical for the process of antibody humanization. Therefore, further systematic studies are essential to uncover the relationships between sequences/structures at the V_L/V_H interfaces and the biophysical properties of antibodies.

Finally, structure-based simulations also serve a unique purpose in identifying promising sites for experimental random mutagenesis, instead of depending solely on computational predictions for specific beneficial mutations.[126,149] This approach can augment existing scoring functions, which are not flawless, by addressing the limited understanding of sequence-function relationships in antibodies and pinpointing advantageous mutations through screening experiments.

6.3.5 Computational Strategy to Improve Functionality of Antibodies through Machine Learning

Although only a few public repositories exist that store experimentally measured binding affinity data,[89,91,151] this availability makes it increasingly feasible to construct ML-based predictive models for antibody-binding affinities.[32,152,153] As an example, Shan et al. developed a predictive model using an attention-based geometric neural network architecture.[150] This model was trained to understand the effects of mutations on protein-protein interactions from the three-dimensional structures of protein complexes. The training dataset was derived from the SKEMPI 2.0 database,[91] which includes data on antibody-antigen-binding affinities. Mutant structures for the training were generated using the Rosetta Cartesian ΔΔG program.[100] This model specifically aims to predict how mutations within CDRs can affect their binding affinity and neutralization activity against various SARS-CoV-2 variants. By integrating this predictive model with experimental validation, Shan et al. successfully optimized an antibody sequence, pinpointing mutations that enhance neutralizing activity against various SARS-CoV-2 strains (Table 6.2). This accomplishment highlights the potential of ML-based methods for the efficient optimization of antibody sequences, leveraging publicly available experimental datasets for training.

However, a caveat in using public data for training ML-based predictive models is the dataset bias. Publicly available data on mutations, associated with experimental ΔΔG values, is often skewed toward mutations involving alanine, due to its popularity in protein engineering experiments. This can lead to overfitting to such biased data. To develop generalizable ML models for antibody design, a comprehensive dataset covering a broad spectrum of mutations is essential. Directed evolution would be the most accurate approach to improve the binding affinity of antibodies.[154] Such library-based approaches can also be used to generate training sets to be used for machine learning.[155]

In addition to sequence generation, deep generative models can also optimize antibody sequences. Starting with existing sequences, display-based panning experiments can directly generate data on sequence-function relationships of antibodies. A significant advantage of learning from these sequence-function fitness landscapes is that, although a few rounds of library-based screening may be necessary to build predictive models, unlike previously mentioned studies, structural information is not required during the optimization process. Interestingly, previous studies have shown that sequence enrichment in library-based selections does not always translate to superior antibody properties,[147,148] highlighting the complexity of sequence-function relationships. ML approaches are poised to unravel these intricate, non-linear relationships, offering a sophisticated understanding that could revolutionize how we predict and enhance antibody functionality. Table 6.3 summarizes such ML-based optimizations of antibody functions reported in the literature.

The studies by Liu et al.[156] and Saka et al.[157] demonstrated the potential of high-throughput experimental data in developing deep learning-based predictive models for antibody binding. Training data from phage display panning experiments facilitated

TABLE 6.3 Machine learning applications for antibody sequence optimization

	ARCHITECTURE	*TRAINING SETS FOR SEQUENCE OPTIMIZATION*	*REGIONS OPTIMIZED IN THE LITERATURE*	*EXPERIMENTAL VALIDATIONS*
Liu et al. (2020)[156]	CNN	~121,485 sequences of CDR-H3 from phage display library	CDR-H3	ELISA (antigen binding)
Saka et al. (2021)[157]	RNN-LSTM	959 $\boldsymbol{V_H}$ sequences from phage display library	CDRs	SPR (antigen binding)
Makowski et al. (2022)[107]	Linear discriminant analysis, two-layer neural network	4,000 sequences of $\boldsymbol{V_H}$ domain from yeast display library	CDR-H2/H3	Polyspecificity particle assay (antigen binding)[171]
Harvey et al. (2022)[160]	Logistic regression, CNN, RNN	Sequences from nanobody synthetic library (65,147 low polyreactive sequences/69,155 high polyreactive sequences)	CDR-Hs	Polyspecificity reagent analytical staining, flow cytometry, ELISA, thermal shift assay, X-ray crystallography
Mason et al. (2021)[162]	CNN	Sequences of CDR-H3 from DMS	CDR-H3	BLI (antigen binding), thermal ramp experiments (thermal stability), T-cell proliferation assay (immunogenicity)
RESP (Parkinson et al., 2023)[158]	Autoencoder, Bayesian neural network	2,725,492 $\boldsymbol{V_H}$ sequences from cAb-Rep[172] Binding data from yeast display library	CDR-Hs	Yeast display library (antigen binding)
Hie et al. (2023)[164]	Transformer/ESM[81]	~27 million sequences (including non-antibody proteins) from UniRef50[173] ~98 million sequences (including non-antibody proteins) from UniRef90	$\boldsymbol{F_V}$	BLI (antigen binding), DSF (thermal stability), polyspecificity particle assay (antigen binding), pseudovirus assay (neutralization)
Biotransfer (Li et al., 2023)[163]	Ensemble of regression models, Gaussian process, Bayesian optimization	26,453 heavy chains and 26,223 light chains with binding affinities from AlphaSeq assay,[174] ~242,612,962 and 32,593,668 sequences from OAS and Pfam, respectively	CDRs	AlphaSeq assays (antigen binding)

(*Continued*)

TABLE 6.3 (*Continued*) Machine learning applications for antibody sequence optimization

	ARCHITECTURE	*TRAINING SETS FOR SEQUENCE OPTIMIZATION*	*REGIONS OPTIMIZED IN THE LITERATURE*	*EXPERIMENTAL VALIDATIONS*
Akbar et al. (2022)[165]	RNN-LSTM	70,000 sequences of CDR-H3 from naive B-cell repertoires[175] with the predicted binding affinities[18] Sequences of CDR-H3 from DMS[162] (11,300 HER2 binder/27,539 HER2 non-binder)	CDR-H3	N/A
F_V Hallucinator (Mahajan et al., 2022)[168]	Hallucination	Predicted structures of an input sequence	CDRs, V_L/V_H interface	N/A
Lim et al. (2022)[167]	GAN	Sequences of CDR3s from yeast display library targeting PD-1 and CTLA-4[176] (~1,800 binders/~6,000 non-binders)	CDR3s	N/A
AbGAN-LMG (Zhao et al., 2023)[166]	GAN	10, 563 sequences in the CoV-AbDab[177]	CDRs	N/A
AntBO (Khan et al., 2023)[169]	Bayesian optimization	Biophysical properties and binding energies of CDR-H3s toward the antigen calculated via Absolut!	CDR-H3	N/A
GLMAb-BO (Wang et al., 2023)[170]	General language model,[178] CNN, Bayesian optimization	Sequences from OAS	CDR-H3	N/A

the creation of CNN and LSTM models that optimize antibody sequences, yielding novel sequences with enhanced affinities. Both studies employed one-hot sequence representations to train their ML models. Notably, these ML-generated sequences were validated experimentally through ELISA and SPR, respectively, demonstrating the effectiveness of machine learning with self-compiled datasets from detailed experiments in advancing antibody design.

The RESP model, developed by Parkinson et al.,[158] employed a three-component approach to rapidly identify high-affinity antibodies. The first component, an autoencoder, was trained on over 3 million B-cell receptor sequences to encode the learned representations of these repertoire sequences. The second component, a variational Bayesian neural network, was trained with the learned representations of a different set of sequences alongside their corresponding binding affinity data from yeast surface display experiments. This network was utilized to predict the binding affinities of antibody sequences. The third component, a modified simulated annealing algorithm, leveraged the predictions of the neural network to guide *in silico*-directed evolution, generating optimized antibody sequences not present in the original mutation libraries. Experimental validation using a yeast display library confirmed the efficacy of the model, revealing a 17-fold improvement in the binding affinity for the PD-L1 antibody atezolizumab, thereby demonstrating the capability of the RESP model to accelerate the discovery of high-affinity antibodies.

In another study, Makowski et al. employed linear discriminant analysis (LDA) models and two-layer neural networks, each trained on 4,000 sequences from a yeast display library followed by deep sequencing, using one-hot encoded sequence features.[107] Their objective was to construct ML-based models to predict antibody-binding affinity and specificity. The findings revealed that the accuracy of the simpler LDA models was comparable to that of the more complex neural network models, suggesting that model complexity does not significantly impact the accuracy of classifying antibody properties based on deep sequencing data. Furthermore, utilizing the LDA models with features derived from a unified representation,[159] their predictive models could explore new sequence spaces that were not covered in the original library. When 29 ML-generated new sequences were experimentally evaluated, one new antibody clearly demonstrated both higher binding affinity and reduced non-specific binding.

The manufacturing quality control of antibodies is significantly challenged by their lack of specificity, necessitating the reduction of polyreactivity as a key focus in the development of therapeutic and diagnostic antibodies to guarantee their safety, efficacy, and specificity. Continuous advancements in antibody engineering and computational methodologies are enhancing our capabilities to achieve these goals. Within this framework, ML models, including logistic regression (LR), CNNs, and RNNs, have been applied to assess polyreactivity from antibody sequences.[160] These models underwent training using sequences from a naive synthetic yeast display library, sorted into high or low polyreactivity categories via fluorescence-activated cell sorting (FACS). Such training empowered the models to accurately predict the polyreactivity of nanobodies. Notably, a LR model, utilizing a one-hot embedding of the CDR sequences, demonstrated the best performance. Beyond mere classification into high or low polyreactivity, this methodology also facilitates the discovery of novel mutations that decrease the polyreactivity of certain antibody clones, a finding substantiated by various biochemical assays, as summarized in Table 6.3.

Besides library-based selection above, another example of using self-compiled training sets to enhance antibody functions is deep mutational scanning (DMS) experiments,[161] offering a comprehensive mutational landscape compared to the less systematic coverage of library-based selections. DMS quantifies the impact of each mutation, providing rich data for training deep learning models. Mason et al. utilized DMS to create a detailed training set, guiding the design of a combinatorial mutagenesis library.[162] After screening 50,000 variants for antigen specificity, they used a CNN, trained with one-hot encoding sequence features, to predict antigen specificity from antibody sequences. The trained networks predicted over 3 million variants with high antigen specificity. Experimental validation of 30 randomly selected variants confirmed their antigen-specific binding. Subsequent *in silico* filtering optimized nearly 8,000 variants for developability, with top candidates showing improved properties over the wild type, including one significantly less immunogenic candidate. This approach demonstrates the synergy of DMS and machine learning in refining antibody design.

Bayesian optimization (BO) is a technique used to optimize "black-box" functions—complex systems where each evaluation is costly—by using a probabilistic model to predict improved outcomes and strategically selecting the next set of parameters. In antibody design, BO maps antibody sequences to posterior probabilities, aiding in the evaluation and optimization of binding affinity. Li et al. integrated language models, Bayesian optimization, and high-throughput experiments to create high-affinity scFv libraries.[163] Their method, establishing a Bayesian fitness landscape of antibody sequence and function, accurately predicted binding affinities for diverse mutants. Notably, this computational approach to library design yielded an scFv with a 28.7-fold increase in predicted binding affinity compared to a control, demonstrating the effectiveness of the model in efficiently optimizing antibody efficacy. These predictions were verified through display-based assays; most experimentally evaluated ML-generated scFvs exhibited better binding affinity than the initial candidate scFv.

Finally, general transformer-based protein language models[81] have also shown the capacity to evolve human antibodies without detailed knowledge of the target antigen or structure.[164] They achieved this by predicting mutations with evolutionary plausibility, informed by patterns in natural protein sequences. These models efficiently suggested mutations that can significantly enhance antibody-binding affinities. Leveraging evolutionary data, the predictive model can pinpoint variants with high evolutionary likelihood for further experimental testing. This approach has successfully evolved antibodies, enhancing binding affinities of clinically relevant antibodies by up to sevenfold and unmatured antibodies by up to 160-fold, as verified by bio-layer interferometry (BLI) measurements. Additionally, some evolved antibodies exhibited increased thermostability and effective viral neutralization against Ebola and SARS-CoV-2 pseudoviruses, demonstrating the potential of general protein language models to guide antibody evolution against various antigens.

Together, these studies have highlighted the potential of ML-based antibody engineering, where training sets are generated through library-based methods like panning experiments and DMS, followed by deep sequencing. Notably, these studies succeeded in generating novel antibody sequences not found in the original libraries. However, experimental validation remains essential for ML-generated antibody sequences to confirm their binding to target antigens and assess their developability. Although several other studies have also employed generative models and ML techniques, including LSTM,[165] GAN,[166] CNN/GAN,[167] hallucination,[168] and Bayesian optimization,[169,170] for optimizing antibody sequences, the evaluation of such ML-generated sequences has frequently been confined to computational analyses. Given the current technological limitations and our partial understanding of sequence-function relationships in antibodies, experimental validation remains an indispensable step in the optimization process.

6.3.6 Computer-Aided Antibody Repositioning

Drug repositioning, also known as drug repurposing, is a strategic approach in pharmaceutical development that seeks to identify new therapeutic uses for existing drugs.[179] This innovative strategy offers a promising pathway to accelerate the drug development process, significantly reducing the time and costs associated with bringing a drug to market. Unlike traditional drug discovery, drug repositioning leverages the existing safety and pharmacokinetic profiles of approved or investigational drugs and explores their potential applications beyond their original intended use, thereby uncovering valuable novel therapeutic benefits. This approach not only enhances the efficiency of drug development but also opens new avenues for treating complex diseases, providing hope for patients with conditions that currently lack effective therapies.

This concept of drug repositioning is equally applicable to antibody drug discovery, where existing antibodies that bind to different antigens may be redesigned to target specific antigens. This process can be seen as an expansion of traditional specificity design. In this context, Nimrod et al. demonstrated the capability to redesign an existing antibody to target a specific epitope.[180] They utilized rigid-body docking programs ZDOCK[181] and HEX,[182] molecular dynamics simulations, and antibody sequence design calculations through the Discovery Studio suite, followed by screening with a yeast surface display experiment. This study demonstrated the successful engineering of a functional antibody targeting the cytokine interleukin-17A through computational "re-epitoping" of an existing antibody.[180] The crystal structure of the antibody-antigen complex confirms the targeted epitope, validating the accuracy of the computational design. A ML classifier based on a random forest algorithm was also employed to predict specific residue-residue contacts between the antigen and antibody, contributing to the rational design of the functional antibody.

The study by Nimrod et al.[180] utilized a yeast surface display experiment to mature the initial hits of the repurposed antibody. In a more computationally focused study, Tam et al. repurposed nanobodies found in the PDB to bind to ELMO1-RBD, a non-cognate antigen of the original nanobodies.[183] They employed the shape-based, rigid-body docking program PatchDock[184] to select initial binding poses for subsequent design calculations. The initial docking poses were further filtered based on their resemblance to the binding modes of known nanobodies and the quality of their docking energy landscapes. Following this, rigid-body docking and design calculations with Rosetta,[100] based on the selected poses, identified a weakly binding nanobody. This was followed by a purely computational affinity maturation process, which involved rigid-body docking and design calculations with Rosetta using parameters different from those in the initial design step. This process enabled further refinement of the docking poses and sequences. The final selection was based on MM/PBSA calculations through MD simulations and Flex-ddG calculations[185] in Rosetta. The best binder, as computationally identified, was experimentally confirmed to exhibit improved binding affinity toward the target antigen and inhibit the target interaction effectively.

These results emphasize the innovative and impactful nature of the computational repositioning approach, offering new possibilities for the development of targeted antibody therapies. In a potentially related approach, Schneider et al. proposed a deep learning framework, DLAB, for screening existing antibodies.[186] DLAB utilizes a structure-based deep learning approach, specifically a CNN model trained on rigid-body docking decoys of antibody-antigen complex structures. It includes components such as DLAB-Re (rescoring) and DLAB-VS (virtual screening) to improve pose selection in antibody-antigen docking simulations and enable the classification of antibody-antigen pairings as binders or non-binders. Demonstrated to enrich binders against non-binding sequences in realistic scenarios of predicting antibody escape in SARS-CoV-2 variants, DLAB effectively discriminated binding antibodies from non-binders, enhancing the virtual screening process in antibody drug discovery. Although antibody-non-cognate antigen pairs would exhibit poor scores, making it more difficult to discriminate potential binders from non-binders, computational screening approaches like DLAB may still be useful for antibody repositioning studies.

6.4 GEOMETRIC AND COMPUTATIONAL CONSIDERATIONS IN THE DESIGN OF MULTISPECIFIC BIOLOGICS

Multispecific antibodies represent a significant advancement in antibody engineering, designed to target two or more distinct antigens.[187,188] This capability offers a broader therapeutic potential than traditional monospecific therapies. This section explores various antibody formats used in creating multispecific biologics and highlights the role of computational methods in their design.

6.4.1 Antibody Formats in Multispecific Biologics

The modular nature of antibodies enables the creation of various formats beyond traditional IgG structures, illustrating the versatility and innovation in antibody engineering. However, existing bispecific antibody (BsAb) formats encounter limitations, such as altered antibody geometry and the need for extensive engineering. Formats like diabodies, IgG-single-chain F_V (scF_V), and dual-variable domain (DVD)-Ig modify the native structure of antibodies to target multiple antigens, but this may affect stability and solubility. Furthermore, antibody fragments often require substantial modification to stabilize variable domains outside their native context, adding complexity to the design process.

Among various immunotherapeutic agents, bispecific T-cell engagers (BiTEs) are innovative formats that harness the immune system to target cancer.[189] They function by linking two scF_Vs with a peptide linker: one targeting CD3 on T cells and the other targeting tumor cell antigens. This dual binding facilitates an immunological synapse, enabling direct cytotoxic signaling from T cells to tumor cells and resulting in targeted cancer cell destruction. BiTEs, therefore, offer a potent and specific strategy to enhance the immune system's tumor cell recognition and elimination capabilities. A notable example of BiTEs in clinical use is blinatumomab,[190] which targets CD3 on T cells and CD19 on B-cell malignancies. Future advancements of BiTEs may involve substituting the scFv component targeting CD19 with one aimed at different tumor-associated antigens, potentially even employing T-cell receptor (TCR)-like antibodies. While TCRs and antibodies share structural similarities in antigen recognition through their CDRs, the specifics of their interaction mechanisms and domain orientations—V_α/V_β in TCRs and V_L/V_H in antibodies—are distinctively different.[191] This distinction illustrates the complexity of computationally designing TCR-like antibodies and further BiTEs, presenting a challenging yet promising frontier in antibody engineering.

While computational methods could offer insights, their application in multispecific antibody design is nascent. MD simulations with explicit solvent can trace the dynamics of these modified formats, although their use is somewhat constrained by the large sizes of multispecific antibody formats. A noteworthy application is the study by Aertker et al., which utilized all-atom MD simulations to elucidate the *in vitro* pharmacokinetic profiles of a BsAb.[192] In this study, the C-terminal ends of an IgG1 were covalently linked to a single-chain F_V format targeting a distinct antigen via a peptide linker. The computational analysis began with a hypothesized complex between the BsAb and two neonatal Fc receptors (FcRn). This study unveiled interactions between the Fab and scF_V regions of the BsAb with FcRn, potentially influencing its pharmacokinetic behavior. Notably, these atomic contacts were not apparent in the initial model; they emerged only after 300 ns of simulation, underscoring the importance of extended simulations to capture the dynamic interactions that influence the mechanism of action of BsAbs. Such interactions, revealed through all-atom MD simulations, could serve as a foundation for enhancing the pharmacokinetic profiles of BsAbs. The computational design strategies discussed in this chapter could then be applied to modify these interactions without compromising the antibody's antigen recognition capabilities.

Another bispecific format is IgG BsAbs, which provide several advantages over conventional BsAb formats. They maintain the biophysical behavior and pharmacokinetic properties of native IgG, potentially enhancing stability and efficacy. Additionally, IgG BsAbs can be produced by directly combining parental mAbs, simplifying production compared to methods requiring separate expression and recombination. Consequently, IgG BsAbs represent a promising strategy for creating therapeutic molecules with enhanced properties and clinical potential.

However, the production of fully IgG BsAbs introduces its own set of challenges. A primary concern is the highly heterogeneous pairing of heavy and light chains when produced in mammalian cells, as achieving the desired quaternary structure requires specific combinations of these chains. Moreover, fully IgG BsAbs often rely on heterodimeric Fc designs to facilitate the heterodimerization of different antibody heavy chains, further complicating the production process. Additionally, the design of fully IgG BsAbs that maintain desired specificity and affinity for both targets, while also preserving IgG-like pharmacokinetics and effector functions, necessitates precise optimization and validation. Overcoming these obstacles through sophisticated computational design and protein engineering is pivotal for the successful development of fully IgG BsAbs with optimal therapeutic properties.

6.4.2 Computational Multistate Design of Bispecific IgG Antibodies

The challenge of heterodimerization in IgG BsAb production, recognized nearly three decades ago,[193] has been the persistent issue of light chain mispairing. Only recently have significant advancements been made toward resolving this. Within this framework, multistate design (MSD) presents itself as a forward-thinking approach.[129] It aims to simultaneously minimize the free energy across multiple protein conformations, each recognizing different targets, moving beyond the conventional focus on designing proteins for a single, static structure.

The primary goal in the design of IgG BsAb is to engineer an orthogonal Fab interface that promotes heavy-chain heterodimerization while minimizing undesired by-products. This involves modifying a constant domain to facilitate heavy-chain heterodimerization and creating an orthogonal heavy chain-light chain interface to ensure correct pairings. For example, Lewis et al. employed a MSD strategy to re-engineer the C_H1-C_L interface of an IgG1 antibody, utilizing multistate design application in Rosetta.[194] They identified sequences that prefer mutant-mutant pairing over mutant-wild type, leading to parental monoclonal antibodies that, when co-expressed, assemble into IgG BsAb with enhanced heavy chain-light chain pairing. The resulting IgG BsAb retained the pharmacokinetic properties of native IgG while binding to target antigens monovalently.

In another study, Leaver-Fay et al. suggested the explicit inclusion of negative state repertoires during antibody sequence optimization, employing an iterative method that alternates between sequence design and protein docking.[195] Incorporating negative state repertoires allows for the addition of fixed-backbone conformations that simulate real docking trajectories, offering deeper insights into the structural impact of mutations on negative states and leading to more specific and stable bispecific antibodies.

Froning et al. focused on developing novel solutions for the heavy chain-light chain pairing issue through computational and rational engineering methods.[196] Their work in producing and characterizing multiple fully IgG BsAbs underscores the critical role of specificity engineering in both the variable and constant domains for achieving robust heavy chain-light chain specificity across all BsAbs. The collective efforts in employing these advanced computational strategies signify a pivotal shift in optimizing bispecific antibodies, emphasizing the synergy between precision engineering and computational antibody design.

6.5 CONCLUSIONS AND PERSPECTIVES

This chapter focused on antibody sequence generation and optimization, utilizing structure-based simulations and machine learning. While AlphaFold2 has made significant strides in structural biology and drug discovery, accurately predicting antibody structures, particularly the highly variable CDR-H3, remains challenging. The extensive conformational and sequence space of CDR-H3 is not fully represented by existing experimental structures or B-cell repertoire data, highlighting the need for innovative computational or experimental methods to bridge this gap. The current limitations in accurately modeling long CDR-H3 make structure-based simulations challenging.

In the absence of structures, how do we proceed with drug discovery? Advances in B-cell repertoire sequencing have significantly enhanced our understanding of the immune system, impacting research in infectious diseases, vaccine development, and antibody therapeutics. High-throughput technologies have rapidly expanded the sequence data available, allowing the exploration of sequence-function relationships in antibodies. Consequently, optimizing antibody functions based solely on sequence information is becoming a viable approach. However, computationally predicted model structures still play a crucial role in rationalizing the sequences generated and optimized through such sequence-based methods. The concept of "developability" in antibody drug discovery, akin to the "rule of five" for small molecules,[197] further underscores the importance of structural information. Structural features are invaluable in assessing the developability of antibody therapeutics.

While machine learning models present a promising path for antibody sequence generation and optimization, experimental validation is still vital to confirm their effectiveness. These models offer an alternative to structure-based approaches, potentially lowering costs and increasing efficiency. Nevertheless, the functionality and developability of computationally generated antibodies warrant further investigation.

With the evolution of computational antibody design and optimization, several future directions emerge. Miniaturizing antibodies is a logical progression, enhancing their therapeutic applicability. Single-domain antibodies and peptide-based drugs are promising for their potential to penetrate tissues and treat respiratory diseases, with computational technologies playing a crucial role in their development.

In conclusion, this chapter highlighted the synergy between machine learning, structure-based simulations, and high-throughput experimental techniques in advancing antibody engineering. It pointed toward a future where machine-made therapeutics could play a pivotal role in healthcare, demonstrating the evolving landscape of antibody design and its implications for medical science.

ACKNOWLEDGMENTS

The author acknowledges support from the Japan Agency for Medical Research and Development (JP23wm0325047), the Japan Society for the Promotion of Science (JP19H04202 and JP21K18310), and the Okawa Foundation for Information and Telecommunications (20-10).

REFERENCES

1. Makowski EK, Chen H-T, Tessier PM. Simplifying complex antibody engineering using machine learning. *Cell Syst* 2023; 14:667–75. Available from: https://linkinghub.elsevier.com/retrieve/pii/S2405471223001187
2. Mieczkowski C, Zhang X, Lee D, Nguyen K, Lv W, Wang Y, Zhang Y, Way J, Gries J-M. Blueprint for antibody biologics developability. *MAbs* 2023; 15:2185924. Available from: https://www.tandfonline.com/doi/full/10.1080/19420862.2023.2185924
3. Marks C, Deane CM. How repertoire data are changing antibody science. *J Biol Chem* 2020; 295:9823–37. Available from: https://linkinghub.elsevier.com/retrieve/pii/S0021925817489265
4. Wilman W, Wróbel S, Bielska W, Deszynski P, Dudzic P, Jaszczyszyn I, Kaniewski J, Młokosiewicz J, Rouyan A, Satława T, et al. Machine-designed biotherapeutics: opportunities, feasibility and advantages of deep learning in computational antibody discovery. *Brief Bioinform* 2022; 23(4):bbac267. Available from: https://academic.oup.com/bib/article/doi/10.1093/bib/bbac267/6643456
5. Chungyoun MF, Gray JJ. AI models for protein design are driving antibody engineering. *Curr Opin Biomed Eng* 2023; 28:100473. Available from: https://linkinghub.elsevier.com/retrieve/pii/S2468451123000296
6. Kumar S, Singh SK. *Developability of Biotherapeutics.* CRC Press; 2015. Available from: https://www.taylorfrancis.com/books/9781482246155
7. Akbar R, Bashour H, Rawat P, Robert PA, Smorodina E, Cotet T-S, Flem-Karlsen K, Frank R, Mehta BB, Vu MH, et al. Progress and challenges for the machine learning-based design of fit-for-purpose monoclonal antibodies. *MAbs* 2022; 14(1):2008790. Available from: https://www.tandfonline.com/doi/full/10.1080/19420862.2021.2008790
8. Krizhevsky A, Sutskever I, Hinton GE. ImageNet classification with deep convolutional neural networks. In: F. Pereira and C.J. Burges and L. Bottou and K.Q. Weinberger (eds.), *Advances in Neural Information Processing Systems 25*(NIPS 2012).
9. Graves A, Mohamed A, Hinton G. Speech recognition with deep recurrent neural networks. In: *2013 IEEE International Conference on Acoustics, Speech and Signal Processing.* IEEE; 2013. page 6645–9. Available from: https://ieeexplore.ieee.org/document/6638947/
10. He K, Zhang X, Ren S, Sun J. Deep residual learning for image recognition. In: *2016 IEEE Conference on Computer Vision and Pattern Recognition (CVPR).* IEEE; 2016. page 770–8.Available from: https://ieeexplore.ieee.org/document/7780459/
11. Hochreiter S, Schmidhuber J. Long short-term memory. *Neural Comput* 1997; 9:1735–80. Available from: https://direct.mit.edu/neco/article/9/8/1735-1780/6109

12. Goodfellow IJ, Pouget-Abadie J, Mirza M, Xu B, Warde-Farley D, Ozair S, Courville A, Bengio Y. Generative adversarial nets. In: Z. Ghahramani and M. Welling and C. Cortes and N. Lawrence, K.Q. Weinberger (eds.), *Advances in Neural Information Processing Systems 27.* (NIPS 2014).
13. Kingma DP, Welling M. Auto-encoding variational bayes. In: *International Conference on Learning Representations (*ICLR) 2014. 2013. Available from: https://doi.org/10.48550/arXiv.1312.6114
14. Vaswani A, Shazeer N, Parmar N, Uszkoreit J, Jones L, Gomez AN, Kaiser Ł, Polosukhin I. Attention is all you need. In: I. Guyon and U. Von Luxburg and S. Bengio and H. Wallach and R. Fergus and S. Vishwanathan and R. Garnett (eds.), *Advances in Neural Information Processing Systems 30.* (NIPS 2017).
15. Kuroda D, Shirai H, Jacobson MP, Nakamura H. Computer-aided antibody design. Protein Eng Design Select 2012; 25:507–21. Available from: https://doi.org/10.1093/protein/gzs024
16. Childers MC, Daggett V. Molecular dynamics methods for antibody design. In: Kouhei Tsumoto, Daisuke Kuroda (eds.), *Methods in Molecular Biology.* 2023. page 109–24. New York: Humana New York. Available from: https://link.springer.com/10.1007/978-1-0716-2609-2_5
17. Hummer AM, Abanades B, Deane CM. Advances in computational structure-based antibody design. *Curr Opin Struct Biol* 2022; 74:102379. Available from: https://doi.org/10.1016/j.sbi.2022.102379
18. Robert PA, Akbar R, Frank R, Pavlović M, Widrich M, Snapkov I, Slabodkin A, Chernigovskaya M, Scheffer L, Smorodina E, et al. Unconstrained generation of synthetic antibody–antigen structures to guide machine learning methodology for antibody specificity prediction. *Nat Comput Sci* 2022; 2:845–65. Available from: https://www.nature.com/articles/s43588-022-00372-4
19. Weitzner BD, Jeliazkov JR, Lyskov S, Marze N, Kuroda D, Frick R, Adolf-Bryfogle J, Biswas N, Dunbrack RL, Gray JJ. Modeling and docking of antibody structures with Rosetta. *Nat Protoc* 2017; 12:401–16. Available from: https://www.nature.com/doifinder/10.1038/nprot.2016.180
20. Schritt D, Li S, Rozewicki J, Katoh K, Yamashita K, Volkmuth W, Cavet G, Standley DM. Repertoire Builder: high-throughput structural modeling of B and T cell receptors. *Mol Syst Des Eng* 2019; 4:761–8. Available from: https://xlink.rsc.org/?DOI=C9ME00020H
21. Ruffolo JA, Sulam J, Gray JJ. Antibody structure prediction using interpretable deep learning. *Patterns* 2022; 3:100406. Available from: https://doi.org/10.1016/j.patter.2021.100406
22. Abanades B, Georges G, Bujotzek A, Deane CM. ABlooper: fast accurate antibody CDR loop structure prediction with accuracy estimation. *Bioinformatics* 2022; 38:1877–80. Available from: https://academic.oup.com/bioinformatics/article/38/7/1877/6517780
23. Ruffolo JA, Chu L-S, Mahajan SP, Gray JJ. Fast, accurate antibody structure prediction from deep learning on massive set of natural antibodies. *Nat Commun* 2023; 14:2389. Available from: https://www.nature.com/articles/s41467-023-38063-x
24. Abanades B, Wong WK, Boyles F, Georges G, Bujotzek A, Deane CM. ImmuneBuilder: deep-learning models for predicting the structures of immune proteins. *Commun Biol* 2023; 6:575. Available from: https://www.nature.com/articles/s42003-023-04927-7
25. Jin W, Barzilay R, Jaakkola T. Antibody-antigen docking and design via hierarchical structure refinement. In: *Proceedings of the 39th International Conference on Machine Learning. (PMLR)* 2022; 162:10217-27
26. Xu Z, Davila A, Wilamowski J, Teraguchi S, Standley DM. Improved antibody-specific epitope prediction using AlphaFold and AbAdapt. *ChemBioChem* 2022; 23(18):e202200303. Available from: https://chemistry-europe.onlinelibrary.wiley.com/doi/10.1002/cbic.202200303

27. Ambrosetti F, Jandova Z, Bonvin AMJJ. Information-driven antibody–antigen modelling with HADDOCK. In: Kouhei Tsumoto, Daisuke Kuroda (eds.), *Methods in Molecular Biology*. 2023. page 267–82. New York: Humana New York. Available from: https://link.springer.com/10.1007/978-1-0716-2609-2_14
28. Gaudreault F, Corbeil CR, Sulea T. Enhanced antibody-antigen structure prediction from molecular docking using AlphaFold2. *Sci Rep* 2023; 13:15107. Available from: https://www.nature.com/articles/s41598-023-42090-5
29. Liberis E, Veličković P, Sormanni P, Vendruscolo M, Liò P. Parapred: antibody paratope prediction using convolutional and recurrent neural networks. *Bioinformatics* 2018; 34:2944–50. Available from: https://academic.oup.com/bioinformatics/article/34/17/2944/4972995
30. Leem J, Mitchell LS, Farmery JHR, Barton J, Galson JD. Deciphering the language of antibodies using self-supervised learning. *Patterns* 2022; 3:100513. Available from: https://doi.org/10.1016/j.patter.2022.100513
31. Wang M, Cang Z, Wei G-W. A topology-based network tree for the prediction of protein–protein binding affinity changes following mutation. *Nat Mach Intell* 2020; 2:116–23. Available from: https://www.nature.com/articles/s42256-020-0149-6
32. Myung Y, Pires DE V, Ascher DB. mmCSM-AB: guiding rational antibody engineering through multiple point mutations. *Nucleic Acids Res* 2020; 48:W125–31. Available from: https://academic.oup.com/nar/article/48/W1/W125/5841134
33. Kuroda D, Tsumoto K. Engineering stability, viscosity, and immunogenicity of antibodies by computational design. *J Pharm Sci* 2020; 109:1631–51. Available from: https://linkinghub.elsevier.com/retrieve/pii/S0022354920300162
34. Marks C, Hummer AM, Chin M, Deane CM. Humanization of antibodies using a machine learning approach on large-scale repertoire data. *Bioinformatics* 2021; 37:4041–7. Available from: https://academic.oup.com/bioinformatics/article/37/22/4041/6295884
35. Liu X, Luo Y, Li P, Song S, Peng J. Deep geometric representations for modeling effects of mutations on protein-protein binding affinity. *PLoS Comput Biol* 2021; 17:e1009284. Available from: https://dx.plos.org/10.1371/journal.pcbi.1009284
36. Feng J, Jiang M, Shih J, Chai Q. Antibody apparent solubility prediction from sequence by transfer learning. *iScience* 2022; 25:105173. Available from: https://linkinghub.elsevier.com/retrieve/pii/S2589004222014456
37. Yuan Y, Chen Q, Mao J, Li G, Pan X. DG-Affinity: predicting antigen-antibody affinity with language models from sequences. *BMC Bioinform* 2023; 24:430. Available from: https://www.ncbi.nlm.nih.gov/pubmed/37957563
38. Jia L, Sun Y. In silico prediction method for protein asparagine deamidation. In: Kouhei Tsumoto, Daisuke Kuroda (eds.), *Methods in Molecular Biology*. 2023. page 199–217. New York: Humana New York. Available from: https://link.springer.com/10.1007/978-1-0716-2609-2_10
39. Liang S, Zhang C. PITHA: a webtool to predict immunogenicity for humanized and fully human therapeutic antibodies. In: Kouhei Tsumoto, Daisuke Kuroda (eds.), *Methods in Molecular Biology*. 2023. page 143–50. New York: Humana New York. Available from: https://link.springer.com/10.1007/978-1-0716-2609-2_7
40. Ramon A, Ali M, Atkinson M, Saturnino A, Didi K, Visentin C, Ricagno S, Xu X, Greenig M, Sormanni P. Assessing antibody and nanobody nativeness for hit selection and humanization with AbNatiV. *Nat Mach Intell* 2024; 6:74–91. Available from: https://www.nature.com/articles/s42256-023-00778-3
41. Bozhanova NG, Sangha AK, Sevy AM, Gilchuk P, Huang K, Nargi RS, Reidy JX, Trivette A, Carnahan RH, Bukreyev A, et al. Discovery of Marburg virus neutralizing antibodies from virus-naïve human antibody repertoires using large-scale structural predictions. *Proc Natl Acad Sci U S A* 2020; 117:31142–8. Available from: https://pnas.org/doi/full/10.1073/pnas.1922654117

42. Magar R, Yadav P, Barati Farimani A. Potential neutralizing antibodies discovered for novel corona virus using machine learning. *Sci Rep* 2021; 11:5261. Available from: https://www.nature.com/articles/s41598-021-84637-4
43. Zhang J, Du Y, Zhou P, Ding J, Xia S, Wang Q, Chen F, Zhou M, Zhang X, Wang W, et al. Predicting unseen antibodies' neutralizability via adaptive graph neural networks. *Nat Mach Intell* 2022; 4:964–76. Available from: https://www.nature.com/articles/s42256-022-00553-w
44. Bozhanova NG, Flyak AI, Brown BP, Ruiz SE, Salas J, Rho S, Bombardi RG, Myers L, Soto C, Bailey JR, et al. Computational identification of HCV neutralizing antibodies with a common HCDR3 disulfide bond motif in the antibody repertoires of infected individuals. *Nat Commun* 2022; 13:3178. Available from: https://www.nature.com/articles/s41467-022-30865-9
45. Saksena SD, Liu G, Banholzer C, Horny G, Ewert S, Gifford DK. Computational counterselection identifies nonspecific therapeutic biologic candidates. *Cell Rep Methods* 2022; 2:100254. Available from: https://linkinghub.elsevier.com/retrieve/pii/S2667237522001278
46. Vu MH, Akbar R, Robert PA, Swiatczak B, Sandve GK, Greiff V, Haug DTT. Linguistically inspired roadmap for building biologically reliable protein language models. *Nat Mach Intell* 2023; 5:485–96. Available from: https://www.nature.com/articles/s42256-023-00637-1
47. Gomes T, Teichmann SA, Talavera-López C. Immunology driven by large-scale single-cell sequencing. *Trends Immunol* 2019; 40:1011–21. Available from: https://linkinghub.elsevier.com/retrieve/pii/S1471490619301929
48. Morgan D, Tergaonkar V. Unraveling B cell trajectories at single cell resolution. *Trends Immunol* 2022; 43:210–29. Available from: https://linkinghub.elsevier.com/retrieve/pii/S1471490622000035
49. Rawlings DJ, Metzler G, Wray-Dutra M, Jackson SW. Altered B cell signalling in autoimmunity. *Nat Rev Immunol* 2017; 17:421–36. Available from: https://www.nature.com/articles/nri.2017.24
50. Parameswaran P, Liu Y, Roskin KM, Jackson KKL, Dixit VP, Lee J-Y, Artiles KL, Zompi S, Vargas MJ, Simen BB, et al. Convergent antibody signatures in human dengue. *Cell Host Microbe* 2013; 13:691–700. Available from: https://linkinghub.elsevier.com/retrieve/pii/S1931312813001911
51. Jackson KJL, Liu Y, Roskin KM, Glanville J, Hoh RA, Seo K, Marshall EL, Gurley TC, Moody MA, Haynes BF, et al. Human responses to influenza vaccination show seroconversion signatures and convergent antibody rearrangements. *Cell Host Microbe* 2014; 16:105–14. Available from: https://linkinghub.elsevier.com/retrieve/pii/S193131281400184X
52. Galson JD, Schaetzle S, Bashford-Rogers RJM, Raybould MIJ, Kovaltsuk A, Kilpatrick GJ, Minter R, Finch DK, Dias J, James LK, et al. Deep sequencing of B cell receptor repertoires from COVID-19 patients reveals strong convergent immune signatures. *Front Immunol* 2020; 11. Available from: https://www.frontiersin.org/articles/10.3389/fimmu.2020.605170/full
53. Liu R-X, Wen C, Ye W, Li Y, Chen J, Zhang Q, Li W, Liang W, Wei L, Zhang J, et al. Altered B cell immunoglobulin signature exhibits potential diagnostic values in human colorectal cancer. *iScience* 2023; 26:106140. Available from: https://linkinghub.elsevier.com/retrieve/pii/S2589004223002171
54. Galson JD, Trück J, Fowler A, Clutterbuck EA, Münz M, Cerundolo V, Reinhard C, van der Most R, Pollard AJ, Lunter G, et al. Analysis of B cell repertoire dynamics following hepatitis B vaccination in humans, and enrichment of vaccine-specific antibody sequences. *EBioMedicine* 2015; 2:2070–9. Available from: https://linkinghub.elsevier.com/retrieve/pii/S2352396415302176

55. Forgacs D, Abreu RB, Sautto GA, Kirchenbaum GA, Drabek E, Williamson KS, Kim D, Emerling DE, Ross TM. Convergent antibody evolution and clonotype expansion following influenza virus vaccination. *PLoS One* 2021; 16:e0247253. Available from: https://dx.plos.org/10.1371/journal.pone.0247253
56. Kwong PD, DeKosky BJ, Ulmer JB. Antibody-guided structure-based vaccines. *Semin Immunol* 2020; 50:101428. Available from: https://linkinghub.elsevier.com/retrieve/pii/S1044532320300440
57. Cao Y, Su B, Guo X, Sun W, Deng Y, Bao L, Zhu Q, Zhang X, Zheng Y, Geng C, et al. Potent neutralizing antibodies against SARS-CoV-2 identified by high-throughput single-cell sequencing of convalescent patients' B cells. *Cell* 2020; 182:73–84.e16. Available from: https://linkinghub.elsevier.com/retrieve/pii/S0092867420306206
58. Yaari G, Kleinstein SH. Practical guidelines for B-cell receptor repertoire sequencing analysis. *Genome Med* 2015; 7:121. Available from: https://genomemedicine.biomedcentral.com/articles/10.1186/s13073-015-0243-2
59. Hederman AP, Ackerman ME. Leveraging deep learning to improve vaccine design. *Trends Immunol* 2023; 44:333–44. Available from: https://linkinghub.elsevier.com/retrieve/pii/S1471490623000467
60. Horst A, Smakaj E, Natali EN, Tosoni D, Babrak LM, Meier P, Miho E. Machine learning detects anti-DENV signatures in antibody repertoire sequences. *Front Artif Intell* 2021; 4. Available from: https://www.frontiersin.org/articles/10.3389/frai.2021.715462/full
61. Ostmeyer J, Christley S, Rounds WH, Toby I, Greenberg BM, Monson NL, Cowell LG. Statistical classifiers for diagnosing disease from immune repertoires: a case study using multiple sclerosis. *BMC Bioinform* 2017; 18:401. Available from: https://bmcbioinformatics.biomedcentral.com/articles/10.1186/s12859-017-1814-6
62. Yang KK, Wu Z, Bedbrook CN, Arnold FH. Learned protein embeddings for machine learning. *Bioinformatics* 2018; 34:2642–8. Available from: https://academic.oup.com/bioinformatics/article/34/15/2642/4951834
63. Wang M, Patsenker J, Li H, Kluger Y, Kleinstein SH. Language model-based B cell receptor sequence embeddings can effectively encode receptor specificity. *Nucleic Acids Res* 2024; 52:548–57. Available from: https://doi.org/10.1093/nar/gkad1128
64. Ostrovsky-Berman M, Frankel B, Polak P, Yaari G. Immune2vec: embedding B/T cell receptor sequences in RN using natural language processing. *Front Immunol* 2021; 12:680687. Available from: https://www.frontiersin.org/articles/10.3389/fimmu.2021.680687/full
65. Lin Z, Akin H, Rao R, Hie B, Zhu Z, Lu W, Smetanin N, Verkuil R, Kabeli O, Shmueli Y, et al. Evolutionary-scale prediction of atomic-level protein structure with a language model. *Science (1979)* 2023; 379:1123–30. Available from: https://www.science.org/doi/10.1126/science.ade2574
66. Elnaggar A, Heinzinger M, Dallago C, Rehawi G, Wang Y, Jones L, Gibbs T, Feher T, Angerer C, Steinegger M, et al. ProtTrans: toward understanding the language of life through self-supervised learning. *IEEE Trans Pattern Anal Mach Intell* 2022; 44:7112–27. Available from: https://ieeexplore.ieee.org/document/9477085/
67. Ruffolo JA, Gray JJ, Sulam J. Deciphering antibody affinity maturation with language models and weakly supervised learning. In: *Machine Learning for Structural Biology Workshop* at *NeurIPS 2021*. 2021. Available from: https://doi.org/10.48550/arXiv.2112.07782
68. Wu R, Ding F, Wang R, Shen R, Zhang X, Luo S, Su C, Wu Z, Xie Q, Berger B, et al. High-resolution de novo structure prediction from primary sequence. *bioRxiv* 2022. Available from: https://doi.org/10.1101/2022.07.21.500999
69. Wang W, Peng Z, Yang J. Single-sequence protein structure prediction using supervised transformer protein language models. *Nat Comput Sci* 2022; 2:804–14. Available from: https://www.nature.com/articles/s43588-022-00373-3

70. Moussad B, Roche R, Bhattacharya D. The transformative power of transformers in protein structure prediction. *Proc Natl Acad Sci U S A* 2023; 120(32):e2303499120. Available from: https://pnas.org/doi/10.1073/pnas.2303499120
71. Zhu Y-H, Zhang C, Yu D-J, Zhang Y. Integrating unsupervised language model with triplet neural networks for protein gene ontology prediction. *PLoS Comput Biol* 2022; 18:e1010793. Available from: https://dx.plos.org/10.1371/journal.pcbi.1010793
72. Yuan Q, Xie J, Xie J, Zhao H, Yang Y. Fast and accurate protein function prediction from sequence through pretrained language model and homology-based label diffusion. *Brief Bioinform* 2023; 24(3): bbad117. Available from: https://academic.oup.com/bib/article/doi/10.1093/bib/bbad117/7085635
73. Yamaguchi H, Saito Y. Evotuning protocols for Transformer-based variant effect prediction on multi-domain proteins. *Brief Bioinform* 2021; 22(6):bbab234. Available from: https://academic.oup.com/bib/article/doi/10.1093/bib/bbab234/6309928
74. Jiang TT, Fang L, Wang K. Deciphering "the language of nature": a transformer-based language model for deleterious mutations in proteins. *Innovation* 2023; 4:100487. Available from: https://linkinghub.elsevier.com/retrieve/pii/S2666675823001157
75. Amimeur T, Shaver JM, Ketchem RR, Taylor JA, Clark RH, Smith J, Citters D Van, Siska CC, Smidt P, Sprague M, et al. Designing feature-controlled humanoid antibody discovery libraries using generative adversarial networks. *bioRxiv* 2020. Available from: https://doi.org/10.1101/2020.04.12.024844
76. McCoy LE, Rutten L, Frampton D, Anderson I, Granger L, Bashford-Rogers R, Dekkers G, Strokappe NM, Seaman MS, Koh W, et al. Molecular evolution of broadly neutralizing llama antibodies to the CD4-binding site of HIV-1. *PLoS Pathog* 2014; 10:e1004552. Available from: https://dx.plos.org/10.1371/journal.ppat.1004552
77. Shin J-E, Riesselman AJ, Kollasch AW, McMahon C, Simon E, Sander C, Manglik A, Kruse AC, Marks DS. Protein design and variant prediction using autoregressive generative models. *Nat Commun* 2021; 12:2403. Available from: https://www.nature.com/articles/s41467-021-22732-w
78. Xu X, Xu T, Zhou J, Liao X, Zhang R, Wang Y, Zhang L, Gao X. AB-Gen: antibody library design with generative pre-trained transformer and deep reinforcement learning. *Genomics Proteomics Bioinform* 2023; 21(5):1043–53. Available from: https://linkinghub.elsevier.com/retrieve/pii/S167202292300092X
79. Olsen TH, Boyles F, Deane CM. Observed antibody space: a diverse database of cleaned, annotated, and translated unpaired and paired antibody sequences. *Protein Sci* 2022; 31:141–6. Available from: https://onlinelibrary.wiley.com/doi/10.1002/pro.4205
80. Olsen TH, Moal IH, Deane CM. AbLang: an antibody language model for completing antibody sequences. *Bioinform Adv* 2022; 2(1):vbac046. Available from: https://doi.org/10.1093/bioadv/vbac046
81. Rives A, Meier J, Sercu T, Goyal S, Lin Z, Liu J, Guo D, Ott M, Zitnick CL, Ma J, et al. Biological structure and function emerge from scaling unsupervised learning to 250 million protein sequences. *Proc Natl Acad Sci U S A* 2021; 118(15):e2016239118. Available from: https://pnas.org/doi/full/10.1073/pnas.2016239118
82. Shuai RW, Ruffolo JA, Gray JJ. IgLM: infilling language modeling for antibody sequence design. *Cell Syst* 2023; 14:979–989.e4. Available from: https://linkinghub.elsevier.com/retrieve/pii/S2405471223002715
83. Nijkamp E, Ruffolo JA, Weinstein EN, Naik N, Madani A. ProGen2: exploring the boundaries of protein language models. *Cell Syst* 2023; 14:968–978.e3. Available from: https://linkinghub.elsevier.com/retrieve/pii/S2405471223002727

84. Chennamsetty N, Voynov V, Kayser V, Helk B, Trout BL. Design of therapeutic proteins with enhanced stability. *Proc Natl Acad Sci U S A* 2009; 106:11937–42. Available from: https://www.pnas.org/content/106/29/11937.short
85. Sormanni P, Aprile FA, Vendruscolo M. The CamSol method of rational design of protein mutants with enhanced solubility. *J Mol Biol* 2015; 427:478–90. Available from: https://linkinghub.elsevier.com/retrieve/pii/S0022283614005312
86. Eguchi RR, Choe CA, Huang P-S. Ig-VAE: generative modeling of protein structure by direct 3D coordinate generation. *PLoS Comput Biol* 2022; 18:e1010271. Available from: https://dx.plos.org/10.1371/journal.pcbi.1010271
87. Ferdous S, Martin ACR. AbDb: antibody structure database—a database of PDB-derived antibody structures. *Database* 2018; 2018:bay040. Available from: https://academic.oup.com/database/article/doi/10.1093/database/bay040/4989324
88. Berman H, Henrick K, Nakamura H, Markley JL. The worldwide Protein Data Bank (wwPDB): ensuring a single, uniform archive of PDB data. *Nucleic Acids Res* 2007; 35:D301–3. Available from: https://academic.oup.com/nar/article-lookup/doi/10.1093/nar/gkl971
89. Sirin S, Apgar JR, Bennett EM, Keating AE. AB-Bind: antibody binding mutational database for computational affinity predictions. *Protein Sci* 2016; 25:393–409. Available from: https://doi.org/10.1002/pro.2829
90. Jain T, Sun T, Durand S, Hall A, Houston NR, Nett JH, Sharkey B, Bobrowicz B, Caffry I, Yu Y, et al. Biophysical properties of the clinical-stage antibody landscape. *Proc Natl Acad Sci U S A* 2017; 114:944–9. Available from: https://www.pnas.org/lookup/doi/10.1073/pnas.1616408114
91. Jankauskaitė J, Jiménez-García B, Dapkūnas J, Fernández-Recio J, Moal IH. SKEMPI 2.0: an updated benchmark of changes in protein–protein binding energy, kinetics and thermodynamics upon mutation. *Bioinformatics* 2019; 35:462–9. Available from: https://academic.oup.com/bioinformatics/article/35/3/462/5055583
92. Okamoto Y. Generalized-ensemble algorithms: enhanced sampling techniques for Monte Carlo and molecular dynamics simulations. *J Mol Graph Model* 2004; 22:425–39. Available from: https://linkinghub.elsevier.com/retrieve/pii/S1093326303001980
93. Ponder JW, Case DA. Force fields for protein simulations. In: *Advances in Protein Chemistry*. 2003. page 27–85.Available from: https://linkinghub.elsevier.com/retrieve/pii/S006532330366002X
94. Mackerell AD. Empirical force fields for biological macromolecules: overview and issues. *J Comput Chem* 2004; 25:1584–604. Available from: https://onlinelibrary.wiley.com/doi/10.1002/jcc.20082
95. Dauparas J, Anishchenko I, Bennett N, Bai H, Ragotte RJ, Milles LF, Wicky BIM, Courbet A, de Haas RJ, Bethel N, et al. Robust deep learning–based protein sequence design using ProteinMPNN. *Science (1979)* 2022; 378:49–56. Available from: https://www.science.org/doi/10.1126/science.add2187
96. Bekker G-J, Kamiya N. Thermal stability estimation of single domain antibodies using molecular dynamics simulations. In: Kouhei Tsumoto, Daisuke Kuroda (eds.), *Methods in Molecular Biology*. 2023. page 151–63. New York: Humana New York. Available from: https://link.springer.com/10.1007/978-1-0716-2609-2_8
97. Zhang C, Dalby PA. Assessing and engineering antibody stability using experimental and computational methods. In: Kouhei Tsumoto, Daisuke Kuroda (eds.), *Methods in Molecular Biology*. 2023. page 165–97. New York: Humana New York. Available from: https://link.springer.com/10.1007/978-1-0716-2609-2_9

98. Sulea T, Deprez C, Corbeil CR, Purisima EO. Optimizing antibody–antigen binding affinities with the ADAPT platform. In: Kouhei Tsumoto, Daisuke Kuroda (eds.), *Methods in Molecular Biology*. 2023. page 361–74. New York: Humana New York. Available from: https://link.springer.com/10.1007/978-1-0716-2609-2_20
99. Medagli B, Soler MA, De Zorzi R, Fortuna S. Antibody affinity maturation using computational methods: from an initial hit to small-scale expression of optimized binders. In: Kouhei Tsumoto, Daisuke Kuroda (eds.), *Methods in Molecular Biology*. 2023. page 333–59. New York: Humana New York. Available from: https://link.springer.com/10.1007/978-1-0716-2609-2_19
100. Leman JK, Weitzner BD, Lewis SM, Adolf-Bryfogle J, Alam N, Alford RF, Aprahamian M, Baker D, Barlow KA, Barth P, et al. Macromolecular modeling and design in Rosetta: recent methods and frameworks. *Nat Methods* 2020; 17:665–80. Available from: https://www.nature.com/articles/s41592-020-0848-2
101. Guerois R, Nielsen JE, Serrano L. Predicting changes in the stability of proteins and protein complexes: a study of more than 1000 mutations. *J Mol Biol* 2002; 320:369–87. Available from: https://linkinghub.elsevier.com/retrieve/pii/S0022283602004424
102. Ikeuchi E, Kuroda D, Nakakido M, Murakami A, Tsumoto K. Delicate balance among thermal stability, binding affinity, and conformational space explored by single-domain VHH antibodies. *Sci Rep* 2021; 11:20624. Available from: https://doi.org/10.1038/s41598-021-98977-8
103. Schueler-Furman O, Wodak SJ. Computational approaches to investigating allostery. *Curr Opin Struct Biol* 2016; 41:159–71. Available from: https://linkinghub.elsevier.com/retrieve/pii/S0959440X16300744
104. Tomimoto Y, Yamazaki R, Shirai H. Increasing the melting temperature of VHH with the in silico free energy score. *Sci Rep* 2023; 13:4922. Available from: https://www.nature.com/articles/s41598-023-32022-8
105. Okazaki K, Kobashigawa Y, Morita H, Yamauchi S, Fukuda N, Liu C, Toyota Y, Sato T, Morioka H. Molecular dynamics-based design and biophysical evaluation of thermostable single-chain Fv antibody mutants derived from pharmaceutical antibodies. *ACS Omega* 2023; 8:22945–54. Available from: https://pubs.acs.org/doi/10.1021/acsomega.3c01948
106. Almagro JC, Fransson J. Humanization of antibodies. *Front Biosci* 2008; 13:1619–33. Available from: https://www.ncbi.nlm.nih.gov/pubmed/17981654
107. Makowski EK, Kinnunen PC, Huang J, Wu L, Smith MD, Wang T, Desai AA, Streu CN, Zhang Y, Zupancic JM, et al. Co-optimization of therapeutic antibody affinity and specificity using machine learning models that generalize to novel mutational space. *Nat Commun* 2022; 13:3788. Available from: https://www.nature.com/articles/s41467-022-31457-3
108. Prihoda D, Maamary J, Waight A, Juan V, Fayadat-Dilman L, Svozil D, Bitton DA. BioPhi: a platform for antibody design, humanization, and humanness evaluation based on natural antibody repertoires and deep learning. *MAbs* 2022; 14:2020203. Available from: https://www.tandfonline.com/doi/full/10.1080/19420862.2021.2020203
109. Devlin J, Chang MW, Lee K, Toutanova K. BERT: pre-training of deep bidirectional transformers for language understanding. In: *Proceedings of the 2019 Conference of the North American Chapter of the Association for Computational Linguistics: Human Language Technologies (NAACL HLT 2019)*. 2019. Kerrville, TX: Association for Computational Linguistics. Available from: https://doi.org/10.48550/arXiv.1810.04805
110. Lai P-K, Gallegos A, Mody N, Sathish HA, Trout BL. Machine learning prediction of antibody aggregation and viscosity for high concentration formulation development of protein therapeutics. *MAbs* 2022; 14:2026208. Available from: https://www.tandfonline.com/doi/full/10.1080/19420862.2022.2026208

111. Harmalkar A, Rao R, Richard Xie Y, Honer J, Deisting W, Anlahr J, Hoenig A, Czwikla J, Sienz-Widmann E, Rau D, et al. Toward generalizable prediction of antibody thermostability using machine learning on sequence and structure features. *MAbs* 2023; 15:2163584. Available from: https://www.tandfonline.com/doi/full/10.1080/19420862.2022.2163584
112. Harmalkar A, Gray JJ. Advances to tackle backbone flexibility in protein docking. *Curr Opin Struct Biol* 2021; 67:178–86. Available from: https://linkinghub.elsevier.com/retrieve/pii/S0959440X20302141
113. Mahajan SP, Meksiriporn B, Waraho-Zhmayev D, Weyant KB, Kocer I, Butler DC, Messer A, Escobedo FA, DeLisa MP. Computational affinity maturation of camelid single-domain intrabodies against the nonamyloid component of alpha-synuclein. *Sci Rep* 2018; 8:17611. Available from: https://www.nature.com/articles/s41598-018-35464-7
114. Cannon DA, Shan L, Du Q, Shirinian L, Rickert KW, Rosenthal KL, Korade M, van Vlerken-Ysla LE, Buchanan A, Vaughan TJ, et al. Experimentally guided computational antibody affinity maturation with de novo docking, modelling and rational design. *PLoS Comput Biol* 2019; 15:e1006980. Available from: https://dx.plos.org/10.1371/journal.pcbi.1006980
115. Jumper J, Evans R, Pritzel A, Green T, Figurnov M, Ronneberger O, Tunyasuvunakool K, Bates R, Žídek A, Potapenko A, et al. Highly accurate protein structure prediction with AlphaFold. *Nature* 2021; 596:583–589. Available from: https://doi.org/10.1038/s41586-021-03819-2
116. Lensink MF, Brysbaert G, Raouraoua N, Bates PA, Giulini M, Honorato R V., van Noort C, Teixeira JMC, Bonvin AMJJ, Kong R, et al. Impact of AlphaFold on structure prediction of protein complexes: the CASP15-CAPRI experiment. *Proteins Struct Funct Bioinform* 2023; 91:1658–83. Available from: https://onlinelibrary.wiley.com/doi/10.1002/prot.26609
117. Google DeepMind AlphaFold Team, Isomorphic Labs Team. Performance and structural coverage of the latest, in-development AlphaFold model. 2023 [cited 2024 Feb 19]. Available from: https://storage.googleapis.com/deepmind-media/DeepMind.com/Blog/a-glimpse-of-the-next-generation-of-alphafold/alphafold_latest_oct2023.pdf
118. Guest JD, Vreven T, Zhou J, Moal I, Jeliazkov JR, Gray JJ, Weng Z, Pierce BG. An expanded benchmark for antibody-antigen docking and affinity prediction reveals insights into antibody recognition determinants. *Structure* 2021; 29:606–621.e5. Available from: https://linkinghub.elsevier.com/retrieve/pii/S0969212621000058
119. Setliff I, Shiakolas AR, Pilewski KA, Murji AA, Mapengo RE, Janowska K, Richardson S, Oosthuysen C, Raju N, Ronsard L, et al. High-throughput mapping of B cell receptor sequences to antigen specificity. *Cell* 2019; 179:1636–1646.e15. Available from: https://linkinghub.elsevier.com/retrieve/pii/S0092867419312243
120. Harmalkar A, Lyskov S, Gray JJ. Reliable protein-protein docking with AlphaFold, Rosetta, and replica-exchange. *Elife* 2024. Available from: https://doi.org/10.7554/eLife.94029.1
121. Farady CJ, Sellers BD, Jacobson MP, Craik CS. Improving the species cross-reactivity of an antibody using computational design. *Bioorg Med Chem Lett* 2009; 19:3744–7.
122. Lippow SM, Wittrup KD, Tidor B. Computational design of antibody-affinity improvement beyond in vivo maturation. *Nat Biotechnol* 2007; 25:1171–6. Available from: https://www.nature.com/doifinder/10.1038/nbt1336
123. Clark LA, Boriack-Sjodin PA, Eldredge J, Fitch C, Friedman B, Hanf KJM, Jarpe M, Liparoto SF, Li Y, Lugovskoy A, et al. Affinity enhancement of an in vivo matured therapeutic antibody using structure-based computational design. *Protein Sci* 2006; 15:949–60. Available from: https://doi.wiley.com/10.1110/ps.052030506
124. Kiyoshi M, Caaveiro JMM, Miura E, Nagatoishi S, Nakakido M, Soga S, Shirai H, Kawabata S, Tsumoto K. Affinity improvement of a therapeutic antibody by structure-based computational design: generation of electrostatic interactions in the transition state stabilizes the antibody-antigen complex. *PLoS One* 2014; 9:e87099. Available from: https://doi.org/10.1371/journal.pone.0087099

125. Willis JR, Sapparapu G, Murrell S, Julien JP, Singh V, King HG, Xia Y, Pickens JA, La Branche CC, Slaughter JC, et al. Redesigned HIV antibodies exhibit enhanced neutralizing potency and breadth. *J Clin Invest* 2015; 125:2523–31. Available from: https://doi.org/10.1172/JCI80693
126. Barderas R, Desmet J, Timmerman P, Meloen R, Casal JI. Affinity maturation of antibodies assisted by in silico modeling. *Proc Nat Acad Sci U S A* 2008; 105:9029–34. Available from: https://www.pnas.org/content/105/26/9029.abstract
127. Fukunaga A, Tsumoto K. Improving the affinity of an antibody for its antigen via long-range electrostatic interactions. *Protein Eng Design Select* 2013; 26:773–80. Available from: https://doi.org/10.1093/protein/gzt053
128. Yanaka S, Moriwaki Y, Tsumoto K, Sugase K. Elucidation of potential sites for antibody engineering by fluctuation editing. *Sci Rep* 2017; 7:9597. Available from: https://www.nature.com/articles/s41598-017-10246-9
129. Davey JA, Chica RA. Multistate approaches in computational protein design. *Protein Sci* 2012; 21:1241–52. Available from: https://onlinelibrary.wiley.com/doi/10.1002/pro.2128
130. Sevy AM, Jacobs TM, Crowe JE, Meiler J. Design of protein multi-specificity using an independent sequence search reduces the barrier to low energy sequences. *PLoS Comput Biol* 2015; 11:e1004300. Available from: https://dx.plos.org/10.1371/journal.pcbi.1004300
131. Sevy AM, Wu NC, Gilchuk IM, Parrish EH, Burger S, Yousif D, Nagel MBM, Schey KL, Wilson IA, Crowe JE, et al. Multistate design of influenza antibodies improves affinity and breadth against seasonal viruses. *Proc Natl Acad Sci U S A* 2019; 116:1597–602. Available from: https://pnas.org/doi/full/10.1073/pnas.1806004116
132. Lawrence MC, Colman PM. Shape complementarity at protein/protein interfaces. *J Mol Biol* 1993; 234:946–50. Available from: https://www.ncbi.nlm.nih.gov/pubmed/8263940
133. Cohen GH, Silverton EW, Padlan EA, Dyda F, Wibbenmeyer JA, Willson RC, Davies DR. Water molecules in the antibody–antigen interface of the structure of the Fab HyHEL-5–lysozyme complex at 1.7 Å resolution: comparison with results from isothermal titration calorimetry. *Acta Crystallogr D Biol Crystallogr* 2005; 61:628–33. Available from: https://scripts.iucr.org/cgi-bin/paper?S0907444905007870
134. Kuroda D, Gray JJ. Shape complementarity and hydrogen bond preferences in protein-protein interfaces: implications for antibody modeling and protein-protein docking. *Bioinformatics* 2016; 32:2451–6. Available from: https://doi.org/10.1093/bioinformatics/btw197
135. Yoshida K, Kuroda D, Kiyoshi M, Nakakido M, Nagatoishi S, Soga S, Shirai H, Tsumoto K. Exploring designability of electrostatic complementarity at an antigen-antibody interface directed by mutagenesis, biophysical analysis, and molecular dynamics simulations. *Sci Rep* 2019; 9:4482. Available from: https://www.nature.com/articles/s41598-019-40461-5
136. Holt GT, Gorman J, Wang S, Lowegard AU, Zhang B, Liu T, Lin BC, Louder MK, Frenkel MS, McKee K, et al. Improved HIV-1 neutralization breadth and potency of V2-apex antibodies by in silico design. *Cell Rep* 2023; 42:112711. Available from: https://doi.org/10.1016/j.celrep.2023.112711
137. Ozawa T, Ikeda Y, Chen L, Suzuki R, Hoshino A, Noguchi A, Kita S, Anraku Y, Igarashi E, Saga Y, et al. Rational in silico design identifies two mutations that restore UT28K SARS-CoV-2 monoclonal antibody activity against Omicron BA.1. *Structure* 2024; 32(3):263–272.e7. Available from: https://doi.org/10.1016/j.str.2023.12.013
138. Moriyama S, Anraku Y, Taminishi S, Adachi Y, Kuroda D, Kita S, Higuchi Y, Kirita Y, Kotaki R, Tonouchi K, et al. Structural delineation and computational design of SARS-CoV-2-neutralizing antibodies against Omicron subvariants. *Nat Commun* 2023; 14:4198. Available from: https://www.nature.com/articles/s41467-023-39890-8

139. Maguire JB, Haddox HK, Strickland D, Halabiya SF, Coventry B, Griffin JR, Pulavarti SVSRK, Cummins M, Thieker DF, Klavins E, et al. Perturbing the energy landscape for improved packing during computational protein design. *Proteins Struct FunctBioinform* 2021; 89:436–49. Available from: https://onlinelibrary.wiley.com/doi/10.1002/prot.26030
140. Jeong B-S, Cha JS, Hwang I, Kim U, Adolf-Bryfogle J, Coventry B, Cho H-S, Kim K-D, Oh B-H. Computational design of a neutralizing antibody with picomolar binding affinity for all concerning SARS-CoV-2 variants. *MAbs* 2022; 14:2021601. Available from: https://www.tandfonline.com/doi/full/10.1080/19420862.2021.2021601
141. Wang T, Wang L, Zhang X, Shen C, Zhang O, Wang J, Wu J, Jin R, Zhou D, Chen S, et al. Comprehensive assessment of protein loop modeling programs on large-scale datasets: prediction accuracy and efficiency. *Brief Bioinform* 2024; 25:1–14. Available from: https://academic.oup.com/bib/article/doi/10.1093/bib/bbad486/7505239
142. Adolf-Bryfogle J, Kalyuzhniy O, Kubitz M, Weitzner BD, Hu X, Adachi Y, Schief WR, Dunbrack RL. RosettaAntibodyDesign (RAbD): a general framework for computational antibody design. *PLoS Comput Biol* 2018; 14:e1006112. Available from: https://www.biorxiv.org/content/early/2017/08/31/183350
143. Chothia C, Lesk AM. Canonical structures for the hypervariable regions of immunoglobulins. *J Mol Biol* 1987; 196:901–17. Available from: https://doi.org/10.1016/0022-2836(87)90412-8
144. Hernandez NE, Jankowski W, Frick R, Kelow SP, Lubin JH, Simhadri V, Adolf-Bryfogle J, Khare SD, Dunbrack RL, Gray JJ, et al. Computational design of nanomolar-binding antibodies specific to multiple SARS-CoV-2 variants by engineering a specificity switch of antibody 80R using RosettaAntibodyDesign (RAbD) results in potential generalizable therapeutic antibodies for novel SARS-CoV-2 virus. *Heliyon* 2023; 9:e15032. Available from: https://linkinghub.elsevier.com/retrieve/pii/S2405844023022399
145. Giudicelli V. IMGT/LIGM-DB, the IMGT(R) comprehensive database of immunoglobulin and T cell receptor nucleotide sequences. *Nucleic Acids Res* 2006; 34:D781–4. Available from: https://academic.oup.com/nar/article-lookup/doi/10.1093/nar/gkj088
146. Corbeil CR, Manenda MS, Sulea T, Baardsnes J, Picard M-È, Hogues H, Gaudreault F, Deprez C, Shi R, Purisima EO. Redesigning an antibody H3 loop by virtual screening of a small library of human germline-derived sequences. *Sci Rep* 2021; 11:21362. Available from: https://www.nature.com/articles/s41598-021-00669-w
147. Tokuriki N, Stricher F, Serrano L, Tawfik DS. How protein stability and new functions trade off. *PLoS Comput Biol* 2008; 4:e1000002. Available from: https://dx.plos.org/10.1371/journal.pcbi.1000002
148. Warszawski S, Borenstein Katz A, Lipsh R, Khmelnitsky L, Ben Nissan G, Javitt G, Dym O, Unger T, Knop O, Albeck S, et al. Optimizing antibody affinity and stability by the automated design of the variable light-heavy chain interfaces. *PLoS Comput Biol* 2019; 15:e1007207. Available from: https://dx.plos.org/10.1371/journal.pcbi.1007207
149. Akiba H, Tamura H, Caaveiro JMM, Tsumoto K. Computer-guided library generation applied to the optimization of single-domain antibodies. *Protein Eng Design Selec* 2019; 32:423–31. Available from: https://academic.oup.com/peds/article/32/9/423/5804847
150. Shan S, Luo S, Yang Z, Hong J, Su Y, Ding F, Fu L, Li C, Chen P, Ma J, et al. Deep learning guided optimization of human antibody against SARS-CoV-2 variants with broad neutralization. *Proc Natl Acad Sci U S A* 2022; 119(11):e2122954119. Available from: https://doi.org/10.1073/pnas.2122954119
151. Sulea T, Vivcharuk V, Corbeil CR, Deprez C, Purisima EO. Assessment of solvated interaction energy function for ranking antibody-antigen binding affinities. *J Chem Inf Model* 2016; 56:1292–303. Available from: https://doi.org/10.1021/acs.jcim.6b00043

152. Kurumida Y, Saito Y, Kameda T. Predicting antibody affinity changes upon mutations by combining multiple predictors. *Sci Rep* 2020; 10:19533. Available from: https://doi.org/10.1038/s41598-020-76369-8
153. Kang Y, Leng D, Guo J, Pan L. Sequence-based deep learning antibody design for in silico antibody affinity maturation. *arXiv* 2021. Available from: https://arxiv.org/abs/2103.03724
154. Rajpal A, Beyaz N, Haber L, Cappuccilli G, Yee H, Bhatt RR, Takeuchi T, Lerner RA, Crea R. A general method for greatly improving the affinity of antibodies by using combinatorial libraries. *Proc Natl Acad Sci U S A* 2005; 102:8466–71. Available from: https://pnas.org/doi/full/10.1073/pnas.0503543102
155. Yang KK, Wu Z, Arnold FH. Machine-learning-guided directed evolution for protein engineering. *Nat Methods* 2019; 16:687–94. Available from: https://www.nature.com/articles/s41592-019-0496-6
156. Liu G, Zeng H, Mueller J, Carter B, Wang Z, Schilz J, Horny G, Birnbaum ME, Ewert S, Gifford DK. Antibody complementarity determining region design using high-capacity machine learning. *Bioinformatics* 2020; 36:2126–33. Available from: https://academic.oup.com/bioinformatics/article/36/7/2126/5645171
157. Saka K, Kakuzaki T, Metsugi S, Kashiwagi D, Yoshida K, Wada M, Tsunoda H, Teramoto R. Antibody design using LSTM based deep generative model from phage display library for affinity maturation. *Sci Rep* 2021; 11:5852. Available from: https://doi.org/10.1038/s41598-021-85274-7
158. Parkinson J, Hard R, Wang W. The RESP AI model accelerates the identification of tight-binding antibodies. *Nat Commun* 2023; 14:454. Available from: https://www.nature.com/articles/s41467-023-36028-8
159. Alley EC, Khimulya G, Biswas S, AlQuraishi M, Church GM. Unified rational protein engineering with sequence-based deep representation learning. *Nat Methods* 2019; 16:1315–22. Available from: https://www.nature.com/articles/s41592-019-0598-1
160. Harvey EP, Shin J-E, Skiba MA, Nemeth GR, Hurley JD, Wellner A, Shaw AY, Miranda VG, Min JK, Liu CC, et al. An in silico method to assess antibody fragment polyreactivity. *Nat Commun* 2022; 13:7554. Available from: https://www.nature.com/articles/s41467-022-35276-4
161. Fowler DM, Fields S. Deep mutational scanning: a new style of protein science. *Nat Methods* 2014; 11:801–7. Available from: https://www.ncbi.nlm.nih.gov/pubmed/25075907
162. Mason DM, Friedensohn S, Weber CR, Jordi C, Wagner B, Meng SM, Ehling RA, Bonati L, Dahinden J, Gainza P, et al. Optimization of therapeutic antibodies by predicting antigen specificity from antibody sequence via deep learning. *Nat Biomed Eng* 2021; 5:600–612. Available from: https://doi.org/10.1038/s41551-021-00699-9
163. Li L, Gupta E, Spaeth J, Shing L, Jaimes R, Engelhart E, Lopez R, Caceres RS, Bepler T, Walsh ME. Machine learning optimization of candidate antibody yields highly diverse sub-nanomolar affinity antibody libraries. *Nat Commun* 2023; 14:3454. Available from: https://www.nature.com/articles/s41467-023-39022-2
164. Hie BL, Shanker VR, Xu D, Bruun TUJ, Weidenbacher PA, Tang S, Wu W, Pak JE, Kim PS. Efficient evolution of human antibodies from general protein language models. *Nat Biotechnol* 2024; 42:275–83. Available from: https://www.nature.com/articles/s41587-023-01763-2
165. Akbar R, Robert PA, Weber CR, Widrich M, Frank R, Pavlović M, Scheffer L, Chernigovskaya M, Snapkov I, Slabodkin A, et al. In silico proof of principle of machine learning-based antibody design at unconstrained scale. *MAbs* 2022; 14:2031482. Available from: https://doi.org/10.1080/19420862.2022.2031482

166. Zhao W, Luo X, Tong F, Zheng X, Li J, Zhao G, Zhao D. Improving antibody optimization ability of generative adversarial network through large language model. *Comput Struct Biotechnol J* 2023; 21:5839–50. Available from: https://linkinghub.elsevier.com/retrieve/pii/S2001037023004555
167. Lim YW, Adler AS, Johnson DS. Predicting antibody binders and generating synthetic antibodies using deep learning. *MAbs* 2022; 14:2069075. Available from: https://www.tandfonline.com/doi/full/10.1080/19420862.2022.2069075
168. Mahajan SP, Ruffolo JA, Frick R, Gray JJ. Hallucinating structure-conditioned antibody libraries for target-specific binders. *Front Immunol* 2022; 13:999034. Available from: https://www.frontiersin.org/articles/10.3389/fimmu.2022.999034/full
169. Khan A, Cowen-Rivers AI, Grosnit A, Deik D-G-X, Robert PA, Greiff V, Smorodina E, Rawat P, Akbar R, Dreczkowski K, et al. Toward real-world automated antibody design with combinatorial Bayesian optimization. *Cell Rep Methods* 2023; 3:100374. Available from: https://linkinghub.elsevier.com/retrieve/pii/S2667237522002764
170. Wang Y, Wang B, Shi T, Fu J, Kong H, Biomap YZ, Zhang Z. Sample-efficient antibody design through protein language model for risk-aware batch Bayesian optimization. In: *NeurIPS 2023 Workshop: New Frontiers of AI for Drug Discovery and Development* 2023. Available from: https://doi.org/10.1101/2023.11.06.565922
171. Makowski EK, Wu L, Desai AA, Tessier PM. Highly sensitive detection of antibody non-specific interactions using flow cytometry. *MAbs* 2021; 13:1951426. Available from: https://www.tandfonline.com/doi/full/10.1080/19420862.2021.1951426
172. Guo Y, Chen K, Kwong PD, Shapiro L, Sheng Z. cAb-Rep: a database of curated antibody repertoires for exploring antibody diversity and predicting antibody prevalence. *Front Immunol* 2019; 10. Available from: https://www.frontiersin.org/article/10.3389/fimmu.2019.02365/full
173. Suzek BE, Huang H, McGarvey P, Mazumder R, Wu CH. UniRef: comprehensive and non-redundant UniProt reference clusters. *Bioinformatics* 2007; 23:1282–8. Available from: https://academic.oup.com/bioinformatics/article/23/10/1282/197795
174. Engelhart E, Emerson R, Shing L, Lennartz C, Guion D, Kelley M, Lin C, Lopez R, Younger D, Walsh ME. A dataset comprised of binding interactions for 104,972 antibodies against a SARS-CoV-2 peptide. *Sci Data* 2022; 9:653. Available from: https://www.nature.com/articles/s41597-022-01779-4
175. Greiff V, Menzel U, Miho E, Weber C, Riedel R, Cook S, Valai A, Lopes T, Radbruch A, Winkler TH, et al. Systems analysis reveals high genetic and antigen-driven predetermination of antibody repertoires throughout B cell development. *Cell Rep* 2017; 19(7):1467–1478. Available from: https://doi.org/10.1016/j.celrep.2017.04.054
176. Asensio MA, Lim YW, Wayham N, Stadtmiller K, Edgar RC, Leong J, Leong R, Mizrahi RA, Adams MS, Simons JF, et al. Antibody repertoire analysis of mouse immunization protocols using microfluidics and molecular genomics. *MAbs* 2019; 11:870–83. Available from: https://www.tandfonline.com/doi/full/10.1080/19420862.2019.1583995
177. Raybould MIJ, Kovaltsuk A, Marks C, Deane CM. CoV-AbDab: the coronavirus antibody database. *Bioinformatics* 2021; 37:734–5. Available from: https://academic.oup.com/bioinformatics/article/37/5/734/5893556
178. Du Z, Qian Y, Liu X, Ding M, Qiu J, Yang Z, Tang J. GLM: general language model pretraining with autoregressive blank infilling. In: Smaranda Muresan, Preslav Nakov, Aline Villavicencio (eds.), *Proceedings of the 60th Annual Meeting of the Association for Computational Linguistics (Volume 1: long papers)*. Kerrville, TX: Association for Computational Linguistics; 2022. page 320–35.Available from: https://aclanthology.org/2022.acl-long.26
179. Chong CR, Sullivan DJ. New uses for old drugs. *Nature* 2007; 448:645–6. Available from: https://www.nature.com/articles/448645a

180. Nimrod G, Fischman S, Austin M, Herman A, Keyes F, Leiderman O, Hargreaves D, Strajbl M, Breed J, Klompus S, et al. Computational design of epitope-specific functional antibodies. *Cell Rep* 2018; 25:2121–2131.e5. Available from: https://linkinghub.elsevier.com/retrieve/pii/S2211124718316851
181. Chen R, Li L, Weng Z. ZDOCK: an initial-stage protein-docking algorithm. *Proteins Struct Funct Genetics* 2003; 52:80–7. Available from: https://doi.wiley.com/10.1002/prot.10389
182. Ritchie DW, Venkatraman V. Ultra-fast FFT protein docking on graphics processors. *Bioinformatics* 2010; 26:2398–405. Available from: https://academic.oup.com/bioinformatics/article/26/19/2398/229220
183. Tam C, Kukimoto-Niino M, Miyata-Yabuki Y, Tsuda K, Mishima-Tsumagari C, Ihara K, Inoue M, Yonemochi M, Hanada K, Matsumoto T, et al. Targeting Ras-binding domain of ELMO1 by computational nanobody design. *Commun Biol* 2023; 6:284. Available from: https://www.nature.com/articles/s42003-023-04657-w
184. Schneidman-Duhovny D, Inbar Y, Nussinov R, Wolfson HJ. PatchDock and SymmDock: servers for rigid and symmetric docking. *Nucleic Acids Res* 2005; 33:W363–7. Available from: https://academic.oup.com/nar/article-lookup/doi/10.1093/nar/gki481
185. Barlow KA, Ó Conchúir S, Thompson S, Suresh P, Lucas JE, Heinonen M, Kortemme T. Flex ddG: Rosetta ensemble-based estimation of changes in protein–protein binding affinity upon mutation. *J Phys Chem B* 2018; 122:5389–99. Available from: https://pubs.acs.org/doi/10.1021/acs.jpcb.7b11367
186. Schneider C, Buchanan A, Taddese B, Deane CM. DLAB: deep learning methods for structure-based virtual screening of antibodies. *Bioinformatics* 2022; 38:377–83. Available from: https://academic.oup.com/bioinformatics/article/38/2/377/6373413
187. Kontermann RE, Brinkmann U. Bispecific antibodies. *Drug Discov Today* 2015; 20:838–47. Available from: https://linkinghub.elsevier.com/retrieve/pii/S135964461500077X
188. Sawant MS, Streu CN, Wu L, Tessier PM. Toward drug-like multispecific antibodies by design. *Int J Mol Sci* 2020; 21:7496. Available from: https://www.mdpi.com/1422-0067/21/20/7496
189. Tian Z, Liu M, Zhang Y, Wang X. Bispecific T cell engagers: an emerging therapy for management of hematologic malignancies. *J Hematol Oncol* 2021; 14:75. Available from: https://jhoonline.biomedcentral.com/articles/10.1186/s13045-021-01084-4
190. Wu J, Fu J, Zhang M, Liu D. Blinatumomab: a bispecific T cell engager (BiTE) antibody against CD19/CD3 for refractory acute lymphoid leukemia. *J Hematol Oncol* 2015; 8:104. Available from: https://www.jhoonline.org/content/8/1/104
191. Dunbar J, Knapp B, Fuchs A, Shi J, Deane CM. Examining variable domain orientations in antigen receptors gives insight into TCR-like antibody design. *PLoS Comput Biol* 2014; 10:e1003852. Available from: https://dx.plos.org/10.1371/journal.pcbi.1003852
192. Aertker KMJ, Pilvankar MR, Prass TM, Blech M, Higel F, Kasturirangan S. Exploring molecular determinants and pharmacokinetic properties of IgG1-scFv bispecific antibodies. *MAbs* 2024; 16:2318817. Available from: https://doi.org/10.1080/19420862.2024.2318817
193. Carter P, Ridgway JBB, Presta LG. 'Knobs-into-holes' provides a rational design strategy for engineering antibody CH3 domains for heavy chain heterodimerization. *Immunotechnology* 1996; 2:73. Available from: https://linkinghub.elsevier.com/retrieve/pii/1380293396806853
194. Lewis SM, Wu X, Pustilnik A, Sereno A, Huang F, Rick HL, Guntas G, Leaver-Fay A, Smith EM, Ho C, et al. Generation of bispecific IgG antibodies by structure-based design of an orthogonal Fab interface. *Nat Biotechnol* 2014; 32:191–8. Available from: https://www.nature.com/articles/nbt.2797

195. Leaver-Fay A, Froning KJ, Atwell S, Aldaz H, Pustilnik A, Lu F, Huang F, Yuan R, Hassanali S, Chamberlain AK, et al. Computationally designed bispecific antibodies using negative state repertoires. *Structure* 2016; 24:641–51. Available from: https://linkinghub.elsevier.com/retrieve/pii/S0969212616000721
196. Froning KJ, Leaver-Fay A, Wu X, Phan S, Gao L, Huang F, Pustilnik A, Bacica M, Houlihan K, Chai Q, et al. Computational design of a specific heavy chain/κ light chain interface for expressing fully IgG bispecific antibodies. *Protein Sci* 2017; 26:2021–38. Available from: https://onlinelibrary.wiley.com/doi/10.1002/pro.3240
197. Lipinski CA, Lombardo F, Dominy BW, Feeney PJ. Experimental and computational approaches to estimate solubility and permeability in drug discovery and development settings. *Adv Drug Deliv Rev* 1997; 23:3–25. Available from: https://linkinghub.elsevier.com/retrieve/pii/S0169409X96004231

Computational Biophysical Analyses of Antibody Structure-Function Relationships with Emphasis on Therapeutic Antibody-Based Biologics 7

Puneet Rawat, Eva Smorodina, Divya Sharma, R. Prabakaran, Jack Wade, Rahmad Akbar, Amrinder Singh, Sandeep Kumar, Victor Greiff, and M. Michael Gromiha

7.1 INTRODUCTION

Antibodies or immunoglobulins (IGs) are a key component of the adaptive immune response. They play a pivotal role in recognizing and binding to a foreign molecule, followed by triggering an immune response against the antigen by recruiting other cells and molecules. Antibodies are heavy proteins with an approximate size of 10 nm and a weight of 150 kDa (Reth, 2013). These molecules are roughly Y shape, composed

DOI: 10.1201/9781003300311-7

of two identical heavy and light chains connected by disulfide bonds (Glockshuber et al., 1992). The recognition of any foreign molecule called "antigen" mainly involves the variable region of the antibody (upper tips of the Y shape including both heavy and light chains), and the activation of the immune response is regulated by the constant region of the antibody (Segal et al., 1974; Sela-Culang et al., 2013; Sinclair et al., 1968). Antibodies possess a high level of diversity, allowing them to target a wide range of antigens, particularly in the antibody variable regions. The diversity in the antibody chains is generated through V(D)J recombination, insertion/deletion during the recombination process, and somatic hypermutation. The variable region of the antibody further consists of four framework regions (FRs) and three complementarity-determining regions (CDRs) on both light and heavy chains. The CDRs on the antibody predominantly interact with the antigen (Padlan et al., 1995). However, the highly diverse CDR3 on the heavy chain (CDRH3) contributes significantly to the antibody–antigen interactions, in most cases (Akbar et al., 2021; Chothia & Lesk, 1987; Xu & Davis, 2000). The interacting residues in the antibody–antigen complex are called "paratopes" on the antibody side and "epitopes" on the antigen side (Akbar et al., 2021; Sela-Culang et al., 2013). Therefore the main features of antibodies can be classified as (i) specificity toward the antigens due to high diversity on the CDRs, (ii) antibody diversity through V(D)J recombination and somatic hypermutations to recognize wide range of antigens, (iii) tolerance toward host proteins/cells, (iv) optimized biophysical properties to work effectively at the biological conditions, (v) immunological memory to build immunity upon reinfection, (vi) trigger Fc mediated effector functions by neutralization (direct binding to the pathogen to prevent infection), opsonization (activation of phagocytic cells), complement activation (activation of downstream cascade to neutralize the pathogens), and antibody-dependent cellular cytotoxicity (ADCC; lysis of infected cells activated by antibodies) (Lu et al., 2018).

There are two major aspects related to antibody–antigen interaction, namely, "affinity and avidity" (Rudnick & Adams, 2009; Yin et al., 2021). Affinity measures the strength of the epitope binding to an antibody and is often represented by the dissociation constant K_D. The paratope and epitope residues interact through various types of non-covalent interactions, such as hydrogen bonds, ionic bonds, Van der Waals, and hydrophobic interactions. Avidity measures the overall strength of the antibody–antigen complex. Avidity includes the valency of the protein, the binding affinity of the antibody–antigen complex, as well as the structural arrangement of the antibody(s) (Evans & Thurber, 2022).

Antibodies are also attractive therapeutic candidates due to their biological role in the immune response and high specificity toward the antigen (Lu et al., 2020). Monoclonal antibodies or mAbs are widely used as therapeutic candidates because they are highly specific to particular antigens. Currently, there are approximately 100 Food and Drug Administration (FDA)-approved mAbs with an estimated market size of ~185.50 billion USD, which is expected to reach more than 500 billion USD by 2030 (Mullard, 2021). Antibody-based therapeutics development also faces several challenges. For example, antibody repertoire sizes for an individual range from 10^8 to 10^{10} in humans (Elhanati et al., 2014; Glanville et al., 2009). Although antibody repertoires may show convergence based on post-exposure to similar antigens, there is still vast diversity in the repertoires to be analyzed experimentally (Greiff et al., 2017).

For example, Wardemann et al. used single-cell cloning strategies and antibody expression to exhibit self-expression of newly generated B-cells in the bone marrow (Wardemann et al., 2003). This opened an arena to several immunological insights and furthered the isolation of antibodies to neutralize clinical pathogens like SARS-CoV, influenza, Human immunodeficiency viruses (HIV), and many others using these B-cell cloning strategies. However, the main limitation posed by this technique is that it provides only a sliver of information on the full antibody repertoire and is usually limited to a subset of antigen binding activity. Further, Ig-sequencing limitations include identification of a suitable source of DNA sequence and quantification errors; ability to distinguish which V_H genes pair with V_L genes in each B-cell; use of appropriate data and visualization tools to detangle the large amounts of information furnished post-analysis; cross-reactivity of antibodies with host protein; and lack of structural information about the antibodies (Brown et al., 2019; Georgiou et al., 2014; Greiff et al., 2015). Once the binding with the target antigen is established, there are several other developability challenges for naturally occurring antibodies to be used as therapeutic antibodies, which include folding stability, aggregation, viscosity, etc. (Ahmed et al., 2021; Akbar et al., 2022; Jain et al., 2017; Młokosiewicz et al., 2022; Pérez et al., 2022; Raybould et al., 2019; Xu et al., 2019). Therefore, experimental validation of potential therapeutic antibody leads is challenging due to high production costs and a lack of large-scale and high-throughput methods for developability prediction (Schlander et al., 2021).

With the advent of computational sciences, researchers have probed machine-learning (ML) methods and other informatics technologies to study plausible relationships between an antibody's structure and function. For example, there are several computational resources available for antibody structure modeling, *in silico* screening of epitope/paratope regions, docking of antibodies with antigen, estimation of binding energies, and calculation of biophysical properties of antibodies such as aggregation propensity, solubility, and melting temperature (Ahmed et al., 2021; Chiu et al., 2019a; Harmalkar et al., 2022; Narayanan et al., 2021a; Sankar et al., 2022). However, even in these methods, there are certain limitations wherein predicting models of disordered or post-translationally modified (e.g., glycosylated) proteins is not possible. The docking methods require prior information on epitope regions for better prediction of the antibody–antigen complex, and binding energy prediction methods include inaccuracies in the calculation of absolute binding free energies. Moreover, the sequence-based approach to study binding prediction is unreliable because the structural conformation of the CDRs can significantly affect the binding (Akbar et al., 2022).

Conventionally, antibody repurposing and optimization of the therapeutic antibodies have been widely used by *in silico* researchers due to limited resources and long downstream validation processes (Mason et al., 2021; Rawat et al., 2021; Rodriguez-Quijada et al., 2020; Wang, Gallolu Kankanamalage, et al., 2021). However, there have been significant advances in the computational approaches for the design and development of antibodies in recent years (Hummer et al., 2022; Norman et al., 2020; Tiller & Tessier, 2015). In this chapter, we will be focusing on the computational resource developed to curate antibody-related information, antibody–antigen binding (docking and binding affinity prediction), and biophysical parameters affecting antigen design (Figure 7.1). We have also highlighted the role of language models and molecular dynamics (MD) simulations in antibody structure-function relationship prediction.

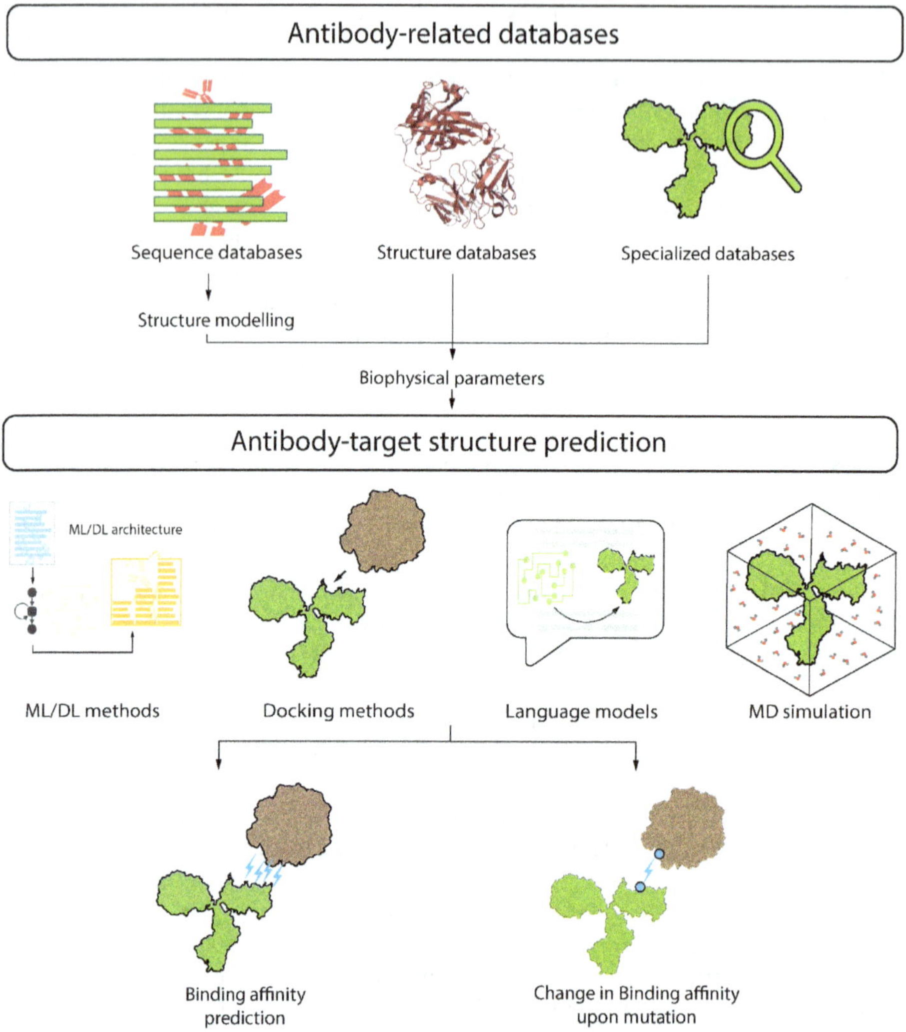

FIGURE 7.1 Overview of the topics related to antibody structure-function considered in the book chapter. There are several antibody-related databases that contain sequence, structure, and other relevant information related to antibodies (e.g., interacting pathogen, residue-level interaction, binding affinity, epitope, developability-related information, and so on). There are several possible approaches for the antibody–antigen structure prediction. Docking (faster computation time) and MD simulations (very high computation time) are classical approaches widely used to predict antibody–antigen structures, whereas ML/DL methods and language models are more recently developed approaches showing much better speed and accuracy compared to the classical approaches. Several binding affinity prediction methods are also developed, which can use the sequence/structure information of the antibody–antigen complex to predict (i) the binding affinity of the complex or (ii) the change in binding affinity upon point mutation at the interaction interface.

7.2 COMPUTATIONAL RESOURCES FOR ANTIBODY STRUCTURE AND FUNCTION

7.2.1 Antibody-Related Online Resources and Databases

As antibodies become an increasingly interesting topic in the field of biotherapeutics, the demand for antibody-specific data in public repositories is growing (Norman et al., 2020). The online resource and databases on antibodies can be divided into two major categories: (i) primary sequence/structure databases and (ii) derived databases, which contain the secondary information obtained from the antibody sequence/structure or biological activity (such as antibody–antigen interaction, binding affinity/neutralization activity, and epitope/antigen information) (Table 7.1).

7.2.1.1 Primary databases

Primary databases contain the sequence/structure of antibodies or immune repertoire sequences from B-cells. The international ImMunoGeneTics information system (IMGT) is one of the comprehensive resources on IGs or antibodies, T-cell receptors (TCRs), and major histocompatibility (MH) of human and other vertebrate species. It consists of sequence databases, genome databases, structure databases, and mAbs' databases, which are embedded into several web resources and interactive tools (Ehrenmann et al., 2010). Antibody sequences and structures are also curated in the abYsis database (Swindells et al., 2017), providing an inbuilt analysis platform. The AntiBodies Chemically Defined (ABCD) database is a manually curated repository of sequenced antibodies (Lima et al., 2020). The Protein Data Bank (PDB) is a general resource for experimentally determined protein structures, which also include structures of antibodies/antibody complexes (Rose et al., 2021). Several antibody-specific structure databases were also developed using PDB, which contains PDB structures as well as related annotated information. For example, SAbDab (Dunbar et al., 2014) contains the antibody structures from PDB, which are annotated with several details, including experimental details, antibody nomenclature (e.g., heavy-light pairings), curated affinity data, and sequence annotations; abYbank contains sequences (EMBLIG, Kabat, and AbPDBSeq databases) and renumbered experimental structures (Kabat, Chothia, and Martin antibody numbering scheme in the AbDb database) of antibodies from PDB (Ferdous & Martin, 2018). The B-cell repertoire-specific databases include observed antibody space (OAS) (Olsen et al., 2022a), VBASE2 (Retter et al., 2005), cAb-Rep (Guo et al., 2019), Pan Immune Repertoire Database (PIRD) (Zhang et al., 2020), and VDJbase (Omer et al., 2020). These immune repertoire databases can be used as benchmarking datasets for humanness and developability parameters of therapeutic antibodies.

TABLE 7.1 List of antibody-related sequence, structure, and other specialized databases

SR. NO.	*DATABASE*	*LINK*	*DESCRIPTION*	*REFERENCE*
Sequence/Structure Database				
1	IMGT	https://www.imgt.org/	Comprehensive sequence/structure/genome/ monoclonal antibody database	Ehrenmann et al. (2010)
2	SAbDab	https://opig.stats.ox.ac.uk/ webapps/newsabdab/sabdab/	Structure database for antibodies	Dunbar et al. (2014)
3	PDB	https://www.rcsb.org/	A generalized structure database which also contains antibodies	Rose et al. (2021)
4	abYbank	http://www.abybank.org/	Antibody sequence and renumbered structures data	Ferdous and Martin (2018)
5	cAb-Rep	https://cab-rep.c2b2.columbia.edu/	Database of curated antibody repertoires	(Guo et al., 2019)
6	OAS	http://opig.stats.ox.ac.uk/webapps/ oas/	Annotated immune repertoire database	Olsen et al. (2022a)
7	VDJbase	https://vdjbase.org/	Database of adaptive immune receptor genes, genotypes, and haplotypes	Omer et al. (2020)
8	PIRD	https://db.cngb.org/pird/	Database of raw and processed sequences of IGs and T-cell receptors (TCRs) of human and other vertebrate species	Zhang et al. (2020)
9	VBASE2	http://www.vbase2.org/	Database of human germline variable region sequences	Retter et al. (2005)
10	ABCD	https://web.expasy.org/abcd/	Database is a manually curated depository of sequenced antibodies	Lima et al. (2020)
11	abYsis	http://www.abysis.org/abysis/	Integrated database of antibody sequence and structure data	Swindells et al. (2017)

(*Continued*)

TABLE 7.1 (*Continued*) List of antibody-related sequence, structure, and other specialized databases

SR. NO.	DATABASE	LINK	DESCRIPTION	REFERENCE
12	iReceptor	http://ireceptor.irmacs.sfu.ca/	NGS sequence data on B-cell receptors	Corrie et al. (2018)
Specialized Sequence/Structure Database				
1	Thera-SAbDab	http://opig.stats.ox.ac.uk/webapps/newsabdab/therasabdab/	Sequence database for approved or clinical-stage therapeutic antibodies	Raybould et al. (2020)
2	CoV-Ab-Dab	http://opig.stats.ox.ac.uk/webapps/covabdab/	Sequence database for coronavirus-related antibodies	Raybould et al. (2021)
3	Ab-CoV	https://web.iitm.ac.in/bioinfo2/ab-cov/home	Experimental neutralization profile of coronavirus-related antibodies	Rawat et al. (2022)
4	IEDB	https://www.iedb.org/	Antibody epitope database	Vita et al. (2019)
5	bNAber	http://bnaber.org/*	Database of broadly neutralizing HIV antibodies	Eroshkin et al. (2014)
6	AgAbDb	http://bioinfo.net.in/AgAbDb.htm*	Antibody–antigen interaction database	Kulkarni-Kale et al. (2014)
7	CPAD 2.0	https://web.iitm.ac.in/bioinfo2/cpad2/	Experimental protein aggregation information which also includes antibodies	Rawat et al. (2020)
8	AL-Base	https://wwwapp.bumc.bu.edu/BEDAC_ALBase/	Experimental amyloidogenic antibody light chain database	Bodi et al. (2009)
9	AB-Bind	https://github.com/sarahsirin/AB-Bind-Database	Database of experimentally determined changes in binding free energies	Sirin et al. (2016)
10	SKEMPI 2.0	https://life.bsc.es/pid/skempi2/	Database of kinetics and energetics information upon mutation and includes antibody–antigen complexes	Jankauskaitė et al. (2018)
11	PROXiMATE	https://www.iitm.ac.in/bioinfo/PROXiMATE/	A mutant protein–protein interaction kinetics and thermodynamics database	Jemimah et al. (2017)

The links that are not active (as checked on Dec 2022) are denoted with "*" sign.

7.2.1.2 Specialized sequence/structure databases

Specialized databases contain a variety of secondary information derived from the primary databases. There are several specialized sequence databases such as Thera-SAbDab (Raybould et al., 2020) for approved and clinical-stage antibodies, CoV-AbDab (Raybould et al., 2021) for coronavirus-related antibodies, broadly neutralizing antibodies electronic resource (bNAber) for broadly neutralizing HIV antibodies (Eroshkin et al., 2014), Amyloid Light Chain Database (AL-Base) for antibody light chains with aggregation capability (Bodi et al., 2009), and so on. AgAbDb is a unique database, which contains the antibody–antigen interaction details at the residue level along with other parameters such as interaction type and accessible surface area (Kulkarni-Kale et al., 2014). The AB-Bind database contains the experimentally determined change in binding free energy values (Sirin et al., 2016). SKEMPI 2.0 and PROXiMATE databases contain changes in thermodynamic parameters and kinetic rate constants upon point mutations for protein–protein interaction, which also includes antibody–antigen interactions (Jankauskaitė et al., 2018; Jemimah et al., 2017). The curated protein aggregation database (CPAD) 2.0 database provides comprehensive experimentally determined information on aggregation-prone regions and aggregation kinetics for all proteins, including antibodies (Rawat et al., 2020). The Ab-CoV database is a coronavirus-specific antibody database that contains the antibodies' neutralization profile (IC50 and EC50) and binding affinity (KD), as well as computationally predicted changes in stability and binding affinity upon epitope/paratope residue mutation (Rawat et al., 2022). The Immune Epitope Database (IEDB) considers the antigen-side information and collects epitope information (Vita et al., 2019).

7.2.2 Computational Methods for Investigating Structure-Function Relationship

7.2.2.1 Docking tools for antibody–antigen complex structure prediction

Docking is a molecular modeling technique, which predicts the conformation of one molecule (ligand) on the surface of another static and larger molecule (receptor). In the case of antibody–antigen docking, antibodies are usually considered receptors, while antigen is considered a ligand. Molecular docking allows us to see residue-level interactions between the receptor and ligand, which is crucial for drug discovery (Pagadala et al., 2017; Pinzi & Rastelli, 2019). Most docking methods generate several docked structures and provide docking scores to rank each pose. These scores can be unrelated to real estimation of the binding strengths of complex structures, but they allow the comparison of different conformations of one binder or different binders between each other within one docking tool. Lower scores usually represent better binder conformations. Recent docking methods prefer ensemble of protein structures (usually from MD simulations) to identify the correct pose (Amaro et al., 2018). Docking poses can also be evaluated through an affinity scoring function representing electrostatic and Van der Waals interactions (Pagadala et al., 2017; Rawat et al., 2021). Docking programs typically rely on an estimated binding site to accurately predict the binding interfaces, and they often lack precision in predicting binding energies (Wang et al., 2003). MD simulations provide a more precise estimation of binding energies (Fernández-Quintero et al., 2022; Kralj et al., 2021; Salmaso & Moro, 2018). Some docking models allow the ligand to be treated as a rigid body object (all atoms and residues are static and immovable) or flexible (when some amino acids are allowed to move during the docking procedure). The docking procedure that requires many ligands to bind to one receptor is called virtual screening. It's a very common approach in the early stages of drug discovery (Schneider et al., 2022). Docking is the widely used approach to investigate the function or binding of the antibody against an antigen using protein structural information (Brooks et al., 2020; Chaves et al., 2020; Guest et al., 2021). ZDOCK (Pierce et al., 2014), Haddock (Dominguez et al., 2003), ClusPro 2.0 (Comeau et al., 2004), LightDock (Jiménez-García et al., 2018), and Rosetta (Schoeder et al., 2021) are highly used for protein–protein docking tasks, including antibody–antigen docking (Ambrosetti et al., 2020). Antibody–antigen-specific docking models include Antibody i-Patch and Absolut!. Antibody i-Patch defines antibody residues which are most likely in contact with antigen (Krawczyk et al., 2013) and Absolut! generates coarse-grained synthetic antibody–antigen complexes with information about paratope and epitope conformations and affinity (Robert et al., 2022). A list of the most used docking tools for antibody–antigen docking is provided in Table 7.2.

ML techniques can outperform classical molecular docking results (Ganea et al., 2021). One of the models is based on graph neural networks (EquiDock) and predicts rotations and translations of molecules in rigid docking (Ganea et al., 2021). Another model is a diffusion generative model (DiffDock), which refines random docking poses to reach the best complex conformation via translations, rotations, and torsion angles (Corso et al., 2022). The recently introduced architecture called Hierarchical Equivariant Refinement Network (HERN) in "abdockgen" allows not only rigid docking but also refines side chains after pose generation (Jin et al., 2022).

7.2.3 Computational Approaches to Predict Antibody–Antigen Interaction

In this section, we have summarized different methods for antibody–antigen binding and affinity prediction tools that do not involve docking approaches. The details regarding the tools predicting only paratope or epitope regions can be found elsewhere and not discussed here (Akbar et al., 2022; Chinery et al., 2023; Lo et al., 2021). The initial methods for antibody–antigen complex prediction were based on amino acid usage. For example, Bepar uses a sliding window of amino acids between the antigen and CDR regions of antibodies to predict the epitope residues using sequence information (Zhao & Li, 2010). EpiPred is another method that predicts the epitope regions for antibodies using conformational matching and specific antibody–antigen scores (Krawczyk et al., 2014). The methods developed later utilized machine learning (ML) or deep learning (DL) approaches for the prediction of epitope–paratope residues. The random forest-based ML model "PEASE" (Sela-Culang et al., 2014, 2015) calculates the "residue-score" for both antibody and antigen to identify interacting epitope–paratope

TABLE 7.2 List of molecular docking tools widely used for antibody–antigen complex prediction

SR. NO.	*DOCKING TOOL*	*LINK*	*DESCRIPTION*	*REFERENCE*
1	ZDOCK	https://zdock.umassmed.edu/	Fast Fourier Transform-based protein docking	Pierce et al. (2014)
2	Haddock	https://wenmr.science.uu.nl/haddock2.4/	Information-driven flexible docking approach	Dominguez et al. (2003)
3	ClusPro 2.0	https://cluspro.org/login.php	Fast Fourier Transform-based rigid docking	Comeau et al. (2004)
4	LightDock	https://lightdock.org/	Docking protocol based on the Glowworm Swarm Optimization (GSO) algorithm	Jiménez-García et al. (2018)
5	Rosetta (SnugDock)	https://www.rosettacommons.org/software	Simulates the induced-fit mechanism	Schoeder et al. (2021)
6	PatchDock	http://bioinfo3d.cs.tau.ac.il/PatchDock/	Geometry-based molecular docking algorithm	Schneidman-Duhovny et al. (2005)
7	AbAdapt	https://sysimm.org/abadapt/	From sequence to docked structure generation pipeline	Davila et al. (2022)
8	EquiDock	https://github.com/octavian-ganea/equidock_public	Pairwise-independent SE(3)-equivariant graph matching network-based rigid docking	Ganea et al. (2021)
9	DiffDock	https://github.com/gcorso/DiffDock	A diffusion generative model over the non-Euclidean manifold of ligand poses	Corso et al. (2022)
10	abdockgen	https://github.com/wengong-jin/abdockgen	Antibody–antigen docking and design via hierarchical equivariant refinement	Jin et al. (2022)
11	Antibody i-Patch	http://opig.stats.ox.ac.uk/webapps/newsabdab/sabpred/antibodyipatch	Contact likelihood score to each residue	Krawczyk et al. (2013)
12	Absolut!	https://github.com/csi-greifflab/Absolut	Unconstrained lattice-based antibody–antigen bindings generator	Robert et al. (2022)

residues using 120 properties calculated from the antibody–antigen complexes. It also allows an optional additional step of finding surface patches on antigens with high scores (termed patch scores). Another study combined statistical and ML algorithms to predict the antibody-specific epitopes. They calculated several geometric and physicochemical features related to interacting regions of antibody–antigen complexes. These features were used in Monte Carlo algorithms to generate putative epitope–paratope pairs, used as training datasets in the ML model (Jespersen et al., 2019). A DL-based framework, "Paratope and Epitope prediction with graph Convolution Attention Network" (PECAN), uses a protein–protein interaction dataset for training and utilizes transfer learning to predict the binding interface of antibody–antigen complexes. The local residues at close spatial proximity of the interfaces were captured using graph convolutions, while an attention layer was employed to encode antibody–antigen pair interactions (Pittala & Bailey-Kellogg, 2020).

A protein–protein interaction prediction method, "Molecular Surface Interaction Fingerprints" (MaSIF), applies geometric deep learning (GDL), which can incorporate geometric features such as structure and symmetry of the input to improve the quality of the predictions (Gainza et al., 2020). MaSIF also uses antibody–antigen complexes as training data. MaSIF converts the protein surface into a mesh representation, where each vertex contains two geometric features (shape and distance-dependent curvature) and three chemical features (hydropathy, continuum electrostatics, and location of free electrons/proton donors). Furthermore, a set of geodesic filters generates an output with a fixed dimension. The performance of MaSIF showed an ROC AUC of 0.77 with one geodesic convolutional layer and 0.86 with three layers. A simple version of this GDL model, differentiable molecular surface interaction fingerprinting (dMaSIF), is also now available for large-scale analysis with simplified surface representation (Sverrisson et al., 2020). Del Vecchio et al. (2021) argued that paratope and epitope prediction require asymmetric treatment. Therefore, they developed a separate paratope model (Para-EPMP) and epitope model (Epi-EPMP) for joint paratope and epitope prediction. Rangel et al. utilized a fragment-based approach for the combinatorial design of antibody binding loops (CDRs) and grafted them onto antibody scaffolds (Rangel et al., 2021). The designed CDRs were also computationally optimized for solubility and conformational solubility.

7.2.4 In Silico Prediction of Binding Affinity

Understanding protein–protein interactions is crucial in the investigation of biological systems, as these play a critical role in almost all cellular processes (Gromiha, 2020; Jones & Thornton, 1996; Perkins et al., 2010). Binding affinity is defined as the strength of interaction between proteins and/or peptides. However, higher binding affinity does not always lead to the best therapeutic response (Yu et al., 2023). The binding affinity of an interaction is described through the equilibrium dissociation constant K_D, or, in thermodynamic terms, the Gibbs free energy ΔG (ΔG=-RT ln K_D) (Kastritis & Bonvin, 2013). Experimentally measuring K_D values is a time-consuming and expensive process (Jarmoskaite et al., 2020). Therefore, many computational methods have been developed for predicting the binding affinity (Table 7.3). Binding affinity prediction is important because it not only allows to control interactions and develop innovative therapeutics but

TABLE 7.3 List of computational resources available for the prediction of the absolute value of binding affinity and change in binding affinity upon mutation(s)

BINDING AFFINITY PREDICTION TOOLS					
METHOD	*FEATURES*	*PERFORMANCE*	*MODEL TYPE*	*TRAINED ON AB-AG DATA?*	*URL/REFERENCE*
1. Sequence-Based Prediction Methods					
PPA-Pred	Sequence-based affinity prediction using functional information	r=0.90 on 135 complexes selected from structure-based benchmark	Regression	Yes (11.1%; 15 out of 135)	https://www.iitm.ac.in/bioinfo/PPA_Pred/ (Yugandhar & Michael Gromiha, 2014)
ISLAND	Kernel representation	r=0.44 on structure-based benchmark	Support vector machine (SVM)	Yes (11.1%; 15 out of 135)	https://sites.google.com/view/wajidarshad/software (Abbasi et al., 2020)
PIPR	Pre-trained embeddings	r=0.87 on SKEMPI	RRCNN	Yes (11.02%; 781 out of 7,085)	https://github.com/muhaochen/seq_ppi (Chen et al., 2019)
RAPPPID	Pairs of amino acid sequences	r=0.97 on STRING	Neural network	Yes (2.6%; 242 out of 9,340)	https://github.com/jszym/rapppid (Szymborski & Emad, 2022)
TcellMatch	Sequence embedding	r=0.63 on 10x dataset	Neural network	Yes (100%; 4,812)	(Fischer et al., 2020)

(Continued)

TABLE 7.3 (*Continued*) List of computational resources available for the prediction of the absolute value of binding affinity and change in binding affinity upon mutation(s)

BINDING AFFINITY PREDICTION TOOLS					
METHOD	*FEATURES*	*PERFORMANCE*	*MODEL TYPE*	*TRAINED ON AB-AG DATA?*	*URL/REFERENCE*
2. Structure-Based Prediction Methods					
PRODIGY	Inter-residue contacts and noninteracting surface	r=0.73 on benchmark of 81 protein–protein complexes	Linear regression	Yes (14.8%; 12 out of 81)	https://wenmr.science.uu.nl/prodigy/ (Xue et al., 2016)
FoldX	Empirical function	64%	Empirical function	No	https://foldxsuite.crg.eu/ (Delgado et al., 2019)
PPI-Affinity	ProtDcal	r=0.77 on the PDBbind database	SVM	Yes (9.8%; 82 out of 833)	https://protdcal.zmb.uni-due.de/PPIAffinity* (Romero-Molina et al., 2022)
CSM-AB	Graph-based signatures	r=0.64 on PDBbind, SabDab, RCSB PDB	Regression	Yes (100%; 472)	http://biosig.unimelb.edu.au/csm_ab/prediction (Myung et al., 2022)

(*Continued*)

TABLE 7.3 (*Continued*) List of computational resources available for the prediction of the absolute value of binding affinity and change in binding affinity upon mutation(s)

BINDING AFFINITY PREDICTION TOOLS					
METHOD	*FEATURES*	*PERFORMANCE*	*MODEL TYPE*	*TRAINED ON AB-AG DATA?*	*URL/REFERENCE*
Change in Binding Affinity Upon Mutation Prediction Tools					
1. Sequence-Based Prediction Methods					
ProAffiMuSeq	Sequence-based features and functional class	r=0.73 on mutation data from PROXiMATE	Regression	Yes (16.1%; 189 out of 1,173)	https://web.iitm.ac.in/bioinfo2/proaffimuseq/ (Jemimah et al., 2019)
PANDA	Sequence-based	r=0.52 on SKEMPI 2.0	Regression	Yes (11.02%; 781 out of 7,085)	https://github.com/wajidarshad/panda (Abbasi et al., 2021)
SAAMBE-SEQ	Sequence-based, physical properties	r=0.83 on SKEMPI 2.0	Gradient-boosting decision tree	Yes (11.02%; 781 out of 7,085)	http://compbio.clemson.edu/saambe_webserver/indexSEQ.php#started (Li et al., 2021)
2. Structure-Based Prediction Methods					
BeAtMuSic	Statistical potentials	r=0.4 on SKEMPI	Regression	Yes (1.8%; 55 out of 3,047)	http://babylone.ulb.ac.be/beatmusic/index.php (Dehouck et al., 2013)
BindProfX	Interface profile score, shape complementarity, and sequence-based features	r=0.68 on SKEMPI	Profile score	Yes (1.8%; 55 out of 3,047)	https://zhanggroup.org/BindProfX/ (Xiong et al., 2017)

(*Continued*)

TABLE 7.3 (*Continued*) List of computational resources available for the prediction of the absolute value of binding affinity and change in binding affinity upon mutation(s)

BINDING AFFINITY PREDICTION TOOLS					
METHOD	*FEATURES*	*PERFORMANCE*	*MODEL TYPE*	*TRAINED ON AB-AG DATA?*	*URL/REFERENCE*
SAAMBE	Van der Waals, solvation and Coulomb energy, entropy, hydrophobicity, solvent accessible surface area, hydrogen bonds and interface area	$r=0.82$ on SKEMPI 2.0	XGBoost	Yes (11.02%; 781 out of 7,085)	http://compbio.clemson.edu/saambe_webserver/ (Li et al., 2021)
MutaBind	Van der Waals energy, solvation energy, free energy change due to unfolding, solvent accessible surface area	$r=0.68$ on SKEMPI	Molecular mechanics force fields, statistical potentials, and fast side-chain optimization algorithms	Yes (1.8%; 55 out of 3,047)	https://lilab.jysw.suda.edu.cn/research/mutabind2// (Li et al., 2016)
mmCSM-PPI	Graph-based signatures and complementary features	$r=0.75$ on SKEMPI 2.0	Extra trees	Yes (11.02%; 781 out of 7,085)	https://biosig.lab.uq.edu.au/mmcsm_ppi/ (Rodrigues et al., 2021)

(*Continued*)

TABLE 7.3 (*Continued*) List of computational resources available for the prediction of the absolute value of binding affinity and change in binding affinity upon mutation(s)

BINDING AFFINITY PREDICTION TOOLS					
METHOD	*FEATURES*	*PERFORMANCE*	*MODEL TYPE*	*TRAINED ON AB-AG DATA?*	*URL/REFERENCE*
GeoPPI	Graph neural network	r=0.52 on SKEMPI 2.0	Gradient-boosting tree	Yes (11.02%; 781 out of 7,085)	https://github.com/Liuxg16/GeoPPI (Liu et al., 2021)
TopNetTree	CNN, persistent homology	r=0.79 on SKEMPI 2.0	Gradient-boosting tree	Yes (11.02%; 781 out of 7,085)	(Wang et al., 2020)
PerSpect-EL	Physical properties, persistent homology	r=0.85 on SKEMPI 2.0	CNN+gradient-boosting tree	Yes (11.02%; 781 out of 7,085)	https://github.com/ExpectozJJ/PerSpect-Ensemble-Learning (Wee & Xia, 2022)
mCSM-AB	Graph-based signatures	r=0.53 on 29 Ab-Ag complexes	Regression	Yes (100%; 645)	https://biosig.lab.uq.edu.au/mcsm_ab/prediction (Pires & Ascher, 2016)
FoldX	Empirical function	64%	Empirical function	No	https://foldxsuite.crg.eu/ (Delgado et al., 2019)

The links that are not active (as checked on Dec 2024) are denoted with "*" sign.

also for other applications such as protein engineering, computational mutagenesis, and docking (Ben-Shimon & Eisenstein, 2010; Keskin et al., 2005; Kortemme et al., 2004; Vangone & Bonvin, 2015).

7.2.4.1 Binding affinity prediction methods

There are several sequence-based binding affinity prediction methods available for protein–protein complexes. Protein-Protein Affinity Predictor (PPA-Pred), a tool for predicting the real value of binding affinity from amino acid sequences, is based on a multiple regression model (Yugandhar & Michael Gromiha, 2014). The sequence-based features include predicted binding site residues and property values of 20 amino acids from the AAindex database (Kawashima, 2000). The training data included antibody–antigen complexes as well and showed a correlation from 0.74 to 0.99 for different classes of complexes. ISLAND (*In SiLico* protein AffiNity preDictor), another sequence-based tool, combined a kernel representation of protein sequences with the support vector regression to predict the binding affinity. The correlation between the experimental and predicted ΔG was 0.44, and the structure-based benchmark for protein–protein binding affinity data was used (Abbasi et al., 2020). Another tool based on the recurrent convolutional neural network (RCNN) that takes amino acid sequence as the input for prediction of protein–protein binding affinity was developed by Chen et al. (2019). The correlation of 0.87 was obtained from a Siamese residual RCNN with a pre-trained embedding representation of protein sequences. A similar model known as DPPI was developed by Hashemifar et al., which was based on DL model (Hashemifar et al., 2018). Another end-to-end DL framework that learns both robust local features and contextualized information from sequences is Protein–Protein Interaction Prediction Based on Siamese Residual RCNN (PIPR), which was able to predict the binding affinities between the interacting proteins (Chen et al., 2019). Moreover, a method named regularized automatic prediction of PPIs using deep learning (RAPPPID) was trained by considering pairs of amino acid sequences of interacting proteins and allowing better distinctiveness between the interacting motifs and other parts of the proteins. Further, Xue et al. developed a method based on pre-trained embedding; and residual RCNN, structure information, and functions of proteins were used in the pre-training stage to generate sequence embeddings. However, the performance of the model was poor with a correlation of only 0.26 (Xue et al. 2021). Fischer et al considered the UMI counts in 10x Genomics single-cell immune profiling dataset as binding strength of the TCR-pMHC complex and developed a model named "TcellMatch" with r^2 values of 0.63 (Fischer et al., 2020). Makowski et al. combined high-throughput experimental methods including deep sequencing and ML to identify therapeutic antibody variants with superior combinations of affinity and non-specific binding (Makowski et al., 2022). The model is trained on binary datasets for affinity and specificity and does not consider real binding affinity prediction, yet it correlates with continuous affinity values. A similar approach is also used in the pipeline called RESP that is trained on over 3 million human B-cell receptor sequences. The pipeline efficiently identifies the high-affinity antibodies but is not designed to predict binding affinity (Parkinson et al., 2023). Bachas et al. (2022) used deep contextual language models trained on high-throughput affinity data to quantitatively predict binding of unseen antibody sequence variants and included a metric to score antibody variants for similarity to natural IGs. It is important to note that sequence-based methods are unable to perform predictions for different binding poses of the interacting proteins and do not take conformational changes into account (Gromiha et al., 2017).

The structure-based methods have significant advantages over sequence-based prediction. However, they usually lack the high-quality structural data. The first study to relate binding affinities with a set of structures was by Horton and Lewis who used 15 ΔG values from literature as training data and used interface polar and non-polar groups as features to obtain a linear regression coefficient of $r=0.96$, and a mean absolute difference of 0.8 kcal/mol between the calculated and observed ΔG values (Horton & Lewis, 1992). Kastritis et al. (2011) benchmarked the protein–protein binding affinity data for 144 protein–protein complexes (including 19 antibody–antigen complexes) with varying biological functions and observed that the performance was poor on a validation set because of noise in the experimental data (Kastritis & Bonvin, 2011). Faster methods based on empirical functions (empirical, force-field-based potentials, statistical potentials, and scoring functions used in docking) could be successful on small training sets (Audie & Scarlata, 2007; Horton & Lewis, 1992) but most of them fail to predict binding affinity accurately (Rosato, 2010) for large datasets or discriminate between binders and non-binders (Sacquin-Mora et al., 2008). Vangone and Bonvin (2015) worked on relating the interfacial contacts (ICs) and noninteracting surface (NIS) residues with the experimental binding affinity and obtained a Pearson correlation of –0.73. They used a training dataset of 81 protein–protein complexes (including ten antibody–antigen complexes) and developed a method, PROtein binDIng enerGY prediction (PRODIGY), that can predict the binding affinity of protein–protein complexes from their 3D structure with a correlation of 0.73 between experimental and predicted ΔG values (Vangone & Bonvin, 2015; Xue et al., 2016). Vangone and Bonvin (2015) further analyzed 122 complexes with binding affinity data and observed that structure-based methods such as free energy perturbation and thermodynamics integration could be very accurate, but due to their computational costs, their application is extremely limited. Apart from regression models, QSAR models were also used for relating structural descriptors with binding affinity of protein–protein complexes using structure-based benchmark datasets compiled by Kastritis et al. (2011) and Zhou et al. (2013). Using the same dataset Marillet et al. (2016) utilized 12 features which account for enthalpic and entropic changes upon binding and devised protein–protein affinity prediction models. Wang et al. used the knowledge-based potentials and reformulated the binding affinity based on the Monte Carlo algorithm to obtain a Pearson correlation of 0.7 for the prediction (Wang, Su, et al., 2021). FoldX by Delgado et al. (2019) uses an empirical function for predicting the binding free energy between the protein–protein complexes.

In recent years, ML methods have been developed which are faster and more accurate for predicting protein–protein binding affinity (Li et al., 2022). PPI-Affinity is a web-based tool that predicts the binding affinity using support vector machines and other classic ML models (Romero-Molina et al., 2022). The ML model showed a performance of $r=0.77$ on the SKEMPI dataset (Romero-Molina et al., 2022). In addition, a few antibody–antigen-specific binding affinity prediction methods have also been developed in the past few years. CSM-AB developed by Myung et al. (2022) is a ML method capable of predicting antibody–antigen binding affinity by modeling interaction interfaces as graph-based signatures. It obtained a correlation of up to 0.64 on a blind test dataset. Yang et al. carried out a ML analysis based on interface and surface areas for antibody–antigen complexes. They constructed different models to predict antibody–antigen binding using area-based and contacts-based descriptors through constructing and training different predictive models. They obtained the best correlation of 0.85 (with 33 antibody–antigen complexes) and 0.74 (with 262 antibody–antigen complexes). Their results showed that the area-based descriptors are slightly better than the contacts-based descriptors in terms of predictive power; the new models specific for antibody–protein antigen binding affinity prediction are superior to the previously used general models for predicting the protein–protein binding affinities; and the performances of the best area-based and contacts-based models are better than the performances of the graph-based model (i.e., CSM-AB) specific for antibody–antigen binding affinity prediction (Yang et al., 2023). Recently, Sharma et al. developed a model for predicting the binding affinity for SARS-CoV-2 spike protein and neutralizing antibodies using 29 antibody–antigen complexes. They obtained a correlation of 0.90 for the jack-knife test on SARS-CoV-2 protein data bank structures (Sharma et al., 2022).

7.2.4.2 Change in binding affinity upon mutation prediction methods

Mutations in a protein cause changes in its structure, function, interactions, and binding affinity, which can lead to disease (Gromiha et al., 2016). Several methods have been developed over the years for the prediction of change in binding affinity upon mutation utilizing sequence, structure, and energy-based features as well as a combination of them. BeAtMuSiC is a coarse-grained predictor of the changes in binding free energy induced by point mutations. It is based on a set of statistical potentials derived from known protein structures and integrates the mutation's effect on: (i) the strength of the interactions at the interface, and (ii) the overall stability of the complex (Dehouck et al., 2013). The correlation obtained by BeAtMuSiC with 90% of the SKEMPI dataset (including antibody–antigen complexes) was 0.70. Brender et al. used random forest training and combined interface structure profile scores with residue-level coarse-grained potentials to develop a composite predictive model. They obtained a correlation of >0.8 between the predicted and observed binding free energy changes upon mutation using the SKEMPI database (Jankauskaitė et al., 2018). The single amino acid mutation-based change in binding free energy (SAAMBE) method developed by Petukh et al. (2015) took advantage of both sequence and structure-based methods and utilized structure minimization, statistical energy scoring functions, and modified molecular mechanics energies combined with the Poisson–Boltzmann surface area continuum solvation (MM-PBSA) for estimating the effect of single and multiple mutations on binding affinity. The mmCSM-PPI, developed by Rodrigues et al. (2021), predicts binding affinity change upon mutation using graph-based signatures, which describe the distance patterns between atoms on the binding interface (Rodrigues et al., 2021). Another method that depends on graph-based signatures is mCSM-AB, which is specific for predicting antibody–antigen affinity changes upon mutation (Pires & Ascher, 2016). Methods such as GeoPPI, TopNetTree, and PerSpect-EL are based on neural networks and persistent homology (Liu et al., 2021; Wang et al., 2020; Wee & Xia, 2022). All these methods used the SKEMPI dataset and showed a correlation of up to 0.85 between predicted and experimental data. FoldX software can also evaluate the effect of mutations on protein stability, interaction, folding, and dynamics using structures (Delgado et al., 2019). Jemimah et al. developed ProAffiMuSeq, a method that predicts protein–protein binding affinity change upon mutation using sequence-based features and functional class. It shows a correlation of 0.73 and a mean absolute error (MAE) of 0.86 kcal/mol in cross-validation (Jemimah et al., 2019). Another sequence-based predictor is PANDA that could predict a change in protein binding affinity upon mutation with a correlation coefficient of 0.52 (Abbasi et al., 2021). SAAMBE-SEQ is based on a gradient-boosting decision tree ML algorithm. It utilized 80 features representing evolutionary information, sequence-based features, and change of physical properties upon mutation at the mutation site and achieved a Pearson correlation coefficient (PCC) of 0.83 (Li et al., 2021).

7.2.5 Biophysical Parameters Affecting Antibody Design

There are certain biophysical parameters which need consideration for the optimization of binding affinity, specificity, and developability of an antibody (Khetan et al., 2022). These parameters include variations of protein features such as charge, pH/isoelectric point, hydrophobicity, and CDR length and are calculated for either the whole antibody or a specific region of the antibody. The first developability guidelines were defined using 137 clinical-stage antibodies, which empirically define boundaries of antibody drug-like behavior (Jain et al., 2017). Further, Raybould et al. presented five biophysical parameters for antibodies similar to the "Lipinski rule of fives" for orally active drugs, which include CDR length, surface hydrophobicity of the near CDR region, and three charge-based metrics (patches of positive charge, patches of negative charge near the CDR region, and charge symmetry of the surface exposed residues) (Raybould et al., 2019). More recently, *in silico* analyses of the variable regions of 77 marketed antibody-based biotherapeutics have revealed five non-redundant physicochemical descriptors, which represent stability, isoelectric point, and molecular surface characteristics of Fv regions (Ahmed et al., 2021). Another study on the same dataset found that antibody specificity is dependent on the net charge of CDR regions, with positively charged CDRs having a higher risk of low specificity than negatively charged antibodies (Rabia et al., 2018). Negatively charged CDRs were also linked with the poor biophysical properties of the antibody. Bashour et al. (2024) conducted a computational assessment of 40 sequence-based and 46 structure-based developability parameters (DPs) across more than two million native and human-engineered single-chain antibody sequences. Their analysis revealed that structure-based DPs exhibited lower redundancy compared to sequence-based DPs, suggesting that sequence DPs are more predictable and operate within a more constrained design space. Sharma et al. looked into the viscosity and clearance of antibodies (Sharma et al., 2014). They observed that antibody viscosity increases with the increase in charge dipole distribution and hydrophobicity, and decreases with net charge. On the other hand, antibody clearance correlates with high hydrophobicity of CDR regions and highly positive/negative net charge. An analysis of a small set of FDA-approved antibodies revealed that the shift in isoelectric point, and interaction between negatively charged cell membranes and predominantly positive surface charge of antibodies can significantly influence the kinetics of blood clearance and tissue deposition (Boswell et al., 2010).

The conformational stability of an antibody not only depends on intramolecular interactions but also on interactions between the protein and its surrounding solvent. Formulation is one way to stabilize antibodies and prevent them from aggregation or degradation (Narayanan et al., 2021b; Strickley & Lambert, 2021). Antibodies have also been engineered to have binding affinity conditional to changes in pH. Two applications exist: either binding is strengthened at mildly acidic pH relative to neutral pH, or high affinity is maintained at neutral pH but lowered at acidic pH. Antibodies engineered to have stronger binding at acidic pH are more selective for acidified environments, such as the tumor microenvironment (Johnston et al., 2019). In contrast, lower affinity at acidic pH increases the release of antigen during the endosome within the antibody recycling pathway, which can confer more than one neutralization cycle per antibody (Igawa et al., 2013). As the histidine amino acid residue naturally experiences a change in charge within the biological pH range relevant to these applications, it has been used for engineering pH switches into antibody CDR loops using *in vitro* display technologies (Schröter et al., 2015). pH-dependent structural changes that affect antibody biophysical properties have been reported. A histidine located in the antibody light chain was presumed to be important in the trans-isomerization of a proline residue in the CDRH3 loop of an anti-HIV antibody. A comparison of structures determined in neutral and acidic pH environments revealed that the lower affinity was due to a change in CDRH3 conformation at acidic pH (Masiero et al., 2020). Additionally, the change in the environment of a pair of aspartate residues was hypothesized as being important in affecting the CDRH3 loop dynamics of a therapeutic antibody (Lan et al., 2020) at different pHs, although the effect on affinity at acidic pH wasn't measured. Determinants outside of the paratope that affect CDRH3 conformational dynamics may be a consideration in the design of antibodies with pH-dependent binding properties, in addition to the epitope–paratope interface.

Several other biophysical parameters can influence the design of antibodies on a case-by-case basis (Tiller & Tessier, 2015). For example, a study analyzed the role of the "DE loop," which sits adjacent to CDR1 and CDR2 and joins the D and E strands on the antibody v-type fold mainly in the HIV-related antibodies. These loops are treated as FRs, although mutations in these regions can affect CDR conformations and also sometimes contact antigens. The study observes insertion in the DE loop in many antibodies, which are related to broadly neutralizing HIV-1 antibodies (bNabs) (Kelow et al., 2020). Evolutionary information calculated as a position-specific scoring matrix (PSSM) has also been used in antibody design (Finn et al., 2020; Petersen et al., 2021; Schmitz et al., 2022; Warszawski et al., 2019). In residue-level mutations, optimization of electrostatic interaction and H-bond can potentially improve antibody binding affinity (Lippow et al., 2007). Researchers have also introduced intramolecular disulfide bonds within antibodies to improve folding stability (Hagihara & Saerens, 2014; Kim et al., 2012; Nakamura et al., 2021; Saerens et al., 2008). Post-translational modifications (PTMs) of constant regions modulate the binding and activity of an antibody. PTMs allow the generation of antibody variants with improved activity and potency without changing the sequence/structure of the antibody. PTMs include covalently adding functional groups or proteins, proteolysis of regulatory subunits, or degradation of the full protein. The wide variety of PTMs includes chain additions (N- and O-linked glycosylation, glycation, cysteinylation, and sulfation) (Kayser et al., 2011; Qasba, 2015; Reusch & Tejada, 2015), chain trimming (C-terminal lysine clipping) (Faid et al., 2021; van den Bremer et al., 2015), and amino acid modifications (cyclization into an N-terminal pyro-glutamic acid, deamidation, oxidation, isomerization, and carbonylation) (Gupta et al., 2022; Liu et al., 2019; Lu et al., 2019; Xie et al., 2009; Yang et al., 2014). Computational approaches and databases related to PTMs have been discussed in detail previously (Jefferis, 2016; Ramazi & Zahiri, 2021; Vatsa, 2022), and the effect of each PTM on the antibody is summarized by Chiu et al. (2019b).

7.2.6 Role of Mutational Scanning in Antibody–Antigen Interaction Prediction

Mutational scanning is transforming the prediction of binding affinity and stability for the ag-ab complexes. It helps generate a large amount of data, which leads to more accurate AI prediction. Assessing all the mutations at once also accelerates the analysis of the interactions for the protein–protein complex. Mutational scanning could be done by both computational and experimental methods. The experimental deep mutational scanning study by Greaney et al. assessed all possible single amino acid variants in the spike protein of SARS-CoV-2 and provided immune escape maps for mutations in the presence of antibodies (Greaney et al., 2021). Computationally scanning the antibody–antigen interface residues for all 20 amino acid mutations for change in binding affinity has led to the identification of antibody escape mutations for the SARS-CoV-2 spike protein (Greaney et al., 2021; Sharma et al., 2022). Moreover, mutations that increase the binding affinity and stability of the SARS-CoV-2 to the human cell receptor ACE-2 could be identified by mutational scanning (Teng et al., 2021). Furthermore, analyzing each residue for all the amino acid mutations can help identify its role in the binding and stability of the complex and assist in deriving physiological mechanisms. It will assist in antibody engineering, leading to better antibody design, which in turn may lead to the development of better therapeutics for diseases such as COVID-19 and cancer.

7.2.7 Language Models for Antibody Structure and Function Prediction

During the last several years, ML and DL have made a breakthrough in various fields of research, including natural language processing (NLP) that is responsible for text and language analysis. One of the pillars of recent NLP progress is language models (LMs) (Ofer et al., 2021). LM is an ML term representing the probability distribution over sentences of tokens such as characters, words, or subwords. In the biological context, such sentences are amino acid sequences (Vu, Akbar, et al., 2022). Protein sequences similar to natural languages contain contextual information in inter-residue relationships (An & Weng, 2022). Deep LMs can reveal molecular characteristics based on large-scale protein sequence data. Protein LMs encode amino acid sequences into vector representations that reflect their structural, functional, and evolutionary properties (Bepler & Berger, 2021).

There are several LM models developed for proteins (Bepler & Berger, 2021; Ofer et al., 2021; Unsal et al., 2022), which also include a few antibody-specific models

TABLE 7.4 List of language models developed for protein/antibody structure and function prediction

LM MODELS/ REFERENCE	*ARCHITECTURE*	*APPLICATIONS*
ProtGPT2 (Ferruz et al., 2022)	Transformer decoder	Generation of de novo protein sequences based on the natural sequence rules
AlphaFold2 (Jumper et al., 2021)	Transformer encoder	Protein structure prediction
ProteinBERT (Brandes et al., 2022)	Transformer encoder	Prediction of protein characteristics (function, structure, PTMs, and biophysical properties)
AntiBERTa (Leem et al., 2022)	Transformer encoder	Prediction of antibody function (paratope prediction)
AbLang (Olsen et al., 2022b)	Transformer encoder	Restoration of missing residues in antibody sequence data
OpenFold (Ahdritz et al., 2022)	Transformer encoder	Protein structure prediction
OmegaFold (Wu et al., 2022)	Transformer encoder	Protein structure prediction
ESMFold (Lin et al., 2022)	Transformer encoder	Protein structure prediction
IgFold (Ruffolo et al., 2022a)	Transformer encoder	Antibody structure prediction
ImmuneBuilder (Abanades et al., 2022)	Transformer encoder	Antibody structure prediction
AlphaFold-Multimer (Evans et al., 2022)	Transformer encoder	Multi-chain protein complex structure prediction
DeepAb (Ruffolo et al., 2022b)	Bidirectional long short-term memory (biLSTM) encoder–decoder	Antibody structure prediction
RoseTTAFold (Baek et al., 2021)	Transformer encoder	Protein structure prediction
IgLM (Shuai et al., 2022)	Transformer decoder	Generation of antibody synthetic sequence libraries (re-designing variable-length spans)
ProGen2 (Nijkamp et al., 2022)	Transformer decoder	Generation of protein sequences, protein fitness prediction

(Table 7.4) (Vu, Robert, et al., 2022). Most of the current protein and antibody LMs are responsible for structure prediction. Most deep LMs have similar architecture types, following either the evolutionary-focused AlphaFold2 (Jumper et al., 2021) model or classical NLP models such as BERT (Devlin et al. 2018) or GPT2 (Radford et al., 2019). A detailed overview of the LM models is provided by Hu et al. (2022).

7.2.8 Role of MD Simulations in Antibody Structure-Function Prediction

MD is a simulation technique that predicts the system's evolution in time, based on the physical laws of atomic interactions. The system can be any molecular object, such as small-molecule ligands, nucleic acid molecules, proteins, and macro-complexes (Hollingsworth & Dror, 2018). MD simulations generate atomic trajectories describing the system's behavior. The scale of the simulation can vary from femtoseconds for atomic vibration investigations to milliseconds for protein folding studies. The accuracy and correctness of MD simulations depend on a description of how the molecules will interact, called a force field (FF), which defines the parameter sets used to calculate the potential energy of a system (Lopes et al., 2015). In the context of antibodies, MD simulations are utilized for (i) prediction of antibodies structure dynamics (Chen et al., 2016; Tucs et al., 2023), (ii) determination of CDR loops conformational ensemble (Fernández-Quintero et al., 2018, 2019; Fernández-Quintero, Heiss, et al., 2020; Löhr et al., 2022), (iii) investigation of paratope–epitope binding mechanism and structure (Bekker et al., 2020; Brandt et al., 2021; Fernández-Quintero et al., 2022; Fernández-Quintero, Hoerschinger, et al., 2020; Huang et al., 2022; Ieong et al., 2015; Lees et al., 2017), (iv) exploration of antibody–antigen interface conformation (Wong et al., 2022; Yamashita, 2018), (v) understanding of affinity maturation process (Fernández-Quintero, Loeffler, et al., 2020), (vi) improvement of binding affinity (Conti et al., 2022; Kralj et al., 2021; Wong et al., 2022; Yamashita et al., 2019), (vii) influence of substitutions on antibody–antigen binding (Lees et al., 2017), and (viii) benchmarking of antibody MD parameters (Al Qaraghuli et al., 2018).

Classical MD simulations are time-consuming and computationally expensive (Ciccotti et al., 2022). Some MD simulations can take months to be finished. Hence, there are many variations of MD simulations, making conformational sampling faster and easier. Such approaches are called enhanced sampling and include replica-exchange MD simulation (REMD), accelerated MD (aMD), metadynamics (MetaD), and Markov state (MC) (Lazim et al., 2020; Qing et al., 2022). Most of the approaches are available within MD software tools such as GROMACS+PLUMED, AMBER, or NAMD+VMD. Apart from enhanced sampling techniques, another way to speed up MD simulation is to use ML. ML-MD uses reference data from the electronic structure calculations to parametrize interatomic potentials and calculate the energies and forces of a system (Ciccotti et al., 2022). ML models, mostly based on neural networks or kernel-based approaches, are able to interpolate the reference data and predict force fields with equivalent accuracy of quantum *ab initio* calculations with much higher speed (Unke et al., 2021).

7.3 CONCLUSION

In this chapter, we have summarized the databases for the antibody sequence and structure; different tools for predicting antibody–antigen complex structure and binding affinity; LMs; and MD simulations for antibody structure-function prediction. The recent computational resources for the above-mentioned tasks have increasingly used ML or DL approaches and have shown much better performance than the conventional methods (Wilman et al., 2022). However, the limited availability of data has been a major challenge for the robustness, transferability, and generalization of such models. To counter these challenges, especially in the case of antibodies, the paradigm is shifting toward: (i) either developing models on large-scale synthetic data, which can represent real-world data (Robert et al., 2022) or (ii) training models on large-scale protein–protein interaction data and utilizing transfer learning for antibody–antigen interaction data (Pittala & Bailey-Kellogg, 2020). Recently, significant advances in the field of protein/antibody structure prediction have opened the possibility of reimagining epitope/paratope prediction. However, several challenges, including cross-reactivity, polyspecificity, and sensitivity toward minor changes in sequence/structure/pH, are yet to be answered. Understanding the complex rules of antigen and antibody interactions can be the starting point for the end-to-end dry lab-based early-stage development of antibody therapeutics.

COMPETING INTERESTS

V.G. declares advisory board positions in aiNET GmbH, Enpicom B.V, Absci, Omniscope, and Diagonal Therapeutics. VG is a consultant for Adaptyv Biosystems, Specifica Inc., Roche/Genentech, immunai, and LabGenius, Proteinea and FairJourney Biologics.

ACKNOWLEDGMENTS

The project has received partial funding from Leona M. and Harry B. Helmsley Charitable Trust (#2019PG-T1D011), UiO World-Leading Research Community, UiO: LifeScience Convergence Environment Immunolingo, EU Horizon 2020 iReceptorplus (#825821), a Norwegian Cancer Society Grant (#215817), Research Council of Norway projects (#300740, #331890), the European Union (ERC, AB-AG-INTERACT, 101125630, to VG), and a Research Council of Norway IKTPLUSS project (#311341) to VG. This project has received funding from the European Union's Horizon 2020 research and innovation program under the Marie Skłodowska-Curie grant agreement

No 801133 to PR. We thank the Department of Biotechnology and the Indian Institute of Technology Madras for their computational facilities, as well as the Ministry of Human Resource and Development (MHRD) for the HTRA scholarship to DS.

REFERENCES

Abanades, B., Wong, W. K., Boyles, F., Georges, G., Bujotzek, A., & Deane, C. M. (2023). ImmuneBuilder: Deep-learning models for predicting the structures of immune proteins. *Communications Biology*, *6*(1), 575.

Abbasi, W. A., Abbas, S. A., & Andleeb, S. (2021). PANDA: predicting the change in proteins binding affinity upon mutations by finding a signal in primary structures. *Journal of Bioinformatics and Computational Biology*, *19*(4), 2150015.

Abbasi, W. A., Yaseen, A., Hassan, F. U., Andleeb, S., & Minhas, F. U. A. A. (2020). ISLAND: in-silico proteins binding affinity prediction using sequence information. *BioData Mining*, *13*(1), 20.

Ahdritz, G., Bouatta, N., Floristean, C., Kadyan, S., Xia, Q., Gerecke, W., … AlQuraishi, M. (2024). OpenFold: retraining AlphaFold2 yields new insights into its learning mechanisms and capacity for generalization. *Nature Methods, 21*(8), 1514–1524.

Ahmed, L., Gupta, P., Martin, K., Scheer, J. M., Nixon, A. E., & Kumar, S. (2021). Intrinsic physicochemical profile of marketed antibody-based biotherapeutics. Proceedings of the National Academy of Sciences of the United States of America, *118*(37). https://doi.org/10.1073/pnas.2020577118

Akbar, R., Bashour, H., Rawat, P., Robert, P. A., Smorodina, E., Cotet, T.-S., Flem-Karlsen, K., Frank, R., Mehta, B. B., Vu, M. H., Zengin, T., Gutierrez-Marcos, J., Lund-Johansen, F., Andersen, J. T., & Greiff, V. (2022). Progress and challenges for the machine learning-based design of fit-for-purpose monoclonal antibodies. *mAbs*, *14*(1), 2008790.

Akbar, R., Robert, P. A., Pavlović, M., Jeliazkov, J. R., Snapkov, I., Slabodkin, A., Weber, C. R., Scheffer, L., Miho, E., Haff, I. H., Haug, D. T. T., Lund-Johansen, F., Safonova, Y., Sandve, G. K., & Greiff, V. (2021). A compact vocabulary of paratope-epitope interactions enables predictability of antibody-antigen binding. *Cell Reports*, *34*(11), 108856.

Al Qaraghuli, M. M., Kubiak-Ossowska, K., & Mulheran, P. A. (2018). Thinking outside the laboratory: analyses of antibody structure and dynamics within different solvent environments in molecular dynamics (MD) simulations. *Antibodies (Basel, Switzerland), 7*(3). https://doi.org/10.3390/antib7030021

Amaro, R. E., Baudry, J., Chodera, J., Demir, Ö., McCammon, J. A., Miao, Y., & Smith, J. C. (2018). Ensemble Docking in Drug Discovery. *Biophysical Journal*, *114*(10), 2271–2278.

Ambrosetti, F., Jiménez-García, B., Roel-Touris, J., & Bonvin, A. M. J. J. (2020). Modeling Antibody-Antigen complexes by information-driven docking. *Structure*, *28*(1), 119–129.e2.

An, J., & Weng, X. (2022). Collectively encoding protein properties enriches protein language models. *BMC Bioinformatics*, *23*(1), 467.

Audie, J., & Scarlata, S. (2007). A novel empirical free energy function that explains and predicts protein-protein binding affinities. *Biophysical Chemistry*, *129*(2–3), 198–211.

Bachas, S., Rakocevic, G., Spencer, D., Sastry, A. V., Haile, R., Sutton, J. M., Kasun, G., Stachyra, A., Gutierrez, J. M., Yassine, E., Medjo, B., Blay, V., Kohnert, C., Stanton, J. T., Brown, A., Tijanic, N., McCloskey, C., Viazzo, R., Consbruck, R., … Spreafico, R. (2022). Antibody optimization enabled by artificial intelligence predictions of binding affinity and naturalness. bioRxiv, 2022.08.16.504181. https://doi.org/10.1101/2022.08.16.504181

Baek, M., DiMaio, F., Anishchenko, I., Dauparas, J., Ovchinnikov, S., Lee, G. R., Wang, J., Cong, Q., Kinch, L. N., Schaeffer, R. D., Millán, C., Park, H., Adams, C., Glassman, C. R., DeGiovanni, A., Pereira, J. H., Rodrigues, A. V., van Dijk, A. A., Ebrecht, A. C., … Baker, D. (2021). Accurate prediction of protein structures and interactions using a three-track neural network. *Science*, *373*(6557), 871–876.

Bashour, H., Smorodina, E., Pariset, M., Zhong, J., Akbar, R., Chernigovskaya, M., Lê Quý, K., Snapkow, I., Rawat, P., Krawczyk, K., Sandve, G. K., Gutierrez-Marcos, J., Gutierrez, D. N.-Z., Andersen, J. T., & Greiff, V. (2024). Biophysical cartography of the native and human-engineered antibody landscapes quantifies the plasticity of antibody developability. *Communications Biology, 7*(1), 922.

Bekker, G.-J., Fukuda, I., Higo, J., & Kamiya, N. (2020). Mutual population-shift driven antibody-peptide binding elucidated by molecular dynamics simulations. *Scientific Reports*, *10*(1), 1406.

Ben-Shimon, A., & Eisenstein, M. (2010). Computational mapping of anchoring spots on protein surfaces. *Journal of Molecular Biology*, *402*(1), 259–277.

Bepler, T., & Berger, B. (2021). Learning the protein language: evolution, structure, and function. *Cell Systems*, *12*(6), 654–669.e3.

Bodi, K., Prokaeva, T., Spencer, B., Eberhard, M., Connors, L. H., & Seldin, D. C. (2009). AL-Base: a visual platform analysis tool for the study of amyloidogenic immunoglobulin light chain sequences. *Amyloid: The International Journal of Experimental and Clinical Investigation*, *16*(1), 1–8.

Boswell, C. A., Tesar, D. B., Mukhyala, K., Theil, F.-P., Fielder, P. J., & Khawli, L. A. (2010). Effects of charge on antibody tissue distribution and pharmacokinetics. *Bioconjugate Chemistry*, *21*(12), 2153–2163.

Brandes, N., Ofer, D., Peleg, Y., Rappoport, N., & Linial, M. (2022). ProteinBERT: a universal deep-learning model of protein sequence and function. *Bioinformatics*, *38*(8), 2102–2110.

Brandt, A. A. M. L., Rodrigues-da-Silva, R. N., Lima-Junior, J. C., Alves, C. R., & de Souza-Silva, F. (2021). Combining well-tempered metadynamics simulation and SPR assays to characterize the binding mechanism of the universal T-lymphocyte tetanus toxin Epitope TT830–843. *BioMed Research International*, 2021, 5568980.

Brooks, B. D., Closmore, A., Yang, J., Holland, M., Cairns, T., Cohen, G. H., & Bailey-Kellogg, C. (2020). Characterizing epitope binding regions of entire antibody panels by combining experimental and computational analysis of antibody: antigen binding competition. *Molecules*, *25*(16). https://doi.org/10.3390/molecules25163659

Brown, A. J., Snapkov, I., Akbar, R., Pavlović, M., Miho, E., Sandve, G. K., & Greiff, V. (2019). Augmenting adaptive immunity: progress and challenges in the quantitative engineering and analysis of adaptive immune receptor repertoires. *Molecular Systems Design & Engineering*, *4*(4), 701–736.

Chaves, B., Sartori, G. R., Vasconcelos, D. C. A., Savino, W., Caffarena, E. R., Cotta-de-Almeida, V., & da Silva, J. H. M. (2020). Guidelines to predict binding poses of antibody-integrin complexes. *ACS Omega*, *5*(27), 16379–16385.

Chen, M., Ju, C. J. T., Zhou, G., Chen, X., Zhang, T., Chang, K.-W., Zaniolo, C., & Wang, W. (2019). Multifaceted protein–protein interaction prediction based on Siamese residual RCNN. *Bioinformatics*, 35(14), i305–i314. https://doi.org/10.1093/bioinformatics/btz328

Chen, M., Wang, J., & Zhu, W. (2016). Molecular mechanism and energy basis of conformational diversity of antibody SPE7 revealed by molecular dynamics simulation and principal component analysis. *Scientific Reports*, *6*, 36900.

Chinery, L., Wahome, N., Moal, I., & Deane, C. M. (2023). Paragraph—antibody paratope prediction using graph neural networks with minimal feature vectors. *Bioinformatics*, 39(1). https://doi.org/10.1093/bioinformatics/btac732

Chiu, M. L., Goulet, D. R., Teplyakov, A., & Gilliland, G. L. (2019a). Antibody structure and function: the basis for engineering therapeutics. *Antibodies*, *8*(4), 55.

Chiu, M. L., Goulet, D. R., Teplyakov, A., & Gilliland, G. L. (2019b). Antibody structure and function: the basis for engineering therapeutics. *Antibodies (Basel, Switzerland)*, *8*(4). https://doi.org/10.3390/antib8040055

Chothia, C., & Lesk, A. M. (1987). Canonical structures for the hypervariable regions of immunoglobulins. *Journal of Molecular Biology*, *196*(4), 901–917.

Ciccotti, G., Dellago, C., Ferrario, M., Hernández, E. R., & Tuckerman, M. E. (2022). Molecular simulations: past, present, and future (a topical issue in EPJB). *The European Physical Journal. B*, *95*(1), 3.

Comeau, S. R., Gatchell, D. W., Vajda, S., & Camacho, C. J. (2004). ClusPro: a fully automated algorithm for protein-protein docking. *Nucleic Acids Research*, 32(Web Server issue), W96–W99.

Conti, S., Lau, E. Y., & Ovchinnikov, V. (2022). On the rapid calculation of binding affinities for antigen and antibody design and affinity maturation simulations. *Antibodies (Basel, Switzerland)*, *11*(3). https://doi.org/10.3390/antib11030051

Corrie, B. D., Marthandan, N., Zimonja, B., Jaglale, J., Zhou, Y., Barr, E., Knoetze, N., Breden, F. M. W., Christley, S., Scott, J. K., Cowell, L. G., & Breden, F. (2018). iReceptor: a platform for querying and analyzing antibody/B-cell and T-cell receptor repertoire data across federated repositories. *Immunological Reviews*, *284*(1), 24–41.

Corso, G., Stärk, H., Jing, B., Barzilay, R., & Jaakkola, T. (2022). DiffDock: diffusion steps, twists, and turns for molecular docking. arXiv [q-bio.BM], arXiv. https://arxiv.org/abs/2210.01776

Davila, A., Xu, Z., Li, S., Rozewicki, J., Wilamowski, J., Kotelnikov, S., Kozakov, D., Teraguchi, S., & Standley, D. M. (2022). AbAdapt: an adaptive approach to predicting Antibody–Antigen complex structures from sequence. *Bioinformatics Advances*, *2*(1), vbac015.

Dehouck, Y., Kwasigroch, J. M., Rooman, M., & Gilis, D. (2013). BeAtMuSiC: prediction of changes in protein–protein binding affinity on mutations. *Nucleic Acids Research*, 41(W1), W333–W339). https://doi.org/10.1093/nar/gkt450

Delgado, J., Radusky, L. G., Cianferoni, D., & Serrano, L. (2019). FoldX 5.0: working with RNA, small molecules and a new graphical interface. *Bioinformatics*, *35*(20), 4168–4169.

Del Vecchio, A., Deac, A., Liò, P., & Veličković, P. (2021). Neural message passing for joint paratope-epitope prediction. arXiv [q-bio.QM], arXiv. https://arxiv.org/abs/2106.00757

Devlin, J., Chang, M.-W., Lee, K., & Toutanova, K. (2018). BERT: Pre-training of deep bidirectional Transformers for language understanding. In arXiv [cs.CL]. https://arxiv.org/abs/1810.04805v2http://arxiv.org/abs/1810.04805

Dominguez, C., Boelens, R., & Bonvin, A. M. J. J. (2003). HADDOCK: a protein-protein docking approach based on biochemical or biophysical information. *Journal of the American Chemical Society*, *125*(7), 1731–1737.

Dunbar, J., Krawczyk, K., Leem, J., Baker, T., Fuchs, A., Georges, G., Shi, J., & Deane, C. M. (2014). SAbDab: the structural antibody database. *Nucleic Acids Research*, *42*(Database issue), D1140–D1146.

Ehrenmann, F., Kaas, Q., & Lefranc, M.-P. (2010). IMGT/3Dstructure-DB and IMGT/DomainGapAlign: a database and a tool for immunoglobulins or antibodies, T cell receptors, MHC, IgSF and MhcSF. *Nucleic Acids Research*, *38*(Database issue), D301–D307.

Elhanati, Y., Murugan, A., Callan, C. G., Jr, Mora, T., & Walczak, A. M. (2014). Quantifying selection in immune receptor repertoires. *Proceedings of the National Academy of Sciences of the United States of America*, *111*(27), 9875–9880.

Eroshkin, A. M., LeBlanc, A., Weekes, D., Post, K., Li, Z., Rajput, A., Butera, S. T., Burton, D. R., & Godzik, A. (2014). bNAber: database of broadly neutralizing HIV antibodies. *Nucleic Acids Research*, *42*(Database issue), D1133–D1139.

Evans, R., O'Neill, M., Pritzel, A., Antropova, N., Senior, A., Green, T., Žídek, A., Bates, R., Blackwell, S., Yim, J., Ronneberger, O., Bodenstein, S., Zielinski, M., Bridgland, A., Potapenko, A., Cowie, A., Tunyasuvunakool, K., Jain, R., Clancy, E., … Hassabis, D. (2022). Protein complex prediction with AlphaFold-Multimer. bioRxiv, 2021.10.04.463034. https://doi.org/10.1101/2021.10.04.463034

Evans, R., & Thurber, G. M. (2022). Design of high avidity and low affinity antibodies for in situ control of antibody drug conjugate targeting. *Scientific Reports*, *12*(1), 7677.

Faid, V., Leblanc, Y., Berger, M., Seifert, A., Bihoreau, N., & Chevreux, G. (2021). C-terminal lysine clipping of IgG1: impact on binding to human FcγRIIIa and neonatal Fc receptors. *European Journal of Pharmaceutical Sciences*, *159*, 105730.

Ferdous, S., & Martin, A. C. R. (2018). AbDb: antibody structure database—a database of PDB-derived antibody structures. *Database: The Journal of Biological Databases and Curation*, 2018, bay040.

Fernández-Quintero, M. L., Heiss, M. C., Pomarici, N. D., Math, B. A., & Liedl, K. R. (2020). Antibody CDR loops as ensembles in solution vs. canonical clusters from X-ray structures. *mAbs*, *12*(1), 1744328.

Fernández-Quintero, M. L., Hoerschinger, V. J., Lamp, L. M., Bujotzek, A., Georges, G., & Liedl, K. R. (2020). VH -VL interdomain dynamics observed by computer simulations and NMR. *Proteins*, *88*(7), 830–839.

Fernández-Quintero, M. L., Kraml, J., Georges, G., & Liedl, K. R. (2019). CDR-H3 loop ensemble in solution - conformational selection upon antibody binding. *mAbs*, *11*(6), 1077–1088.

Fernández-Quintero, M. L., Loeffler, J. R., Bacher, L. M., Waibl, F., Seidler, C. A., & Liedl, K. R. (2020). Local and global rigidification upon antibody affinity maturation. *Frontiers in Molecular Biosciences*, *7*, 182.

Fernández-Quintero, M. L., Loeffler, J. R., Kraml, J., Kahler, U., Kamenik, A. S., & Liedl, K. R. (2018). Characterizing the diversity of the CDR-H3 loop conformational ensembles in relationship to antibody binding properties. *Frontiers in Immunology*, *9*, 3065.

Fernández-Quintero, M. L., Vangone, A., Loeffler, J. R., Seidler, C. A., Georges, G., & Liedl, K. R. (2022). Paratope states in solution improve structure prediction and docking. *Structure*, *30*(3), 430–440.e3.

Ferruz, N., Schmidt, S., & Höcker, B. (2022). ProtGPT2 is a deep unsupervised language model for protein design. *Nature Communications*, *13*(1), 4348.

Finn, J. A., Dong, J., Sevy, A. M., Parrish, E., Gilchuk, I., Nargi, R., Scarlett-Jones, M., Reichard, W., Bombardi, R., Voss, T. G., Meiler, J., & Crowe, J. E., Jr. (2020). Identification of structurally related antibodies in antibody sequence databases using Rosetta-derived position-specific scoring. *Structure*, *28*(10), 1124–1130.e5.

Fischer, D. S., Wu, Y., Schubert, B., & Theis, F. J. (2020). Predicting antigen specificity of single T cells based on TCR CDR 3 regions. *Molecular Systems Biology*, 16(8). https://doi.org/10.15252/msb.20199416

Gainza, P., Sverrisson, F., Monti, F., Rodolà, E., Boscaini, D., Bronstein, M. M., & Correia, B. E. (2020). Deciphering interaction fingerprints from protein molecular surfaces using geometric deep learning. *Nature Methods*, *17*(2), 184–192.

Ganea, O.-E., Huang, X., Bunne, C., Bian, Y., Barzilay, R., Jaakkola, T., & Krause, A. (2021). Independent SE(3)-equivariant models for end-to-end rigid protein docking. arXiv [cs.AI], arXiv. https://arxiv.org/abs/2111.07786

Georgiou, G., Ippolito, G. C., Beausang, J., Busse, C. E., Wardemann, H., & Quake, S. R. (2014). The promise and challenge of high-throughput sequencing of the antibody repertoire. *Nature Biotechnology*, *32*(2), 158–168.

Glanville, J., Zhai, W., Berka, J., Telman, D., Huerta, G., Mehta, G. R., Ni, I., Mei, L., Sundar, P. D., Day, G. M. R., Cox, D., Rajpal, A., & Pons, J. (2009). Precise determination of the diversity of a combinatorial antibody library gives insight into the human immunoglobulin repertoire. *Proceedings of the National Academy of Sciences of the United States of America*, *106*(48), 20216–20221.

Glockshuber, R., Schmidt, T., & Plückthun, A. (1992). The disulfide bonds in antibody variable domains: effects on stability, folding in vitro, and functional expression in Escherichia coli. *Biochemistry*, *31*(5), 1270–1279.

Greaney, A. J., Loes, A. N., Crawford, K. H. D., Starr, T. N., Malone, K. D., Chu, H. Y., & Bloom, J. D. (2021). Comprehensive mapping of mutations in the SARS-CoV-2 receptor-binding domain that affect recognition by polyclonal human plasma antibodies. *Cell Host & Microbe*, 29(3), 463–476.e6.

Greiff, V., Menzel, U., Miho, E., Weber, C., Riedel, R., Cook, S., Valai, A., Lopes, T., Radbruch, A., Winkler, T. H., & Reddy, S. T. (2017). Systems analysis reveals high genetic and antigen-driven predetermination of antibody repertoires throughout B cell development. *Cell Reports*, *19*(7), 1467–1478.

Greiff, V., Miho, E., Menzel, U., & Reddy, S. T. (2015). Bioinformatic and statistical analysis of adaptive immune repertoires. *Trends in Immunology*, *36*(11), 738–749.

Gromiha. (2020). *Protein Interactions: Computational Methods, Analysis and Applications*. World Scientific.

Gromiha, M. M., Anoosha, P., & Huang, L.-T. (2016). Applications of protein thermodynamic database for understanding protein mutant stability and designing stable mutants. *Methods in Molecular Biology*, *1415*, 71–89.

Gromiha, M. M., Michael Gromiha, M., Yugandhar, K., & Jemimah, S. (2017). Protein–protein interactions: scoring schemes and binding affinity. *Current Opinion in Structural Biology*, *44*, 31–38. https://doi.org/10.1016/j.sbi.2016.10.016

Guest, J. D., Vreven, T., Zhou, J., Moal, I., Jeliazkov, J. R., Gray, J. J., Weng, Z., & Pierce, B. G. (2021). An expanded benchmark for antibody-antigen docking and affinity prediction reveals insights into antibody recognition determinants. *Structure*, *29*(6), 606–621.e5.

Guo, Y., Chen, K., Kwong, P. D., Shapiro, L., & Sheng, Z. (2019). cAb-Rep: a database of curated antibody repertoires for exploring antibody diversity and predicting antibody prevalence. *Frontiers in Immunology*, *10*, 2365.

Gupta, S., Jiskoot, W., Schöneich, C., & Rathore, A. S. (2022). Oxidation and deamidation of monoclonal antibody products: potential impact on stability, biological activity, and efficacy. *Journal of Pharmaceutical Sciences*, *111*(4), 903–918.

Hagihara, Y., & Saerens, D. (2014). Engineering disulfide bonds within an antibody. *Biochimica et Biophysica Acta*, *1844*(11), 2016–2023.

Harmalkar, A., Rao, R., Richard Xie, Y., Honer, J., Deisting, W., Anlahr, J., … Wei, K. Y. (2023). Toward generalizable prediction of antibody thermostability using machine learning on sequence and structure features. mAbs, 15(1), 2163584.

Hashemifar, S., Neyshabur, B., Khan, A. A., & Xu, J. (2018). Predicting protein–protein interactions through sequence-based deep learning. *Bioinformatics*, 34(17), i802–i810. https://doi.org/10.1093/bioinformatics/bty573

Hollingsworth, S. A., & Dror, R. O. (2018). Molecular dynamics simulation for all. *Neuron*, *99*(6), 1129–1143.

Horton, N., & Lewis, M. (1992). Calculation of the free energy of association for protein complexes. *Protein Science*, 1(1), 169–181. https://doi.org/10.1002/pro.5560010117

Huang, Y., Li, Z., Hong, Q., Zhou, L., Ma, Y., Hu, Y., Xin, J., Li, T., Kong, Z., Zheng, Q., Chen, Y., Zhao, Q., Gu, Y., Zhang, J., Wang, Y., Yu, H., Li, S., & Xia, N. (2022). A stepwise docking molecular dynamics approach for simulating antibody recognition with substantial conformational changes. *Computational and Structural Biotechnology Journal*, *20*, 710–720.

Hu, B., Xia, J., Zheng, J., Tan, C., Huang, Y., Xu, Y., & Li, S. Z. (2022). Protein language models and structure prediction: connection and progression. arXiv [q-bio.QM], arXiv. https://arxiv.org/abs/2211.16742

Hummer, A. M., Abanades, B., & Deane, C. M. (2022). Advances in computational structure-based antibody design. *Current Opinion in Structural Biology*, *74*, 102379.

Ieong, P., Amaro, R. E., & Li, W. W. (2015). Molecular dynamics analysis of antibody recognition and escape by human H1N1 influenza hemagglutinin. *Biophysical Journal*, *108*(11), 2704–2712.

Igawa, T., Maeda, A., Haraya, K., Tachibana, T., Iwayanagi, Y., Mimoto, F., Higuchi, Y., Ishii, S., Tamba, S., Hironiwa, N., Nagano, K., Wakabayashi, T., Tsunoda, H., & Hattori, K. (2013). Engineered monoclonal antibody with novel antigen-sweeping activity in vivo. PloS One, 8(5), e63236.

Jain, T., Sun, T., Durand, S., Hall, A., Houston, N. R., Nett, J. H., Sharkey, B., Bobrowicz, B., Caffry, I., Yu, Y., Cao, Y., Lynaugh, H., Brown, M., Baruah, H., Gray, L. T., Krauland, E. M., Xu, Y., Vásquez, M., & Wittrup, K. D. (2017). Biophysical properties of the clinical-stage antibody landscape. Proceedings of the National Academy of Sciences of the United States of America, *114*(5), 944–949.

Jankauskaitė, J., Jiménez-García, B., Dapkūnas, J., Fernández-Recio, J., & Moal, I. H. (2018). SKEMPI 2.0: an updated benchmark of changes in protein–protein binding energy, kinetics and thermodynamics upon mutation. *Bioinformatics*, *35*(3), 462–469.

Jarmoskaite, I., AlSadhan, I., Vaidyanathan, P. P., & Herschlag, D. (2020). How to measure and evaluate binding affinities. eLife, 9. https://doi.org/10.7554/elife.57264

Jefferis, R. (2016). Posttranslational modifications and the immunogenicity of biotherapeutics. *Journal of Immunology Research*, 2016, 5358272.

Jemimah, S., Sekijima, M., & Gromiha, M. M. (2019). ProAffiMuSeq: sequence-based method to predict the binding free energy change of protein–protein complexes upon mutation using functional classification. *Bioinformatics*, *36*(6), 1725–1730.

Jemimah, S., Yugandhar, K., & Michael Gromiha, M. (2017). PROXiMATE: a database of mutant protein–protein complex thermodynamics and kinetics. *Bioinformatics*, *33*(17), 2787–2788.

Jespersen, M. C., Mahajan, S., Peters, B., Nielsen, M., & Marcatili, P. (2019). Antibody specific B-Cell epitope predictions: leveraging information from antibody-antigen protein complexes. *Frontiers in Immunology*, 10. https://doi.org/10.3389/fimmu.2019.00298

Jiménez-García, B., Roel-Touris, J., Romero-Durana, M., Vidal, M., Jiménez-González, D., & Fernández-Recio, J. (2018). LightDock: a new multi-scale approach to protein-protein docking. *Bioinformatics*, *34*(1), 49–55.

Jin, W., Barzilay, R., & Jaakkola, T. (2022). Antibody–Antigen docking and design via hierarchical equivariant refinement. arXiv [q-bio.BM], arXiv. https://arxiv.org/abs/2207.06616

Johnston, R. J., Su, L. J., Pinckney, J., Critton, D., Boyer, E., Krishnakumar, A., Corbett, M., Rankin, A. L., Dibella, R., Campbell, L., Martin, G. H., Lemar, H., Cayton, T., Huang, R. Y.-C., Deng, X., Nayeem, A., Chen, H., Ergel, B., Rizzo, J. M., … Korman, A. J. (2019). VISTA is an acidic pH-selective ligand for PSGL-1. *Nature*, *574*(7779), 565–570.

Jones, S., & Thornton, J. M. (1996). Principles of protein-protein interactions. *Proceedings of the National Academy of Sciences*, 93(1), 13–20. https://doi.org/10.1073/pnas.93.1.13

Jumper, J., Evans, R., Pritzel, A., Green, T., Figurnov, M., Ronneberger, O., Tunyasuvunakool, K., Bates, R., Žídek, A., Potapenko, A., Bridgland, A., Meyer, C., Kohl, S. A. A., Ballard, A. J., Cowie, A., Romera-Paredes, B., Nikolov, S., Jain, R., Adler, J., … Hassabis, D. (2021). Highly accurate protein structure prediction with AlphaFold. *Nature*, *596*(7873), 583–589.

Kastritis, P. L., & Bonvin, A. M. J. (2011). Are scoring functions in protein–protein docking ready to predict interactomes? Clues from a novel binding affinity benchmark. *Journal of Proteome Research*, 10(2), 921–922. https://doi.org/10.1021/pr101118t

Kastritis, P. L., & Bonvin, A. M. J. J. (2013). On the binding affinity of macromolecular interactions: daring to ask why proteins interact. *Journal of the Royal Society, Interface/the Royal Society*, *10*(79), 20120835.

Kastritis, P. L., Moal, I. H., Hwang, H., Weng, Z., Bates, P. A., Bonvin, A. M. J., & Janin, J. (2011). A structure-based benchmark for protein-protein binding affinity. *Protein Science*, 20(3), 482–491. https://doi.org/10.1002/pro.580

Kawashima, S. (2000). AAindex: amino acid index database. *Nucleic Acids Research*, 28(1), 374–374. https://doi.org/10.1093/nar/28.1.374

Kayser, V., Chennamsetty, N., Voynov, V., Forrer, K., Helk, B., & Trout, B. L. (2011). Glycosylation influences on the aggregation propensity of therapeutic monoclonal antibodies. *Biotechnology Journal*, *6*(1), 38–44.

Kelow, S. P., Adolf-Bryfogle, J., & Dunbrack, R. L. (2020). Hiding in plain sight: structure and sequence analysis reveals the importance of the antibody DE loop for antibody-antigen binding. *mAbs*, *12*(1), 1840005.

Keskin, O., Ma, B., Rogale, K., Gunasekaran, K., & Nussinov, R. (2005). Protein–protein interactions: organization, cooperativity and mapping in a bottom-up systems biology approach. *Physical Biology*, 2(2), S24–S35. https://doi.org/10.1088/1478-3975/2/2/s03

Khetan, R., Curtis, R., Deane, C. M., Hadsund, J. T., Kar, U., Krawczyk, K., Kuroda, D., Robinson, S. A., Sormanni, P., Tsumoto, K., Warwicker, J., & Martin, A. C. R. (2022). Current advances in biopharmaceutical informatics: guidelines, impact and challenges in the computational developability assessment of antibody therapeutics. *mAbs*, *14*(1), 2020082.

Kim, D. Y., Kandalaft, H., Ding, W., Ryan, S., van Faassen, H., Hirama, T., Foote, S. J., MacKenzie, R., & Tanha, J. (2012). Disulfide linkage engineering for improving biophysical properties of human VH domains. *Protein Engineering, Design & Selection: PEDS*, *25*(10), 581–589.

Kortemme, T., Joachimiak, L. A., Bullock, A. N., Schuler, A. D., Stoddard, B. L., & Baker, D. (2004). Computational redesign of protein-protein interaction specificity. *Nature Structural & Molecular Biology*, *11*(4), 371–379.

Kralj, S., Hodošček, M., Podobnik, B., Kunej, T., Bren, U., Janežič, D., & Konc, J. (2021). Molecular dynamics simulations reveal interactions of an IgG1 antibody with selected Fc receptors. *Frontiers in Chemistry*, *9*, 705931.

Krawczyk, K., Baker, T., Shi, J., & Deane, C. M. (2013). Antibody i-Patch prediction of the antibody binding site improves rigid local antibody-antigen docking. *Protein Engineering, Design & Selection: PEDS*, *26*(10), 621–629.

Krawczyk, K., Liu, X., Baker, T., Shi, J., & Deane, C. M. (2014). Improving B-cell epitope prediction and its application to global antibody-antigen docking. *Bioinformatics*, 30(16), 2288–2294. https://doi.org/10.1093/bioinformatics/btu190

Kulkarni-Kale, U., Raskar-Renuse, S., Natekar-Kalantre, G., & Saxena, S. A. (2014). Antigen–antibody interaction database (AgAbDb): a compendium of antigen–antibody interactions. In R. K. De & N. Tomar (Eds.), *Immunoinformatics* (pp. 149–164). Springer New York.

Lan, W., Valente, J. J., Ilott, A., Chennamsetty, N., Liu, Z., Rizzo, J. M., Yamniuk, A. P., Qiu, D., Shackman, H. M., & Bolgar, M. S. (2020). Investigation of anomalous charge variant profile reveals discrete pH-dependent conformations and conformation-dependent charge states within the CDR3 loop of a therapeutic mAb. *mAbs*, 12(1). https://doi.org/10.1080/19420862.2020.1763138

Lazim, R., Suh, D., & Choi, S. (2020). Advances in molecular dynamics simulations and enhanced sampling methods for the study of protein systems. *International Journal of Molecular Sciences*, *21*(17). https://doi.org/10.3390/ijms21176339

Leem, J., Mitchell, L. S., Farmery, J. H. R., Barton, J., & Galson, J. D. (2022). Deciphering the language of antibodies using self-supervised learning. Patterns (New York, N.Y.), *3*(7), 100513.

Lees, W. D., Stejskal, L., Moss, D. S., & Shepherd, A. J. (2017). Investigating substitutions in antibody–antigen complexes using molecular dynamics: a case study with broad-spectrum, influenza A antibodies. Frontiers in Immunology, *8*. https://doi.org/10.3389/fimmu.2017.00143

Lima, W. C., Gasteiger, E., Marcatili, P., Duek, P., Bairoch, A., & Cosson, P. (2020). The ABCD database: a repository for chemically defined antibodies. *Nucleic Acids Research*, *48*(D1), D261–D264.

Li, M., Simonetti, F. L., Goncearenco, A., & Panchenko, A. R. (2016). MutaBind estimates and interprets the effects of sequence variants on protein–protein interactions. *Nucleic Acids Research*, *44*(W1), W494–W501. https://doi.org/10.1093/nar/gkw374

Lin, Z., Akin, H., Rao, R., Hie, B., Zhu, Z., Lu, W., … Rives, A. (2023). Evolutionary-scale prediction of atomic-level protein structure with a language model. Science (New York, N.Y.), 379(6637), 1123–1130.

Li, Pahari, S., Murthy, A. K., Liang, S., Fragoza, R., Yu, H., & Alexov, E. (2021). SAAMBE-SEQ: a sequence-based method for predicting mutation effect on protein-protein binding affinity. *Bioinformatics*, *37*(7), 992–999.

Lippow, S. M., Dane Wittrup, K., & Tidor, B. (2007). Computational design of antibody-affinity improvement beyond in vivo maturation. *Nature Biotechnology*, *25*(10), 1171–1176. https://doi.org/10.1038/nbt1336

Liu, Luo, Y., Li, P., Song, S., & Peng, J. (2021). Deep geometric representations for modeling effects of mutations on protein-protein binding affinity. *PLoS Computational Biology*, *17*(8), e1009284.

Liu, Z., Valente, J., Lin, S., Chennamsetty, N., Qiu, D., & Bolgar, M. (2019). Cyclization of N-terminal glutamic acid to pyro-Glutamic acid impacts monoclonal antibody charge heterogeneity despite its appearance as a neutral transformation. *Journal of Pharmaceutical Sciences*, *108*(10), 3194–3200.

Li, S., Wu, S., Wang, L., Li, F., Jiang, H., & Bai, F. (2022). Recent advances in predicting protein-protein interactions with the aid of artificial intelligence algorithms. *Current Opinion in Structural Biology*, *73*, 102344.

Löhr, T., Sormanni, P., & Vendruscolo, M. (2022). Conformational entropy as a potential liability of computationally designed antibodies. *Biomolecules, 12*(5). https://doi.org/10.3390/biom12050718

Lopes, P. E. M., Guvench, O., & MacKerell, A. D., Jr. (2015). Current status of protein force fields for molecular dynamics simulations. *Methods in Molecular Biology*, *1215*, 47–71.

Lo, Y.-T., Shih, T.-C., Pai, T.-W., Ho, L.-P., Wu, J.-L., & Chou, H.-Y. (2021). Conformational epitope matching and prediction based on protein surface spiral features. *BMC Genomics*, *22*(Suppl 2), 116.

Lu, R.-M., Hwang, Y.-C., Liu, I.-J., Lee, C.-C., Tsai, H.-Z., Li, H.-J., & Wu, H.-C. (2020). Development of therapeutic antibodies for the treatment of diseases. *Journal of Biomedical Science*, *27*(1), 1.

Lu, L. L, Suscovich, T. J., Fortune, S. M., & Alter, G. (2018). Beyond binding: antibody effector functions in infectious diseases. *Nature Reviews. Immunology*, *18*(1), 46–61.

Lu, X., Nobrega, R. P., Lynaugh, H., Jain, T., Barlow, K., Boland, T., Sivasubramanian, A., Vásquez, M., & Xu, Y. (2019). Deamidation and isomerization liability analysis of 131 clinical-stage antibodies. *mAbs*, *11*(1), 45–57.

Makowski, E. K., Kinnunen, P. C., Huang, J., Wu, L., Smith, M. D., Wang, T., Desai, A. A., Streu, C. N., Zhang, Y., Zupancic, J. M., Schardt, J. S., Linderman, J. J., & Tessier, P. M. (2022). Co-optimization of therapeutic antibody affinity and specificity using machine learning models that generalize to novel mutational space. *Nature Communications*, *13*(1), 3788.

Marillet, S., Boudinot, P., & Cazals, F. (2016). High-resolution crystal structures leverage protein binding affinity predictions. *Proteins*, *84*(1), 9–20.

Masiero, A., Nelly, L., Marianne, G., Christophe, S., Florian, L., Ronan, C., Claire, B., Cornelia, Z., Grégoire, B., Eric, L., Ludovic, L., Dominique, B., Sylvie, A., Marie, G., Francis, D., Fabienne, S., Cécile, C., Isabelle, A., Jacques, D., … Catherine, P. (2020). The impact of proline isomerization on antigen binding and the analytical profile of a trispecific anti-HIV antibody. *mAbs*, 12(1), 1698128.

Mason, D. M., Friedensohn, S., Weber, C. R., Jordi, C., Wagner, B., Meng, S. M., Ehling, R. A., Bonati, L., Dahinden, J., Gainza, P., Correia, B. E., & Reddy, S. T. (2021). Optimization of therapeutic antibodies by predicting antigen specificity from antibody sequence via deep learning. *Nature Biomedical Engineering*. https://doi.org/10.1038/s41551-021-00699-9

Młokosiewicz, J., Deszyński, P., Wilman, W., Jaszczyszyn, I., Ganesan, R., Kovaltsuk, A., Leem, J., Galson, J. D., & Krawczyk, K. (2022). AbDiver: a tool to explore the natural antibody landscape to aid therapeutic design. *Bioinformatics*, 38(9), 2628–2630. https://doi.org/10.1093/bioinformatics/btac151

Mullard, A. (2021). FDA approves 100th monoclonal antibody product. *Nature Reviews. Drug Discovery*, *20*(7), 491–495.

Myung, Y., Pires, D. E. V., & Ascher, D. B. (2022). CSM-AB: graph-based antibody–antigen binding affinity prediction and docking scoring function. *Bioinformatics*, 38(4), 1141–1143. https://doi.org/10.1093/bioinformatics/btab762

Nakamura, H., Yoshikawa, M., Oda-Ueda, N., Ueda, T., & Ohkuri, T. (2021). A comprehensive analysis of novel disulfide bond introduction site into the constant domain of human Fab. *Scientific Reports*, *11*(1), 12937.

Narayanan, H., Dingfelder, F., Condado Morales, I., Patel, B., Heding, K. E., Bjelke, J. R., Egebjerg, T., Butté, A., Sokolov, M., Lorenzen, N., & Arosio, P. (2021a). Design of biopharmaceutical formulations accelerated by machine learning. *Molecular Pharmaceutics*, *18*(10), 3843–3853.

Narayanan, H., Dingfelder, F., Condado Morales, I., Patel, B., Heding, K. E., Bjelke, J. R., Egebjerg, T., Butté, A., Sokolov, M., Lorenzen, N., & Arosio, P. (2021b). Design of biopharmaceutical formulations accelerated by machine learning. *Molecular Pharmaceutics*, *18*(10), 3843–3853.

Nijkamp, E., Ruffolo, J. A., Weinstein, E. N., Naik, N., & Madani, A. (2023). ProGen2: Exploring the boundaries of protein language models. Cell Systems, 14(11), 968–978.e3.

Norman, R. A., Ambrosetti, F., Bonvin, A. M. J. J., Colwell, L. J., Kelm, S., Kumar, S., & Krawczyk, K. (2020). Computational approaches to therapeutic antibody design: established methods and emerging trends. *Briefings in Bioinformatics*, *21*(5), 1549–1567.

Ofer, D., Brandes, N., & Linial, M. (2021). The language of proteins: NLP, machine learning & protein sequences. *Computational and Structural Biotechnology Journal*, *19*, 1750–1758.

Olsen, T. H., Boyles, F., & Deane, C. M. (2022a). Observed antibody space: a diverse database of cleaned, annotated, and translated unpaired and paired antibody sequences. *Protein Science: A Publication of the Protein Society*, *31*(1), 141–146.

Olsen, T. H., Moal, I. H., & Deane, C. M. (2022). AbLang: an antibody language model for completing antibody sequences. Bioinformatics Advances, 2(1), vbac046.

Omer, A., Shemesh, O., Peres, A., Polak, P., Shepherd, A. J., Watson, C. T., Boyd, S. D., Collins, A. M., Lees, W., & Yaari, G. (2020). VDJbase: an adaptive immune receptor genotype and haplotype database. *Nucleic Acids Research*, *48*(D1), D1051–D1056.

Padlan, E. A., Abergel, C., & Tipper, J. P. (1995). Identification of specificity-determining residues in antibodies. *FASEB Journal*, *9*(1), 133–139.

Pagadala, N. S., Syed, K., & Tuszynski, J. (2017). Software for molecular docking: a review. *Biophysical Reviews*, *9*(2), 91–102.

Parkinson, J., Hard, R., & Wang, W. (2023). The RESP AI model accelerates the identification of tight-binding antibodies. *Nature Communications*, *14*(1), 454.

Pérez, A.-M. W., Lorenzen, N., Vendruscolo, M., & Sormanni, P. (2022). Assessment of therapeutic antibody developability by combinations of in vitro and in silico methods. Therapeutic Antibodies, 57–113. https://doi.org/10.1007/978-1-0716-1450-1_4

Perkins, J. R., Diboun, I., Dessailly, B. H., Lees, J. G., & Orengo, C. (2010). Transient protein-protein interactions: structural, functional, and network properties. *Structure*, *18*(10), 1233–1243.

Petersen, B. M., Ulmer, S. A., Rhodes, E. R., Gutierrez-Gonzalez, M. F., Dekosky, B. J., Sprenger, K. G., & Whitehead, T. A. (2021). Regulatory approved monoclonal antibodies contain framework mutations predicted from human antibody repertoires. Frontiers in Immunology, 12, 728694.

Petukh, M., Li, M., & Alexov, E. (2015). Predicting binding free energy change caused by point mutations with knowledge-modified MM/PBSA method. *PLoS Computational Biology*, 11(7), e1004276. https://doi.org/10.1371/journal.pcbi.1004276

Pierce, B. G., Wiehe, K., Hwang, H., Kim, B.-H., Vreven, T., & Weng, Z. (2014). ZDOCK server: interactive docking prediction of protein-protein complexes and symmetric multimers. *Bioinformatics*, *30*(12), 1771–1773.

Pinzi, L., & Rastelli, G. (2019). Molecular docking: shifting paradigms in drug discovery. *International Journal of Molecular Sciences*, *20*(18). https://doi.org/10.3390/ijms20184331

Pires, D. E. V., & Ascher, D. B. (2016). mCSM-AB: a web server for predicting antibody–antigen affinity changes upon mutation with graph-based signatures. *Nucleic Acids Research*, *44*(W1), W469–W473.

Pittala, S., & Bailey-Kellogg, C. (2020). Learning context-aware structural representations to predict antigen and antibody binding interfaces. *Bioinformatics*, *36*(13), 3996–4003.

Qasba, P. K. (2015). Glycans of antibodies as a specific site for drug conjugation using glycosyltransferases. *Bioconjugate Chemistry*, *26*(11), 2170–2175.

Qing, R., Hao, S., Smorodina, E., Jin, D., Zalevsky, A., & Zhang, S. (2022). Protein design: from the aspect of water solubility and stability. *Chemical Reviews*, *122*(18), 14085–14179.

Rabia, L. A., Zhang, Y., Ludwig, S. D., Julian, M. C., & Tessier, P. M. (2018). Net charge of antibody complementarity-determining regions is a key predictor of specificity. *Protein Engineering, Design & Selection*, *31*(11), 409–418.

Radford, A., Wu, J., Child, R., Luan, D., Amodei, D., & Sutskever, I. (2019). Language Models are Unsupervised Multitask Learners. https://life-extension.github.io/2020/05/27/GPT%E6%8A%80%E6%9C%AF%E5%88%9D%E6%8E%A2/language-models.pdf

Ramazi, S., & Zahiri, J. (2021). Posttranslational modifications in proteins: resources, tools and prediction methods. *Database: The Journal of Biological Databases and Curation*, 2021. https://doi.org/10.1093/database/baab012

Rangel, M. A., Bedwell, A., Costanzi, E., Taylor, R. J., Russo, R., Bernardes, G. J. L., … Sormanni, P. (2022). Fragment-based computational design of antibodies targeting structured epitopes. Science Advances, 8(45), eabp9540.

Rawat, P., Prabakaran, R., Sakthivel, R., Mary Thangakani, A., Kumar, S., & Gromiha, M. M. (2020). CPAD 2.0: a repository of curated experimental data on aggregating proteins and peptides. *Amyloid: The International Journal of Experimental and Clinical Investigation: The Official Journal of the International Society of Amyloidosis*, *27*(2), 128–133.

Rawat, P., Sharma, D., Prabakaran, R., Ridha, F., Mohkhedkar, M., Janakiraman, V., & Gromiha, M. M. (2022). Ab-CoV: a curated database for binding affinity and neutralization profiles of coronavirus related antibodies. *Bioinformatics*. https://doi.org/10.1093/bioinformatics/btac439

Rawat, P., Sharma, D., Srivastava, A., Janakiraman, V., & Gromiha, M. M. (2021). Exploring antibody repurposing for COVID-19: beyond presumed roles of therapeutic antibodies. *Scientific Reports*, *11*(1), 10220.

Raybould, M. I. J., Kovaltsuk, A., Marks, C., & Deane, C. M. (2021). CoV-AbDab: the coronavirus antibody database. *Bioinformatics*, *37*(5), 734–735.

Raybould, M. I. J., Marks, C., Krawczyk, K., Taddese, B., Nowak, J., Lewis, A. P., Bujotzek, A., Shi, J., & Deane, C. M. (2019). Five computational developability guidelines for therapeutic antibody profiling. *Proceedings of the National Academy of Sciences of the United States of America*, *116*(10), 4025–4030.

Raybould, M. I. J., Marks, C., Lewis, A. P., Shi, J., Bujotzek, A., Taddese, B., & Deane, C. M. (2020). Thera-SAbDab: the therapeutic structural antibody database. *Nucleic Acids Research*, *48*(D1), D383–D388.

Reth, M. (2013). Matching cellular dimensions with molecular sizes. *Nature Immunology*, *14*(8), 765–767.

Retter, I., Althaus, H. H., Münch, R., & Müller, W. (2005). VBASE2, an integrative V gene database. *Nucleic Acids Research*, *33*(Database issue), D671–D674.

Reusch, D., & Tejada, M. L. (2015). Fc glycans of therapeutic antibodies as critical quality attributes. *Glycobiology*, *25*(12), 1325–1334.

Robert, P. A., Akbar, R., Frank, R., Pavlović, M., Widrich, M., Snapkov, I., Slabodkin, A., Chernigovskaya, M., Scheffer, L., Smorodina, E., Rawat, P., Mehta, B. B., Vu, M. H., Mathisen, I. F., Prósz, A., Abram, K., Olar, A., Miho, E., Haug, D. T. T., … Greiff, V. (2022). Unconstrained generation of synthetic antibody–antigen structures to guide machine learning methodology for antibody specificity prediction. *Nature Computational Science*, *2*(12), 845–865.

Rodrigues, C. H. M., Pires, D. E. V., & Ascher, D. B. (2021). mmCSM-PPI: predicting the effects of multiple point mutations on protein–protein interactions. *Nucleic Acids Research*, 49(W1), W417–W424. https://doi.org/10.1093/nar/gkab273

Rodriguez-Quijada, C., Gomez-Marquez, J., & Hamad-Schifferli, K. (2020). Repurposing old antibodies for new diseases by exploiting cross-reactivity and multicolored nanoparticles. *ACS Nano*, *14*(6), 6626–6635.

Romero-Molina, S., Ruiz-Blanco, Y. B., Mieres-Perez, J., Harms, M., Münch, J., Ehrmann, M., & Sanchez-Garcia, E. (2022). PPI-Affinity: a web tool for the prediction and optimization of protein-peptide and protein-protein binding affinity. *Journal of Proteome Research*, *21*(8), 1829–1841.

Rosato, A. (2010). Faculty opinions recommendation of are scoring functions in protein-protein docking ready to predict interactomes? Clues from a novel binding affinity benchmark. *Faculty Opinions – Post-Publication Peer Review of the Biomedical Literature*. https://doi.org/10.3410/f.3437978.3131081

Rose, Y., Duarte, J. M., Lowe, R., Segura, J., Bi, C., Bhikadiya, C., Chen, L., Rose, A. S., Bittrich, S., Burley, S. K., & Westbrook, J. D. (2021). RCSB protein data bank: architectural advances towards integrated searching and efficient access to macromolecular structure data from the PDB archive. *Journal of Molecular Biology*, *433*(11), 166704.

Rudnick, S. I., & Adams, G. P. (2009). Affinity and avidity in antibody-based tumor targeting. *Cancer Biotherapy & Radiopharmaceuticals*, *24*(2), 155–161.

Ruffolo, J. A., Chu, L.-S., Mahajan, S. P., & Gray, J. J. (2023). Fast, accurate antibody structure prediction from deep learning on massive set of natural antibodies. Nature Communications, 14(1), 2389.

Ruffolo, J. A., Sulam, J., & Gray, J. J. (2022b). Antibody structure prediction using interpretable deep learning. Patterns (New York, N.Y.), 3(2), 100406.

Sacquin-Mora, S., Carbone, A., & Lavery, R. (2008). Identification of protein interaction partners and protein–protein interaction sites. *Journal of Molecular Biology*, 382(5), 1276–1289. https://doi.org/10.1016/j.jmb.2008.08.002

Saerens, D., Conrath, K., Govaert, J., & Muyldermans, S. (2008). Disulfide bond introduction for general stabilization of immunoglobulin heavy-chain variable domains. *Journal of Molecular Biology*, *377*(2), 478–488.

Salmaso, V., & Moro, S. (2018). Bridging molecular docking to molecular dynamics in exploring ligand-protein recognition process: an overview. Frontiers in Pharmacology, *9*. https://doi.org/10.3389/fphar.2018.00923

Sankar, K., Trainor, K., Blazer, L. L., Adams, J. J., Sidhu, S. S., Day, T., Meiering, E., & Maier, J. K. X. (2022). A descriptor set for quantitative structure-property relationship prediction in biologics. Molecular Informatics, *41*, e2100240.

Schlander, M., Hernandez-Villafuerte, K., Cheng, C.-Y., Mestre-Ferrandiz, J., & Baumann, M. (2021). How much does it cost to research and develop a new drug? A systematic review and assessment. *PharmacoEconomics*, *39*(11), 1243–1269.

Schmitz, S., Schmitz, E. A., Crowe, J. E., Jr, & Meiler, J. (2022). The human antibody sequence space and structural design of the V, J regions, and CDRH3 with Rosetta. mAbs, *14*(1), 2068212.

Schneider, C., Buchanan, A., Taddese, B., & Deane, C. M. (2022). DLAB: deep learning methods for structure-based virtual screening of antibodies. *Bioinformatics*, *38*(2), 377–383. https://doi.org/10.1093/bioinformatics/btab660

Schneidman-Duhovny, D., Inbar, Y., Nussinov, R., & Wolfson, H. J. (2005). PatchDock and SymmDock: servers for rigid and symmetric docking. *Nucleic Acids Research*, 33(Web Server issue), W363–W367.

Schoeder, C. T., Schmitz, S., Adolf-Bryfogle, J., Sevy, A. M., Finn, J. A., Sauer, M. F., Bozhanova, N. G., Mueller, B. K., Sangha, A. K., Bonet, J., Sheehan, J. H., Kuenze, G., Marlow, B., Smith, S. T., Woods, H., Bender, B. J., Martina, C. E., Del Alamo, D., Kodali, P., … Moretti, R. (2021). Modeling immunity with Rosetta: methods for antibody and antigen design. *Biochemistry*, *60*(11), 825–846.

Schröter, C., Günther, R., Rhiel, L., Becker, S., Toleikis, L., Doerner, A., Becker, J., Schönemann, A., Nasu, D., Neuteboom, B., Kolmar, H., & Hock, B. (2015). A generic approach to engineer antibody pH-switches using combinatorial histidine scanning libraries and yeast display. mAbs, *7*(1), 138–151.

Segal, D. M., Padlan, E. A., Cohen, G. H., Rudikoff, S., Potter, M., & Davies, D. R. (1974). The three-dimensional structure of a phosphorylcholine-binding mouse immunoglobulin Fab and the nature of the antigen binding site. *Proceedings of the National Academy of Sciences of the United States of America*, *71*(11), 4298–4302.

Sela-Culang, I., Ashkenazi, S., Peters, B., & Ofran, Y. (2015). PEASE: predicting B-cell epitopes utilizing antibody sequence. *Bioinformatics*, 31(8), 1313–1315. https://doi.org/10.1093/bioinformatics/btu790

Sela-Culang, I., Benhnia, M. R.-E.-I., Matho, M. H., Kaever, T., Maybeno, M., Schlossman, A., Nimrod, G., Li, S., Xiang, Y., Zajonc, D., Crotty, S., Ofran, Y., & Peters, B. (2014). Using a combined computational-experimental approach to predict antibody-specific B cell epitopes. *Structure*, *22*(4), 646–657.

Sela-Culang, I., Kunik, V., & Ofran, Y. (2013). The structural basis of antibody-antigen recognition. *Frontiers in Immunology*, *4*, 302.

Sharma, D., Rawat, P., Janakiraman, V., & Gromiha, M. M. (2022). Elucidating important structural features for the binding affinity of spike - SARS-CoV-2 neutralizing antibody complexes. *Proteins*, *90*(3), 824–834.

Sharma, Patapoff, T. W., Kabakoff, B., Pai, S., Hilario, E., Zhang, B., Li, C., Borisov, O., Kelley, R. F., Chorny, I., Zhou, J. Z., Dill, K. A., & Swartz, T. E. (2014). In silico selection of therapeutic antibodies for development: viscosity, clearance, and chemical stability. *Proceedings of the National Academy of Sciences of the United States of America*, *111*(52), 18601–18606.

Shuai, R. W., Ruffolo, J. A., & Gray, J. J. (2023). IgLM: Infilling language modeling for antibody sequence design. Cell Systems, 14(11), 979–989.e4.

Sinclair, N. R., Lees, R. K., & Elliott, E. V. (1968). Role of the Fc fragment in the regulation of the primary immune response. *Nature*, *220*(5171), 1048–1049.

Sirin, S., Apgar, J. R., Bennett, E. M., & Keating, A. E. (2016). AB-Bind: antibody binding mutational database for computational affinity predictions. *Protein Science: A Publication of the Protein Society*, *25*(2), 393–409.

Strickley, R. G., & Lambert, W. J. (2021). A review of formulations of commercially available antibodies. *Journal of Pharmaceutical Sciences*, *110*(7), 2590–2608.e56.

Sverrisson, F., Feydy, J., Correia, B. E., & Bronstein, M. M. (2021). Fast end-to-end learning on protein surfaces. 2021 IEEE/CVF Conference on Computer Vision and Pattern Recognition (CVPR), 15267–15276. IEEE.

Swindells, M. B., Porter, C. T., Couch, M., Hurst, J., Abhinandan, K. R., Nielsen, J. H., Macindoe, G., Hetherington, J., & Martin, A. C. R. (2017). abYsis: integrated antibody sequence and structure—management, analysis, and prediction. *Journal of Molecular Biology*, *429*(3), 356–364.

Szymborski, J., & Emad, A. (2022). RAPPPID: towards generalizable protein interaction prediction with AWD-LSTM twin networks. *Bioinformatics*, *38*(16), 3958–3967.

Teng, S., Sobitan, A., Rhoades, R., Liu, D., & Tang, Q. (2021). Systemic effects of missense mutations on SARS-CoV-2 spike glycoprotein stability and receptor-binding affinity. *Briefings in Bioinformatics*, *22*(2), 1239–1253.

Tiller, K. E., & Tessier, P. M. (2015). Advances in antibody design. *Annual Review of Biomedical Engineering*, *17*, 191–216.

Tucs, A., Tsuda, K., & Sljoka, A. (2023). Probing conformational dynamics of antibodies with geometric simulations. *Methods in Molecular Biology*, *2552*, 125–139.

Unke, O. T., Chmiela, S., Sauceda, H. E., Gastegger, M., Poltavsky, I., Schütt, K. T., Tkatchenko, A., & Müller, K.-R. (2021). Machine learning force fields. *Chemical Reviews*, *121*(16), 10142–10186.

Unsal, S., Atas, H., Albayrak, M., Turhan, K., Acar, A. C., & Doğan, T. (2022). Learning functional properties of proteins with language models. *Nature Machine Intelligence*, *4*(3), 227–245.

van den Bremer, E. T. J., Beurskens, F. J., Voorhorst, M., Engelberts, P. J., de Jong, R. N., van der Boom, B. G., Cook, E. M., Lindorfer, M. A., Taylor, R. P., van Berkel, P. H. C., & Parren, P. W. (2015). Human IgG is produced in a pro-form that requires clipping of C-terminal lysines for maximal complement activation. *mAbs*, 7(4), 672–680. https://doi.org/10.1080/19420862.2015.1046665

Vangone, A., & Bonvin, A. M. (2015). Contacts-based prediction of binding affinity in protein–protein complexes. eLife, 4. https://doi.org/10.7554/elife.07454

Vatsa, S. (2022). In silico prediction of post-translational modifications in therapeutic antibodies. *mAbs*, *14*(1), 2023938.

Vita, R., Mahajan, S., Overton, J. A., Dhanda, S. K., Martini, S., Cantrell, J. R., Wheeler, D. K., Sette, A., & Peters, B. (2019). The immune epitope database (IEDB): 2018 update. *Nucleic Acids Research*, *47*(D1), D339–D343.

Vu, M. H., Akbar, R., Robert, P. A., Swiatczak, B., Sandve, G. K., Greiff, V., & Haug, D. T. T. (2023). Linguistically inspired roadmap for building biologically reliable protein language models. Nature Machine Intelligence, 5(5), 485–496.

Vu, M. H., Robert, P. A., Akbar, R., Swiatczak, B., Sandve, G. K., Haug, D. T. T., & Greiff, V. (2022). ImmunoLingo: linguistics-based formalization of the antibody language. arXiv [q-bio.QM], arXiv. https://arxiv.org/abs/2209.12635

Wang, Gallolu Kankanamalage, S., Dong, J., & Liu, Y. (2021). Optimization of therapeutic antibodies. *Antibody Therapeutics*, *4*(1), 45–54.

Wang, Lu, Y., & Wang, S. (2003). Comparative evaluation of 11 scoring functions for molecular docking. *Journal of Medicinal Chemistry*, *46*(12), 2287–2303.

Wang, M., Cang, Z., & Wei, G.-W. (2020). A topology-based network tree for the prediction of protein-protein binding affinity changes following mutation. *Nature Machine Intelligence*, *2*(2), 116–123.

Wang, Su, Z., & Wu, Y. (2021). Computational assessment of protein-protein binding affinity by reversely engineering the energetics in protein complexes. *Genomics, Proteomics & Bioinformatics*, *19*(6), 1012–1022.

Wardemann, H., Yurasov, S., Schaefer, A., Young, J. W., Meffre, E., & Nussenzweig, M. C. (2003). Predominant autoantibody production by early human B cell precursors. *Science*, *301*(5638), 1374–1377.

Warszawski, S., Katz, A. B., Lipsh, R., Khmelnitsky, L., Ben Nissan, G., Javitt, G., Dym, O., Unger, T., Knop, O., Albeck, S., Diskin, R., Fass, D., Sharon, M., & Fleishman, S. J. (2019). Optimizing antibody affinity and stability by the automated design of the variable light-heavy chain interfaces. *PLoS Computational Biology*, 15(8), e1007207. https://doi.org/10.1371/journal.pcbi.1007207

Wee, J., & Xia, K. (2022). Persistent spectral based ensemble learning (PerSpect-EL) for protein-protein binding affinity prediction. *Briefings in Bioinformatics*, *23*(2). https://doi.org/10.1093/bib/bbac024

Wilman, W., Wróbel, S., Bielska, W., Deszynski, P., Dudzic, P., Jaszczyszyn, I., Kaniewski, J., Młokosiewicz, J., Rouyan, A., Satława, T., Kumar, S., Greiff, V., & Krawczyk, K. (2022). Machine-designed biotherapeutics: opportunities, feasibility and advantages of deep learning in computational antibody discovery. *Briefings in Bioinformatics*, *23*(4). https://doi.org/10.1093/bib/bbac267

Wong, M. T. Y., Kelm, S., Liu, X., Taylor, R. D., Baker, T., & Essex, J. W. (2022). Higher Affinity antibodies bind with lower hydration and flexibility in large scale simulations. *Frontiers in Immunology*, *13*. https://doi.org/10.3389/fimmu.2022.884110

Wu, R., Ding, F., Wang, R., Shen, R., Zhang, X., Luo, S., Su, C., Wu, Z., Xie, Q., Berger, B., Ma, J., & Peng, J. (2022). High-resolution de novo structure prediction from primary sequence. bioRxiv, 2022.07.21.500999. https://doi.org/10.1101/2022.07.21.500999

Xie, H., Gilar, M., & Gebler, J. C. (2009). Analysis of deamidation and oxidation in monoclonal antibody using peptide mapping with UPLC/MSE. Waters Application Note. https://www.waters.com/webassets/cms/library/docs/720002897en.pdf

Xiong, P., Zhang, C., Zheng, W., & Zhang, Y. (2017). BindProfX: assessing mutation-induced binding affinity change by protein interface profiles with pseudo-counts. *Journal of Molecular Biology*, *429*(3), 426–434.

Xu, & Davis, M. M. (2000). Diversity in the CDR3 region of VH is sufficient for most antibody specificities. *Immunity*, *13*(1), 37–45.

Xue, L. C., Rodrigues, J. P., Kastritis, P. L., Bonvin, A. M., & Vangone, A. (2016). PRODIGY: a web server for predicting the binding affinity of protein–protein complexes. Bioinformatics, btw514. https://doi.org/10.1093/bioinformatics/btw514

Xue, Y., Liu, Z., Fang, X., & Wang, F. (2021). Multimodal pre-training model for sequence-based prediction of protein-protein interaction. *ArXiv*, https://arxiv.org/abs/2112.04814.

Xu, Y., Wang, D., Mason, B., Rossomando, T., Li, N., Liu, D., Cheung, J. K., Xu, W., Raghava, S., Katiyar, A., Nowak, C., Xiang, T., Dong, D. D., Sun, J., Beck, A., & Liu, H. (2019). Structure, heterogeneity and developability assessment of therapeutic antibodies. *mAbs*, *11*(2), 239–264. https://doi.org/10.1080/19420862.2018.1553476

Yamashita, T. (2018). Toward rational antibody design: recent advancements in molecular dynamics simulations. *International Immunology*, *30*(4), 133–140.

Yamashita, T., Mizohata, E., Nagatoishi, S., Watanabe, T., Nakakido, M., Iwanari, H., Mochizuki, Y., Nakayama, T., Kado, Y., Yokota, Y., Matsumura, H., Kawamura, T., Kodama, T., Hamakubo, T., Inoue, T., Fujitani, H., & Tsumoto, K. (2019). Affinity improvement of a cancer-targeted antibody through alanine-induced adjustment of antigen-antibody interface. *Structure*, *27*(3), 519–527.e5.

Yang, Wang, P., & Zhu, B. T. (2023). Binding affinity prediction for antibody–protein antigen complexes: a machine learning analysis based on interface and surface areas. *Journal of Molecular Graphics and Modelling*, 118, 108364. https://doi.org/10.1016/j.jmgm.2022.108364

Yang, Y., Stella, C., Wang, W., Schöneich, C., & Gennaro, L. (2014). Characterization of oxidative carbonylation on recombinant monoclonal antibodies. *Analytical Chemistry*, *86*(10), 4799–4806.

Yin, V., Lai, S.-H., Caniels, T. G., Brouwer, P. J. M., Brinkkemper, M., Aldon, Y., Liu, H., Yuan, M., Wilson, I. A., Sanders, R. W., van Gils, M. J., & Heck, A. J. R. (2021). Probing affinity, avidity, anticooperativity, and competition in antibody and receptor binding to the SARS-CoV-2 spike by single particle mass analyses. *ACS Central Science*, *7*(11), 1863–1873.

Yugandhar, K., & Michael Gromiha, M. (2014). Protein–protein binding affinity prediction from amino acid sequence. *Bioinformatics*, 30(24), 3583–3589. https://doi.org/10.1093/bioinformatics/btu580

Yu, X., Orr, C. M., Chan, H. T. C., James, S., Penfold, C. A., Kim, J., Inzhelevskaya, T., Mockridge, C. I., Cox, K. L., Essex, J. W., Tews, I., Glennie, M. J., & Cragg, M. S. (2023). Reducing affinity as a strategy to boost immunomodulatory antibody agonism. Nature. https://doi.org/10.1038/s41586-022-05673-2

Zhang, W., Wang, L., Liu, K., Wei, X., Yang, K., Du, W., Wang, S., Guo, N., Ma, C., Luo, L., Wu, J., Lin, L., Yang, F., Gao, F., Wang, X., Li, T., Zhang, R., Saksena, N. K., Yang, H., … Liu, X. (2020). PIRD: pan immune repertoire database. *Bioinformatics*, *36*(3), 897–903.

Zhao, L., & Li, J. (2010). Mining for the antibody-antigen interacting associations that predict the B cell epitopes. *BMC Structural Biology*, 10(Suppl 1), S6. https://doi.org/10.1186/1472-6807-10-s1-s6

Zhou, P., Wang, C., Tian, F., Ren, Y., Yang, C., & Huang, J. (2013). Biomacromolecular quantitative structure-activity relationship (BioQSAR): a proof-of-concept study on the modeling, prediction and interpretation of protein-protein binding affinity. *Journal of Computer-Aided Molecular Design*, *27*(1), 67–78.

Use of Molecular Simulations to Understand Structural Dynamics of Antibodies

8

Daniel A. Nissley, Matthew I. J. Raybould, Charlotte M. Deane, and Sandeep Kumar

8.1 WHY RUN MOLECULAR SIMULATIONS ON ANTIBODIES?

Gaining insight into the structure-function relationships that underlie antibody-antigen binding is a critical step in lead development. While static structures provided by computational predictions[1–8] or experimental methods like macromolecular crystallography[9,10] (MX) provide a wealth of information to direct, for example, mutational studies to optimize antigen binding and limit off-target interactions, they only ever provide single snapshots of the ensemble of structures that an antibody populates in solution. Furthermore, the MX structures themselves might contain deviations from the structure that would be found in the solution state, as they are obtained under conditions that are dissimilar to the environment they would encounter in a patient or in transport.[9]

DOI: 10.1201/9781003300311-8

For example, almost all MX structures are collected at cryogenic temperatures (~100 K) to minimize radiation damage.[11] Cryogenic electron microscopy (cryo-EM) is another experimental method of imaging biomolecules under more native-like conditions.[12] However, the low throughput of cryo-EM experiments and a practical resolution threshold of ~3 Å[13] currently limit their application in the drug development pipeline.

Diverse structural questions may be of interest during antibody drug development. Frequently, these are related to the antibody-antigen binding event: which of the contacts present in the MX structure remain when it is heated to room or body temperature? Of those contacts that remain, which contribute most strongly to the overall free-energy change upon binding? Over time, how do the two binding arms reposition with respect to one another in solution? How does a given mutation in a complementarity-determining region (CDR) loop alter the kinetics of binding? Each of these questions requires dynamic information about antibody structures that is difficult or impossible to access from the experimental structures alone.

Other questions may relate to the developability of the antibody, such as predicting its propensity to aggregate or self-associate, or its response to changes in salt content made during formulation. These solution properties would ideally be assessed at an early development stage before they could become problematic. Complicating the issue, experimental and theoretical studies indicate that factors difficult to predict from either sequence or structure alone, including the anisotropy of electrostatic interactions and the characteristic oligomeric state of the antibody in solution, strongly influence these properties.

As described in the following sections, the careful application of molecular simulation techniques can provide suggestions or even answers to these and other common issues that arise throughout the development of an antibody therapeutic. Before we consider these case studies, we first step through the most common molecular simulation techniques and important considerations when designing a simulation study of antibody biophysics.

8.2 COMMON TYPES OF MOLECULAR SIMULATIONS FOR BIOMOLECULES

8.2.1 Molecular Dynamics (MD) Simulations

For biomolecules, molecular dynamics (MD) simulations are perhaps the most widely used molecular simulation technique today. MD is a mature, broadly used technique as recognized by the 2013 Nobel Prize in Chemistry awarded to Karplus, Levitt, and Warshel for the development of multi-scale modeling techniques for complex chemical systems, including MD. In this section, we describe in general terms the theory of MD simulations.

The main goal of MD is the prediction of the time evolution of a molecular system.[14,15] In our application, the system likely consists of an antibody and perhaps its antigen in aqueous solution in a patient. We choose to model the system using the rules of classical (Newtonian) mechanics and represent each particle as a sphere and each bond as a spring. At the beginning of the simulation, each particle in the system is initialized with some starting position and velocity. Initial particle velocities are often randomly selected from a Maxwell-Boltzmann distribution generated at the user-defined absolute temperature of the system, T. A mathematical function, termed a forcefield or Hamiltonian, E, determines the forces different particles in the simulation exert on one another. Despite their name, forcefields are typically written as expressions that, given system particle coordinates X, compute the *potential energy* of the system $E = E(X)$. The specific terms within the forcefield are chosen to match quantum-mechanical calculations and, in most cases, experimental data. Typically, potential energy expressions contain terms that constrain particle motions through both bonded terms, E_{bonded}, representing constraints on particle motions due to bonds, angles, and torsions, and non-bonded terms, $E_{\text{non-bonded}}$, representing non-covalent interactions like Van der Waals (VDW) and electrostatics:

$$E = E_{\text{bonded}} + E_{\text{non-bonded}} \tag{8.1}$$

$$E_{\text{bonded}} = E_{\text{bonds}} + E_{\text{angles}} + E_{\text{torsions}} \tag{8.2}$$

$$E_{\text{non-bonded}} = E_{\text{VDW}} + E_{\text{el}} \tag{8.3}$$

Equations (8.2 and 8.3) provide a typical decomposition of the E_{bonded} and $E_{\text{non-bonded}}$ terms in Equation (8.1) into their component parts. Some typical mathematical functions used to represent bonds, angles, torsion angles, VDW, and electrostatic interactions are shown in Figure 8.1a–e.

With positions and velocities determined, the forces acting on particles are then computed as the negative gradient, $F = -\nabla E$, of the potential energy. Having determined the forces, Newton's equations of motion are then used to predict the positions and velocities at some future time of the system. To reduce numerical integration errors, the time step must be small; values on the order of 1 to 10 femtoseconds are typical for MD simulations of biomolecules. By stepping through time and updating the positions and velocities of all particles in the system according to the forcefield, we produce a movie in 3D space of the time evolution of the system. Various methods for maintaining the system temperature during the simulation by updating or rescaling particle velocities have been developed.[16–19] Analysis of the resulting "trajectory" of the system based on the coordinates collected during the run can then be carried out.

Several different types of MD are frequently encountered in the academic literature, all of which fall under the umbrella of MD techniques. In Langevin dynamics,[18] additional drag and random collision terms are added to Newton's equations of motion to approximate the influence of solvent. In Brownian Dynamics,[20] which may be considered a simplified form of Langevin dynamics that applies to certain particles in solution, the equations of motion are modified such that there is no average acceleration in the system.

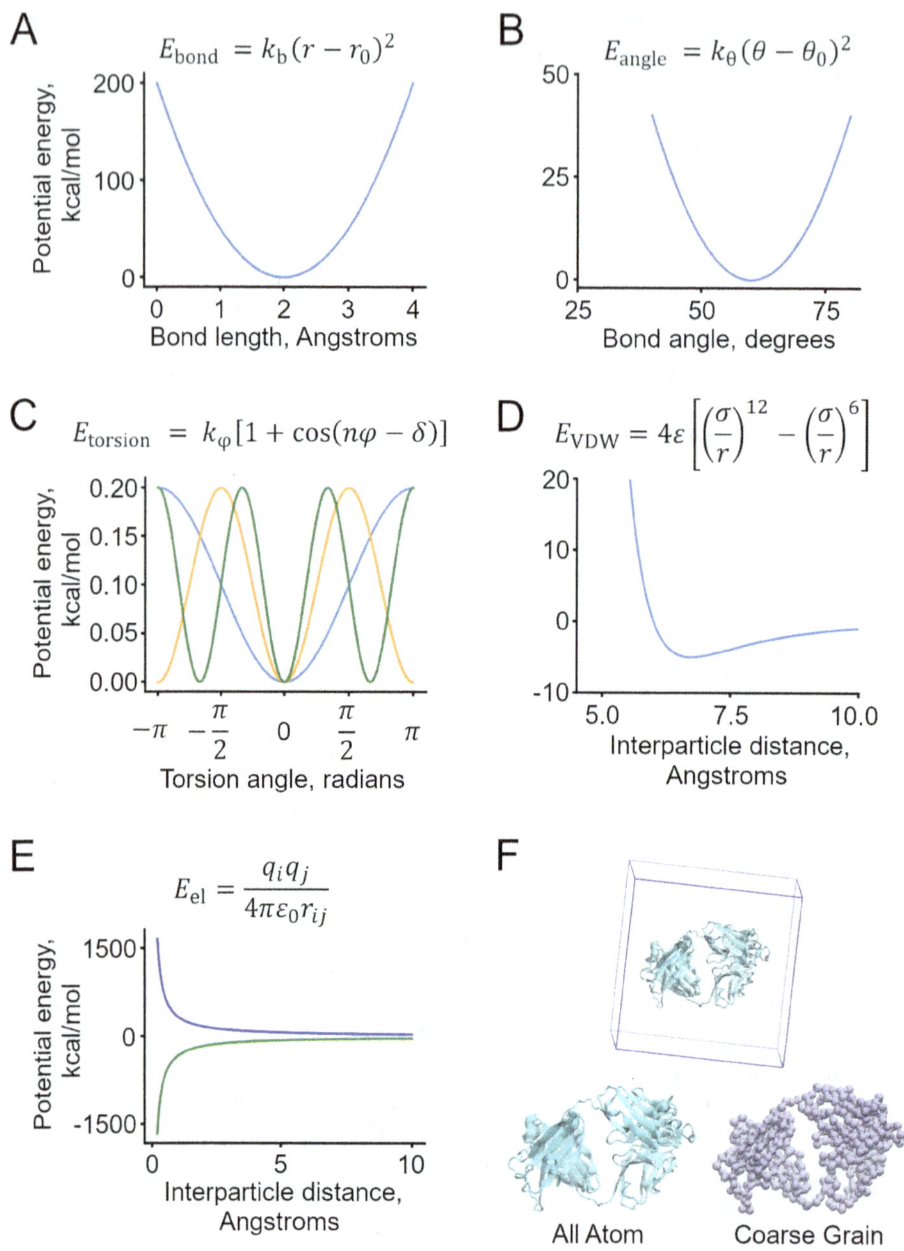

FIGURE 8.1 Common mathematical forms of MD forcefield terms are shown in panels A through E.[87] Each equation is accompanied by a plot generated with sample forcefield parameters.

(Continued)

(Continued)

(a) In the expression for E_{bond}, k_b is a force constant and r and r_0 are, respectively, the current and equilibrium bond lengths. The plot shows results for a bond with $k_b = 50\frac{kcal}{\left[mol \times \AA^2\right]}$ and $r_0 = 2\AA$. (b) Angles are computed as a function of the force constant k_θ and the squared deviation between the equilibrium (θ_0) and current (θ) bond angles. The energy as a function of θ is shown for $k_\theta = 0.1\frac{kcal}{mol}$ and $\theta_0 = 60°$. (c) $E_{torsion}$ is expressed as a function of the force constant k_φ, the multiplicity n (equal to the integer number of minima in the period $[-\pi, \pi]$), the torsion angle φ, and its phase δ. The displayed plots were generated using $k_\varphi = 0.1\frac{kcal}{mol}$, $\delta = \pi$ radians, and $n = \{1, 2, 3\}$. (d) The Lennard-Jones potential or a slightly modified form is frequently used for VDW interactions; it computes the non-bonded interaction between two particles as a function of the distance between the particles, r, the collision diameter for the interaction, σ, and the depth of the potential well, ε. Results are displayed for $\sigma = 6$ Å and $\varepsilon = 5\frac{kcal}{mol}$. (e) Electrostatic interactions may be treated with a Coulombic term; the charges of particles i and j are given by q_i and q_j, respectively, ε_0 is the permittivity of vacuum, and r_{ij} is the current distance between particles i and j. Results are shown for $q_i = +1$ and $q_j = +1$ (blue) and for $q_i = +1$ and $q_j = -1$ (green). (f) (Top) Cartoon model of an all-atom antibody Fab in a periodic simulation box. (Bottom) All-atom and CG models of the same Fab are shown. In the CG model, each amino acid is reduced to one representative interaction site.

8.2.2 Monte Carlo (MC) Simulations

Monte Carlo (MC) simulation refers to a broad range of simulation tools in which random numbers are used to perform a search across some phase space.[18] The goal of an MC simulation is typically the same as an MD simulation: the prediction of a quantity of interest from the system. In the case of biomolecules, we are typically concerned with exploring the conformational and energetic landscape of our system of interest to predict its conformations or energies. In many cases, a forcefield much like those used in MD simulations will be used to assess the energy of molecular conformations.

MC and MD simulations may also be used within the same protocol, such as in the enhanced sampling technique of replica exchange.[21] In replica exchange, a set of identical replicas of a chemical system are initialized, each at a different temperature. After a short MD simulation, exchanges are attempted between "neighboring" replicas that are at adjacent temperatures. The probability that the exchange is accepted, w, is calculated based on the energy and temperature differences between the two replica conformations, and a random number, r, is generated on the interval [0,1]. If $r \leq w$, the exchange is accepted and the replicas are swapped between temperatures. By performing many thousands of these exchanges, the replicas undergo a random walk in temperature space. This random walk helps the simulation avoid spending most of its runtime in local potential energy minima and thereby reduces the total time required to fully explore the energy landscape relative to a single-temperature simulation.[22] This is but one example of the multitude of MC methods that may be applied to molecular simulations.

We note that the key difference between MC and MD simulations is that subsequent structures generated by an MC method are not time-correlated with one another as they are in an MD trajectory.

8.2.3 Challenges of Molecular Simulations

As with all models, molecular simulations have certain limitations that must be considered. As described above, the goal of molecular simulation is normally to predict the time evolution or ensemble average of some property for a chemical system of interest. Due to the use of random numbers when generating particle velocities (as well as numerous other technical factors[23]), individual MD trajectories, even when initiated from identical starting coordinates, quickly diverge from one another. The use of random numbers in MC simulations likewise leads to divergent behavior between runs. This leads to questions such as: which of these simulations can be considered correct? Are they all reasonable predictions of behavior?

The random nature of the simulations means that each can be thought of as similar to one observation from a single-molecule experiment – each is individually valid, but the average (ensemble) behavior of the system only becomes clear in the limit of many simulations/observations. In practice, this means that multiple simulations are run and the conformations sampled across them are averaged together. If we are trying to compute the average radius of gyration, we compute the radius of gyration of all of the different conformations of the molecule from each simulation and average them. Determining if enough statistically independent simulations of sufficient length have been run to collect a representative sample of possible conformations is a key issue in the practical application of molecular simulations. Obtaining sufficient sampling over the different possible conformational states/trajectories of the system to achieve converged results is time consuming and expensive, especially for larger systems.[23] Some systems are simply too large to achieve converged results,[24] meaning that reliable answers cannot be obtained without reducing the complexity of the calculation. Below we describe methods that have been developed to overcome these sampling problems.

Molecular simulations also require starting structures. In most cases, these are MX structures deposited in the Protein Data Bank[25] (PDB). However, many experimental structures have residues with missing side chains or sections in which entire residues could not be resolved, which require careful rebuilding before simulation. Most proteins solved by MX are small globular proteins or single domains of multi-domain proteins crystallized in isolation. This means that, in many cases, multiple PDB models must be merged to build a complete model for simulation. With the advent of more accurate structure prediction tools such as AlphaFold[1–3] and various antibody-specific protein structure prediction tools,[4–8] the reliance on experimental structures is beginning to ease. This explosion in predicted structures has opened up exciting new opportunities for simulations by providing a wealth of starting structures for previously inaccessible systems.

One shortcoming of classical MD simulations is that they cannot model the formation and breakage of chemical bonds, meaning that they cannot be used to model enzymatic reactions. Several methods are available to overcome this shortcoming. Multi-scale modeling methods in which most of the biomolecular system is represented classically and the catalytic region is modeled using quantum-mechanical methods have been developed,[26] and some specialized forcefields can model bond breakage and formation.[27] However, in simulations of antibodies, we tend to be interested in either antibody-antibody or antibody-antigen non-covalent interactions, meaning that the majority of simulations run are classical.

8.3 MODELING PERSPECTIVE: WHY WE CANNOT SIMULATE EVERYTHING IN THE REAL SYSTEM

The goal of a molecular simulation is to predict a property of interest for an antibody. In practice, this is achieved by initializing a simulation with some set of particles representing the molecules in the system, defining how they interact with one another, and then by some method sampling the different accessible conformational states of the simulation.

It makes intuitive sense that the most accurate simulation would be one in which every molecule present in the real system and their interactions with one another are represented as accurately as possible. However, the cellular or test tube environment is far too large and complex to be explicitly simulated with current computational power. For example, a 1-mL aliquot of a 150-mg/mL full-length immunoglobulin G 1 (IgG1) monoclonal antibody (mAb) solution contains on the order of 10^{17} antibody molecules (assuming a molecular weight of 150 kDa[28]), each of which is composed of ~20,000 atoms. Without even considering the need to add water and cosolutes to our simulation, we can see that the number of atoms exceeds 10^{21}. Complete simulations of a cell-like environment would be necessarily even more complex, containing each of the thousands of unique macromolecules and small molecules composing the cellular milieu.

Current computational power limits molecular simulations to an upper bound of 10^9 particles, though this feat required utilizing 65,000 processors.[29] Simplifying assumptions must be applied to reduce the size and complexity of the real system by many orders of magnitude to make up the difference between the $\leq 10^9$-atom systems we can simulate and the real system of 1 mL of 150-mg/mL mAb with $\gg 10^{21}$ atoms that we cannot. Some of the most important and frequently employed simplifying assumptions are considered below before we discuss different resolution molecular simulations in detail.

8.3.1 Periodic Boundary Conditions

To avoid needing to explicitly model large volumes of solution, molecular simulations typically apply periodic boundary conditions around a single copy of the biomolecule of interest. These boundary conditions define an infinitely mirrored simulation space, such that when a particle's trajectory causes it to exit the system from one end, it reappears at the other end with the same velocity (Figure 8.1f). The volume within the periodic boundary cell is chosen to be sufficiently large that the protein cannot interact with itself through the periodic wall during the simulation. Thus, these simulations can be thought of as being run at infinite dilution. Limiting the size of the simulation in this way massively decreases the number of particles to within the realm of feasibility.

8.3.2 Inclusion versus Exclusion of Constant Domains

Structurally, antibodies can be broken into the Fab regions at the ends of the two binding arms that recognize and bind antigens and the Fc region that is involved in signaling. Due to the strong emphasis on understanding antibody-antigen binding, as well as the inherent difficulty in crystallizing full-length antibodies (and larger proteins in general), experimental structures predominantly contain only Fab segments. For example, as of the 1-Nov-2022 update, the SAbDab database[30,31] of antibody structures contains 6,029 Fvs, 5,084 Fabs, and 17 full-length antibodies (all sequence non-redundant). These numbers indicate that while the entire Fab is present in 84% of structures, intact antibodies are rarely crystallized. Molecular simulation studies are frequently performed using only the Fv and antigen to reduce computational expense. This approximation appears to make intuitive sense for cases in which only the energetics of antigen/antibody interactions are of interest, as the entire interface is resolved. However, published MD simulations indicate that including the full Fab rather than just the Fv does, in fact, influence results,[32] suggesting that using full Fabs whenever possible in simulations should be considered a matter of best practice. Simulations of the properties of solutions of mAbs, however, require a representation of the Fc to account for, at a minimum, the effects of its steric bulk. Simulations of full-length antibodies have been published,[33–38] though they are far less common than studies using a reduced section of the molecule.

8.3.3 All-Atom versus Coarse-Grain (CG) Simulations

Simulations in which each atom of the chosen reduced chemical system is explicitly represented are the most common type of molecular simulations performed. All-atom simulations of this type must use notably short integration time steps of 1 or 2 fs in order to maintain stability during numerical integration. The magnitude of the integration time step is limited by the highest-frequency vibrations in the system, which for atomic systems described classically are the bond vibrations.[39] By constraining covalent bonds containing hydrogen atoms, a time step of up to 2–3 fs may be used. All-atom methods are considered the standard for accuracy in MD simulations. All-atom MD simulations also benefit from a plethora of well-used and documented tools for preparing, running, and analyzing simulations, including Amber,[40] GROMACS,[41] CHARMM,[42] OpenMM,[43] and NAMD.[44] The popular forcefields like Amber also have many iterations that are best applied in different situations, making the choice of forcefield an important question. A benefit of all-atom methods is that all of the mainstream protein forcefields (*e.g.*, CHARMM, Amber) are transferable, meaning that they can be applied to any protein system without needing to generate custom parameters. As we will see below, reduced-resolution models frequently include non-transferable terms that must be parameterized on a case-by-case basis. All-atom simulations are typically run with explicit representations of solvent molecules and ions. Various water models exist, and some are designed to work in concert with specific biomolecular forcefields (for example, the Amber FF14SB protein forcefield[45] is optimized to run simulations with the TIP3P water model[46]).

Reducing the computational expense of all-atom simulations to allow the simulation of larger systems for longer timescales while preserving as much of their predictive power as possible is highly desirable. One method to reduce the computational cost of MD simulations is to coarse-grain (CG) the system by reducing groups of atoms to representative interaction sites[47,48] (Figure 8.1f). In most cases, this is achieved by starting with an atomistic structure of the system and using a CG mapping function that determines how groups of atoms are reduced to interaction sites.[48] Once a CG mapping has been determined, a forcefield that describes the interactions between CG sites can be used to calculate the forces between them and their time evolution is then predicted with Newton's equations of motion, just as in all-atom MD.

Simulations of CG models are faster than all-atom simulations for several reasons. First, by reducing the number of particles in the system, they significantly reduce the number of forcefield terms that must be computed at every time step. Returning to our example of a ~20,000-atom full-length IgG1 mAb, if we CG it to one interaction site per residue, we reduce the number of particles by an order of magnitude to ~1,400 interaction sites. Reducing the number of degrees of freedom in the system not only reduces the number of forcefield terms to compute but also reduces the roughness of the free-energy landscape,[47] tending to accelerate dynamic processes like protein folding. By increasing the length scale of the system and reducing the frequency of the highest-frequency vibrations, CG models can also typically be run with integration timesteps up to 10-fold larger than all-atom MD, allowing them to take larger leaps through time and simulate longer timescales faster. Finally, CG models are frequently run without explicit solvent representations, opting instead to represent the solvent implicitly to further reduce the number of particles. Together, these effects mean that CG simulations are often several orders of magnitude faster than all-atom simulations.

The selection of model resolution is a critical modeling decision that is typically made at the earliest steps of a molecular simulation project. In general, the selection of model resolution should be motivated by the time and length scales of the process of interest. In fact, it is always worth considering whether the time and length scales of the property of interest put it outside of the practical realm of MD simulations. In terms of timescale, all-atom simulations can routinely access timescales on the order of 10^{-9} to 10^{-6} seconds (nano- to microsecond), with longer simulations (up to millisecond) possible for small systems or with the heroic application of computational power. For example, the benchmarks[43,49] for OpenMM v7.7 on a single A100 GPU indicate that for a ~10^5-atom system (apolipoprotein A1) one can expect to achieve 429 ns of simulation per day, but with a ~10^6-atom system (satellite tobacco mosaic virus), that performance drops to just 32 ns/day on the same hardware. At the far upper limit of the computational performance curve, the state-of-the-art Anton 3 supercomputer, which is custom-built for high-speed MD simulations, boasts a speed of 100,000 ns/day using 512 nodes to simulate a 10^6-atom system.[50]

The speed-up of a CG simulation relative to an equivalent all-atom system is not frequently reported in the literature. General estimates suggest CG models tend to be 10^3 to 10^4 times more efficient than all-atom simulations.[51] In some cases, however, the improvement can be more drastic; a set of 19 CG models of globular proteins were found to fold on average 4×10^6 times faster than in experiments.[52] The specific sources of acceleration in CG simulations are many and differ between models; we direct interested readers to the discussion in Refs. 47 and 48.

While CG models enjoy favorable increases in simulation speed, they suffer from a loss of spatial resolution. CG models with one interaction site per amino acid placed at the coordinates of the C_α atom are reduced to a spatial resolution on the order of the average C_α–C_α bond length, 3.8 Å. In many cases when simulating mAbs, resolution is reduced much further, with the mAb represented by perhaps 6, 10, or 12 interaction sites. These simulations, which may be considered "ultra-coarse-grained" as they reduce hundreds of amino acids to single interaction sites, can be used to simulate protein-protein interactions and predict experimental solution properties. However, a six-bead model of antibodies is not suitable for the investigation of which residues contribute to antibody-antigen binding as the individual residues at the interface are not modeled. Given that an infinite number of different CG representations may be designed for a given protein, methods to determine the "best" representation have emerged.[53] However, in most cases, researchers employ a combination of intuition for the chemical system under investigation and trial and error to achieve a realistic model for the system and parameter of interest. Despite these issues that would appear to limit the accuracy of CG models, when carefully parameterized and used within their limitations, they can accurately predict the dependence of protein properties on osmolyte concentration,[54] the influence of pH on antibody viscosity,[55] and various other properties.[56,57]

8.4 USES OF MOLECULAR SIMULATION IN ANTIBODY DRUG DEVELOPMENT

8.4.1 Predicting and Understanding Protein-Protein Interactions in mAb Solutions

Most antibody drugs, such as mAbs, are delivered by sub-cutaneous injection. The comfort of the patient receiving the injection places certain limits on the volume (1–2 mL[58]) that can be administered and the viscosity (practical delivery threshold ~20 cP[59]) of the solution, the latter of which determines the injection force required. The injection volume limit and typical dosages dictate mAb solutions with concentrations of >100 mg/mL.[58]

At these high concentrations, protein-protein interactions between mAbs drive self-association and determine viscosity.[60,61] As a result of differences in their self-interaction profiles, some mAbs may have poor solubility and not reach the required concentration for effective delivery; some may demonstrate prohibitively high viscosity for comfortable injection; yet others may aggregate irreversibly.

In the development pipeline, experimental methods are used to assess a mAb's physical and chemical properties and response to environmental conditions like salt concentration[55,62,63] to judge its "developability". These methods are, however, time consuming and consumptive of mAb material. In response, computational ways to both predict and understand what mAb structures give rise to the properties of concentrated mAb solutions have been developed in the academic literature. Some of these methods leverage information from available experimental structures of known therapeutic antibodies to define developability guidelines.[64] These methods do not, however, provide information about what structures these antibodies populate that give rise to their behavior in solution. Simulation methods have also been designed to address this shortcoming.[55,59,65] Some of these methods take a small amount of experimental data as input and extrapolate to diverse solution conditions. Yet other methods are designed to work in tandem with experimental methods to answer the inverse question of what structures give rise to an observed experimental signal. As their goal is predicting protein-protein interaction-based parameters, these methods typically consider multiple copies of the mAb of interest at a CG resolution to control computational costs. In the following three **case studies**, we describe important lessons learned from CG models of concentrated antibody solutions.

Case Study 8.1: Predicting the response of mAb solutions to salt and pH

The propensity of mAbs to self-interact in low-concentration regimes can be assessed experimentally through the second osmotic virial coefficient, B_{22}.[55] Its values are a commonly used proxy for protein-protein interactions in a dilute solution, with positive values indicating repulsive interactions and negative values indicating attractive interactions. In situations in which a mAb displays strong protein-protein interactions, such as at higher concentrations, higher-order virial coefficients that report on the formation of higher-order oligomers may also be computed. Shahfar and co-workers[55] compared the ability of four different CG models (Figure 8.2a) to predict B_{22} as a function of the total ionic strength (TIS) for mAb1, mAbB, mAb2, and mAbC.

Diverse qualitative behavior in experimental B_{22} versus TIS curves is observed; mAb1, mAbB, and mAb2 can serve as exemplars of the different behaviors displayed by mAb solutions as TIS changes (Figure 8.2b). mAb1 at pH 5 displays large positive B_{22} values at low values of TIS in the solution, with a monotonic decrease observed as TIS increases (Figure 8.2b, blue). mAbB at pH 5 displays values near zero at low TIS, negative values at intermediate TIS, and then increases monotonically with increasing TIS to a plateau near zero (Figure 8.2b, green). Finally, mAb2 at pH 6.5 displays large magnitude negative values of B_{22} at low TIS that increase monotonically toward zero as TIS increases (Figure 8.2b, yellow). In each of these examples, B_{22} versus TIS curves

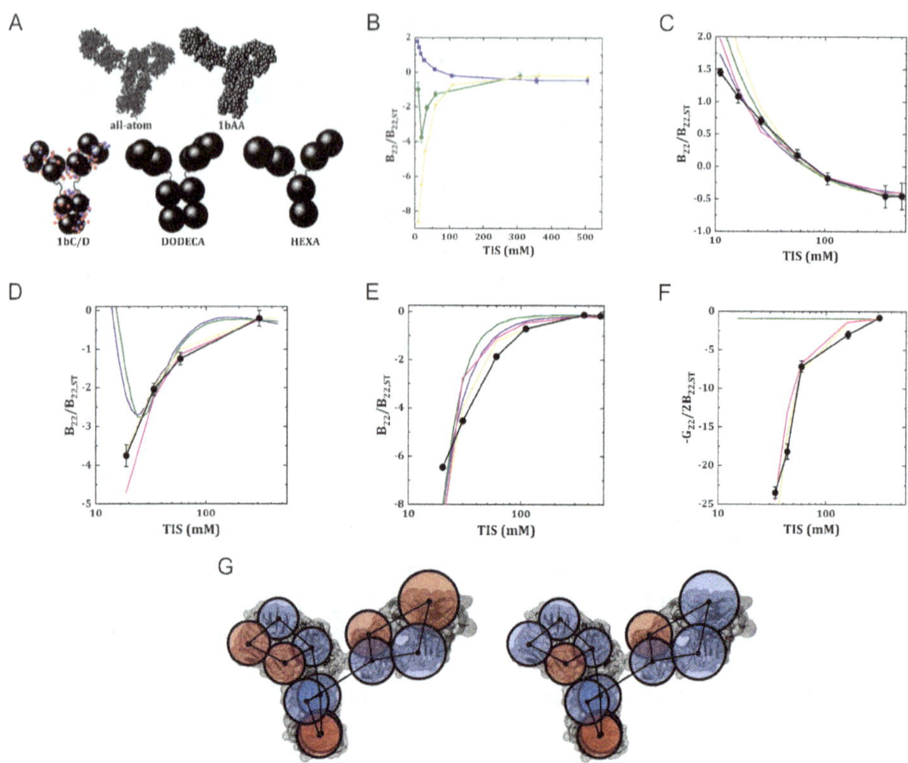

FIGURE 8.2 (a) All-atom and CG models from Shahfar et al. (2021). (b) Experimental second virial coefficient versus TIS curves for mAb1 at pH 5 (blue), mAbB at pH 5 (green), and mAb2 at pH 6.5 (yellow). (c) Experimental (black symbols) and CG model predicted second virial coefficient versus TIS curves for the HEXA (blue), DODECA (green), 1bC/D (yellow), and 1bAA (purple) for mAb1 at pH 5. (d) Same as (c) but for mAbB. (e) Same as (c) but for mAb2 at pH 6.5. (f) Same as (c) but for mAbC. B_{22} is the second virial coefficient and $B_{22,ST}$ is the second virial coefficient under the assumption of steric-only interactions. The y-axis in panel F is denoted $-G_{22} / 2B_{22,\,ST}$ because the large values suggest the formation of higher-order oligomers. (g) DODECA models used by Wang et al. for mAb1 (left) and mAb2 (right). Red and blue beads indicate negative and positive net charges, respectively. Panels A-F reprinted with permission from Shahfar, H.; Forder, J. K.; Roberts, C. J. *J. Phys. Chem. B* **2021**, *125* (14), 3574–3588. Copyright 2023 American Chemical Society. Panel G reprinted with permission from Wang, G. et al. *J. Phys. Chem. B* **2018**, *122* (11), 2867–2880. Copyright 2023 American Chemical Society.

can be understood in terms of the interactions that predominate in each mAb solution as TIS changes. For mAb1, the behavior is as predicted for a colloid-like system in which high net surface charge leads to repulsive interactions at low ionic strength that become screened as TIS increases, leading to a plateau near zero at high TIS. mAbB contains both positive and negative charges that can interact attractively, leading to attractive interactions at intermediate TIS when screening is at a favorable level. In the case of mAb2, the results are consistent with a mAb with a low net charge but with charged patches that can strongly attract one another at low TIS and become screened as TIS increases.[55] These experimental curves indicate that a CG model must be able to account for various behaviors in order to accurately predict the different behaviors of mAbs.

To predict the B_{22} versus TIS curves for mAb1, mAbB, mAb2, and mAbC, Shahfar and co-workers constructed CG models with (i) one bead per amino acid (1bAA), (ii) 12 beads total with one each per domain in the mAb and the net charge of its constituent amino acids applied to each site (DODECA), (iii) 12 beads total with charged amino acids represented as charged patches on the beads (1bC/D), and (iv) six beads total with two for each Fab and the Fc and the net charge applied to each bead (HEXA). Simulations were run using the Mayer sampling method[66] with the overlap sampling algorithm to compute values of B_{22} and higher-order virial coefficients; this amounts to a specialized MC sampler that biases the simulations to preferentially explore states that contribute most to the final value of a given virial coefficient. Within each of these CG models, solvent is considered implicitly, and similar functions are used to represent their short-range non-electrostatic and electrostatic interactions as functions of ε_{SR} and ψ, respectively. ε_{SR} sets the maximum strength (well-depth) of non-electrostatic interactions between sites and ψ is an adjustment factor that scales the theoretical charges of CG sites. Values of ε_{SR} and ψ were selected based on test simulations for each CG model until pairs of parameters that minimize the average relative deviation from the experimental B_{22} versus TIS curves were determined.

For mAb1, which displays standard colloid-like behavior, each of the four CG models provides a reasonably good fit at pH 5, with the 1bAA model displaying the smallest deviation from the experiment (Figure 8.2c). In the case of mAbB at pH 5, the DODECA and HEXA models fail to predict the correct shape of B_{22} at low TIS, while the 1bC/D and 1bAA models both provide strong predictions of the experimental data (Figure 8.2d). With mAb2 at pH 6.5, each of the four models again makes reasonable predictions, while for mAbC at pH 5, the HEXA and DODECA models completely fail to capture the shape of the experimental curve (Figure 8.2e and f). These results indicate that a bespoke modeling process for understanding the influence of salt is required, with some mAbs proving to be less tractable than others and requiring more expensive, higher-resolution CG models. Shahfar et al. provide a table of CG model parameters for their different mAbs that can help choose reasonable parameters for simulations even in the absence of experimental data (see Ref. 55 Table 1).

This MC investigation of the influence of pH on low-concentration mAb solutions highlights the challenges and future directions for such predictive models. Strong fits to experimental data tend to be obtained when at least 12 CG interaction sites are considered with amino-acid-specific charge representations rather than domain-level net charge electrostatics (*i.e.*, 1bC/D). Models using one CG interaction site per amino acid also provide strong predictions, though at a higher computational cost. Even in some more difficult cases, CG simulations can reproduce the fine structure of experimental B_{22} curves.

Case Study 8.2: Predictive models for the properties of high-concentration mAb solutions

Significant interest in recent years has been focused on the development of CG models coupled with either MD or MC sampling methods to predict mAb solution viscosity. Wang and co-workers[67] ran Brownian Dynamics simulations using 12-bead CG representations of two mAbs, mAb1 and mAb2, that have 92% sequence similarity. Despite their highly similar sequences, these mAbs have disparate rheological behavior, with the viscosity of mAb1 varying much more strongly in response to changes in its concentration, pH, and the ionic strength of the solution.[61] The initial simulations of Wang and co-workers were in poor agreement with experimental relative viscosity measurements. In these initial simulations, CG beads within a single mAb molecule interact through bond, angle, dihedral, and Urey-Bradley terms (additional forcefield terms to account for angle bending), and each CG bead is assigned a charge equal to the net charge of the particles it represents (Figure 8.2g). The spring constants and equilibrium values for the CG bead interactions were computed from all-atom MD simulations of a single mAb molecule.[68]

Wang and co-workers noted that in scattering experiments, mAb1 has been found to form reversible dimers and higher-order oligomers.[60,69] These dynamically forming and breaking groups of self-associated mAbs are referred to as clusters. Upon adding additional constraints to their CG simulations that cause groups of self-associated CG mAbs to move together rigidly, simulation results for both mAb1 and mAb2 significantly improve. In these new results, mAb2's viscosity as a function of concentration is predicted well, while the prediction of mAb1's viscosity is improved but still inaccurate at high concentrations. These simulation results indicate the importance of considering soluble clusters of mAbs in simulations for the prediction of mAb viscosity.

In a follow-up to the work of Wang et al., Lai and co-workers attempted to improve this 12-bead CG model's ability to capture changes in viscosity with mAb concentration by making a key modification to interactions within the model.[70] Rather than using a single constant interaction energy between all CG beads regardless of the part of the mAb they represent, they incorporate separate Fv and Fc interaction terms. Interaction strengths between Fv regions are determined using the high-viscosity index (HVI), a parameter relative to viscosity developed using a machine-learning approach,[71] and then scaled by a fitting parameter α. Interactions between mAbs depend, then, on an Fv-Fv specific term $A_H^V = \text{HVI} * \alpha$, an Fc-Fc specific term set to a constant for all Fcs, and electrostatic forces dependent on the net charge of the CG interaction site. Equilibrium values and force constants for CG interactions were determined from all-atom simulations of each mAb, as in the work of Wang et al. Values for α and the strength of Fc interactions were determined using a grid search to find the pair of parameters that best predicts the viscosity of a 150-mg/mL mAb solution over all 20 mAbs studied. This simple fitting procedure generates models that provide accurate relative viscosity predictions for [mAb] = {50, 100, 125, 150} mg/mL for a set of 20 mAbs. In comparison to the original model of Wang et al., the root-mean-square deviation improves from 1 to 0.68 and the correlation coefficient improves from 0.63 to 0.87. Despite better performance on the data set overall, the original model of Wang and co-workers performs better than that of Lai and co-workers on mAb2.

Lai and co-workers also, for the first time, explored the influence of the size of the simulated system on the calculated relative viscosity. In principle, an infinite number of simulation boxes with different volumes can be designed and then filled with an appropriate number of mAb representations to reach a desired concentration. All publications investigating mAb solution viscosity prior to this simulated only a single box size. Lai and co-workers, however, demonstrated that changing the size of the simulation box influences the calculated relative viscosity. With $A_H^V = 1$, the simulations predict a 3-fold larger viscosity when run with a box containing N=4,096 monomers versus a smaller box with N=512 monomers. This result breaks one of our basic intuitions for how a chemical system should behave: if we have 10 mL of 1.0 M NaCl and split the solution into two 5-mL aliquots, we expect each of them to have identical intensive properties like viscosity to the initial solution before it was split. In the case at hand, this illogical result is a simulation artifact that arises due to the dependence of viscosity on the size of self-associated clusters formed in the finite simulation box. The calculation of the viscosity can only consider the clusters formed within the simulation box—for large cluster sizes and small periodic boxes, however, the effective cluster extends through the periodic boundary conditions. This leads to computed viscosities that depend on the size of the simulation box with mAb concentration held constant. When interactions between mAbs are weak and clusters tend to be small, this dependence disappears.

These studies of high-concentration mAb solutions highlight that Fv-specific interactions are crucial for predicting protein-protein interactions in high-concentration mAb solutions. The dependence of the cluster-size distribution on the simulation box dimensions is a particularly troubling simulation challenge.

Case Study 8.3: Tandem experimental and simulation studies of mAb solutions

While the previous two **case studies** have focused on using CG methods to predict the solution properties of mAbs with minimal reliance on experimental data in their parameterization procedures, other CG models have been designed to work exclusively in tandem with experiments.[59,62] In such studies, experimental small-angle X-ray scattering (SAXS) or small-angle neutron scattering (SANS) data are used as a target during CG model optimization and parameters chosen that minimize the deviation between simulated and experimental scattering data.[59,62] The scattering data may also be refined into an all-atom model, from which a CG representation is then constructed.[59] While these methods are not, strictly speaking, predictive models for viscosity, the best-fit simulations can provide a wealth of information about cluster-size distributions and help understand the dynamic oligomeric states populated by mAbs in solution.

In the work of Dear and co-workers,[62] SAXS experiments at [mAb] = 5 mg/mL were used to assign a mAb shape by fitting to $P(q)$, the normalized form factor, as a function of $q = \frac{4\pi}{\lambda}\sin\left(\frac{\theta}{2}\right)$ where λ is the wavelength of the incident X-ray and

θ is the scattering angle. The scattering pattern represented by the form factor is related to the size and shape of the protein in the solution.[72] In general, low concentrations of 1–10 mg/mL are used to provide a compromise between the increase in signal-to-noise ratio and the decrease in interparticle distances that both accompany increased solution concentration. While the former is always favorable, the latter can influence the scattering pattern. In practice, profiles collected at different concentrations may be merged into a single representative curve. Solutions of mAb2 at [NaCl] = 0, 50, 250 mM and with [Arg] = 250 mM and mAb4 with either 250-mM NaCl or Arg were all assayed with SAXS, and the resulting $P(q)$ curves used to build 12-bead CG models with various interaction profiles. While each of the CG models for a particular mAb and solution condition have identical bead sizes and locations, they differ in terms of how many beads experience attractive interactions and where these attractive beads are placed within the CG model. MD simulations were run with 6,000 identical mAb CG representations in a periodic box with a constant number of particles, system volume, and system temperature for each mAb and CG model. They find that models with only three attractive beads per mAb are frequently detected bound to more than three neighbors. This result appears to invalidate theoretical models that only allow one interaction partner per binding site.[73,74] In a follow-up paper, Chowdhury and co-workers investigated the influence of short-range, non-electrostatic attractive and electrostatic repulsive interactions. Using a similar fitting and simulation procedure to Dear et al., Chowdhury and coworkers[59] manipulated the parameters within the CG model to best fit the experimental structure factor curves under various conditions for mAb2. These simulations found that including uniform VDW interactions between all model beads allows predictions of structure factor curves to become more accurate.

These simulation results highlight the ability of CG simulations to aid in understanding the difficult inverse question that accompanies spectroscopic assays of biomolecules: what ensemble of structures gives rise to the observed signal? In this case, exploring various CG models and rationalizing which models provide the best predictions for which system provides a wealth of additional information beyond the experiment alone. As computational power increases and model accuracy continues to improve, the use of CG simulations to understand the fine structure of experimental scattering curves may become more commonplace.

8.4.2 Predicting and Understanding Binding Mechanisms, Energetics, and Aggregation

The following three **case studies** highlight the use of all-atom simulations to understand and predict the mechanisms and energetics of mAbs interacting with their targets as well as the aggregation propensity of mAbs. While all-atom simulations have become more common in the academic literature, their high computational cost limits their application across the development pipeline. As enhanced sampling techniques and simulation speeds continue to improve, however, we expect to see an ever-increasing reliance on the predictions of all-atom MD.

Case Study 8.4: Predicting the binding mechanism and oligomeric preference of solanezumab

Bekker et al. (2020) ran multicanonical MD simulations, in which structures from trajectories run at different temperatures may be reweighted to provide better sampling at a specific temperature of interest, of solanezumab binding to the monomeric form of its target peptide amyloid-β (Aβ, Figure 8.3a).[75] In these simulations, the Fv region of solanezumab was held partially restrained, and the center of mass of Aβ was then allowed to explore conformations within a cylinder positioned normal to the solanezumab binding site. This combination of multicanonical MD with positional restraints increases the speed of the conformational search relative to single-temperature simulations and reduces the number of accessible conformations for the solanezumab/Aβ system, enhancing sampling.

A free-energy landscape projected onto the two principal components (PC1 and PC2) of this simulation suggested that the solanezumab Fv explores a diverse conformational landscape at 300 K around the experimental structure (Figure 8.3b, experimental structure indicated by a white X). Bekker et al. 2020 then selected a set of ten structures from the free-energy landscape representing steps along the binding reaction and performed path-sampling simulations to characterize the binding mechanism. These path-sampling simulations revealed the mechanism of interaction between solanezumab and Aβ. As the peptide approaches the antibody, transient non-specific interactions begin to form. Once it reaches 8–10 Å from the binding pocket center of mass, the hydrophobic core of the pocket begins to interact via CDRH1 and CDRL1 (observed as a metastable plateau in the potential of mean force, Figure 8.3c). Many individual native contacts are made as the peptide nears the core, with non-specific interactions gradually becoming stabilized by salt bridges. Bekker et al. 2020 conclude by simulating Aβ in isolation, revealing (consistent with previous work[76]) that Aβ explores disordered structures in the unbound state.

Comparing their results to the three typical binding models of (i) lock and key, (ii) population shift, and (iii) induced fit, Bekker and co-workers determined that their results are most consistent with what they term a "mutual population shift". Each molecule undergoes random structural perturbations until it is in a native-like state, which allows it to rapidly bind if its partner is nearby and also in a native-like state. In other words, if a molecule of both the mAb and the peptide are both in native-like states in close proximity, they form a bound complex. Their structure and mechanism of interaction also provide an explanation for the substrate selection of solanezumab: only the monomeric form of Aβ is able to fit deep enough within the solanezumab binding pocket to induce a conformational shift away from the random states it populates in solution. These all-atom MD simulations highlight the strengths of this technique, providing a detailed binding mechanism along the chosen order parameter.

A
Aβ
V_L
V_H
B
PC1
PC2
Free energy (kcal/mol)
C
PMF (kcal/mol)
λ' (Å)
D
α1
α2
peptide
E
CDR1
CDR2
CDR3
VH/VB
VL/VA
Wilm's Tumor 1
RMFPNAPYL
NY-ESO-1
SLLMWITQV
p53 R175H
HMTEVVRHC
F
Solubility from CamSol:
<1.7
<1.6
<1.5
<1.4
Temperature (K)
bevacizumab
rituximab
PPI01
PPI17
PPI10
PPI02
PPI03
trastuzumab
adalimumab
PPI13
G
Δaggregates (%)
≥ 0.5%
≤ 0.5%
PPI04
PPI08
bevacizumab
PPI01
rituximab
PPI11
PPI03
trastuzumab
adalimumab
PPI02
PPI17
PPI13
PPI10

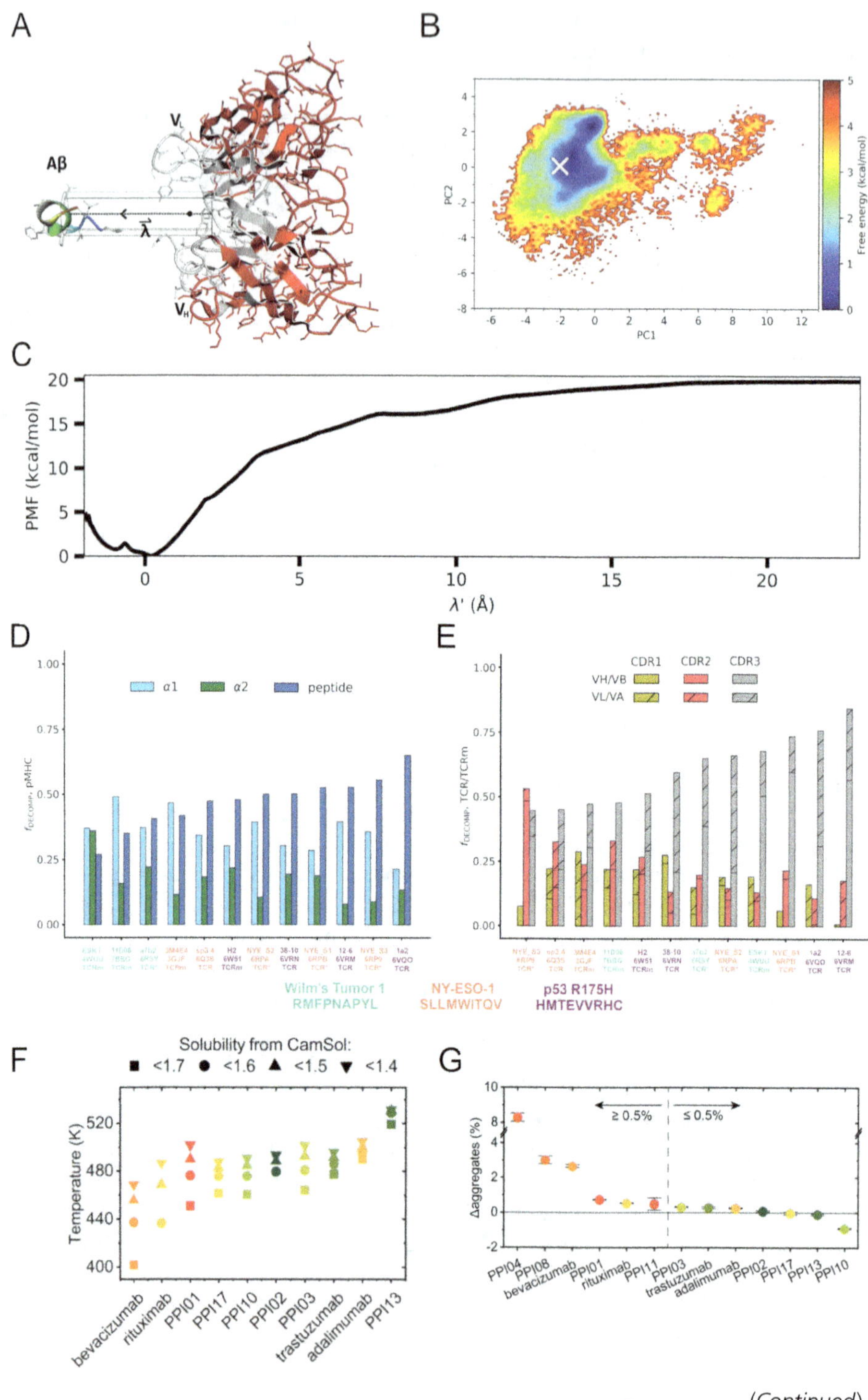

(Continued)

FIGURE 8.3 (*Continued*) (a) Simulation setup of Bekker et al. 2020. (b) Free-energy landscape from all-atom simulations of solanezumab (labeled V_L and V_H for variable light and variable heavy) and Aβ peptide at 300 K. (c) Potential of mean force over the order parameter λ', the distance between the peptide and binding pocket. (d) Fractional contributions to total receptor free energy from the peptide, α1, and α2 helices. (e) Fractional contributions of CDR loops to total ligand-free energy. (f) The temperature at which a set of ten Fvs pass below solubility thresholds during temperature-ramp all-atom MD. (g) Percent change in aggregation for 13 mAb solutions after 3 months of storage at 40°C. Panels a–c: Reprinted with permission from Bekker, G. J. et al. Sci. Rep. 2020, *10* (1), 1–9 (Creative Commons Attribution 4.0, https://creativecommons.org/licenses/by/4.0/). Panels f and g: Reprinted with permission from Berner, C. et al. Mol. Pharm. 2021, 18 (6), 2242–2253. Copyright 2023 American Chemical Society.

Case Study 8.5: Predicting and understanding the energetics of interactions between antibodies and their targets

All-atom simulations allow for the energetic interactions between molecules to be investigated at levels of detail inaccessible by experiment. This advantage has been highlighted by two recent studies of the differences between how T-cell receptors (TCRs) and TCR-mimetic antibodies (TCRms) bind their targets. TCRs are the natural immune system's method of identifying peptide fragments presented by human leukocyte antigens (pHLAs), in which the peptide binds the groove between the $\alpha 1$ and $\alpha 2$ helices of the HLA. The TCR variable region recognizes pHLAs exposing non-self peptides, while the membrane-bound constant region, alongside crucial co-receptors, enables downstream immune activation. The pharmaceutical industry is increasingly interested in translating TCR specificity to solubilized drug formats, which require stronger target binding. This necessitates engineering efforts to "affinity enhance" the TCR variable domain, or the use of immune recognition domains that were not evolutionarily selected to bind pHLAs, such as antibodies (TCRms), but which can routinely achieve binding in the nM-pM range.

In the work of Holland et al. (2020), all-atom simulations were used to generate ensembles of conformations for a set of TCRs, TCRms, and affinity-enhanced TCRs.[77] The molecular mechanic Poisson-Boltzmann surface area (MMPBSA) method was then used to predict the free energy difference between the bound and unbound states for each complex, and the result decomposed into the per-residue contributions.[78] In brief, this method uses all-atom MD structures in concert with implicit solvent calculations to predict free energy differences between states. Decomposition of the free energy revealed that while affinity-enhanced TCRs tend to have a broad binding signature characterized by strong interactions with up to three residues in the peptide antigen, TCRms tend to have only one or two strong energetic contacts with the peptide.

Raybould et al. (2022), leveraging the increasing numbers of MX structures for TCRms, performed a similar MD study comparing TCR, TCRm, and affinity-enhanced TCR interactions with their targets.[79] To provide the best comparisons possible, Raybould and co-workers simulated three different peptide antigens, Wilms' Tumor 1 (WT1), New York esophageal squamous cell carcinoma 1 (NY-ESO-1), and p53_R175H neoantigen (p53 R175H) bound by the same HLA in contact with various TCR, TCRm, and affinity-enhanced TCRs. The contributions of the peptide, $\alpha 1$ helix, and $\alpha 2$ helix to the total free-energy change of the pHLA were computed using the molecular mechanic Generalized-Born surface area approach, a technique related to MMPBSA but using a different representation of the protein for surface area calculations. These calculations revealed that natural and affinity-enhanced TCRs tend to gain a larger proportion of their binding energy from the peptide than any of the TCRms (Figure 8.3d). Further, these calculations suggested that some peptide antigens are easier to bind than others, with WT1 peptides notably all ranking at the bottom in terms of fraction contribution to the total free energy change in the receptor. Similar calculations carried out on the residues in the CDRs (Figure 8.3e) indicated that while the CDR[H/B]3 and CDR[L/A]3 predominate across the board, the pattern changes depending on the context. While TCRms tend to gain more energy from CDRH3 than CDRL3, in TCRs, there is no clear preference.

These studies into the energetics of antibody-antigen and TCR-antigen interactions highlight one of the key uses of MD and molecular simulations in general: gaining more detailed analysis than is possible experimentally. By determining which residues contribute most strongly to binding, studies like these can serve as the beginnings of a computational basis for mutational studies.

Case Study 8.6: Predicting aggregation-prone antibody therapeutics with all-atom MD

The formation of aggregates within mAb solutions during processing and storage can lead to the rejection of otherwise promising candidate molecules; aggregation has therefore been called the "most common and troubling manifestation of protein instability, encountered in almost all stages of protein drug development".[80] Predicting this behavior with small amounts of mAb material at an early stage of development is therefore highly desirable, and computational methods have been developed to identify and interrogate regions prone to aggregation.[81–83] Aggregation-prone regions are specific short-sequence motifs that appear to be able to modulate aggregation, leading to extensive computational investigations of their dynamics. For example, a recent study of aggregation-prone pentapeptides revealed that they can begin to form aggregates within 20 ns at a concentration of 0.1 M, with mutations strongly influencing the kinetics of aggregate assembly.[83] In a recent paper, Berner and co-workers used a combination of two experimental unfolding reversibility methods and one simulation technique to interrogate a set of 13 candidate mAbs, including some aggregation-prone examples.[82] They note that their work is inspired by the observation that antibodies with domains that undergo reversible thermal unfolding also tend to be resistant to aggregation and that antibodies that refold to monomers after unfolding with denaturant aggregate less in formulations. In the following section, we describe the insights into aggregation they gained from all-atom simulations of antibody Fv regions.

Berner and co-workers[82] performed all-atom MD simulations of the Fv segments of the ten antibodies for which they had access to experimental structures (Figure 8.3f) and experimental assay data. These data show the temperatures at which each of the ten antibodies passed below certain solubility thresholds (calculated with CamSol[84]) during temperature-ramp simulations, in which the Fv was heated over the course of the simulation from 300 to 540 K with a temperature increment of 10–20 K and 20 ns of dynamics simulated at each temperature. In general, decreased solubility is observed as temperature increases, indicating partial unfolding that increases the solvent exposure of aggregation-prone regions.

Compared to experimental data showing the percent change in aggregates detected by size-exclusion chromatography after 3 months of storage at 40°C (Figure 8.3g), we see that the all-atom simulations are strongly predictive of aggregation. In the case of trastuzumab, a tight transition to less soluble states is seen from ~480–500 K; these simulations suggest trastuzumab contains a stable Fv that should only begin to lose solubility at high temperatures, consistent with the experimental data showing that trastuzumab indeed has effectively zero aggregation change over a 3-month period. In contrast, bevacizumab has a broad solubility transition from ~400 to 470 K, and in size-exclusion chromatography experiments was found to have a ~3% increase in aggregation over 3 months. These results are consistent with analyses showing that the proportion of trastuzumab in a β-sheet conformation in the all-atom simulations is effectively constant at 0.45 until 500 K, when it transitions sharply to an unfolded state by 525 K. Bevacizumab, in contrast, begins to lose its secondary structure at around 400 K and unfolds partially over the course of over 100 K before losing all structure by 525 K.

By correlating their simulation results with experimental assays on aggregation, Berner et al. suggest an intriguing path forward using simple temperature-ramp simulations to predict solubility. One of the greatest strengths of their method is its simplicity – temperature-ramp simulations are easy to perform, and their analysis relies on standard solubility methods and straightforward calculations of secondary structure content. Future studies considering more of the overall antibody structure (*i.e.*, full Fab or even full-length antibody sequences) in the simulations could improve the information content of such studies as computational speed continues to improve.

8.5 CONCLUSION

In this chapter, we have described the background and modeling philosophy for simulations of antibodies and described some example **case studies** in more detail. The applications of MD simulations for antibodies are quickly evolving, undergoing developments in both all-atom and CG methodologies as well as analysis techniques. The set of studies highlighted here is a small fraction of the complete literature on this topic. For example, in addition to the CG results we described for the prediction of mAb solution properties, all-atom simulations have also been employed to predict the influence of pH and temperature stress on mAbs.[85,86] CG models appear poised to enter the development pipeline as a prime tool to predict the behavior of concentrated mAb solutions. All-atom simulations, on the other hand, will likely be slower to enter industrial use due to their slow speed and high computational cost.

REFERENCES

(1) Senior, A. W.; Evans, R.; Jumper, J.; Kirkpatrick, J.; Sifre, L.; Green, T.; Qin, C.; Žídek, A.; Nelson, A. W. R.; Bridgland, A.; Penedones, H.; Petersen, S.; Simonyan, K.; Crossan, S.; Kohli, P.; Jones, D. T.; Silver, D.; Kavukcuoglu, K.; Hassabis, D. Improved Protein Structure Prediction Using Potentials from Deep Learning. *Nature* **2020**, *577* (7792), 706–710. https://doi.org/10.1038/s41586-019-1923-7.

(2) Jumper, J.; Evans, R.; Pritzel, A.; Green, T.; Figurnov, M.; Ronneberger, O.; Tunyasuvunakool, K.; Bates, R.; Žídek, A.; Potapenko, A.; Bridgland, A.; Meyer, C.; Kohl, S. A. A.; Ballard, A. J.; Cowie, A.; Romera-Paredes, B.; Nikolov, S.; Jain, R.; Adler, J.; Back, T.; Petersen, S.; Reiman, D.; Clancy, E.; Zielinski, M.; Steinegger, M.; Pacholska, M.; Berghammer, T.; Bodenstein, S.; Silver, D.; Vinyals, O.; Senior, A. W.; Kavukcuoglu, K.; Kohli, P.; Hassabis, D. Highly Accurate Protein Structure Prediction with AlphaFold. *Nature* **2021**, *596* (7873), 583–589. https://doi.org/10.1038/s41586-021-03819-2.

(3) Evans, R.; O'Neill, M.; Pritzel, A.; Antropova, N.; Senior, A.; Green, T.; Zidek, A.; Bates, R.; Blackwell, S.; Yim, J.; Ronneberger, O.; Bodenstein, S.; Zielinski, M.; Bridgland, A.; Potapenko, A.; Cowie, A.; Tunyasuvunakool, K.; Jain, R.; Clancy, E.; Kohli, P.; Jumper, J.; Hassabis, D. Protein Complex Prediction with AlphaFold-Multimer. *bioRxiv* **2022**. https://doi.org/10.1007/978-1-61779-361-5_16.

(4) Leem, J.; Dunbar, J.; Georges, G.; Shi, J.; Deane, C. M. ABodyBuilder: Automated Antibody Structure Prediction with Data–Driven Accuracy Estimation. *MAbs* **2016**, *8* (7), 1259–1268. https://doi.org/10.1080/19420862.2016.1205773.

(5) Ruffolo, J. A.; Sulam, J.; Gray, J. J. Antibody Structure Prediction Using Interpretable Deep Learning. *Patterns* **2022**, *3* (2), 100406. https://doi.org/10.1016/j.patter.2021.100406.

(6) Weitzner, B. D.; Jeliazkov, J. R.; Lyskov, S.; Marze, N.; Kuroda, D.; Frick, R.; Adolf-Bryfogle, J.; Biswas, N.; Dunbrack, R. L.; Gray, J. J. Modeling and Docking of Antibody Structures with Rosetta. *Nat. Protoc.* **2017**, *12* (2), 401–416. https://doi.org/10.1038/nprot.2016.180.

(7) Abanades, B.; Wong, W. K.; Boyles, F.; Georges, G.; Bujotzek, A.; Deane, C. M. ImmuneBuilder: Deep-Learning Models for Predicting the Structures of Immune Proteins. *bioRxiv* **2022**, 1–12.

(8) Abanades, B.; Georges, G.; Bujotzek, A.; Deane, C. M. ABlooper: Fast Accurate Antibody CDR Loop Structure Prediction with Accuracy Estimation. *Bioinformatics* **2022**, *38* (7), 1877–1880. https://doi.org/10.1093/bioinformatics/btac016.

(9) Wlodawer, A.; Minor, W.; Dauter, Z.; Jaskolski, M. Protein Crystallography for Aspiring Crystallographers or How to Avoid Pitfalls and Traps in Macromolecular Structure Determination. *FEBS J.* **2013**, *280* (22), 5705–5736. https://doi.org/10.1111/febs.12495.

(10) Mcpherson, A.; Gavira, J. A. Introduction to Protein Crystallization. *Acta Crystallogr.* **2013**, *F70*, 2–20. https://doi.org/10.1107/S2053230X13033141.

(11) Garman, E. "Cool" Crystals: Macromolecular Cryocrystallography and Radiation Damage. *Curr. Opin. Struct. Biol.* **2003**, *13* (5), 545–551. https://doi.org/10.1016/j.sbi.2003.09.013.

(12) Murata, K.; Wolf, M. Cryo-Electron Microscopy for Structural Analysis of Dynamic Biological Macromolecules. *Biochim. Biophys. Acta - Gen. Subj.* **2018**, *1862* (2), 324–334. https://doi.org/10.1016/j.bbagen.2017.07.020.

(13) Lyumkis, D. Challenges and Opportunities in Cryo-EM Single-Particle Analysis. *J. Biol. Chem.* **2019**, *294* (13), 5181–5197. https://doi.org/10.1074/jbc.REV118.005602.

(14) Hollingsworth, S. A.; Dror, R. O. Molecular Dynamics Simulation for All. *Neuron* **2018**, *99* (6), 1129–1143. https://doi.org/10.1016/j.neuron.2018.08.011.

(15) Karplus, M.; McCammon, J. A. Molecular Dynamics Simulations of Biomolecules. *Nat. Struct. Mol. Biol.* 20002, *9*, 646–652. https://doi.org/10.1299/jsmemag.116.1131_78.

(16) Berendsen, H. J. C.; Postma, J. P. M.; Van Gunsteren, W. F.; Dinola, A.; Haak, J. R. Molecular Dynamics with Coupling to an External Bath. *J. Chem. Phys.* **1984**, *81* (8), 3684–3690. https://doi.org/10.1063/1.448118.
(17) Nosé, S. A Unified Formulation of the Constant Temperature Molecular Dynamics. *J. Chem. Phys.* **1984**, *81*, 511–519. https://doi.org/10.1063/1.447334.
(18) Paquet, E.; Viktor, H. L. Molecular Dynamics, Monte Carlo Simulations, and Langevin Dynamics: A Computational Review. *Biomed Res. Int.* **2015**, *2015*. https://doi.org/10.1155/2015/183918.
(19) Zhang, Z.; Liu, X.; Yan, K.; Tuckerman, M. E.; Liu, J. Unified Efficient Thermostat Scheme for the Canonical Ensemble with Holonomic or Isokinetic Constraints via Molecular Dynamics. *J. Phys. Chem. A* **2019**, *123* (28), 6056–6079. https://doi.org/10.1021/acs.jpca.9b02771.
(20) Erban, R. From Molecular Dynamics to Brownian Dynamics. *Proc. R. Soc. A Math. Phys. Eng. Sci.* **2014**, *470* (2167). https://doi.org/10.1098/rspa.2014.0036.
(21) Sugita, Y.; Okamoto, Y. Replica Exchange Molecular Dynamics Method for Protein Folding Simulation. *Chem. Phys. Lett.* **1999**, *314*, 141–151. https://doi.org/10.1385/1-59745-189-4:205.
(22) Lingenheil, M.; Denschlag, R.; Mathias, G.; Tavan, P. Efficiency of Exchange Schemes in Replica Exchange. *Chem. Phys. Lett.* **2009**, *478* (1–3), 80–84. https://doi.org/10.1016/j.cplett.2009.07.039.
(23) Knapp, B.; Ospina, L.; Deane, C. M. Avoiding False Positive Conclusions in Molecular Simulation: The Importance of Replicas. *J. Chem. Theory Comput.* **2018**, *14* (12), 6127–6138. https://doi.org/10.1021/acs.jctc.8b00391.
(24) Fritch, B.; Kosolapov, A.; Hudson, P.; Nissley, D. A.; Woodcock, H. L.; Deutsch, C.; O'Brien, E. P. Origins of the Mechanochemical Coupling of Peptide Bond Formation to Protein Synthesis. *J. Am. Chem. Soc.* **2018**, *140* (15), 5077–5087. https://doi.org/10.1021/jacs.7b11044.
(25) Berman, H. M.; Westbrook, J.; Feng, Z.; Gilliland, G.; Bhat, T. N.; Weissig, H.; Shindyalov, I. N.; Bourne, P. E. The Protein Data Bank. *Nucleic Acids Res.* **2000**, *28* (1), 235–242.
(26) Van Der Kamp, M. W.; Mulholland, A. J. Combined Quantum Mechanics/Molecular Mechanics (QM/MM) Methods in Computational Enzymology. *Biochemistry* **2013**, *52* (16), 2708–2728. https://doi.org/10.1021/bi400215w.
(27) Senftle, T. P.; Hong, S.; Islam, M. M.; Kylasa, S. B.; Zheng, Y.; Shin, Y. K.; Junkermeier, C.; Engel-Herbert, R.; Janik, M. J.; Aktulga, H. M.; Verstraelen, T.; Grama, A.; Van Duin, A. C. T. The ReaxFF Reactive Force-Field: Development, Applications and Future Directions. *npj Comput. Mater.* **2016**, *2* (**November 2015**). https://doi.org/10.1038/npjcompumats.2015.11.
(28) Janeway, C. A.; Travers, P.; Walport, M.; Shlomchick, M. J. Immunobiology: The Immune System in Health and Disease. *J. Allergy Clin. Immunol.* **2001**. https://doi.org/10.1016/s0091-6749(95)70025-0.
(29) Jung, J.; Nishima, W.; Daniels, M.; Bascom, G.; Kobayashi, C.; Adedoyin, A.; Wall, M.; Lappala, A.; Phillips, D.; Fischer, W.; Tung, C. S.; Schlick, T.; Sugita, Y.; Sanbonmatsu, K. Y. Scaling Molecular Dynamics beyond 100,000 Processor Cores for Large-Scale Biophysical Simulations. *J. Comput. Chem.* **2019**, *40* (21), 1919–1930. https://doi.org/10.1002/jcc.25840.
(30) Dunbar, J.; Krawczyk, K.; Leem, J.; Baker, T.; Fuchs, A.; Georges, G.; Shi, J.; Deane, C. M. SAbDab: The Structural Antibody Database. *Nucleic Acids Res.* **2014**, *42* (D1), 1140–1146. https://doi.org/10.1093/nar/gkt1043.
(31) Schneider, C.; Raybould, M. I. J.; Deane, C. M. SAbDab in the Age of Biotherapeutics: Updates Including SAbDab-Nano, the Nanobody Structure Tracker. *Nucleic Acids Res.* **2021**, *50* (D1), D1368–D1372. https://doi.org/10.1093/nar/gkab1050.
(32) Knapp, B.; Dunbar, J.; Alcala, M.; Deane, C. M. Variable Regions of Antibodies and T-Cell Receptors May Not Be Sufficient in Molecular Simulations Investigating Binding. *J. Chem. Theory Comput.* **2017**, *13* (7), 3097–3105. https://doi.org/10.1021/acs.jctc.7b00080.

(33) Saporiti, S.; Parravicini, C.; Pergola, C.; Guerrini, U.; Rossi, M.; Centola, F.; Eberini, I. IgG1 Conformational Behavior: Elucidation of the N-Glycosylation Role via Molecular Dynamics. *Biophys. J.* **2021**, *120* (23), 5355–5370. https://doi.org/10.1016/j.bpj.2021.10.026.
(34) Tomar, D. S.; Licari, G.; Bauer, J.; Singh, S. K.; Li, L.; Kumar, S. Stress-Dependent Flexibility of a Full-Length Human Monoclonal Antibody: Insights from Molecular Dynamics to Support Biopharmaceutical Development. *J. Pharm. Sci.* **2022**, *111* (3), 628–637. https://doi.org/10.1016/j.xphs.2021.10.039.
(35) Blech, M.; Hörer, S.; Kuhn, A. B.; Kube, S.; Göddeke, H.; Kiefer, H.; Zang, Y.; Alber, Y.; Kast, S. M.; Westermann, M.; Tully, M. D.; Schäfer, L. V.; Garidel, P. Structure of a Therapeutic Full-Length Anti-NPRA IgG4 Antibody: Dissecting Conformational Diversity. *Biophys. J.* **2019**, *116* (9), 1637–1649. https://doi.org/10.1016/j.bpj.2019.03.036.
(36) Wang, X.; Kumar, S.; Buck, P. M.; Singh, S. K. Impact of Deglycosylation and Thermal Stress on Conformational Stability of a Full Length Murine IgG2a Monoclonal Antibody: Observations from Molecular Dynamics Simulations. *Proteins Struct. Funct. Bioinforma.* **2013**, *81* (3), 443–460. https://doi.org/10.1002/prot.24202.
(37) Kortkhonjia, E.; Brandman, R.; Zhou, J. Z.; Voelz, V. A.; Chorny, I.; Kabakoff, B.; Patapoff, T. W.; Dill, K. A.; Swartz, T. E. Probing Antibody Internal Dynamics with Fluorescence Anisotropy and Molecular Dynamics Simulations. *MAbs* **2013**, *5* (2), 306–322. https://doi.org/10.4161/mabs.23651.
(38) Al Qaraghuli, M. M.; Kubiak-Ossowska, K.; Mulheran, P. A. Thinking Outside the Laboratory: Analyses of Antibody Structure and Dynamics within Different Solvent Environments in Molecular Dynamics (MD) Simulations. *Antibodies* **2018**, *7* (3). https://doi.org/10.3390/antib7030021.
(39) Braun, E.; Gilmer, J.; Mayes, Heather, B.; Mobley, D. L.; Monroe, J. I.; Prasad, S.; Zuckerman, D. M. Best Practices for Foundations in Molecular Simulations. *Living J. Comput. Mol. Sci.* **2019**, *1* (1). https://doi.org/10.33011/livecoms.1.1.5957.
(40) Case, D. A.; Cheatham, T. E.; Darden, T.; Gohlke, H.; Luo, R.; Merz, K. M.; Onufriev, A.; Simmerling, C.; Wang, B.; Woods, R. J. The Amber Biomolecular Simulation Programs. *J. Comput. Chem.* **2005**, *26* (16), 1668–1688. https://doi.org/10.1002/jcc.20290.
(41) Abraham, J. M.; Murtola, T.; Schulz, R.; Pall, S.; Smith, J. C.; Hess, B.; Lindahl, E. GROMACS: High Performance Molecular Simulations through Multi-Level Parallelism from Laptops to Supercomputers. *SoftwareX* **2015**, *2*, 19–25. https://doi.org/10.1016/j.softx.2015.06.001.
(42) Brooks, B. R.; III, C. L. B.; A. D. Mackerell, J.; Nilsson, L.; Petrella, R. J.; Roux, B.; Won, Y.; Archontis, G.; Bartels, C.; Boresch, S.; Caflisch, A.; Caves, L.; Cui, Q.; Dinner, A. R.; Feig, M.; Fischer, S.; Gao, J.; Hodoscek, M. W. I.; Karplus, M. CHARMM: The Biomolecular Simulation Program. *J. Comput. Chem.* **2009**, *30* (10), 1545–1614. https://doi.org/10.1002/jcc.
(43) Eastman, P.; Swails, J.; Chodera, J. D.; McGibbon, R. T.; Zhao, Y.; Beauchamp, K. A.; Wang, L. P.; Simmonett, A. C.; Harrigan, M. P.; Stern, C. D.; Wiewiora, R. P.; Brooks, B. R.; Pande, V. S. OpenMM 7: Rapid Development of High Performance Algorithms for Molecular Dynamics. *PLoS Comput. Biol.* **2017**, *13* (7), 1–17. https://doi.org/10.1371/journal.pcbi.1005659.
(44) Phillips, J. C.; Hardy, D. J.; Maia, J. D. C.; Stone, J. E.; Ribeiro, J. V.; Bernardi, R. C.; Buch, R.; Fiorin, G.; Hénin, J.; Jiang, W.; McGreevy, R.; Melo, M. C. R.; Radak, B. K.; Skeel, R. D.; Singharoy, A.; Wang, Y.; Roux, B.; Aksimentiev, A.; Luthey-Schulten, Z.; Kalé, L. V.; Schulten, K.; Chipot, C.; Tajkhorshid, E. Scalable Molecular Dynamics on CPU and GPU Architectures with NAMD. *J. Chem. Phys.* **2020**, *153* (4), 1–33. https://doi.org/10.1063/5.0014475.
(45) Maier, J. A.; Martinez, C.; Kasavajhala, K.; Wickstrom, L.; Hauser, K. E.; Simmerling, C. Ff14 SB: Improving the Accuracy of Protein Side Chain and Backbone Parameters from Ff99SB. *J. Chem. Theory Comput.* **2015**, *11* (8), 3696–3713. https://doi.org/10.1021/acs.jctc.5b00255.

(46) Jorgensen, W. L.; Chandrasekhar, J.; Madura, J. D.; Impey, R. W.; Klein, M. L. Comparison of Simple Potential Functions for Simulating Liquid Water. *J. Chem. Phys.* **1983**, *79*, 926–935. https://doi.org/10.1063/1.445869.

(47) Kmiecik, S.; Gront, D.; Kolinski, M.; Wieteska, L.; Dawid, A. E.; Kolinski, A. Coarse-Grained Protein Models and Their Applications. *Chem. Rev.* **2016**, *116* (14), 7898–7936. https://doi.org/10.1021/acs.chemrev.6b00163.

(48) Noid, W. G. Perspective: Coarse-Grained Models for Biomolecular Systems. *J. Chem. Phys.* **2013**, *139* (9). https://doi.org/10.1063/1.4818908.

(49) https://openmm.org/benchmarks.

(50) Shaw, D. E.; Adams, P. J.; Azaria, A.; Bank, J. A.; Batson, B.; Bell, A.; Bergdorf, M.; Bhatt, J.; Adam Butts, J.; Correi, T.; Dirks, R. M.; Dror, R. O.; Eastwoo, M. P.; Edwards, B.; Even, A.; Feldmann, P.; Fenn, M.; Fenton, C. H.; Forte, A.; Gagliardo, J.; Gill, G.; Gorlatova, M.; Greskamp, B.; Grossman, J. P.; Gullingsrud, J.; Harper, A.; Hasenplaugh, W.; Heily, M.; Heshmat, B. C.; Hunt, J.; Ierardi, D. J.; Iserovich, L.; Jackson, B. L.; Johnson, N. P.; Kirk, M. M.; Klepeis, J. L.; Kuskin, J. S.; Mackenzie, K. M.; Mader, R. J.; McGowen, R.; McLaughlin, A.; Moraes, M. A.; Nasr, M. H.; Nociolo, L. J.; O'Donnell, L.; Parker, A.; Peticolas, J. L.; Pocina, G.; Predescu, C.; Quan, T.; Salmon, J. K.; Schwink, C.; Shim, K. S.; Siddique, N.; Spengler, J.; Szalay, T.; Tabladillo, R.; Tartler, R.; Taube, A. G.; Theobald, M.; Towles, B.; Vick, W.; Wang, S. C.; Wazlowski, M.; Weingarten, M. J.; Williams, J. M.; Yuh, K. A. Anton 3: Twenty Microseconds of Molecular Dynamics Simulation Before Lunch. *Int. Conf. High Perform. Comput. Networking, Storage Anal. SC* **2021**. https://doi.org/10.1145/3458817.3487397.

(51) Fritz, D.; Herbers, C. R.; Kremer, K.; Van Der Vegt, N. F. A. Hierarchical Modeling of Polymer Permeation. *Soft Matter* **2009**, *5* (22), 4556–4563. https://doi.org/10.1039/b911713j.

(52) Leininger, S. E.; Trovato, F.; Nissley, D. A.; O'Brien, E. P. Domain Topology, Stability, and Translation Speed Determine Mechanical Force Generation on the Ribosome. *Proc. Natl. Acad. Sci. U.S.A.* **2019**, *116* (12), 5523–5532. https://doi.org/10.1073/pnas.1813003116.

(53) Foley, T. T.; Kidder, K. M.; Scott Shell, M.; Noid, W. G. Exploring the Landscape of Model Representations. *Proc. Natl. Acad. Sci. U. S. A.* **2020**, *117* (39), 24061–24068. https://doi.org/10.1073/pnas.2000098117.

(54) O'Brien, E. P.; Ziv, G.; Haran, G.; Brooks, B. R.; Thirumalai, D. Effects of Denaturants and Osmolytes on Proteins Are Accurately Predicted by the Molecular Transfer Model. *Proc. Natl. Acad. Sci. U. S. A* **2008**, *105* (36), 13403–13408. https://doi.org/10.1073/pnas.0802113105.

(55) Shahfar, H.; Forder, J. K.; Roberts, C. J. Toward a Suite of Coarse-Grained Models for Molecular Simulation of Monoclonal Antibodies and Therapeutic Proteins. *J. Phys. Chem. B* **2021**, *125* (14), 3574–3588. https://doi.org/10.1021/acs.jpcb.1c01903.

(56) Nissley, D. A.; O'Brien, E. P. Structural Origins of FRET-Observed Nascent Chain Compaction on the Ribosome. *J. Phys. Chem. B* **2018**, *122* (43), 9927–9937. https://doi.org/10.1021/acs.jpcb.8b07726.

(57) Jiang, Y.; Neti, S. S.; Sitarik, I.; Pradhan, P.; To, P.; Xia, Y.; Fried, S. D.; Booker, S. J.; Brien, E. P. O. How Synonymous Mutations Alter Enzyme Structure and Function over Long Timescales. *Nat. Chem.* **2022**. https://doi.org/10.1038/s41557-022-01091-z.

(58) Sánchez-Félix, M.; Burke, M.; Chen, H. H.; Patterson, C.; Mittal, S. Predicting Bioavailability of Monoclonal Antibodies after Subcutaneous Administration: Open Innovation Challenge. *Adv. Drug Deliv. Rev.* **2020**, *167*, 66–77. https://doi.org/10.1016/j.addr.2020.05.009.

(59) Chowdhury, A.; Bollinger, J. A.; Dear, B. J.; Cheung, J. K.; Johnston, K. P.; Truskett, T. M. Coarse-Grained Molecular Dynamics Simulations for Understanding the Impact of Short-Range Anisotropic Attractions on Structure and Viscosity of Concentrated Monoclonal Antibody Solutions. *Mol. Pharm.* **2020**, *17* (5), 1748–1756. https://doi.org/10.1021/acs.molpharmaceut.9b00960.

(60) Lilyestrom, W. G.; Yadav, S.; Shire, S. J.; Scherer, T. M. Monoclonal Antibody Self-Association, Cluster Formation, and Rheology at High Concentrations. *J. Phys. Chem. B* **2013**, *117* (21), 6373–6384. https://doi.org/10.1021/jp4008152.

(61) Liu, J.; Nguyen, M. D. H.; Andya, J. D.; Shire, S. J. Reversible Self-Association Increases the Viscosity of a Concentrated Monoclonal Antibody in Aqueous Solution. *J. Pharm. Sci.* **2005**, *94* (9), 1928–1940. https://doi.org/10.1002/jps.20347.

(62) Dear, B. J.; Bollinger, J. A.; Chowdhury, A.; Hung, J. J.; Wilks, L. R.; Karouta, C. A.; Ramachandran, K.; Shay, T. Y.; Nieto, M. P.; Sharma, A.; Cheung, J. K.; Nykypanchuk, D.; Godfrin, P. D.; Johnston, K. P.; Truskett, T. M. X-Ray Scattering and Coarse-Grained Simulations for Clustering and Interactions of Monoclonal Antibodies at High Concentrations. *J. Phys. Chem. B* **2019**. https://doi.org/10.1021/acs.jpcb.9b04478.

(63) Ferreira, G. M.; Calero-Rubio, C.; Sathish, H. A.; Remmele, R. L.; Roberts, C. J. Electrostatically Mediated Protein-Protein Interactions for Monoclonal Antibodies: A Combined Experimental and Coarse-Grained Molecular Modeling Approach. *J. Pharm. Sci.* **2019**, *108* (1), 120–132. https://doi.org/10.1016/j.xphs.2018.11.004.

(64) Raybould, M. I. J.; Marks, C.; Krawczyk, K.; Taddese, B.; Nowak, J.; Lewis, A. P.; Bujotzek, A.; Shi, J.; Deane, C. M. Five Computational Developability Guidelines for Therapeutic Antibody Profiling. *Proc. Natl. Acad. Sci. U. S. A.* **2019**, *116* (10), 4025–4030. https://doi.org/10.1073/pnas.1810576116.

(65) Mahapatra, S.; Polimeni, M.; Gentiluomo, L.; Roessner, D.; Frieß, W.; Peters, G. H. J.; Streicher, W. W.; Lund, M.; Harris, P. Self-Interactions of Two Monoclonal Antibodies: Small-Angle X-Ray Scattering, Light Scattering, and Coarse-Grained Modeling. *Mol. Pharm.* **2022**, *19* (2), 508–519. https://doi.org/10.1021/acs.molpharmaceut.1c00627.

(66) Singh, J. K.; Kofke, D. A. Mayer Sampling: Calculation of Cluster Integrals Using Free-Energy Perturbation Methods. *Phys. Rev. Lett.* **2004**, *92* (22), 22–25. https://doi.org/10.1103/PhysRevLett.92.220601.

(67) Wang, G.; Varga, Z.; Hofmann, J.; Zarraga, I. E.; Swan, J. W. Structure and Relaxation in Solutions of Monoclonal Antibodies. *J. Phys. Chem. B* **2018**, *122* (11), 2867–2880. https://doi.org/10.1021/acs.jpcb.7b11053.

(68) Chaudhri, A.; Zarraga, I. E.; Kamerzell, T. J.; Brandt, J. P.; Patapoff, T. W.; Shire, S. J.; Voth, G. A. Coarse-Grained Modeling of the Self-Association of Therapeutic Monoclonal Antibodies. *J. Phys. Chem. B* **2012**, *116* (28), 8045–8057. https://doi.org/10.1021/jp301140u.

(69) Yearley, E. J.; Godfrin, P. D.; Perevozchikova, T.; Zhang, H.; Falus, P.; Porcar, L.; Nagao, M.; Curtis, J. E.; Gawande, P.; Taing, R.; Zarraga, I. E.; Wagner, N. J.; Liu, Y. Observation of Small Cluster Formation in Concentrated Monoclonal Antibody Solutions and Its Implications to Solution Viscosity. *Biophys. J.* **2014**, *106* (8), 1763–1770. https://doi.org/10.1016/j.bpj.2014.02.036.

(70) Lai, P. K.; Swan, J. W.; Trout, B. L. Calculation of Therapeutic Antibody Viscosity with Coarse-Grained Models, Hydrodynamic Calculations and Machine Learning-Based Parameters. *MAbs* **2021**, *13* (1). https://doi.org/10.1080/19420862.2021.1907882.

(71) Lai, P. K.; Fernando, A.; Cloutier, T. K.; Gokarn, Y.; Zhang, J.; Schwenger, W.; Chari, R.; Calero-Rubio, C.; Trout, B. L. Machine Learning Applied to Determine the Molecular Descriptors Responsible for the Viscosity Behavior of Concentrated Therapeutic Antibodies. *Mol. Pharm.* **2021**, *18* (3), 1167–1175. https://doi.org/10.1021/acs.molpharmaceut.0c01073.

(72) Kikhney, A. G.; Svergun, D. I. A Practical Guide to Small Angle X-Ray Scattering (SAXS) of Flexible and Intrinsically Disordered Proteins. *FEBS Lett.* **2015**, *589* (19), 2570–2577. https://doi.org/10.1016/j.febslet.2015.08.027.

(73) Schmit, J. D.; He, F.; Mishra, S.; Ketchem, R. R.; Woods, C. E.; Kerwin, B. A. Entanglement Model of Antibody Viscosity. *J. Phys. Chem. B* **2014**, *118* (19), 5044–5049. https://doi.org/10.1021/jp500434b.

(74) Kastelic, M.; Dill, K. A.; Kalyuzhnyi, Y. V.; Vlachy, V. Controlling the Viscosities of Antibody Solutions through Control of Their Binding Sites. *J. Mol. Liq.* **2018**, *270*, 234–242. https://doi.org/10.1016/j.molliq.2017.11.106.
(75) Bekker, G. J.; Fukuda, I.; Higo, J.; Kamiya, N. Mutual Population-Shift Driven Antibody-Peptide Binding Elucidated by Molecular Dynamics Simulations. *Sci. Rep.* **2020**, *10* (1), 1–9. https://doi.org/10.1038/s41598-020-58320-z.
(76) Baumketner, A. Amyloid Beta-Protein Monomer Structure: A Computational and Experimental Study. *Protein Sci.* **2006**, *15* (3), 420–428. https://doi.org/10.1110/ps.051762406.
(77) Holland, C. J.; Crean, R. M.; Pentier, J. M.; de Wet, B.; Lloyd, A.; Srikannathasan, V.; Lissin, N.; Lloyd, K. A.; Blicher, T. H.; Conroy, P. J.; Hock, M.; Pengelly, R. J.; Spinner, T. E.; Cameron, B.; Potter, E. A.; Jeyanthan, A.; Molloy, P. E.; Sami, M.; Aleksic, M.; Liddy, N.; Robinson, R. A.; Harper, S.; Lepore, M.; Pudney, C. R.; van der Kamp, M. W.; Rizkallah, P. J.; Jakobsen, B. K.; Vuidepot, A.; Cole, D. K. Specificity of Bispecific T Cell Receptors and Antibodies Targeting Peptide-HLA. *J. Clin. Invest.* **2020**, *130* (5), 2673–2688. https://doi.org/10.1172/JCI130562.
(78) Miller, B. R.; Mcgee, T. D.; Swails, J. M.; Homeyer, N.; Gohlke, H.; Roitberg, A. E. MMPBSA.Py : An Efficient Program for End-State Free Energy Calculations. *J. Chem. Theory Comput.* **2012**, *8*, 3314–3321.
(79) Raybould, M. I. J.; Nissley, D. A.; Kumar, S.; Deane, C. M. Computationally Profiling Peptide: MHC Recognition by T-Cell Receptors and T-Cell Receptor-Mimetic Antibodies. *Front. Immunol.* **2023**, 1–12. https://doi.org/10.3389/fimmu.2022.1080596.
(80) Wang, W. Protein Aggregation and Its Inhibition in Biopharmaceutics. *Int. J. Pharm.* **2005**, *289* (1–2), 1–30. https://doi.org/10.1016/j.ijpharm.2004.11.014.
(81) Wang, X.; Das, T. K.; Singh, S. K.; Kumar, S. Potential Aggregation Prone Regions in Biotherapeutics: A Survey of Commercial Monoclonal Antibodies. *MAbs* **2009**, *1* (3), 254–267. https://doi.org/10.4161/mabs.1.3.8035.
(82) Berner, C.; Menzen, T.; Winter, G.; Svilenov, H. L. Combining Unfolding Reversibility Studies and Molecular Dynamics Simulations to Select Aggregation-Resistant Antibodies. *Mol. Pharm.* **2021**, *18* (6), 2242–2253. https://doi.org/10.1021/acs.molpharmaceut.1c00017.
(83) Prabakaran, R.; Rawat, P.; Yasuo, N.; Sekijima, M.; Kumar, S.; Gromiha, M. M. Effect of Charged Mutation on Aggregation of a Pentapeptide: Insights from Molecular Dynamics Simulations. *Proteins Struct. Funct. Bioinforma.* **2022**, *90* (2), 405–417. https://doi.org/10.1002/prot.26230.
(84) Sormanni, P.; Aprile, F. A.; Vendruscolo, M. The CamSol Method of Rational Design of Protein Mutants with Enhanced Solubility. *J. Mol. Biol.* **2015**, *427* (2), 478–490. https://doi.org/10.1016/j.jmb.2014.09.026.
(85) Zhang, C.; Codina, N.; Tang, J.; Yu, H.; Chakroun, N.; Kozielski, F.; Dalby, P. A. Comparison of the PH- and Thermally-Induced Fluctuations of a Therapeutic Antibody Fab Fragment by Molecular Dynamics Simulation. *Comput. Struct. Biotechnol. J.* **2021**, *19*, 2726–2741. https://doi.org/10.1016/j.csbj.2021.05.005.
(86) Licari, G.; Martin, K. P.; Crames, M.; Mozdzierz, J.; Marlow, M. S.; Karow-Zwick, A. R.; Kumar, S.; Bauer, J. Embedding Dynamics in Intrinsic Physicochemical Profiles of Market-Stage Antibody-Based Biotherapeutics. *Mol. Pharm.* **2022**. https://doi.org/10.1021/acs.molpharmaceut.2c00838.
(87) Wang, J.; Wolf, R. M.; Caldwell, J. W.; Kollman, P. A.; Case, D. A. Development and Testing of a General Amber Force Field. *J. Comput. Chem.* **2004**, *25* (9), 1157–1174. https://doi.org/10.1002/jcc.20035.

Considerations of Developability During the Early Stages of Antibody Drug Discovery and Design

9

Maximiliano Vásquez, Bianka Prinz, Eric Krauland, and Tushar Jain

9.1 INTRODUCTION

Antibodies have become the main class of biotherapeutics, with over a 100 molecules approved for use in the United States and other major markets. Successful antibody therapies require potent and specific engagement of a disease-relevant target. Turning an antibody into a drug, however, demands a series of additional criteria to be met, collectively known as "developability." This has been increasingly appreciated over the last decade, as numerous studies have been published, illustrating various aspects of the developability problem.

Early examples centered around specific case studies, such as the effort to generate potent, affinity-enhanced, versions of the anti-Respiratory Syncytial Virus (RSV) antibody palivizumab.[1,2] A first attempt[1] generated an antibody with a 1500-fold improvement in affinity and a 44-fold enhancement in neutralization of RSV compared to the

DOI: 10.1201/9781003300311-9

parental antibody palivizumab. Subsequent work, however, showed that in vivo potency enhancement was only about 2-fold,[2] which was attributed to unexpectedly poor pharmacokinetics in the form of very fast clearance, and traced to increased nonspecific binding across multiple tissues. A re-examination of the affinity maturation work led to a modified candidate, motavizumab, which had much reduced polyspecificity and showed high potency in vitro and in vivo.[2] Motavizumab entered clinical development progressing to phase III and Food and Drug Administration (FDA) submission after having been administered to thousands of patients[3]; however, it was terminated in late 2010 upon request by the FDA for additional clinical data.

Other work concerned the description of novel assays aimed at evaluating characteristics of antibody therapeutics candidates: cross-interaction chromatography (CIC)[4] and affinity-capture self-interactionnano-spectroscopy (AC-SINS)[5,6] are two such examples. More recently, studies looking at substantial numbers of antibody samples using one or more assays have been published; among them are reports on chemical degradation,[7–9] aggregation propensity,[10] viscosity and pharmacokinetics,[11] multiple biophysical properties,[12] experimental correlates of viscosity and opalescence,[13] and pharmacokinetics.[14,15]

It is also worth mentioning papers describing computational metrics that correlate with general developability qualities; representatives include the therapeutic antibody profiler,[16] the combination of in silico and experimental properties predictive of favorable pharmacokinetics,[17] sequence-based characterization of approved antibody therapeutics as metric to assess candidate antibody sequences,[18] and charge calculations as predictors of viscosity, isoelectric point, pharmacokinetics, and general antibody profiling.[19]

Assessing developability earlier in the discovery process is ideal because there is more opportunity to avoid choosing a poor molecule when down-selecting to a lead. However, the challenge with early developability assessment is two-fold. First, speed, throughput, and minimal use of samples are practical requirements during this stage, where typically large panels of molecules are involved. Secondly, the context within which the antibodies exist is very different than that during clinical development, and therefore, there is a challenge to design assays and then understand and validate their predictive ability. It is crucial to find a balance between the practical considerations and the ability of an assay, or set of such assays, to correlate with relevant metrics important in development. Metrics of polyspecificity and others relating to pharmacokinetics,[14,15,20,21] or the validation at a larger scale that dynamic light scattering can approximately "predict" viscosity,[13,22] are examples of approaches described that can potentially meet this challenge.

This chapter is organized into several sections as follows. First, we offer a historical perspective of how ideas and processes emerged over the last decade in our organization and our collaborators, with reference to mostly published material. Second, we review a set of 137 antibody samples constructed using sequences from clinical-stage molecules[12] and revisit previously published results. Next, we examine developability data for 349 antibodies isolated from human B cells,[20] followed by an analysis of assay data obtained on over 150 antibody samples constructed using sequences from the literature and corresponding to molecules aimed at targets of biomedical interest. These were compiled from antibody controls used across discovery campaigns against over 60 distinct targets. This is followed by a discussion on how these assays have been used in recent antibody discovery campaigns. We also include a section about the assessment of chemical liabilities via undesirable post-translational modifications. Lastly, we offer conclusions and perspectives for future work.

9.2 HISTORICAL PERSPECTIVE

In early years, we dealt mostly, but not exclusively, with antibodies isolated from our synthetic human-like antibody diversities harbored by our engineered yeast host as full-length IgG molecules. Access to yeast-produced, full-length, IgGs was subsequently extended to situations where the source of diversity was from human B cells, or from tissues isolated from immunized animals. With IgG material in hand, we were interested in finding assays and workflows that could address developability concerns earlier in the antibody discovery process than had been incorporated up to that point.

In 2013, we published our first two articles on this general subject.[23,24] In the first article,[24] a polyspecificity assay was described for application in both screening and selection. Briefly, soluble membrane protein (SMP) and soluble cytosolic protein (SCP) are generated from Chinese Hamster Ovary (CHO) cells and biotinylated; a mix of these two constitutes the polyspecificity reagent (PSR), which is then used as a probe to assess binding by a test antibody. In what would become customary in much of our subsequent work, we included data on 24 control samples generated from variable region sequences from clinical development candidates with international nonproprietary name (INN) designations.[25–28] The results indicated that a small number of the 24 "clinical" antibodies had high readouts in the PSR binding assay. The PSR metric also showed a good correlation with those obtained from the CIC[4] and baculovirus particle (BVP) binding[15] assays previously described. This article also demonstrated the ability of the PSR approach to serve as a tool during the selection process utilizing flow cytometry, where millions of antibody variants can be assessed and sorted out of one pot. Lastly, for a subset of the antibody samples studied, a close relation was observed for measurements using the original soluble membrane preparation (SMP) from CHO cells with preparations of membrane or cytosolic proteins from insect *Spodoptera frugiperda* (Sf9) cells.[24] In the second article,[23] a clone self-interaction assay using bio-layer interferometry (CSI-BLI) was introduced and, again, tested with a few named (and some reported as mAb1, mAb2, etc.) antibody samples. The CNTO607 antibody, which had been reported to show poor biophysical properties,[29,30] was used as a positive control, and it showed a high response compared to other controls, such as a sample made with the variable regions of adalimumab.

In later work,[31] we collaborated with the Tessier group, then at Rensselaer Polytechnic, in an adaptation of this group's SINS and AC-SINS assays.[6,32] Here, over 30 named antibodies, including CNTO607, were considered. CNTO607 was confirmed to give high self-interaction in this assay, but, in addition, we observed that three samples made using INN sequences (corresponding to the clinical-stage antibodies ganitumab, tremelimumab, and vesencumab) exhibited evidence for high self-interaction as determined by AC-SINS. We had chosen vesencumab to be in our test set upon inferring that it may correspond to a poorly behaved anti-NRP1 antibody highlighted in the aforementioned BVP assay article[15] after learning that an anti-NRP1 antibody published by a group, also from Genentech, showed unexpected differences in pharmacokinetics in monkey vs. other species.[33]

A proposal to adapt AC-SINS to assess hydrophobicity, by performing the assay under a salt gradient, was reported[34] and evaluated using 32 named antibody samples, including vesencumab. The assay results were shown to correlate with HIC (hydrophobic interaction chromatography) retention times, and like in work with previous assays, a few of the "clinical samples" showed increased hydrophobicity relative to a human polyclonal antibody sample with potential undesirable consequences for purification and aggregation.[35,36]

In collaboration with the Wittrup group at MIT, we explored a potential relationship between a set of high-throughput assays and pharmacokinetics in mice.[21,37] In the first study, 16 samples, including four named (constructed using variable region sequences from INN-assigned, clinical stage, antibodies, and expressed transiently in human embryonic (HEK) cells) were tested for clearance after administration to wild-type mice. In parallel, the corresponding samples were assayed for PSR, AC-SINS, CSI-BLI, CIC, size-exclusion chromatography (SEC), and accelerated stability (AS). Significant correlations to antibody clearance were noted for PSR, CIC, and marginally, AS. In a similar timeframe, a team at Merck put forward a hypothesis tying differential binding to FcRn at neutral pH to pharmacokinetics[38] and was furthered by a team at Roche to help explain the difference in reported clearance in human patients between two antibodies against the same target, ustekinumab and briakinumab, both against the common subunit of IL-12 and IL-23.[39] Ustekinumab, an approved drug, shows normal pharmacokinetics with terminal elimination half-life in the order of 20 days; in contrast, briakinumab, whose development was apparently terminated in 2011 after the sponsor withdrew its regulatory submission, has a reported half-life of under 10 days. In this work,[39] a correlation is observed between FcRn interaction at neutral pH and clearance in humans or animal models. In our work, we aimed to distinguish this mechanism with a simpler explanation involving nonspecific interactions.[37] Simplistically, if an antibody binds to FcRn at neutral pH, it may bind multiple off-target moieties which contribute to the observed poor half-life. First, it was noted that ustekinumab exhibited more favorable scores or results for four metrics associated with polyspecificity and/or self-interaction, namely, AC-SINS, BVP, CIC, and PSR. Second, pharmacokinetics experiments were carried out in FcRn-knockout mice, thus removing FcRn as a factor. As expected, both antibodies presented much faster clearance in comparison to their behavior in wild-type mice (it should be noted that antibodies with human Fc sequences bind mouse FcRn). However, the clearance for the highly polyspecific antibody briakinumab was substantially higher than that for ustekinumab: 207 vs. 125 mL/day/kg, which was statistically significant. Similarly, the terminal half-lives were 9.8 and 13.8 hours, respectively, for briakinumab and ustekinumab, also statistically significant.

At this stage, we had accumulated a series of intriguing results involving a relatively small number of samples where attempts to correlate tended to be limited. Inspired in part by studies involving much larger numbers of antibodies,[11,15] we initiated a project aimed at collecting data on "all" antibodies known to have reached at least phase II of clinical development and with reliably known sequence information (INN designations). This was eventually published in 2017,[12] and some of the results are reviewed and discussed in the next section of this chapter.

9.3 CLINICAL ANTIBODY DATA SET

In this comprehensive analysis,[12] antibody samples were generated using sequences corresponding to 137 molecules with INN designations and at various clinical stages. The antibodies were cloned into a common human IgG1 expression vector, with light chains built using human kappa or lambda constant regions as appropriate and expressed transiently in HEK cells. The analysis of these samples included 12 assays probing the biophysical properties of the resulting molecules. Perhaps, unsurprisingly, there was a trend for more clinically advanced molecules, all the way to approval, to have generally better biophysical profiles than those in earlier stages of clinical development. By looking at the properties of approved antibodies, it was possible to define thresholds for 10 of the 12 assays (see Table 1 of Jain et al.[12]) by which antibodies were flagged for potential aberrant behavior. At the same time, rank-order results between assays were compared to identify clusters of assays measuring redundant biophysical properties. The 12 assays could be grouped into 5 clusters (Figure 9.1), Two clusters, which include six of the assays, could in turn be merged into a larger cluster, group A, by using a different cut point in the cluster analysis. The assays in group A are generally understood as

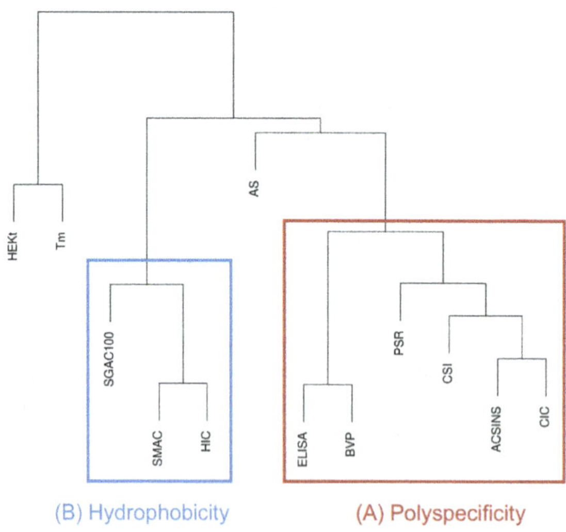

FIGURE 9.1 Clustering of biophysical properties recalculated from published data.[12]

reflecting a measure of polyspecificity while the assays in the other main cluster, termed group B, assess hydrophobicity. By definition, clustering suggests group A (polyspecificity) assays and group B (hydrophobicity) assays are not highly correlated, but beyond that, an interesting observation from this analysis is that very few antibodies in the set of 137 were flagged as having unfavorable measures in both group A assays and group B assays.

As we utilize extensively the PSR binding and HIC assays in our discovery workflow, we focus on them as prototypical group A and group B assays, respectively, to further exemplify this observation.

A bivariate plot of HIC retention time (RT) vs. PSR score with the thresholds defined originally (0.27 for PSR score and 11.7 minutes for HIC RT) reveals a single antibody, with both metrics above the thresholds, among the 137 antibodies used in that study (Figure 9.2). This outlier corresponds to the anti-IGFR1 antibody cixutumumab, which reached phase II of clinical development targeting cancer indications, but further work has been reported as discontinued as of September, 2018. This may well be a target class effect: of nine antibodies to this target with INN designations, seven have been terminated and the surviving ones, including one approved, are aimed at thyroid eye disease. Very similar bivariate plots are obtained when pairing other assays from group A with those from group B. (Data not shown here, but obtainable from the supplemental data in Jain et al.[12]).

The observation that the high PSR, high HIC quadrant is virtually unpopulated by our surrogate clinical antibody set may mean this would be a good metric by which to screen early candidates, or simply that antibodies in general just do not tend to possess these characteristics. Analysis of additional data sets below will distinguish these two possibilities.

It is worth mentioning in this section a handful of published studies that used the same or similar set of samples for analysis of other properties. These include chemical degradation,[8,9] extensional flow,[40] heparin and FcRn chromatography,[41] and interaction with heme.[42]

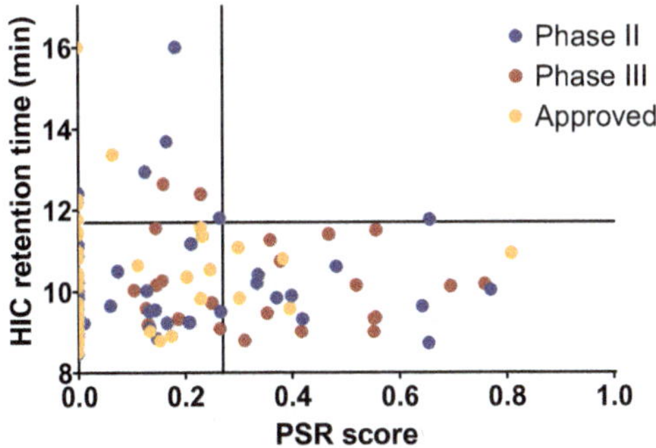

FIGURE 9.2 Bivariate plot of HIC retention time vs. PSR score for 137 samples made with sequences from clinical-stage antibodies, thresholds at 0.27 for PSR and 11.7 minutes for HIC shown as vertical and horizontal lines, respectively. Data from Jain et al.[12] Here and elsewhere, HIC retention time values were set at 16 minutes for cases when the sample did not elute from the column.

TABLE 9.1 Spearman (rank-order) correlations to FcRn and Heparin retention measures[41] with prior work[12]

METRIC	*FCRN RELATIVE RETENTION*	*HEPARIN RELATIVE RETENTION*
CIC	0.86	0.35
AC-SINS	0.83	0.50
CSI_BLI	0.67	0.45
PSR score	0.63	0.44
SGAC-SINS	0.55	−0.07
BVP Score	0.43	0.37
ELISA	0.41	0.40
Heparin relative retention	0.48	1.00
HIC retention time	0.30	−0.34
SMAC	0.30	−0.30
Accelerated stability	0.17	0.12
Estimated Fab pI	0.39	0.78

Since the work of Kraft et al.[41] included data for 130 samples common to the 137 in Jain et al., it is instructive to see how the results of their assays correlate with the previously reported metrics. Rank-order correlation coefficients of their FcRn and heparin relative retention metrics with ten of the assays from Jain et al.[12] are shown in Table 9.1. Notably, very high correlations are observed between FcRn relative retention measures and some assays in group A (polyspecificity category), most prominently CIC retention times, AC-SINS, and the PSR scores. We added to this analysis a simple sequence-based estimation of the isoelectric point for the Fab portion (Fab pI) and showed a high correlation (0.78) with heparin relative retention.

9.4 HUMAN B-CELL-DERIVED ANTIBODIES

In another Adimab study, Shehata et al.[20] isolated 400 antibodies from several B-cell compartments. As described in detail in the original publication, antibody heavy- and light-chain variable regions were PCR-amplified from single B cells, and then cloned and expressed as IgG1 antibodies in an engineered yeast strain. It is important to note that for identical sequences expressed in this system and transiently in CHO cells, we observe a high correlation between PSR score and HIC retention time measurements. For example, in a recent comparative head-to-head set comprising 51 cases, the PSR scores show linear correlation, Pearson R^2, of 0.96, and rank-order correlation, Spearman ρ, of 0.97. For HIC retention time, the correlation coefficients are 0.97 and 0.98, respectively, for R^2 and ρ (our unpublished results). This establishes that for these two metrics, it is possible to compare data obtained with IgG samples made in our engineered yeast with those produced in HEK or CHO cells.

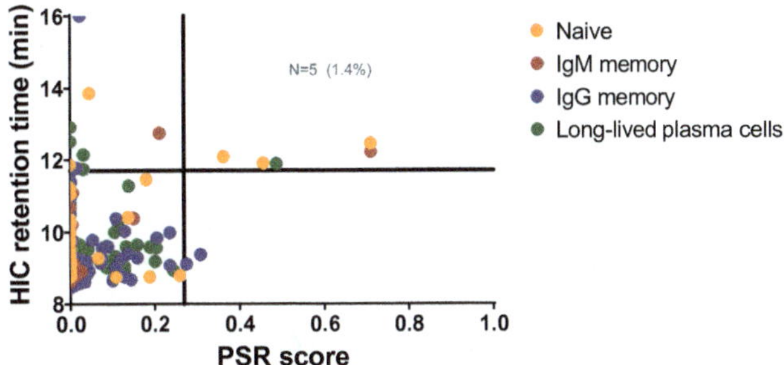

FIGURE 9.3 Bivariate plot of HIC retention time vs. PSR score for 349 samples from human B cells. Thresholds at 0.27 for PSR and 11.7 minutes for HIC are shown as vertical and horizontal lines, respectively. Data from Shehata et al.[20]

Of the 400 antibodies included in the original study, 349 have data reported for both the PSR and HIC assays. The results in Figure 9.3 show that very few antibodies map to the high PSR scores and high HIC RT region defined by the thresholds reported in the clinical antibody study.[12] Specifically, only 5 of the 349 antibodies, or 1.4%, exhibited this behavior, which is virtually absent in the clinical antibody study. In accordance with the general theme of the Shehata et al. investigation, four of these five came from naïve or IgM memory compartments, with only one from plasma cells. As discussed therein, it appears that by these two metrics, natively paired human antibodies have developability characteristics comparable to those of bona fide therapeutic antibodies in clinical development. Of course, this is not a comprehensive developability assessment; for example, in the human-derived antibodies, more mutated sequences (relative to germline) tended to show more favorable polyspecificity scores, but this often came at the expense of reduced thermal stability assessed using differential scanning fluorescence.[43] Nonetheless, antibodies derived directly from human sources via analogous means have been developed for prophylactic and/or therapeutic use for Ebolavirus and SARS-CoV-2 infections.[44,45] More generally, it is notable that of the eight antibodies (corresponding to five distinct products, as three of these are mixtures of two antibodies) receiving emergency use authorization (EUA) in the United States for therapeutic or prophylactic use against SARS2-CoV-2, seven came from human donors[46–51]; the one exception being one component of the two-antibody mix from Regeneron, which came from transgenic mouse immunizations.

9.5 CONTROL ANTIBODIES

More recently, while creating control antibodies for a variety of targets, we have accumulated a subset of 156 cases with measured PSR scores and HIC retention times. To make the samples, genes encoding literature amino acid sequences for the variable regions were sourced and the resulting material was cloned into human IgG1, kappa, or lambda, vectors for expression in our engineered yeast strain, as described in Shehata et al.[20] In addition, these antibodies represent a cross-section of sequences disclosed in the patent literature, most of which, to our knowledge, have not reached clinical development status. They come from a variety of institutions ranging from academia and small biotechnology companies, all the way to large biopharma. As shown in Figure 9.4a, the relative number of antibodies scoring poorly (above the defined thresholds) for both the PSR and HIC metrics is now much higher than in the sets discussed previously: 11 out of 156 or about 7%, compared to about 0.7% and 1.4% for the clinical and human B-cell sets, respectively. Breaking down the results by inferred clinical stage is also informative. This analysis shows that molecules constructed with sequences known or believed to have reached clinical development tend to lie outside the undesirable upper

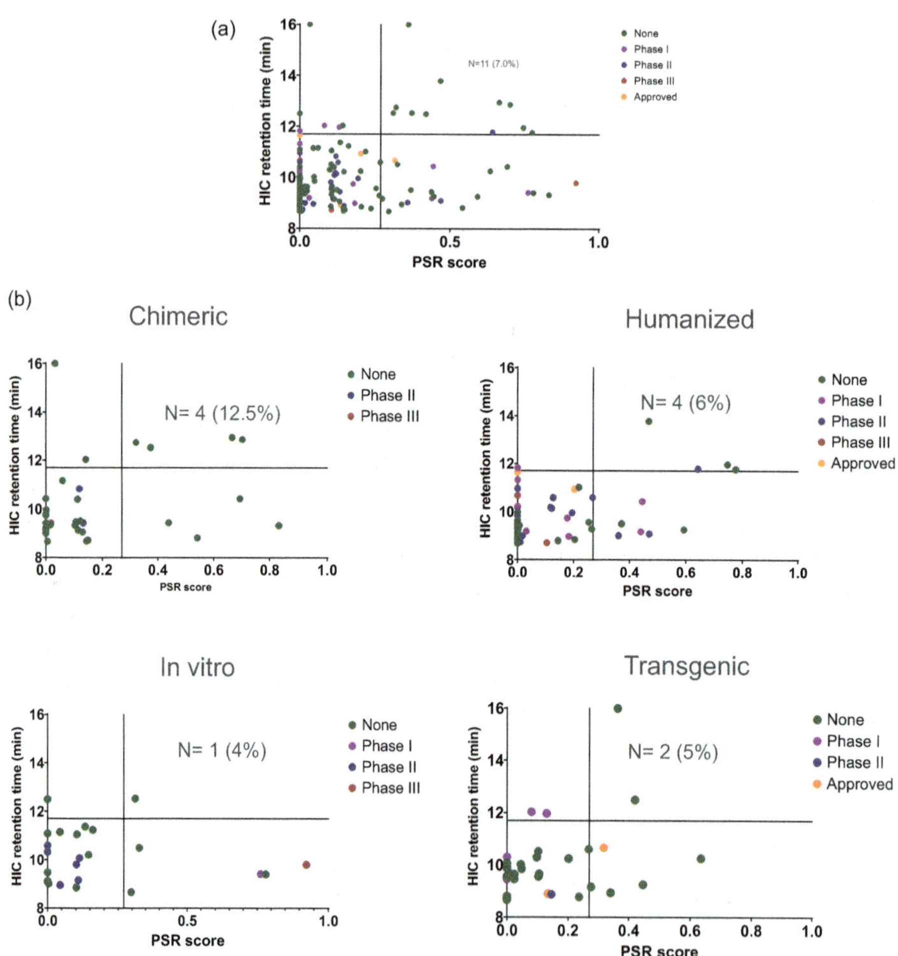

FIGURE 9.4 Bivariate plot of HIC retention time vs. PSR score for 156 samples corresponding to control antibodies, color-coded according to the estimated stage of clinical development (a); same data broken down by antibody type (b). Thresholds at 0.27 for PSR and 11.7 minutes for HIC shown as vertical and horizontal lines, respectively.

right quadrant defined by the PSR and HIC thresholds. The sole exception is a sample made from variable region sequences derived from an antibody believed to be in phase II clinical stage, but not one with an INN designation. It should be noted that the lack of association with an INN creates some uncertainty on the connection between sequence and clinical candidate.

Figure 9.4b breaks out the data by antibody type, defined as chimeric, humanized, in vitro (from libraries), or from humanized transgenic animals. Although numbers are now necessarily smaller, it is notable that antibodies from in vivo sources are not always better, by these metrics, than those emerging from in vitro discovery. Chimeric antibodies have a slightly higher tendency to map to the high PSR, high HIC quadrant than others in this data set: 12.5% vs. 4%–6% for the other antibody types.

One clear result from this set is the observation that the high PSR and high HIC quadrant can be populated by antibodies in general. This suggests that their relative absence in clinical and natively paired human antibodies is the result of direct or indirect filtering away from this dual property.

9.6 ASSAYS RESULTS FOR HUMAN ANTIBODIES FROM DE NOVO DISCOVERY CAMPAIGNS

Before examining the antibody properties, we describe here some basic elements of our discovery process for additional context. The primary aim in early selection rounds is to enrich binders to the target of interest. However, during this process, concomitant enrichment of antibodies with high PSR binding can occur. Therefore, tracking and deselecting such antibodies is essential for identifying target-specific output with low overall polyspecificity. The process is illustrated in Figure 9.5a. As seen in Figure 9.5b, naïve libraries have a small percentage of PSR binders; these can get enriched in the selection process depending on the biophysical properties of the target antigen. After positively selecting for a binding population in the third round of successive enrichment (Figure 9.5a), a subsequent fourth round of selection on the PSR nonbinding population depletes the population of high PSR binders, which can then be observed in round 5 (Figure 9.5b).

Figure 9.5b shows that the percentage of PSR binders in the naïve library, prior to any selection, depends on the germlines present, and it correlates with the output as seen in the right-hand side panels. For example, we see a higher proportion of PSR binders in the unselected VH4 germline family libraries, which would result in more polyreactive (in this work, following historical precedent, we use the terms polyspecific and polyreactive interchangeably, although, a recent review[52] has proposed to ascribe distinct meanings to the terms) antibodies post-selection if no PSR negative round is performed.

To offer a glimpse of sequence correlates of polyspecificity and hydrophobicity, as assessed by the PSR binding and HIC assays, we collected data on several 1,000 samples spread over close to 100 projects and break them down by Variable domain

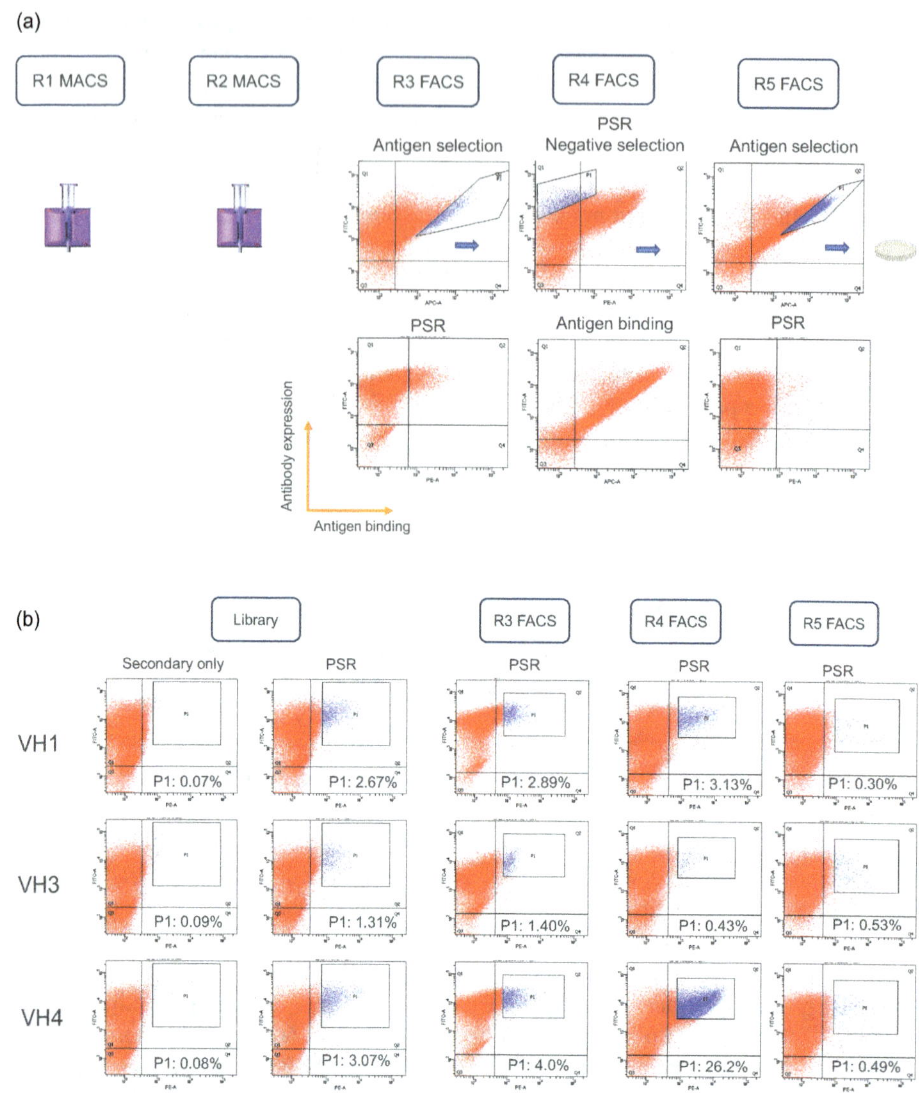

FIGURE 9.5 (a) Representative selection process. Two rounds of magnetic bead enrichment (MACS) with the target antigen, followed by three rounds of selections on FACS. Round 3 (R3): incubation with the target and selection of the binding population. R4: incubation with the polyspecificity reagent (PSR) and negative selection. R5: incubation with the antigen and selection for strong binders for plating and sequencing. (b)Tracking of PSR binding during selection. PSR binding of naïve libraries, and during R3, R4, and R5 of the selection process. The percentage of clones in the P1 gate (PSR binding population) is shown for each plot.

of Heavy chain (VH) or Variable domain of Light chain (VL) germline. They correspond to antibodies from the earliest stages of discovery using input libraries with human-like synthetic diversity. While a detailed description of the design of these

TABLE 9.2 Summary of PSR binding and HIC data for top five occurring VH and VL germlines

GERMLINE	*NUMBER*	*MEAN PSR SCORE*	*PERCENT ABOVE 0.27 THRESHOLD*	*MEAN HIC RETENTION TIME (MIN)*	*PERCENT ABOVE 11.7 THRESHOLD*
VH1–2	572	0.07	5.6%	9.3	0.3%
VH1–69	1,686	0.04	2.7%	11.0	25.8%
VH3–23	1,210	0.04	2.6%	9.4	2.5%
VH3–30	572	0.05	4.0%	9.9	1.9%
VH4–39	1,053	0.07	6.3%	10.1	7.9%
VK1–12	1,211	0.05	3.1%	10.0	8.7%
VK1–33	1,247	0.05	3.4%	9.9	6.4%
VK1–39	1,362	0.07	5.7%	9.9	9.2%
VK3–11	1,764	0.05	4.1%	9.7	5.3%
VK3–20	1,090	0.04	3.0%	9.5	3.2%

libraries is outside the scope of this chapter, it can be noted that the diversity is concentrated almost exclusively on CDRs H3 and L3. Additional details may be found in prior publication.[53]

Table 9.2 summarizes PSR binding and HIC data for the top five VH and VL germlines by occurrence in the set. While the average PSR scores are very close to zero in all cases, a trend for samples with antibodies using the VH4–39 germline to have slightly elevated PSR may be noticed. More than 6% of samples using VH4–39 exhibits PSR scores above the 0.27 threshold, while for the other germlines, this percentage ranges from about 2.5% to 3.9%. In the light-chain analysis, the average PSR scores remain very low, near zero, for all the germlines, but there is a slight trend for VK1–39 to have a higher fraction of samples with PSR readouts above the 0.27 threshold: 5.7% vs. about 3%–4% for the others. We should add here that in many of the selections used to generate the antibodies for this set, negative PSR pressures have been applied, as described above and previously,[24] so that the resulting samples will tend to have lower PSR scores than what would be obtained in a selection process that lacked a PSR depletion step.

Table 9.2 includes a similar analysis and breakdown for the HIC assay. We can see that the VH germline appears to influence the overall hydrophobicity, with VH1–69 showing a higher mean HIC retention time of 11 minutes, while the mean HIC retention time for the other germlines is in the approximate 9–10-minute range. It is also notable that IgGs using the VH1–69 germline have a much larger proportion of cases with HIC retention time above the 11.7-minute threshold, over 25% compared to 0.3 to about 8% for the other germlines. Finally, breaking the results down by light-chain germline does not reveal large differences in averages, but there is a tendency for VK1–39 and VK1–12 to have more antibodies above the 11.7-minute threshold.

In Figure 9.6, we present the PSR-HIC bivariate analysis, with the data parsed by the VH germline gene family. As could be expected from the prior analysis of

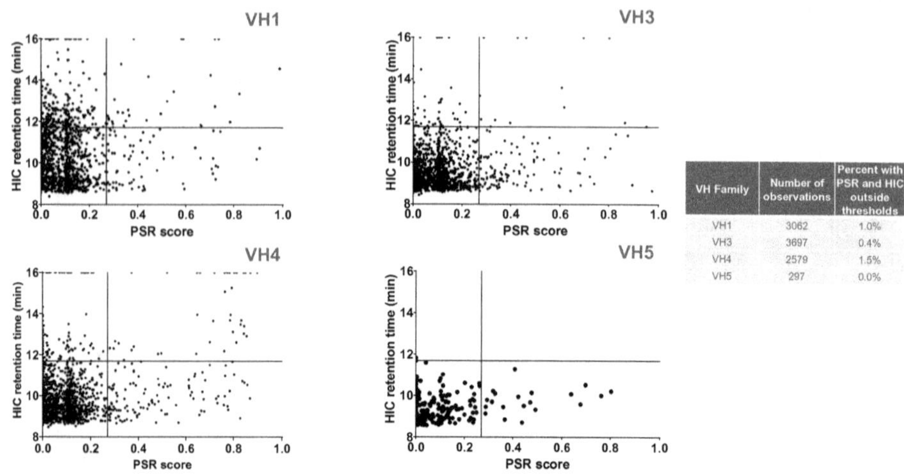

VH Family	Number of observations	Percent with PSR and HIC outside thresholds
VH1	3062	1.0%
VH3	3697	0.4%
VH4	2579	1.5%
VH5	297	0.0%

FIGURE 9.6 Bivariate plot of HIC retention time vs. PSR score for samples from primary discovery; broken down by VH germline family. Thresholds at 0.27 for PSR and 11.7 minutes for HIC are shown as vertical and horizontal lines, respectively.

the individual assays, there is a slight trend for antibodies using genes from the VH1 and VH4 families to have larger percentages of cases with simultaneous high polyspecificity and hydrophobicity metrics, though, of course, these remain low overall (under 2% in all cases).

The observation that certain germlines or even germline families tend to be associated with less favorable properties related to developability raises the question of whether they should be avoided in diversity sources, be they libraries or even choices to construct transgenic mice. We think this would be too radical a solution as, at least for the metrics presented here, the overwhelming majority of molecules using these "less desirable" germlines still can have favorable overall biophysical profiles. Removing or deprioritizing antibodies using, for example, VH4 germlines will result in the loss of unique canonical structures,[54] which will impact overall structural diversity and may compromise the ability to find binders with rare qualities in antibody discovery projects.

To project how these metrics apply to subsequent stages of discovery, we next focus down on a subset of molecules with collected PSR binding and HIC retention time data emerging from a set of 15 discovery efforts that have resulted in lead molecules currently in phase I trials or beyond (Figure 9.7). In contrast with the cumulative campaign data presented earlier, which included only data from the earliest discovery stages, the molecules presented here came from all stages of discovery, including outputs from antibody optimization efforts as required by each project. One can see in Figure 9.7 that the antibodies chosen for clinical development do not populate the undesirable upper right quadrant (high PSR binding and high HIC retention time), and in fact, with very few exceptions, these samples tend to exhibit low PSR binding and low HIC retention times.

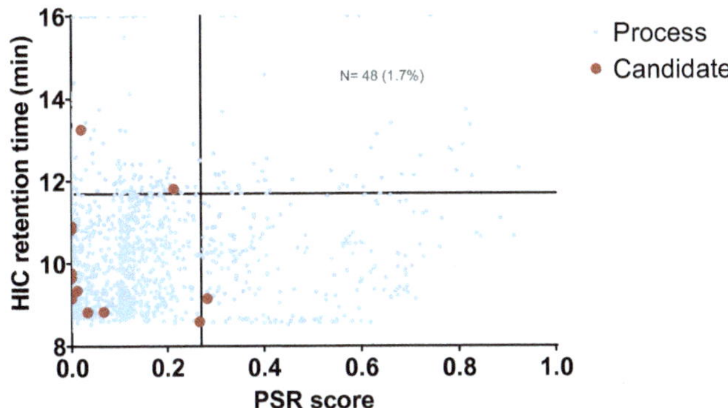

FIGURE 9.7 Bivariate plot of HIC retention time vs. PSR score for samples from 14 antibody discovery campaigns that led to a clinical candidate. The total number of samples assayed was 2,848.

9.7 ASSESSMENT OF CHEMICAL LIABILITIES

An additional important aspect of the developability of antibodies concerns their stability in chemical degradation. Among the most common modes of modification are asparagine deamidation, aspartate isomerization, and oxidation, primarily of methionine residues. As with some of the biophysical properties discussed earlier, there were many early publications of an anecdotal nature, focusing on one or a small handful of antibodies, with the main interest being primarily on method development. One of the first studies that included a relatively large sample of antibodies was published by Sydow et al. in 2014.[7] In this work, the authors examined potential deamidation and isomerization events for 37 mAbs that underwent stress treatment, described as incubation for 2 weeks in a pH 6.0 buffer maintained at 40°C. They observed, for example, that about 67% of sites with the NG motif showed proof of deamidation, while only 36% of DG motifs showed isomerization. Chances of deamidation were even lower for motifs like NS, NN, and NT, as well as isomerization at DS, DT, DD, and DH, motifs. The paper also described a machine learning approach using structural descriptors that could be used to predict deamidation or isomerization from antibody sequences via structural modeling.

A few years later, Lu, Nobrega, and coworkers at Adimab,[9] using a similar set of antibody samples as in the Jain et al. study, examined isomerization and deamidation for 131 mAbs that had been subjected to pH and temperature-related stress. To accelerate potential deamidation, samples were incubated at pH 8.5 for 1 week at 40°C, while for isomerization, incubation occurred at pH 5.5, 40°C for 2 weeks. This study included an examination of 753 asparagine and 1,249 aspartate residues in the variable regions of the 131 antibody samples. A summary of the data parsed out by motif class and by antibody region is presented in Tables 9.3 and 9.4, respectively. These tables also include data (unpublished) from an additional 332 samples that comprise 1,364 asparagine and 3,029 aspartate residues in variable regions.

TABLE 9.3 Summary of asparagine deamidation and aspartate isomerization data by motif class

MOTIF CLASS	*TOTAL*	*% MODIFIED*	*TOTAL, EXPANDED SET*[a]	*% MODIFIED, EXPANDED SET*
NG	30	47%	59	64%
NS	136	5%	376	6%
N [T, D, N, H]	129	10%	268	8%
NX	458	2%	1,414	2%
DG	46	35%	86	41%
DS	88	7%	308	6%
D [T, N, H]	319	2%	1,115	3%
DX	796	0.4%	2,769	0.5%

Source: Data from Lu, Nobrega et al.[9]

[a] Expanded set includes data from an additional 332 antibody samples (unpublished).

TABLE 9.4 Summary of asparagine deamidation and aspartate isomerization data by antibody region

ANTIBODY REGION	*TOTAL ASN*	*% DEAMIDATED*	*TOTAL ASP*	*% ISOMERIZED*
HCDR1	154	3%	92	3%
HCDR2	470	6%	320	4%
HCDR3	66	10%	783	6%
HFR	596	1%	1,008	0.3%
LCDR1	443	16%	155	13%
LCDR2	190	3%	149	5%
LCDR3	129	3%	131	9%
LFR	69	0%	1,640	0%

Source: Data from 463 antibody samples, including those from Lu, Nobrega et al.[9]

Table 9.3 indicates that motifs are modified with percentages similar to what had been reported by Sydow and coworkers[7] on a smaller set, and this is in spite of differences in stress conditions and general methodology. Modifications at "unexpected" motifs, those denoted as NX or DX in Table 9.3, occur at low but observable rates. We break down those by specific amino acid, including looking at the presence of glycine (G) in the position prior to the N or D. It turns out that GN and NA motifs account for most of the cases with detected deamidation at "unexpected" motifs. We observe that 12% and 15%, respectively, of GN and NA sites have evidence of modification. In retrospect, the relative chemical instability of NA should not be too surprising, given this motif having shown potential for deamidation in prior work.[55,56] It happens that in neither of the recent comprehensive studies[7,9] were there examples of deamidation at NA detected, even though they were present in the sequences of some of the antibody samples assessed. All the instances of modification at NA motifs occurred in the new set of samples summarized in Table 9.3. For aspartate isomerization, the motifs in the (initially) "unexpected" group are GD and DE, with 2% and 3% of the cases, respectively, showing detectable modification.

Table 9.4 summarizes the data where modifications are located in the sequence for the consolidated set of 463 antibody samples. Consistent with the literature consensus, we see most of the modifications occurring in CDRs with very few (none in the light chain) observed in framework regions. There are marked differences among the different CDRs, with CDR2 and CDR3 of the heavy chain and CDR1 of the light chain exhibiting the highest relative rates of deamidation. A similar trend is observed for aspartate isomerization, but here, CDR3 of the light chain also contributes to the tally of modifications, with a lesser contribution from CDR2 of the heavy chain.

These experimental studies make it clear that simplistic, purely sequence motif-based approaches to assess deamidation or isomerization potential are insufficient, and more elaborate prediction methods offer improved alternatives.[55,57,58] While avoidance of chemical liability hotspots, reliably predicted or experimentally confirmed, is considered "best practice" in therapeutic antibody development, their relevance to activity and/or safety is hard to assess in advance. An interesting recent example is that of the FDA-approved antibody crizanlizumab, which has been shown to undergo aspartate isomerization at a DG motif within CDR1 of the light chain.[59] This modification had a deleterious impact on the potency of the antibody. However, it was also shown that the change was reversible upon incubation in human serum. The authors of this report[59] conclude that degradation leading to activity loss even under optimized formulation conditions may not always be automatically excluded from development, given that biological activity could be potentially restored under physiological conditions.

In addition to the oxidation of mainly methionine and tryptophane residues, other modifications in antibodies include glycation of lysine residues[60–62] and fragmentation.[63,64] Discussion of other types of post-translational modifications have been reviewed previously.[65,66]

9.8 CONCLUSIONS AND FUTURE PERSPECTIVES

A robust investigation of multiple disparate sets of antibody molecules discussed above supports the idea that a pair of relatively simple and high-throughput assays, PSR binding and HIC, are predictive of general developability behavior. Avoiding antibodies with relatively high readouts in both assays seems a likely prerequisite for eventual successful development in the clinic. There are still several open questions. How unique is the choice of PSR and HIC for this kind of profiling? It is likely that combinations of another assay that assesses "stickiness" or polyreactivity with an assay that measures antibody hydrophobicity will yield similar results. Examples of the first type of assay, in addition to PSR, include CIC,[4] binding to BVP,[15] polyspecificity particle assay,[67] single- and double-stranded DNA, insulin and lipopolysaccharide ELISA,[68] and protein panel profiling.[69] Likewise, assays like Standup Monolayer Adsorption Chromatography (SMAC) or Salt-gradient Affinity-Capture Self-Interaction Nanoparticle Spectroscopy (SGAC-SINS) are potential alternatives in this context to HIC. Parting from the reasonable assumption that avoiding the high PSR, high HIC region, while perhaps necessary, is not sufficient to define an antibody candidate as "developable," at least two more questions may be asked. First, will a more restrictive definition of "acceptable" space in the bivariate plot using more conservative criteria, e.g., asking for both PSR and HIC to be below the thresholds, or even lowering those thresholds further, lead to an enhanced probability of success the development path? Second, are there additional assays that may be routinely deployed in early discovery that could capture important properties or features missing from the simple two-variable analysis? Work in progress in collaboration with other industry investigators aims at finding correlations between high-throughput assays that can be practically used at the earliest stages with more complex behaviors widely recognized to be critical for successful development. An example would be viscosity, where a requirement for low values (an often-used cutoff is 30 cP) exists for high-concentration formulations for subcutaneous administration, and progress has been made on this front, both computationally and experimentally.[13,70–73] Also worth mentioning is the work of Bailly et al., which reports examples of correlations between physicochemical properties and downstream process parameters.[74]

Regarding the correlation of polyspecificity measures with pharmacokinetics,[14,15,17,20,21,75] it is interesting to look at potential outlier cases. The NRP1 antibody, presumably vesencumab,[15,33] is an example where high readouts for the BVP, PSR, and other polyspecificity assays correlate with observed high clearance in humans, but not in cynomolgus monkeys (see Figure 9.5 of the original work[15]). In a separate study, a set of 16 antibodies to infectious agents, seven against bacterial targets and nine against viral targets, with collected terminal half-life in humans[20] is instructive as there is no putative influence from target-mediated drug disposition. Samples were generated from the published variable region amino acid sequences, expressed transiently in HEK, and assayed for several biophysical properties. The sample made with the sequences of the antibody urtoxazumab is shown to have a high PSR score but the clinical molecule seems to have expected pharmacokinetics with a reported half-life of 26 days. When assayed for BVP binding, however, the urtoxazumab sample shows a low score. Conversely, another antibody with a normal terminal half-life (24 days), anti-HIV1 10–1074, gives a low PSR score but a relatively high BVP score (see Figure 9.8). A possible explanation for these seemingly contradictory results is that both assays and others in the polyspecificity class, work by presenting a collection of multiple antigens. The composition and thus the behavior of these collections will depend on the nature of the preparation and on the cell of origin (for example, CHO for PSR, and Sf9 for BVP). It appears that the antigens in CHO-derived PSR that are responsible for the high response for urtoxazumab are likely to be absent in relevant tissue in humans, and those antigens are also absent from the Sf9-based BVP mixture, so urtoxazumab does not appear as a false positive case in the BVP binding assay. The converse situation may be occurring for the 10–1074 anti-HIV1 antibody, which appears as a potential false positive in the BVP but not in the PSR assay. Similarly, for vesencumab, mentioned above, the antigens causing the high BVP binding score (and a high PSR binding score; our unpublished observations with a sample generated from the known variable region sequences) may be absent from cynomolgus monkey tissue, but not from human tissue, so slow clearance is observed in the former, but not in the latter. Progress in understanding important components in PSR particles has been reported,[76] and

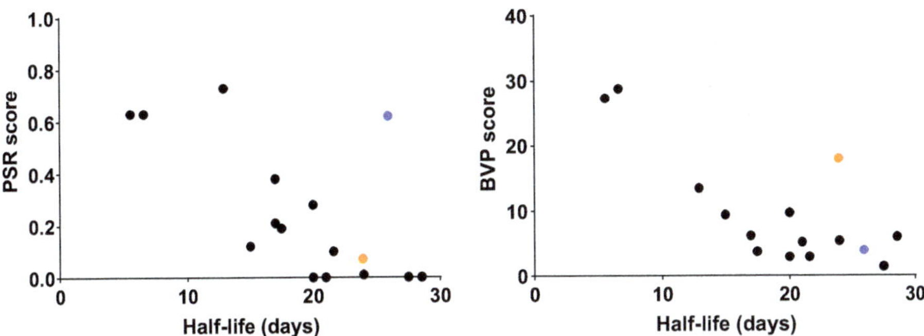

FIGURE 9.8 Correlation of terminal half-life from phase I human trials vs. PSR or BVP scores for samples made from the respective antibody variable region sequences. Points for urtoxazumab and 10–1074 are shown in red and purple, respectively. Data reported originally in Shehata et al.[20]

some of the results have led to alternative assays to probe polyspecificity.[67] Uncovering the underlying interaction in off-target binding can be challenging so it is worth mentioning in this context assays for identifying specific, but off-target binding, such as cell microarray technology.[77] Though limited in throughput, they are gaining traction as means to profile specific on- and off-target binding in antibodies of clinical interest.[78]

It should be noted that the studies reviewed here are concerned almost exclusively with monospecific antibodies in the human IgG1 format. Some of the observations may apply to other isotypes; recent work where isotype effects have been studied systematically includes the publication by Tang et al.[79] Similarly, while the relevance of our observations to bispecific antibodies is not direct, it is the expectation that good behavior by the component monospecific antibodies will be a necessary, but probably not sufficient, condition for good behavior by the bispecific antibody, when built on an IgG-like format. Most of the assays discussed here for IgG molecules can be applied to, especially, Fc-containing bispecific antibodies. However, establishing clear metrics of what ranges constitute concerning behavior, or a "flag" in the nomenclature described earlier,[12] is a work in progress. Some limited data generated internally has suggested that properties like PSR, HIC, and others for bispecific antibodies often have readouts in the range of the averages obtained for the individual monospecific antibodies. More work is required to determine the generality of this preliminary observation. This is an emerging area of research still providing surprises, as exemplified by recent work where the exact arrangement of binding sites on two types of bispecific or bifunctional molecules resulted in different pharmacokinetics.[80,81] In one of the examples,[80] it was observed that an IgG-scFv fusion of antibodies 1 and 2 exhibited poor pharmacokinetics in cynomolgus monkeys when antibody 1 was formatted as the Fab and antibody 2 as the C-terminal scFv fusion. However, when the orientation was reversed and antibody 1 was formatted as the C-terminal scFv fusion, clearance was much slower and in line with expectations of human IgG in cyno. Assessment in vitro of multiple biophysical properties of the respective bispecific antibodies did not readily explain the differences. These, of course, would be even harder, if not impossible, to explain from an assessment of the individual monospecific components.

A final area of the current investigation is the methodology for computational prediction from an amino acid sequence of PSR binding scores, or other metrics of polyspecificity or polyreactivity[82–86] and of HIC retention times.[87,88] Coupled with the kind of analysis presented in this chapter and future correlations that may emerge, such predictions are expected to streamline antibody discovery substantially. Even before such predictions approach quantitative accuracy to the experimental values, we have found them useful to recognize broad trends and in fact, have applied them to help design new antibody repertoires (synthetic libraries) with which to initiate discovery efforts (Jain et al., unpublished research).

ACKNOWLEDGMENTS

We are grateful to the Antibody Engineering, Protein Analytics, Platform Technologies, Core Molecular Biology, and High-Throughput Expression groups at Adimab for sample and data generation. We thank Dr. Xiaojun Lu for access to his group's unpublished chemical degradation data. We also thank Drs. Juergen Nett, James Geoghegan, Arvind Sivasubramanian, and Robert Pejchal for reading the manuscript and for fruitful discussions.

REFERENCES

1. Wu H, Pfarr DS, Tang Y, An LL, Patel NK, Watkins JD, et al. Ultra-potent antibodies against respiratory syncytial virus: effects of binding kinetics and binding valence on viral neutralization. *J Mol Biol* 2005; 350(1):126–44.
2. Wu H, Pfarr DS, Johnson S, Brewah YA, Woods RM, Patel NK, et al. Development of motavizumab, an ultra-potent antibody for the prevention of respiratory syncytial virus infection in the upper and lower respiratory tract. *J Mol Biol* 2007; 368(3):652–65.
3. Cingoz O. Motavizumab. *mAbs* 2009; 1(5):439–42.
4. Jacobs SA, Wu SJ, Feng Y, Bethea D, O'Neil KT. Cross-interaction chromatography: a rapid method to identify highly soluble monoclonal antibody candidates. *Pharm Res* 2010; 27(1):65–71.
5. Sule SV, Sukumar M, Weiss WFt, Marcelino-Cruz AM, Sample T, Tessier PM. High-throughput analysis of concentration-dependent antibody self-association. *Biophys J* 2011; 101(7):1749–57.
6. Sule SV, Dickinson CD, Lu J, Chow CK, Tessier PM. Rapid analysis of antibody self-association in complex mixtures using immunogold conjugates. *Mol Pharm* 2013; 10(4):1322–31.
7. Sydow JF, Lipsmeier F, Larraillet V, Hilger M, Mautz B, Molhoj M, et al. Structure-based prediction of asparagine and aspartate degradation sites in antibody variable regions. *PloS One* 2014; 9(6):e100736.
8. Yang R, Jain T, Lynaugh H, Nobrega RP, Lu X, Boland T, et al. Rapid assessment of oxidation via middle-down LCMS correlates with methionine side-chain solvent-accessible surface area for 121 clinical stage monoclonal antibodies. *mAbs* 2017; 9(4):646–53.

9. Lu X, Nobrega RP, Lynaugh H, Jain T, Barlow K, Boland T, et al. Deamidation and isomerization liability analysis of 131 clinical-stage antibodies. *mAbs* 2019; 11(1):45–57.
10. Obrezanova O, Arnell A, de la Cuesta RG, Berthelot ME, Gallagher TR, Zurdo J, et al. Aggregation risk prediction for antibodies and its application to biotherapeutic development. *mAbs* 2015; 7(2):352–63.
11. Sharma VK, Patapoff TW, Kabakoff B, Pai S, Hilario E, Zhang B, et al. In silico selection of therapeutic antibodies for development: viscosity, clearance, and chemical stability. *Proc Natl Acad Sci U S A* 2014; 111(52):18601–6.
12. Jain T, Sun T, Durand S, Hall A, Houston NR, Nett JH, et al. Biophysical properties of the clinical-stage antibody landscape. *Proc Natl Acad Sci U S A* 2017; 114(5):944–9.
13. Kingsbury JS, Saini A, Auclair SM, Fu L, Lantz MM, Halloran KT, et al. A single molecular descriptor to predict solution behavior of therapeutic antibodies. *Sci Adv* 2020; 6(32):eabb0372.
14. Avery LB, Wade J, Wang M, Tam A, King A, Piche-Nicholas N, et al. Establishing in vitro in vivo correlations to screen monoclonal antibodies for physicochemical properties related to favorable human pharmacokinetics. *mAbs* 2018; 10(2):244–55.
15. Hotzel I, Theil FP, Bernstein LJ, Prabhu S, Deng R, Quintana L, et al. A strategy for risk mitigation of antibodies with fast clearance. *mAbs* 2012; 4(6):753–60.
16. Raybould MIJ, Marks C, Krawczyk K, Taddese B, Nowak J, Lewis AP, et al. Five computational developability guidelines for therapeutic antibody profiling. *Proc Natl Acad Sci U S A* 2019; 116(10):4025–30.
17. Grinshpun B, Thorsteinson N, Pereira JN, Rippmann F, Nannemann D, Sood VD, et al. Identifying biophysical assays and in silico properties that enrich for slow clearance in clinical-stage therapeutic antibodies. *mAbs* 2021; 13(1):1932230.
18. Ahmed L, Gupta P, Martin KP, Scheer JM, Nixon AE, Kumar S. Intrinsic physicochemical profile of marketed antibody-based biotherapeutics. *Proc Natl Acad Sci U S A* 2021; 118(37).
19. Thorsteinson N, Gunn JR, Kelly K, Long W, Labute P. Structure-based charge calculations for predicting isoelectric point, viscosity, clearance, and profiling antibody therapeutics. *mAbs* 2021; 13(1):1981805.
20. Shehata L, Maurer DP, Wec AZ, Lilov A, Champney E, Sun T, et al. Affinity maturation enhances antibody specificity but compromises conformational stability. *Cell Rep* 2019; 28(13):3300–8 e4.
21. Kelly RL, Sun T, Jain T, Caffry I, Yu Y, Cao Y, et al. High throughput cross-interaction measures for human IgG1 antibodies correlate with clearance rates in mice. *mAbs* 2015; 7(4):770–7.
22. Connolly BD, Petry C, Yadav S, Demeule B, Ciaccio N, Moore JM, et al. Weak interactions govern the viscosity of concentrated antibody solutions: high-throughput analysis using the diffusion interaction parameter. *Biophys J* 2012; 103(1):69–78.
23. Sun T, Reid F, Liu Y, Cao Y, Estep P, Nauman C, et al. High throughput detection of antibody self-interaction by bio-layer interferometry. *mAbs* 2013; 5(6):838–41.
24. Xu Y, Roach W, Sun T, Jain T, Prinz B, Yu TY, et al. Addressing polyspecificity of antibodies selected from an in vitro yeast presentation system: a FACS-based, high-throughput selection and analytical tool. *Protein Eng Design Selec PEDS* 2013; 26(10):663–70.
25. Strohl WR. Antibody discovery: sourcing of monoclonal antibody variable domains. *Curr Drug Discov Technol* 2014; 11(1):3–19.
26. Jones TD, Carter PJ, Pluckthun A, Vasquez M, Holgate RG, Hotzel I, et al. The INNs and outs of antibody nonproprietary names. *mAbs* 2016; 8(1):1–9.
27. Guimaraes Koch SS, Thorpe R, Kawasaki N, Lefranc MP, Malan S, Martin ACR, et al. International nonproprietary names for monoclonal antibodies: an evolving nomenclature system. *mAbs* 2022; 14(1):2075078.

28. Wilkinson I, Hale G. Systematic analysis of the varied designs of 819 therapeutic antibodies and Fc fusion proteins assigned international nonproprietary names. *mAbs* 2022; 14(1):2123299.
29. Wu SJ, Luo J, O'Neil KT, Kang J, Lacy ER, Canziani G, et al. Structure-based engineering of a monoclonal antibody for improved solubility. *Protein Eng Design Selec PEDS* 2010; 23(8):643–51.
30. Bethea D, Wu SJ, Luo J, Hyun L, Lacy ER, Teplyakov A, et al. Mechanisms of self-association of a human monoclonal antibody CNTO607. *Protein Eng Design Selec PEDS* 2012; 25(10):531–7.
31. Liu Y, Caffry I, Wu J, Geng SB, Jain T, Sun T, et al. High-throughput screening for developability during early-stage antibody discovery using self-interaction nanoparticle spectroscopy. *mAbs* 2014; 6(2):483–92.
32. Bengali AN, Tessier PM. Biospecific protein immobilization for rapid analysis of weak protein interactions using self-interaction nanoparticle spectroscopy. *Biotechnol Bioeng* 2009; 104(2):240–50.
33. Xin Y, Bai S, Damico-Beyer LA, Jin D, Liang WC, Wu Y, et al. Anti-neuropilin-1 (MNRP1685A): unexpected pharmacokinetic differences across species, from preclinical models to humans. *Pharm Res* 2012; 29(9):2512–21.
34. Estep P, Caffry I, Yu Y, Sun T, Cao Y, Lynaugh H, et al. An alternative assay to hydrophobic interaction chromatography for high-throughput characterization of monoclonal antibodies. *mAbs* 2015; 7(3):553–61.
35. Jarasch A, Koll H, Regula JT, Bader M, Papadimitriou A, Kettenberger H. Developability assessment during the selection of novel therapeutic antibodies. *J Pharm Sci* 2015; 104(6):1885–98.
36. Fekete S, Veuthey JL, Beck A, Guillarme D. Hydrophobic interaction chromatography for the characterization of monoclonal antibodies and related products. *J Pharm Biomed Anal* 2016; 130:3–18.
37. Kelly RL, Yu Y, Sun T, Caffry I, Lynaugh H, Brown M, et al. Target-independent variable region mediated effects on antibody clearance can be FcRn independent. *mAbs* 2016; 8(7):1269–75.
38. Wang W, Lu P, Fang Y, Hamuro L, Pittman T, Carr B, et al. Monoclonal antibodies with identical Fc sequences can bind to FcRn differentially with pharmacokinetic consequences. *Drug Metab Dispos* 2011; 39(9):1469–77.
39. Schoch A, Kettenberger H, Mundigl O, Winter G, Engert J, Heinrich J, et al. Charge-mediated influence of the antibody variable domain on FcRn-dependent pharmacokinetics. *Proc Natl Acad Sci U S A* 2015; 112(19):5997–6002.
40. Willis LF, Kumar A, Jain T, Caffry I, Xu Y, Radford SE, et al. The uniqueness of flow in probing the aggregation behavior of clinically relevant antibodies. *Eng Rep* 2020; 2(5):e12147.
41. Kraft TE, Richter WF, Emrich T, Knaupp A, Schuster M, Wolfert A, et al. Heparin chromatography as an in vitro predictor for antibody clearance rate through pinocytosis. *mAbs* 2020; 12(1):1683432.
42. Lecerf M, Kanyavuz A, Rossini S, Dimitrov JD. Interaction of clinical-stage antibodies with heme predicts their physiochemical and binding qualities. *Commun Biol* 2021; 4(1):391.
43. He F, Woods CE, Becker GW, Narhi LO, Razinkov VI. High-throughput assessment of thermal and colloidal stability parameters for monoclonal antibody formulations. *J Pharm Sci* 2011; 100(12):5126–41.
44. Bornholdt ZA, Herbert AS, Mire CE, He S, Cross RW, Wec AZ, et al. A two-antibody pan-ebolavirus cocktail confers broad therapeutic protection in ferrets and nonhuman primates. *Cell Host Microbe* 2019; 25(1):49–58.

45. Wec AZ, Wrapp D, Herbert AS, Maurer DP, Haslwanter D, Sakharkar M, et al. Broad neutralization of SARS-related viruses by human monoclonal antibodies. *Science* 2020; 369(6504):731–6.
46. Baum A, Ajithdoss D, Copin R, Zhou A, Lanza K, Negron N, et al. REGN-COV2 antibodies prevent and treat SARS-CoV-2 infection in rhesus macaques and hamsters. *Science* 2020; 370(6520):1110–5.
47. Pinto D, Park YJ, Beltramello M, Walls AC, Tortorici MA, Bianchi S, et al. Cross-neutralization of SARS-CoV-2 by a human monoclonal SARS-CoV antibody. *Nature* 2020; 583(7815):290–5.
48. Shi R, Shan C, Duan X, Chen Z, Liu P, Song J, et al. A human neutralizing antibody targets the receptor-binding site of SARS-CoV-2. *Nature* 2020; 584(7819):120–4.
49. Zost SJ, Gilchuk P, Case JB, Binshtein E, Chen RE, Nkolola JP, et al. Potently neutralizing and protective human antibodies against SARS-CoV-2. *Nature* 2020; 584(7821):443–9.
50. Jones BE, Brown-Augsburger PL, Corbett KS, Westendorf K, Davies J, Cujec TP, et al. The neutralizing antibody, LY-CoV555, protects against SARS-CoV-2 infection in nonhuman primates. *Sci Transl Med* 2021; 13(593):eabf1906.
51. Westendorf K, Zentelis S, Wang L, Foster D, Vaillancourt P, Wiggin M, et al. LY-CoV1404 (bebtelovimab) potently neutralizes SARS-CoV-2 variants. *Cell Rep* 2022; 39(7):110812.
52. Cunningham O, Scott M, Zhou ZS, Finlay WJJ. Polyreactivity and polyspecificity in therapeutic antibody development: risk factors for failure in preclinical and clinical development campaigns. *mAbs* 2021; 13(1):1999195.
53. Vasquez M. FM, Gerngross T.U., Wittrup K. D. *Rationally Designed, Synthetic Antibody Libraries and Uses Therefor.* WO 2009/036379, 2009.
54. Chothia C, Lesk AM, Gherardi E, Tomlinson IM, Walter G, Marks JD, et al. Structural repertoire of the human VH segments. *J Mol Biol* 1992; 227(3):799–817.
55. Jia L, Sun Y. In silico prediction method for protein asparagine deamidation. *Methods Mol Biol* 2023; 2552199–217.
56. Robinson NE, Robinson AB. Molecular clocks. *Proc Natl Acad Sci U S A* 2001; 98(3):944–9.
57. Irudayanathan FJ, Zarzar J, Lin J, Izadi S. Deciphering deamidation and isomerization in therapeutic proteins: effect of neighboring residue. *mAbs* 2022; 14(1):2143006.
58. Delmar JA, Wang J, Choi SW, Martins JA, Mikhail JP. Machine learning enables accurate prediction of asparagine deamidation probability and rate. *Mol Ther Methods Clin Dev* 2019; 15264–74.
59. Bickel F, Griaud F, Kern W, Kroener F, Gritsch M, Dayer J, et al. Restoring the biological activity of crizanlizumab at physiological conditions through a pH-dependent aspartic acid isomerization reaction. *mAbs* 2023; 15(1):2151075.
60. Wei B, Berning K, Quan C, Zhang YT. Glycation of antibodies: modification, methods and potential effects on biological functions. *mAbs* 2017; 9(4):586–94.
61. Saleem RA, Affholter BR, Deng S, Campbell PC, Matthies K, Eakin CM, et al. A chemical and computational approach to comprehensive glycation characterization on antibodies. *mAbs* 2015; 7(4):719–31.
62. Liu H, Ponniah G, Neill A, Patel R, Andrien B. Identification and comparative quantitation of glycation by stable isotope labeling and LC-MS. *J Chromatogr B Analyt Technol Biomed Life Sci* 2014; 95890–5.
63. Vlasak J, Ionescu R. Fragmentation of monoclonal antibodies. *mAbs* 2011; 3(3):253–63.
64. Liu H, Gaza-Bulseco G, Lundell E. Assessment of antibody fragmentation by reversed-phase liquid chromatography and mass spectrometry. *J Chromatogr B Analyt Technol Biomed Life Sci* 2008; 876(1):13–23.
65. Xu Y, Wang D, Mason B, Rossomando T, Li N, Liu D, et al. Structure, heterogeneity and developability assessment of therapeutic antibodies. *mAbs* 2019; 11(2):239–64.

66. Haberger M, Bomans K, Diepold K, Hook M, Gassner J, Schlothauer T, et al. Assessment of chemical modifications of sites in the CDRs of recombinant antibodies: susceptibility vs. functionality of critical quality attributes. *mAbs* 2014; 6(2):327–39.
67. Makowski EK, Wu L, Desai AA, Tessier PM. Highly sensitive detection of antibody non-specific interactions using flow cytometry. *mAbs* 2021; 13(1):1951426.
68. Wardemann H, Yurasov S, Schaefer A, Young JW, Meffre E, Nussenzweig MC. Predominant autoantibody production by early human B cell precursors. *Science* 2003; 301(5638):1374–7.
69. Frese K, Eisenmann M, Ostendorp R, Brocks B, Pabst S. An automated immunoassay for early specificity profiling of antibodies. *mAbs* 2013; 5(2):279–87.
70. Agrawal NJ, Helk B, Kumar S, Mody N, Sathish HA, Samra HS, et al. Computational tool for the early screening of monoclonal antibodies for their viscosities. *mAbs* 2016; 8(1):43–8.
71. Tomar DS, Kumar S, Singh SK, Goswami S, Li L. Molecular basis of high viscosity in concentrated antibody solutions: strategies for high concentration drug product development. *mAbs* 2016; 8(2):216–28.
72. Tomar DS, Singh SK, Li L, Broulidakis MP, Kumar S. In silico prediction of diffusion interaction parameter (kD), a key indicator of antibody solution behaviors. *Pharm Res* 2018; 35(10):193.
73. Lai PK, Swan JW, Trout BL. Calculation of therapeutic antibody viscosity with coarse-grained models, hydrodynamic calculations and machine learning-based parameters. *mAbs* 2021; 13(1):1907882.
74. Bailly M, Mieczkowski C, Juan V, Metwally E, Tomazela D, Baker J, et al. Predicting antibody developability profiles through early stage discovery screening. *mAbs* 2020; 12(1):1743053.
75. Sigounas G, Harindranath N, Donadel G, Notkins AL. Half-life of polyreactive antibodies. *J Clin Immunol* 1994; 14(2):134–40.
76. Kelly RL, Geoghegan JC, Feldman J, Jain T, Kauke M, Le D, et al. Chaperone proteins as single component reagents to assess antibody nonspecificity. *mAbs* 2017; 9(7):1036–40.
77. Freeth J, Soden J. New Advances in cell microarray technology to expand applications in target deconvolution and off-target screening. *SLAS Discov* 2020; 25(2):223–30.
78. Finlay WJJ, Coleman JE, Edwards JS, Johnson KS. Anti-PD1 'SHR-1210' aberrantly targets pro-angiogenic receptors and this polyspecificity can be ablated by paratope refinement. *mAbs* 2019; 11(1):26–44.
79. Tang Y, Cain P, Anguiano V, Shih JJ, Chai Q, Feng Y. Impact of IgG subclass on molecular properties of monoclonal antibodies. *mAbs* 2021; 13(1):1993768.
80. Datta-Mannan A, Brown R, Key S, Cain P, Feng Y. Pharmacokinetic developability and disposition profiles of bispecific antibodies: a case study with two molecules. *Antibodies (Basel)* 2021; 11(1):2.
81. Datta-Mannan A, Brown RM, Fitchett J, Heng AR, Balasubramaniam D, Pereira J, et al. Modulation of the biophysical properties of bifunctional antibodies as a strategy for mitigating poor pharmacokinetics. *Biochemistry* 2019; 58(28):3116–32.
82. Boughter CT, Borowska MT, Guthmiller JJ, Bendelac A, Wilson PC, Roux B, et al. Biochemical patterns of antibody polyreactivity revealed through a bioinformatics-based analysis of CDR loops. *Elife* 2020; 9e61393.
83. Harvey EP, Shin J-E, Skiba MA, Nemeth GR, Hurley JD, Wellner A, et al. An in silico method to assess antibody fragment polyreactivity. *Nat Commun* 2022; 13:7554.
84. Ausserwöger H, Schneider MM, Herling TW, Arosio P, Invernizzi G, Knowles TPJ, et al. Non-specificity as the sticky problem in therapeutic antibody development. *Nat Rev Chem* 2022; 6(12):844–861.

85. Zhang Y, Wu L, Gupta P, Desai AA, Smith MD, Rabia LA, et al. Physicochemical rules for identifying monoclonal antibodies with drug-like specificity. *Mol Pharm* 2020; 17(7):2555–69.
86. Hebditch M, Warwicker J. Charge and hydrophobicity are key features in sequence-trained machine learning models for predicting the biophysical properties of clinical-stage antibodies. *PeerJ* 2019; 7e8199.
87. Jain T, Boland T, Lilov A, Burnina I, Brown M, Xu Y, et al. Prediction of delayed retention of antibodies in hydrophobic interaction chromatography from sequence using machine learning. *Bioinformatics* 2017; 33(23):3758–66.
88. Zhou Y, Xie S, Yang Y, Jiang L, Liu S, Li W, et al. SSH2.0: a better tool for predicting the hydrophobic interaction risk of monoclonal antibody. *Front Genet* 2022; 13:13842127.

10 *In Silico* Approaches to Deliver Better Antibodies by Design

The Past, the Present, and the Future

Andreas Evers, Shipra Malhotra, and Vanita D. Sood

ABBREVIATIONS

Anti-drug antibodies	(**ADA**)
Artificial intelligence	(**AI**)
BLOcks SUbstitution Matrix	(**BLOSUM**)
Chemistry, manufacturing, and control	(**CMC**)
Critical quality attribute	(**CQA**)
Complementarity-determining region	(**CDR**)
Deep learning	(**DL**)

DOI: 10.1201/9781003300311-10

Denoising autoencoder	**(DAE)**
Drug metabolism and pharmacokinetics	**(DMPK)**
Fluorescence-activated cell sorting	**(FACS)**
Generative adversarial network	**(GAN)**
Low-density lipoprotein	**(LDL)**
Machine learning	**(ML)**
Major histocompatibility complex class II	**(MHC-II)**
Multiple sequence alignment	**(MSA)**
Next-generation sequencing	**(NGS)**
Post-translational modification	**(PTM)**
Protein Data Bank	**(PDB)**
Modified Probabilistic Neural Network	**(MPNN)**
Sequence-activity-relationship	**(SAR)**
Variational auto-encoder	**(VAE)**
Variable heavy domain of heavy chain	**(VHH)**
Variable heavy chain	**(VH)**
Variable light chain	**(VL)**

10.1 INTRODUCTION

Much ink has been spilled bemoaning the costs, both to drug developers and to society, of poorly developable drugs,[1–4] and rightfully so as every delay and termination of what could have otherwise been an efficacious treatment for patients in need is consequential. Although biologics enjoy a somewhat higher rate of approval than new chemical entities,[3,5–7] there remains substantial room for improvement. Invisible to the general public is the lost potential of many drug candidates that never enter the clinic, due to decisions by drug developers to shelve compounds with poor developability prior to clinical trials, often after most or all of the preclinical discovery and early nonclinical development costs have already been incurred. Yet, the consequences of moving forward with poorly developable drugs are even more harmful, with potential fallout ranging from delays in the progression of clinical trials due to difficulties in producing enough drug supply, or to adverse events resulting from the intrinsic properties of the drug itself, its degradation products, or impurities associated with inconsistencies in manufacturing drugs with poor developability.[8–13] While decisions to terminate trials are typically made for multiple scientific and business reasons that are difficult to deconvolute (for an excellent review, see ref. [14]), there are some illustrative examples where molecular developability characteristics played a major role.

Bococizumab was a promising PCSK9 inhibitory monoclonal antibody with low-density lipoprotein (LDL) lowering activity that was investigated in cardiovascular indications as a possible statin alternative. The clinical development of bococizumab was stopped by the sponsor after considerable investment in eight Phase 3 studies.[15] Anti-drug antibodies (ADA) were observed in only 7% of patients during Phase 2,[16] whereas the larger Phase 3 trial revealed 50% rate of ADA, with 29% of patients developing neutralizing ADA[17] leading to reduced efficacy.[12] While commercial factors certainly played a role in the decision to discontinue (alirocumab and evolocumab, in-class competitors, had already achieved market authorization), as well as variable efficacy even among patients without ADA,[12] developability factors were also cited, specifically the high rate of ADA formation, as a major factor in the termination. ATR-107, an anti-IL21 monoclonal, provides another example of a clinical trial halted, at least in part, due to a high rate of ADA formation.[10]

CMC influencing biophysical properties have been a frequent target of post hoc computational model-informed engineering in the literature. Examples include CNTO607, an anti-IL13 monoclonal[9,13]; an anti-CD3E antibody[18]; and an anti-VEGF antibody[8] for which a dramatic improvement of biophysical properties could be demonstrated. In the case of anti-VEGF, not only were solution properties improved but also expression titer and product quality improved substantially. Stamulumab (MYO-029) provides an additional compelling example; this monoclonal underwent a Phase I/II trial in muscular dystrophy patients who demonstrated a good safety profile, but no statistically significant effect was observed in exploratory efficacy endpoints,[19] likely due to insufficient exposure for efficacy.[20] In addition to intrinsic clearance properties, both dose and route of administration can significantly affect exposure. In retrospect, it seems likely that the low solubility, high viscosity, and aggregation propensity of stamulumab[21] would preclude a formulation and dose that would be needed to support efficacy, given the high free concentration of the drug required to observe pharmacodynamic effects.[20] Initial post hoc developability engineering demonstrated encouraging trends toward improved solubility,[21] and subsequent model-guided optimization yielded remarkable improvements in biophysical properties accompanied by potency increases.[11] Had the optimized variant been available, it is conceivable that higher doses could have been considered for clinical trials.

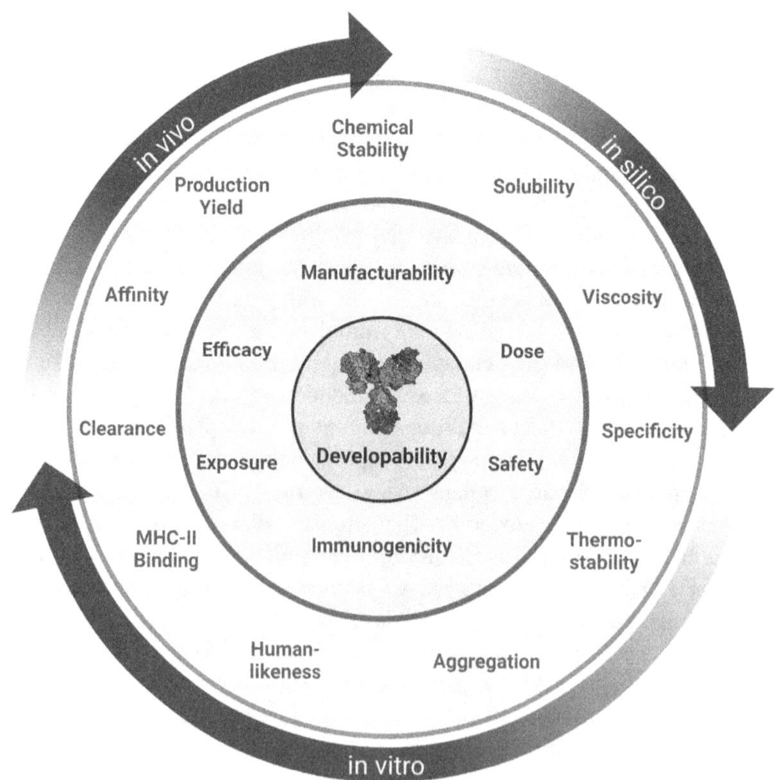

BOX 10.1 TEXT: DEFINING THE ENTITIES FROM HIT TO DEVELOPMENT CANDIDATE

Developability: In a broad sense, developability is defined as the likelihood that an antibody sequence can become a manufacturable, safe, and efficacious drug. To be considered developable, a lead should have a potency, exposure, and formulatability compatible with a reasonable dose; low immunogenicity and low toxicity at the recommended dose; and biophysical properties that ensure a chemistry, manufacturing, and control (CMC) process at reasonable costs within a reasonable timeline.

Hit: A sequence discovered from a screening campaign that displays the desired biochemical activity, target specificity, and mechanism of action (MoA). Typically requires extensive optimization in order to be considered developable.

Optimized Hit: A sequence-optimized hit that meets all the biochemical and cellular screening criteria and in addition is more drug-like in sequence, i.e., displays fewer liabilities and potential liabilities.

Lead: A sequence-optimized hit that has additionally shown *in vivo* pharmacological activity in one or more disease models and demonstrates acceptable pharmacokinetics (PK), nonclinical safety, and manufacturability in preliminary assessments. Typically, a small pool of lead molecules will exist from which a development candidate and a backup candidate will be nominated.

Development Candidate: A stable transfection clone of a lead molecule, which has undergone scale-up, rigorous *in vivo* testing, formal manufacturability assessment, and passed all preclinical pharmacological, PK, nonclinical safety, and CMC criteria to move into early nonclinical development (e.g., pivotal toxicity study) and eventually into clinical development.

These examples of clinical stage developability failures explain the pharmaceutical industry's intense focus on early identification of drug candidates with improved developability characteristics (see Box 10.1 for definitions of developability and candidates, as used in this chapter). As full developability analyses cannot be performed until late program stages, early mitigation is a major challenge; yet, this challenge has been met and overcome in small molecule discovery, providing a heartening example to follow; in the 1990s, most clinical attrition of small molecule drugs was due to poor developability, primarily poor exposure[3]; today, very few small molecules fail due to insufficient exposure in clinic, as modern discovery processes deliver developable candidates with good pharmacokinetics. Improved developability of biologics can in principle be achieved in one of two ways. The first option is to identify and remove from further development any screening hits that are more likely to display developability issues later in development. This option requires the ability to discriminate candidates that have liabilities. Many experimental[22–28] and computational[29–39] methods of candidate assessment are available and have been previously discussed at length.[40–43] The second option is to generate drug candidates de-risked for developability by design. This option requires early hit discovery and optimization methods that prioritize developability properties equally with on-target potency and functional efficacy. In both cases, biopharmaceutical informatics and *in silico* engineering can have a significant impact.

Both options require an understanding of which preclinically measurable properties of the molecule actually predict the clinical developability criteria; the predictivity is not the same for each criterion and for each property. For example, it is well accepted that human and humanized sequences have lower immunogenicity on average than the murine and chimeric antibodies of old.[44] Predicting clinical immunogenicity for a specific drug candidate, however, remains challenging, as it is a complex phenomenon that springs from aspects of sequence (MHC-II presentation and proteasomal processing) and manufacturing process (levels of impurities, high or low molecular weight species, and host cell protein contamination). There is no single *in silico* or *in vitro* assay that is holistically predictive of immunogenicity. The best we can do is to de-risk as much as possible, based on the predictive models that are available to us. The situation is similar for predicting expression yields, clearance, and pharmacokinetics. On the other hand, developability aspects concerned with manufacturing criteria such as viscosity and route of administration are becoming much more predictable preclinically. Furthermore, a growing body of evidence suggests that even for developability properties that are more difficult to predict accurately, optimization on properties that are more easily measured *in silico* and *in vitro*, such as hydrophobicity, polyreactivity, or pI, are likely to positively affect developability.[23,25,29,45–48]

In this chapter, we will focus on proactive and early computational de-risking approaches. We have divided the approaches into three broad categories we have designated as "classical," "contemporary," and "emerging" (Figure 10.1). The classical approach entails applying readily available and standard tools of sequence and structure-based drug design to rapidly & pragmatically re-engineer antibody hits obtained from traditional hit discovery approaches. The contemporary approach is more proactive in that it entails using next-generation sequencing (NGS) and machine learning (ML) to engineer the libraries or repertoires from which hits are obtained for improved developability, thus aiming to ensure that every drug candidate selected on the basis of potency will by default have minimal developability liabilities. Finally, the emerging approach goes one step further, taking advantage of recent advances in deep learning (DL) and artificial intelligence (AI) in the field of protein structure prediction, and utilizes *de novo* computational design to directly generate drug candidates that meet multiple criteria for potency and developability.

10.2 THE CLASSICAL APPROACH – DESIGN OF SPECIFIC SEQUENCE-OPTIMIZATION VARIANTS

A traditional screening cascade for antibody discovery involves screening a repertoire (either synthetic or derived from a naïve or immunized subject) for a few key properties related to the desired on-target effect, e.g., binding to the target and not the off-target(s),

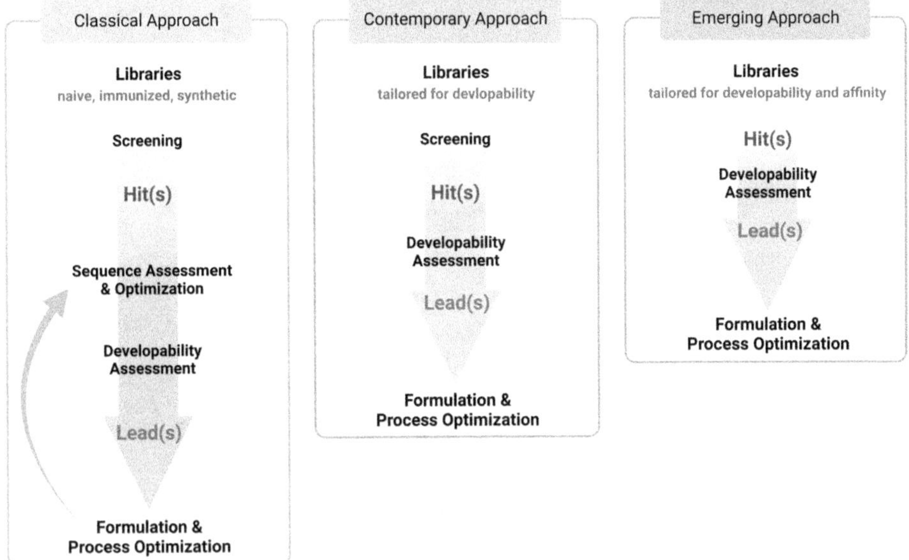

FIGURE 10.1 Illustration of the classical, contemporary, and emerging approaches. The classical approach starts from a pool of potent hits from screening naïve, immunized, or synthetic libraries that typically require cycles of sequence optimization. The contemporary approach utilizes engineered libraries that have been tailored by design toward improved developability properties and will ideally produce hits from the experimental screening campaign that will need minimal to no further sequence optimization. Finally, the emerging approach utilizes AI/ML approaches for a *de novo in silico* design of sequences optimized for developability and manufacturability.

ligand competition, and functional or phenotypic assay (Figure 10.2). After a handful of hits (sometimes as few as 2 or 3) that best meet the criteria are identified, these are prioritized for hit-to-lead optimization, typically including affinity maturation (if needed), humanization (if not derived from a human repertoire), and optimization of CDRs to remove chemically labile residues. The optimized hits are then scaled up and tested *in vivo* for efficacy and clearance, and in early tests of manufacturability to select a lead and backup. With luck, at least one lead and/or backup will continue to demonstrate efficacy, tolerability, and manufacturability as they are tested in increasingly rigorous preclinical studies, and will meet all criteria for a development candidate.

A frequent outcome, however, is that some of the lead candidates may be efficacious and others may be developable, but not all are *both* efficacious and developable. In this case, if the most advanced lead candidates fail to progress for any reason of efficacy or developability, the program may stall as the only alternatives are to return to the pool of unoptimized and partially characterized hits to search for a replacement lead molecule, or to re-engineer the sequence of the most advanced candidate for improved developability, followed by repeating all the downstream preclinical development steps. Either of these remedies entails a delay and an effort at which most organizations balk.

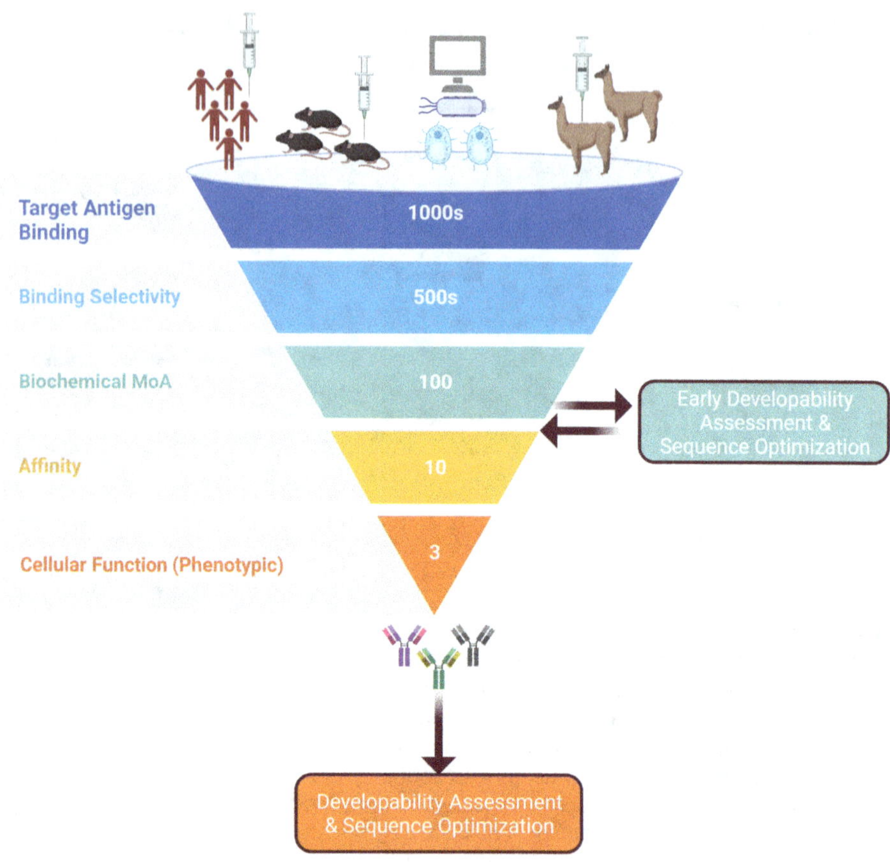

FIGURE 10.2 A generic hit discovery screening funnel. The initial diversity is rapidly restricted by successive high-throughput assays, until the number of hits is sufficiently small that the most relevant, but typically cumbersome and low throughput, phenotypic assay can be performed to identify candidates prioritized for further, more expensive and challenging tests. Once hits with the desired activity are identified (orange), they are intensively optimized and engineered for improved potency and developability while also being scaled up and tested in *in vivo* disease models. An alternative approach is to perform sequence optimization earlier in the screening cascade, thus providing higher quality inputs to critical late screening assays (affinity and phenotypic).

An alternative to the above unhappy scenario is to create a larger pool of fully or partially optimized hits which can serve to safeguard against attrition of the top lead candidates (Figure 10.2, "Early Developability Assessment & Sequence Optimization"). Furthermore, selecting the lead candidates for further characterization from a pool of optimized candidates allows for a fairer comparison between molecules and increases the probability that one or more candidates will meet all criteria for a lead molecule; this is because unoptimized hits may have poor properties (e.g., aggregation propensity or instability) detrimental to their performance in early assays in the screening cascade, obscuring a potential for greatness that becomes obvious only after optimization, as schematically represented in Figure 10.3. We have even seen that sequence optimization based on developability criteria alone can improve affinity up to 15-fold (D. Nannemann, personal communication), thereby increasing the pool of hits that meet affinity criteria and can be assayed for the desired phenotypic effect.

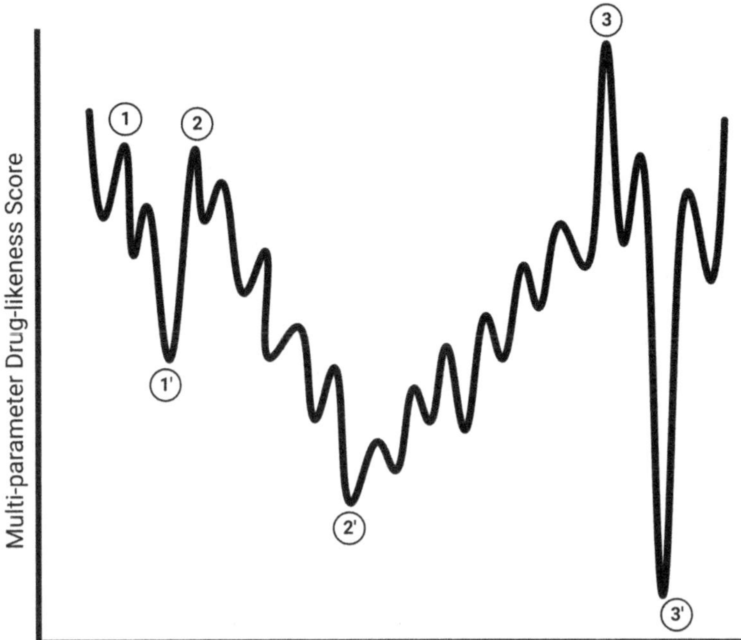

FIGURE 10.3 From confirmed hit to drug-like lead candidate. Three hypothetical hits from a screening cascade are symbolized as 1, 2, and 3. Their position on the y-axis represents how closely each conforms to a multiparameter drug-likeness score that encompasses potency as well as developability and manufacturability aspects such as aggregation propensity, thermostability, hydrophobic or charged patches, viscosity, clearance, and immunogenicity. The sequence-optimized version of each (1′, 2′, and 3′) will have a better drug-like score than its respective parent. Of note, the parental hits 1 and 2, which may score better than hit 3 (perhaps due to higher potency, for example, or fewer chemical instabilities, or less non-human sequence), may not be the ones that yield the most drug-like lead candidate. In certain cases, it may be impossible to fully optimize a certain hit (e.g., if potency and hydrophobicity are correlated). In other cases, a hit (3) might have more liabilities giving a poor score for drug-likeness, but if these can be easily rectified while maintaining potency, the "worst" parental hit may yield the "best" lead candidate (3′).

However, many drug discovery organizations also balk at "investing" in the optimization of hits that are not viewed as lead drug candidates (usually by a potency criterium), creating a sort of paradox. The best way to identify the most promising lead candidates is to compare diverse optimized hits, but optimization is only done for hits that already display the most lead-like attributes. The way out of this paradox is to change the organizational mindset regarding the value of early optimization. The mindset change is arrived at when the organization understands these two truths of drug discovery:

1. Early optimization is an investment that significantly reduces the probability of late-stage attrition of the eventual lead candidate, not a "waste" of effort on hits that will not become leads.
2. The investment in optimization does not have to be costly (in either time or money) if done judiciously and pragmatically.

To convince the organization of truth #1, we suggest referring to concrete examples of later-stage failures as outlined in the introduction, in addition to internal examples from one's own organization; every drug discovery organization will have some confidential examples and it is wise to remember and learn from these. To avoid the fate of Cassandra of Troy,[49] however, one must balance the lessons of past failures by reassuring the organization that a path to success is achievable (truth #2). In a resource-limited setting, it is understandably tempting to adhere to a traditional screening funnel (Figure 10.2) which by its nature not only helps identify the few functional hits, but also limits the number of candidates that will undergo expensive scale-up and *in vivo* testing to very few. However, it is due to the high cost to test the few candidates that it is imperative to ensure that the selected hits are the best possible lead candidates. Similar to the role of cheminformatics in small molecule hit triage,[50] biopharmaceutical informatics can help reconcile the need to select the best hits with the reality of limited resources and time, by helping to select diverse and valid hits for optimization, and by reducing the resource required to fully optimize each selected hit. We outline here a resource-efficient approach to optimization that can be applied to any drug discovery program in any setting, even when advanced computational tools such as those described in Sections 10.3 and 10.4 are not available. This classical approach to optimization relies on access to any structural modeling software,[51–55] general knowledge of antibody optimization, and a tight coupling of computation and experiment.

The first step in a pragmatic and efficient hit optimization procedure is to prospectively plan how many hits can realistically be sequence optimized. This will depend on the priority of the program in the portfolio, the throughput of each assay in the screening funnel, and the resources available to the program. Once the resource allocation is clear, the first critical scientific decision to be made is which hit sequences, of the (hopefully) hundreds that bind to the target, will be triaged and which will be discarded. Biopharmaceutical informatics is indispensable to making the best decision possible. After screening for desired selectivity and MoA, hierarchical clustering (generally on paratope sequence, but also on epitope binning if available) should be used to identify related clusters of hits. From these, one or several representatives from each cluster should be resynthesized and reconfirmed for target antigen binding. If the number of clusters exceeds the resource available for resynthesis and assays, sequence assessment should be used to identify and remove highly hypermutated hits, overly hydrophobic clusters, or clusters with inconsistent structure–activity relationships, as these are likely false positives that need not be tested further. At this point, if the number of clusters still exceeds the capacity for further testing, the number may be reduced by any other criteria such as activity level, high-throughput affinity assessment, protein engineering "intuition," or random selection, until the number conforms to the previously agreed resource available. As long as the diversity is maintained, the program is safeguarded against attrition due to known and unknown causes as it progresses.

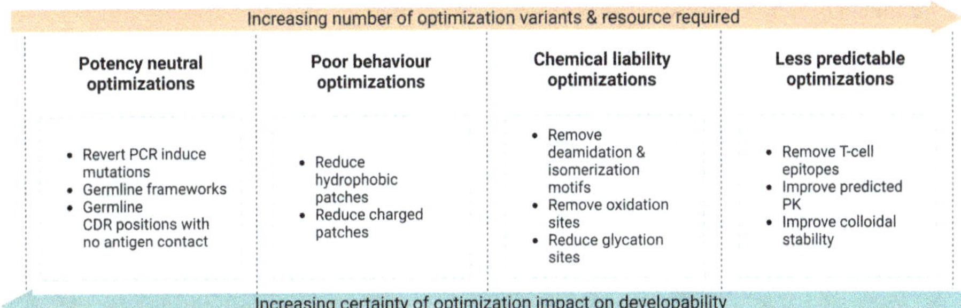

FIGURE 10.4 Tiered approach to prioritizing developability aspects to optimize. Toward the left are "must-have" optimizations that are relatively simple to execute and have a high probability of actually improving developability, as well as reducing false positive or false negative rates in pharmacological tests. The third panel illustrates optimizations that positively impact the CQAs of a candidate; these are "must-have" but may be executed on fewer, prioritized hits. The rightmost panel includes "nice-to-have" optimizations.

If sufficient diversity of biochemically active hits has been obtained in a given discovery project, the number of hits to be optimized will still likely exceed capacity if one attempts to do a comprehensive optimization designed to fully eliminate all risks from each sequence.

Instead of such a "gold-plated" approach, we suggest a tiered approach (Figure 10.4). First, use a structure-based drug design to optimize any liability where the sequence change will not affect potency. This typically entails germline humanization of residues that do not affect the conformation of the paratope[56] and will have the effect of stabilizing the antibody (which reduces the incidence of artifacts in screening assays) as well as reducing downstream immunogenicity risks. Second, include any optimization that will remove liabilities that are likely to interfere with the accuracy of read-outs from decision-making low throughput assays (such as *in vivo* efficacy); beyond the humanization germlining that will be done anyway, this typically should include engineering any overt hydrophobic or charged patches, as these properties are prone to non-specific interactions and induce poor solution behaviors that interfere with assay read-outs and cause artifacts. Third, include any optimization that, should the hit develop into a contender for the program lead molecule, has a high likelihood of becoming a critical quality attribute (CQA), and thus will almost certainly need to be removed from the sequence before manufacturing and clinical development. This will entail engineering the CDRs to remove chemically labile motifs and can be challenging to do in the absence of a co-crystal structure; it is also acceptable to postpone this optimization until a hit is prioritized as a lead candidate. Last, include optimizations that remove less likely theoretical risks or those that remove risk with only moderate certainty of their beneficial effect. These include the risk of immunogenicity which, despite the availability of excellent MHC-II binding predictors,[57] still remains challenging to predict clinically; and the risk of fast clearance or low bioavailability.[25,26,47,58–60] While it is preferable to de-risk as much as possible, it is not always pragmatic to de-risk extensively at the early hit optimization stage. One can safeguard against complete attrition due to less predictable factors by maintaining diversity; thus, allocating resources to partially optimizing more hits is preferable to fully optimizing fewer hits. This decision tree is stopped at the point where the number of sequence variants to be generated matches the capacity of the organization – the "good enough" optimization.

It is also worth mentioning that investing in hit discovery platforms that produce fully human antibodies will obviate the need for humanization, and platforms based on specific immune repertoires frequently yield high-affinity hits that obviate the need for extensive affinity maturation. Platforms such as these maximize the resource allocation for hit optimization as the optimization requirements are minimal and focused on manufacturability. However, if such platforms are not available, more extensive optimization can be done off the critical path, on fewer candidates, and one can focus on the most impactful optimizations for lead candidate selection in the first pass (Figure 10.4).

With agreement on the resources for hit optimization and a philosophy for a "good enough" optimization in place, all that remains is to actually optimize the selected diverse hits. Here again, biopharmaceutical informatics is critical in focusing limited resources on the most relevant and impactful optimizations that will support the best possible decision-making for lead candidates; however, informatics alone cannot and should not guide the optimization. A strong collaboration with analytical sciences is key to a successful optimization in the classical approach. This "lean" approach is exemplified in Figure 10.5.

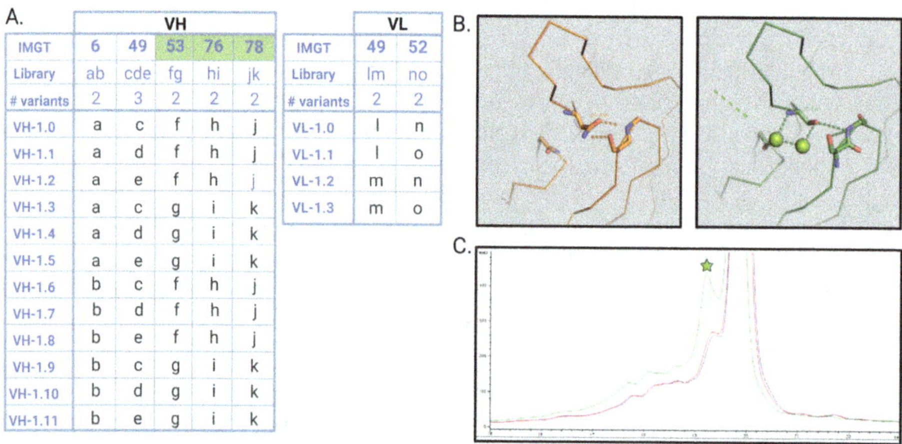

	VH				
IMGT	6	49	53	76	78
Library	ab	cde	fg	hi	jk
# variants	2	3	2	2	2
VH-1.0	a	c	f	h	j
VH-1.1	a	d	f	h	j
VH-1.2	a	e	f	h	j
VH-1.3	a	c	g	i	k
VH-1.4	a	d	g	i	k
VH-1.5	a	e	g	i	k
VH-1.6	b	c	f	h	j
VH-1.7	b	d	f	h	j
VH-1.8	b	e	f	h	j
VH-1.9	b	c	g	i	k
VH-1.10	b	d	g	i	k
VH-1.11	b	e	g	i	k

	VL	
IMGT	49	52
Library	lm	no
# variants	2	2
VL-1.0	l	n
VL-1.1	l	o
VL-1.2	m	n
VL-1.3	m	o

FIGURE 10.5 (a) Design of optimization variants. Upon structural and sequence analysis of a hit antibody, five VH residues and two VL residues are identified as suboptimal, and possible substitutions are designed. The native and designed amino acid suggestions are schematically represented here as lower-case letters. The combinatorial table of designed and native variants is shown. The variants designated as VH-1.0 and VL-1.0 represent the original parental sequence at each optimization position. The three positions shaded in green are covarying residues. For these residues, it is not necessary to sample all combinatorial possibilities. (b) Structural analysis using a homology model of an asparagine residue implicated as a potential deamidation site (left panel) and substitution of the same position by a serine is shown. (c) A chromatogram of a forced deamidation stress test is shown. The blue is the untreated control, red is the profile after 3 days at 37°C, and green is the profile after 3 days at 37°C at an elevated pH. The asterisk indicates the appearance of a well-known deamidation site located in the constant region.

When performing hit optimization, typically one or two mutations should be sampled at each suboptimal sequence position. In order to arrive quickly at a sequence that is fully optimized at each position, it is preferable to express and test all combinatorial variants at once, however, this may exceed the throughput of the screening assays if more residues are selected for optimization (right-hand panels, Figure 10.4). An option to prevent a combinatorial explosion, is to test all heavy chain optimization variants only with the parental light chain, and vice versa; once identified, the optimal variable heavy (VH) and variable light (VL) chain can easily be combined in a transient transfection to yield the fully optimized hit. Additionally, one should use phylogenetic, structural, and empirical analytical knowledge to further limit the number of VH or VL variants to be tested. For example, if residues are known to interact structurally and co-vary (green shaded residues in Figure 10.5a), rather than sampling all combinations it is advisable to restrict to only those combinations that are observed in nature and/or structurally compatible according to a homology model. When attempting to discover substitutions that eliminate the risk of a chemical liability such as asparagine deamidation, structural modeling can aid in reducing the number and type of substitution to be empirically assayed. For example, a homology model might reveal that a susceptible asparagine is involved in a structurally important hydrogen bonding network (Figure 10.5b, left panel). Substitution with glutamine, while conservative from a physicochemical perspective, would disrupt this network whereas substitution with a serine can preserve the network – although it is necessary to model waters in order to predict this (Figure 10.5b, right panel), highlighting the importance of expert computational structural review of the homology models and proposed substitutions. Finally, experimental analytics can also be used to prioritize which substitutions to add to the combinatorial set of variants. Figure 10.5c demonstrates that in a 3-day forced degradation test on an asparagine-containing antibody, no deamidation in the V-region is seen. This does not mean that the residue in question should not be optimized eventually, as it may show chemical instabilities when tested in a more stringent assay; however, it does suggest that resource allocation to optimizing this position should be postponed until downstream tests reveal that residue is truly labile. Additionally, if this hit is not selected as the lead candidate, it will not be necessary to optimize this position at all; the 3-day test already confirms that the antibody will be stable enough to give reliable decision-making data in screening and confirmatory assays. Finally, although it is important to reduce the number of optimization variants to be tested to a lower number suitable to the early phase of the program, it is wise to include at least an alanine mutation at any potentially labile positions identified in the CDRs, as these positions have a higher risk to become CQAs if the selected hit becomes the lead candidate. While alanine substitutions are unlikely to become part of the final optimized lead sequence, they do provide a rapid and reliable way of assessing which CDR residues are directly involved in antigen binding – information that is highly valuable in the absence of a co-crystal structure and will enable a rapid final optimization should the hit become the lead candidate.

This classical approach to developability optimization has the advantage that it is available to any drug discovery organization with a screening team, an analytics expert, and an expert protein modeler. However, in the absence of structural modeling expertise, when under time constraints due to competition, or when the initial hits require more extensive optimization, this approach becomes limiting. Therefore, we describe better contemporary and emerging approaches below which hold the promise of enabling more hits to be extensively de-risked, with fewer resources.

10.3 THE CONTEMPORARY APPROACH – ENGINEERED LIBRARIES TOWARD IMPROVED DEVELOPABILITY AN ALTERNATIVE

An alternative to the design of a few specific sequence optimization variants of antigen-specific antibodies obtained from repertoires of naïve, immunized, or synthetic subjects is the generation of display libraries that are by design engineered to produce hits with improved developability, ideally in combination with NGS and (optionally) deep learning (DL) approaches. Historically, antigen-specific sequences from display libraries are identified using fluorescence-activated cell sorting (FACS "panning"), followed by Sanger sequencing of a few selected clones with increased binding signals. Meanwhile, NGS or "deep" sequencing technologies have been established that reveal how entire sequence spaces and variations of antibody libraries recognize antigens.[61–63] Typically, antigen-binding cells are isolated through multiple rounds of display panning against the target antigen and the binding/non-binding populations are subjected to next-generation sequencing. As an advantage, NGS analysis of sequence pools before and after panning can offer new ways of finding rare but highly enriched potent binders and sequence motifs that might not be identified by random picking and Sanger sequencing.[64] The sequence and sequence-activity-relationship (SAR) spaces obtained from NGS can be used to train AI/ML models to predict new sequences with even further improved binding affinities or developability properties (Figure 10.6).

10.3.1 Design of Specific Hit Optimization Display Libraries in Combination with AI/ML Approaches

Instead of designing specific sequence optimization variants as described in Section 10.2, combinatorial hit optimization display libraries can be produced where sequence diversifications are typically introduced in CDR regions.[65]

As an example, Liu et al.[66] generated a synthetic library containing ~10^{10} ranibizumab variants that were randomized in CDR-H3 for three rounds of phage display panning. Based on the frequency distribution and enrichment analysis obtained from NGS analyses of different panning rounds, ML models were trained with a two-stage approach, by first modeling antibody affinity and specificity in terms of sequence enrichment with an ensemble of neural networks and then optimizing it with gradient-based optimization. These models were able to predict new CDR-H3 sequences of ranibizumab that were superior to the sequences in the training data set. In another

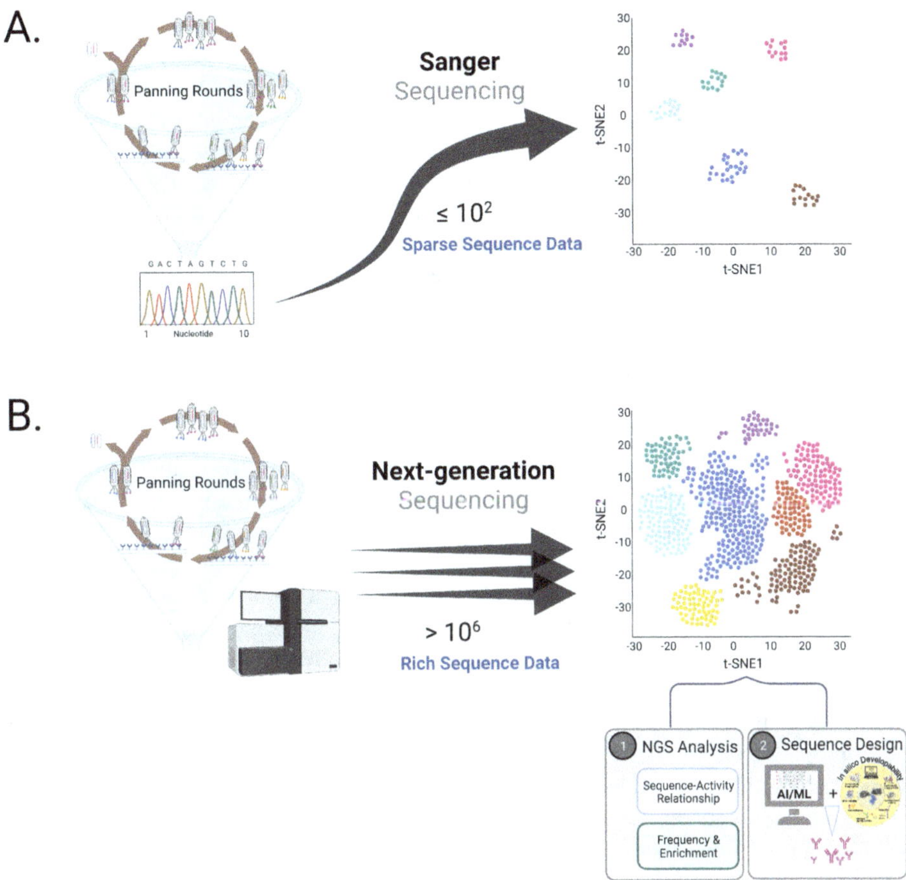

FIGURE 10.6 (a) Process of hit identification from display libraries based on multiple rounds of biopanning and random clone selection for Sanger sequencing. Typically, only a low number of clones can be sequenced, providing the risk that potent sequences with (more) promising sequences or even sequence clusters with favorable developability properties will be missed. (b) NGS analysis of sequence pools before and after panning allows assessment of entire sequence spaces of potent binders. Using SAR, frequency, and enrichment analyses in combination with *in silico* developability predictions, it is possible to rationally identify potent binders with a favorable developability profile. In addition, the sequence pools can be subjected to AI/ML approaches to predict new sequences within even further improved potency and developability properties.

study, Saka et al.[67] employed deep generative models based on NGS-derived sequences from different panning rounds of a combinatorial library diversified in CDR-H1, -H2 and -H3, and FR1 of a kynurenine binding antibody. The affinities of newly designed sequences were over 1,800-fold higher than for the parental clone.

In a further study, Mason et al.[68] performed single-site deep mutational scanning within CDR-H3 of trastuzumab and subjected the antigen-binding vs non-binding samples to NGS to calculate enrichment scores and guide the rational design of a combinatorial mutagenesis library of 5×10^4 variants. This library was screened for specificity to the antigen HER2, again followed by next-generation sequencing. All binding and non-binding variants were used to train deep neural networks that were then applied to design and rank a set of 10^8 virtual sequences. To explicitly account for the aspect of developability, the set of predicted binders was then subjected to sequence-based *in silico* filtering steps to optimize for developability parameters such as viscosity, clearance, solubility, and immunogenicity, resulting in nearly 8,000 antibody sequence variants predicted to have more optimal properties than the starting trastuzumab sequence. Further biophysical characterization of top sequences revealed antibodies with comparable or better properties than trastuzumab for expression and thermal stability, and one candidate that was substantially de-risked for immunogenicity.

Whereas Mason et al. applied *in silico* developability scoring to identify binders with favorable developability properties, Makowski et al. very recently presented an approach where a CDR combinatorial mutagenesis library of ~10^7 emibetuzumab variants was sorted for high and low levels of affinity and non-specific binding to two polyspecificity reagents.[69] Different ML models were then trained on the NGS-enriched libraries for both affinity and (non-)specific binding. In agreement with previous work, the authors demonstrated a high correlation between high-affinity and non-specific binding of emibetuzumab variants. The model predictions were used to identify those antibody mutants in the library that maximize antibody affinity to different extents while minimizing trade-offs due to reduced non-specific binding. Additional models, that used Unified Representation (UniRep) or physicochemical features as antibody descriptors, were built and then applied to design new variants into a novel mutational space that was not covered by the mutations of the training data set.

Production and experimental profiling indeed revealed new sequences with even greater improvements in affinity and specificity than was possible in the experimentally sorted libraries.

The number of reports of such AI/ML approaches in combination with NGS and biopanning is rapidly increasing,[70,71] demonstrating their general potential to transform the biologics drug discovery process. In conclusion, these studies demonstrate how AI/ML models can be built based on sequence enrichments that are observed in antibody libraries through different rounds of biopanning. As demonstrated by Makowski et al., such models can be trained and used for the co-optimization of multiple optimization parameters, such as antigen binding and non-specificity. Of course, these concepts can also be further generalized toward additional optimization parameters, e.g., to optimize antibodies for cross-reactivity against the antigen of relevant pharmacological or toxicological species. Further developability parameters can be considered already in the design of the combinatorial sequence space, for example by excluding sequence combinations that are predicted to show unfavorable predicted physicochemical (e.g., liabilities, PTMs, hydrophobicity, viscosity, charged patches, etc.) or immunogenic (low human-likeness, MHC-II binding) properties. Alternatively, or in addition, such *in silico* predictions can be used to post-filter the ML-predicted variants to select sequences with the best overall affinity and developability profile, as demonstrated by Mason et al.

10.3.2 Design of Diverse Hit Identification Libraries

Different antibody library design approaches toward improved developability have been excellently summarized in the last edition of "Current advances in biopharmaceutical informatics"[33] and in publications cited therein. Here, we briefly summarize the different strategies for the design of diverse libraries for hit identification, and in addition, highlight some very new advances in this field.

Immune and naïve libraries. Libraries derived from antigen-exposed or naïve humans or humanized animals are based entirely on naturally occurring sequence diversity and are generally considered to bias obtained sequence spaces toward high expression and low immunogenicity. However, there is no need or driving force for the immune system to construct antibody sequences toward developability properties that are relevant to the pharmaceutical industry, such as long-term physical and chemical stability. Therefore, antibody libraries obtained from naïve libraries often need further optimization toward required biophysical and chemical stability properties.

Synthetic and semi-synthetic libraries. Synthetic or semi-synthetic libraries combine natural diversity for certain aspects of the library with *in silico* design, often with regard to further improved developability properties,[72] such as high thermal stability, reduced aggregation propensity, and low occurrence of PTM or chemical liability sites. However, synthetic diversity may create artificial complementarity-determining region (CDR) sequences that fold poorly, since the design strategy often includes the combinatorial enumeration of amino acid mutations in specific positions of otherwise fixed antibody scaffolds based on positional frequency analysis (PFA). This procedure, however, ignores how residue types interact to form stabilizing interactions such as hydrophobic, hydrogen, or ionic bonds, potentially resulting in many library members that fold and express poorly.

One recent approach for a rational design of a semi-synthetic library that explicitly considers the developability and functional compatibility of antibody framework and CDR regions was described by Teixeira et al.[73] In their library design, HCDR3s were amplified directly from B cells from ten healthy adult human donors and grafted onto four paired human frameworks. These were derived from a diverse panel of well-behaved ("developable") clinical antibodies, based on known biophysical characteristics[23] and at the same time cover different germline families and thereby assure structural and sequence diversity in the library to improve the ability to select binders against different antigens. Finally, to optimize for developable sequences encoding CDRs able to express well within the chosen scaffolds, natural human CDR-L1–3s and CDR-H1–2s as found in a human NGS dataset were first purged of defined sequence motifs related to chemical instability, PTMs, polyreactivity, and surface hydrophobic/aromatic patches. In contrast to using degenerate oligonucleotides, these sequences were produced using oligonucleotide array-based synthesis, thereby reducing the combinatorial library space and increasing the likelihood of proper folding due to inherent compatibility of the CDRs within the variable light or heavy chain. In the next step, these CDRs were filtered by yeast display as single CDR libraries as a further criterion to eliminate sequences negatively impacting expression, folding, and display. Finally, all CDRs were combinatorially assembled into a single-chain variable (scFv) format.

The general performance of this library approach was successfully demonstrated by the discovery of a large number of unique, highly developable antibodies against four clinically relevant targets with affinities in the sub-nanomolar to low nanomolar range.

Very recently, another novel library approach for the high-throughput *de novo* identification of humanized variable heavy domain of heavy chain antibodies (VHHs) following camelid immunization was implemented.[74] For this, VHH-derived CDR3 regions obtained from a llama (*Lama* glama), immunized against a specific target, are grafted onto a humanized VHH backbone library comprising sequence-diversified CDR1 and CDR2 regions similar to natural immunized and naïve antibody repertoires.

Importantly, these CDRs were tailored toward favorable *in silico* developability properties, by considering human-likeness and excluding potential sequence liabilities and predicted immunogenic motifs. Target-specific humanized VHHs against this specific target were readily obtained by yeast surface display. By exploiting this approach, high-affinity VHHs with an optimized potency and developability profile can be generated that do not require any further sequence optimization. In a further study, the screening pools obtained from different panning rounds were submitted to NGS-derived enrichment analyses.[75] For four highly enriched clusters, deep generative models were trained based on the sequences obtained after panning and used for the sampling of new sequences. Top-ranked sequences were subjected to sequence- and structure-based *in silico* developability assessment to select a set of <10 sequences per cluster for synthesis. As demonstrated by binding measurements and profiling in screening assays for early developability assessment, this procedure might represent a general roadmap for the fast and efficient discovery and design of potent and readily optimized VHH hits with favorable early developability properties directly from screens of immunized lama repertoires.

In future applications, such library spaces might be even further tailored using additional or alternative *in silico* developability descriptors, such as the computed isoelectric point (pI),[48] or other scores that are believed to be predictive for relevant developability properties as described in the next section.

Libraries derived from deep generative models. In recent years, several publications have described the application of deep generative models, a combination of generative models and deep neural networks, for antibody or VHH antibody library design.[76–79] A generative model analyses the distribution of the training data itself, and provides an estimate of how likely a given example (e.g., antibody sequence) is. Deep generative models are neural networks trained to approximate complicated, high-dimensional probability distributions and can be used to create new samples (sequences) from the underlying distribution. Translated to the task of antibody library design, DL offers a route to capture the complex relationships between protein sequence and structure behavior and thereby opens the door to design sequences that are stable and functional. As a concrete example, such models learn from an underlying dataset (e.g., NGS datasets from human donors), which antibody framework and CDR sequences and modifications are (structurally) compatible with each other and will ideally only suggest new sequences that will result in functional antibodies, which will express properly. Since the training data can be derived from huge datasets of human antibody sequences, the DL method implicitly learns the rules for constructing (only) human-like sequences, and the resulting library will ideally retain typical human repertoire characteristics with a low risk of immunogenicity. However, not all predicted sequences will necessarily show favorable properties with regard to further developability parameters, since – as mentioned above – the human immune system does not optimize sequences for all developability properties that are relevant to a pharmaceutical drug product. Thus, the resulting sequences may still contain chemical liability or PTM motifs, or might show non-favorable solution behavior at high concentrations. Options toward the design of more developable libraries might be (i) the "pre-cleaning" of the sequences from the "training" data set based on suited *in silico* developability filters or (ii) the post-filtering of the designed library using the same *in silico* scores. However, such an approach might not be applicable to huge libraries, in particular, in cases where the *in silico* property calculation includes computationally expensive steps, such as 3D-model generation or even conformational sampling. In such cases, it might be feasible to add a further step of so-called "transfer learning" after an initial (purely sequence-based) deep generative learning step. Transfer learning is a continued training of a model with a small subset of sequences with specific desirable characteristics (as for example identified by proper *in silico* filtering using computationally "expensive" methods) and might thereby be suited to bias the properties of the generated antibody sequences toward improved developability properties relevant for the pharmaceutical industry (e.g., physical and chemical stability and immunogenicity).

Such a DL approach, including transfer learning, was described by Amimeur et al.[76]. An initial library that was trained using a generative adversarial network (GAN) to learn the rules of human antibody formation on a set of over 400,000 human antibody sequences. Through transfer learning on small library subsets with desired *in silico* properties, the finally designed library could be biased to generate molecules with key properties of interest such as a defined (*in silico*) pI range, CDR-H3 length, computed negative patch area and lower predicted MHC class II binding.

Whereas these libraries were designed as generic hit identification libraries, applicable for a diverse set of targets, another study reported by Lim et al.[80] demonstrated that DL models can also predict new antibody binders for specific targets when trained on diverse sequences binding against these targets. Sequences that had been obtained from yeast display and NGS analysis against PD-1 and CTLA-4 were classified into binders and nonbinders based on sequence reads and enrichment factors from FACS sorting. The antibody CDR-L3 and CDR-H3 sequences were encoded into antibody images representing their BLOcks SUbstitution Matrix (BLOSUM) substitution scores, which were then used to build and train convolutional neural network models for specific (V gene) sequence clusters to classify binders and nonbinders. The *in silico* generated sequences were similar to the sequences of the training data set, but interestingly also included mutations that were not present in these training sequences. The machine learning (ML) procedure reported in this study was designed to identify potential new binders against the investigated antigens. As a perspective, future studies might combine this approach for library design with the different above-described strategies to tailor such libraries for the prediction of antigen-specific binders with an optimized *in silico* developability profile.

To incorporate these contemporary approaches as a general and essential part of hit discovery toward developable hits, organizations should ensure continuous and close collaboration of protein engineers, deep sequencing experts, data scientists, and ML experts.

10.4 THE EMERGING APPROACH - *DE NOVO* DESIGN OF DEVELOPABLE ANTIBODY THERAPEUTICS

As exciting as the contemporary approach that focuses on engineered libraries with improved developability is, it relies on target-specific datasets (either bespoke or from the public domain), which are then used to train models to optimize libraries. With recent advances in DL technology, neural networks are being applied to tasks such as sequence design, fold recognition, paratope prediction,[81] epitope prediction,[28,82] and structure prediction.[51,83,84] However, the advent of more advanced AI-based approaches for generating new antibodies from scratch, i.e., without relying on the natural immune system or using a template from existing antibodies, has shown immense potential in designing sequences for target antigen binders. These binders are aimed to have properties like high affinity, specificity, stability, and excellent developability profile. This *de novo* antibody design approach bypasses all the time-consuming data generation steps for model training for each discovery campaign for novel and optimized antibody-based therapeutics. Lately, we have seen generative models,[85,86] that approximate the distributions of the data they are trained on. These have garnered interest as a data-driven way to create novel proteins. Unfortunately, most protein-generators create 1D amino acid sequences, making them unsuitable for problems that require structure-based solutions, such as designing protein-protein interfaces like the antigen–antibody complex. In the last few years, two methods shined their capabilities where with known target antigen structure and epitope residues, they either used AI-generated immunoglobulin-like backbones[85] (Figure 10.7 pathway 1a) or used DL to virtually screen the whole structural antibody space for binders[87] (Figure 10.7 pathway 1b).

The first method listed above utilizes variational auto-encoders (VAEs) with a unique loss function for xyz-coordinate generation. This serves as the solution for distance matrix reconstruction and torsion angle inference while preserving the desired invariances. In other words, by constraining and optimizing a protein structure in the VAE latent space, it becomes feasible to specify any desired structural features while the model creates the rest of the molecule. As an example, pathway 1a involves Ig-VAE that can perform constrained loop generation, toward epitope-specific antibody design.[85] After the backbone construction of this model via constrained optimization in the latent space of a generative model, it becomes a protein sequence design problem where given a protein backbone structure of interest, an amino acid sequence that will fold to this structure.

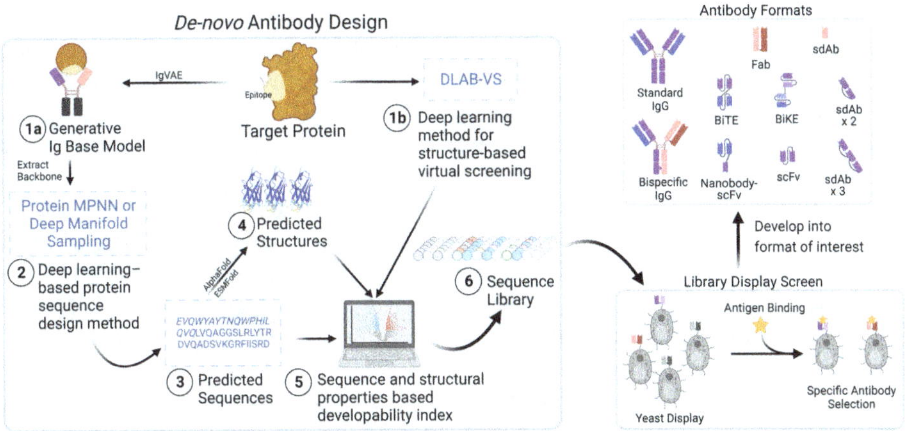

FIGURE 10.7 *De novo* antibody design for known target and epitope. Pathway 1a involves a multistep procedure starting with developing a base structural model of the binder that is complementary to the epitope and is thus expected to bind with high affinity. The backbone of this model is then used by Protein MPNN to generate a set of diverse sequences that can fold into the same backbone while maintaining the affinity for the target. Alphafold2 can be used to fold these predicted sequences. ESMFold can align the predicted structure for the designed sequence with the original backbone. Pathway 1b, on the other hand, uses DLAB, a deep learning-based virtual screening protocol for finding binders based on predicted binding affinity for the target in question. These pathways merge at step 5 where they are evaluated for their developability profile before progressing to antigen binding-based selection. These binders can then be molded into an antibody format of interest.

Historically, physics-based methods such as Rosetta treat sequence design as an energy optimization problem, searching for the combination of amino acid identities and conformations that have the lowest energy for a given input structure.[54] In the past few years, the evolutionary relationship has provided an alternative where the constraints on protein structure are derived from bioinformatics analysis of the evolutionary history of proteins, homology to known structures, and pairwise evolutionary correlations.[54,88] As we know, the protein sequence space is large, discrete, and sparsely functional, where only a small fraction of sequences may fold into stable structural conformations. Adding on to the complexity, immunoglobulins are a class of proteins that follow different evolutionary pressures as compared to other structures in the Protein Data Bank (PDB) and thus re-training on the multiple sequence alignment (MSA) for immunoglobulins is required to predict the structures of CDR loops.[79] Furthermore, the intrinsic flexibility in CDR loops is another reason for this class of proteins to be challenging for automated and efficient exploration of *de novo* design space.

However, recently, autoregressive and non-autoregressive generative model-based approaches have been gaining the limelight. One such example is "deep manifold sampling," which can be used for accelerating function-guided protein design.[89] By combining a sequence denoising autoencoder (DAE) with a function classifier trained on

0.5 M sequences with known function annotations from the Swiss-Prot database, such a sampler can generate diverse sequences of variable length with desired functions. Parallel to this non-autoregressive method, an autoregressive language model Protein Modified Probabilistic Neural Network (MPNN),[90,91] can also be used to generate all possible candidate amino acid sequences given monomeric protein backbones without the need for compute-intensive explicit consideration of side chain rotameric states (Figure 10.7 pathway 1a). These sequences can be folded into respective structural folds using advanced methods like AlphaFold2[88] and ESMFold.[92] Even though these methods are developed for all proteins, these have the potential to be applied to immunoglobulins with ease.

An alternative to abovementioned multistep protocol (pathway 1a) a more concise method, as shown in pathway 1b (Figure 10.7), uses a virtual screening approach to obtain the structure of antibody binders for the target in question. This is done by ranking structural antibody libraries by their predicted binding affinities against the pre-defined target epitope, thereby allowing direct antibody discovery against an epitope region of interest.[87] With great capabilities come great responsibilities. The success of such virtual screening campaigns greatly depends on the prediction accuracy of the docking and scoring algorithms for antibody ranking. As recently reviewed[93] and evaluated,[94] structural prediction of protein-protein complexes using docking and AI-based approaches still leaves room for improvement, especially in antibody–antigen complexes derived from structural antibody homology-based models. Similar to any docking problem, aspects like the accurate prediction of flexibility, the role of water, and further factors might represent potential challenges[95] in this case as well. Therefore, the first success stories about prospective hit identification of new antibodies from AI-based docking approaches of huge libraries are eagerly awaited.

Encouragingly, the first application of deep learning for structure-based virtual screening of antibodies (DLAB) has recently been described in a retrospective study by Schneider et al.[87] DLAB utilizes a convolutional neural network trained on rigid-body docking poses of modeled antibody structures in complex with antigen epitopes to improve the ranking of docking poses from the ZDock algorithm[96] and, in combination with docking scores generated by ZDock, for the prediction of antibody–antigen binding. On datasets of known antibody–antigen pairs, the dataset of known binders could be enriched among the top-ranked scorers, with a significantly higher success rate when docking crystal structure conformations compared to homology models.

Once the sequence and structures of the candidate binders are predicted, their developability needs to be evaluated on several factors like binding affinity, human-likeness, chemical liabilities, PTMs, MHC-II binding, aggregation propensity, thermostability, PK, etc., using suitable *in silico* developability assessment.[29–34,36–39] This focused set of binders can then progress to an antigen-binding screen more confidently and later be developed into an antibody format of interest. The overarching goal for this field is to incorporate all aspects of developability as a learned index at the training stage of the sequence design stage (Stage 2, Figure 10.7) so that all generated sequences can bypass the need for post hoc *in silico* filtering for any liabilities.

10.4.1 Conclusions and Outlook

Historically, antibody discovery and optimization were often pursued following a reactive rather than a proactive approach: Antibody sequences obtained from traditional screening approaches were mainly optimized on binding affinity, human-likeness, and chemical liabilities, followed by developability assessment of the best hits. Since these assessments were typically performed after sequence optimization, they served only to identify and flag suboptimal developability properties in lead molecules. The issues resulting from these suboptimal properties were thrown over the proverbial fence to downstream functions (DMPK, nonclinical safety, and CMC). These functions then attempted to compensate for suboptimal candidate properties through optimization of the downstream process development and dosing regimens, imposing delays in development, increased costs, and finally, a huge risk for the project to achieve approval for First in Human and further clinical studies. Thus, in recent years the pharmaceutical industry has come to the realization that the optimization of antibody-based biologics should be treated as a holistic multiparameter challenge that considers multiple developability aspects (beyond potency and specificity) as defined by the specific target product profile of a project as early design parameters. Prioritizing early optimization for developability or applying emerging options to generate drug candidates de-risked for developability by design, as described in this chapter, should be viewed as investments that will pay dividends in reduced downstream costs and attrition. The contemporary and emerging artificial intelligence and ML (AI/ML) approaches are highly auspicious to transform biologics discovery, using less resources for optimization and promising to deliver truly de-risked candidates with lower attrition rates directly from hit discovery or by *de novo* design. We have to be aware, however, that the general success and impact will critically depend on the general predictivity of the *in silico* predictors. While great progress has been achieved in implementing novel AI algorithms, there remains a gap in the availability of specific relevant experimental data that would allow accurate prediction of the clinical success of antibodies. Whereas there are, for example, large amounts of data from *in vitro* and *in vivo* assays to assess PK properties in preclinical species, the available landscape of antibodies with curated human PK data is currently by far too low to establish and validate robust and truly predictive models for human PK[47]; additionally, the clinical predictivity of *in vitro* assays or even human FcRn transgenic rodent *in vivo* PK assays is not as reliable as the predictivity enjoyed by the equivalent small molecule assays.[97] Furthermore, even the available preclinical data is largely siloed within the "dark matter" of proprietary pharmaceutical company data. Similarly, the amount of experimental viscosity data available to individual pharmaceutical companies from literature or in-house experiments is rather in the range of tens to hundreds instead of thousands or more. Furthermore, viscosity behavior may critically depend on specific assay settings or formulation conditions, thereby increasing the risk that *in silico* models trained on these data are outside the applicability domain of new sequences in specific formulations. Nevertheless, although several "late-stage" parameters cannot yet be accurately predicted, early *in silico* assessment can have a valuable impact in providing an educated rationale for sequence ranking and filtering, triggering relevant and predictive assays, and designing potential backup sequences as early as possible. Such cycles of property prediction and experimental testing will ideally increase the amount of available data and continuously improve model robustness and accuracy. Concomitantly with this continuous improvement of model accuracy, it is essential to ensure each prediction is accompanied by a relevant confidence metric, to guide appropriate use of models. Finally, whereas we have already seen significant progress in property prediction for monospecific antibodies (mostly IgG1s), this progress urgently needs to be extended to *in silico* approaches for developability prediction of different antibody classes (e.g., IgG2, IgG4, and IgM), different multispecific formats, complex fusion proteins, and antibody-drug conjugates as next-generation antibody-based drugs. While the developability data on different biologicals are currently mainly shared through the peer-reviewed literature, we believe that consortia, public–private partnerships, and the establishment of a comprehensive biologics data commons are essential to the future. Through the increasingly open sharing of data the entire industry, as well as patients and payers, will benefit from improved models of developability and the concomitant lower attrition rates.

ACKNOWLEDGMENTS

We thank Jessica Dawson and Navneet Singh for providing representative deamidation data (Figure 10.5c); Lukas Friedrich for valuable discussion on deep learning approaches; Joleen White, David Nannemann, and Sandeep Kumar for suggestions on case studies to illustrate developability challenges and solutions.

REFERENCES

1. Fogel DB. Factors associated with clinical trials that fail and opportunities for improving the likelihood of success: a review. *Contem Clin Trials Commun* 2018; 11:156–64.
2. Hwang TJ, Carpenter D, Lauffenburger JC, Wang B, Franklin JM, Kesselheim AS. Failure of investigational drugs in late-stage clinical development and publication of trial results. *JAMA Int Med* 2016; 176:1826–33.
3. Kola I, Landis J. Can the pharmaceutical industry reduce attrition rates? *Nat Rev Drug Discov* 2004; 3:711–16.
4. Sacks LV, Shamsuddin HH, Yasinskaya YI, Bouri K, Lanthier ML, Sherman RE. Scientific and regulatory reasons for delay and denial of FDA approval of initial applications for new drugs, 2000–2012. *JAMA* 2014; 311:378–84.
5. Adams C, Brantner VV. *New Drug Development: Estimating Entry from Human Clinical Trials* [Internet]. 2003 [cited 30 Dec 2022]; Available from: https://papers.ssrn.com/abstract=428040
6. Ku MS. Recent trends in specialty pharma business model. *J Food Drug Anal* 2015; 23:595–608.

7. Martin KP, Grimaldi C, Grempler R, Hansel S, Kumar S. Trends in industrialization of biotherapeutics: a survey of product characteristics of 89 antibody-based biotherapeutics. *mAbs* 2023; 15:2191301.
8. Bauer J, Mathias S, Kube S, Otte K, Garidel P, Gamer M, Blech M, Fischer S, Karow-Zwick AR. Rational optimization of a monoclonal antibody improves the aggregation propensity and enhances the CMC properties along the entire pharmaceutical process chain. *mAbs* 2020; 12:1787121.
9. Bethea D, Wu S-J, Luo J, Hyun L, Lacy ER, Teplyakov A, Jacobs SA, O'Neil KT, Gilliland GL, Feng Y. Mechanisms of self-association of a human monoclonal antibody CNTO607. *Protein Eng Des Sel* 2012; 25:531–7.
10. Hua F, Comer GM, Stockert L, Jin B, Nowak J, Pleasic-Williams S, Wunderlich D, Cheng J, Beebe JS. Anti-IL21 receptor monoclonal antibody (ATR-107): safety, pharmacokinetics, and pharmacodynamic evaluation in healthy volunteers: a phase I, first-in-human study. *J Clin Pharmacol* 2014; 54:14–22.
11. Kumar S, Roffi K, Tomar DS, Cirelli D, Luksha N, Meyer D, Mitchell J, Allen MJ, Li L. Rational optimization of a monoclonal antibody for simultaneous improvements in its solution properties and biological activity. *Protein Eng Design Selec* 2018; 31:313–25.
12. Ridker PM, Tardif J-C, Amarenco P, Duggan W, Glynn RJ, Jukema JW, Kastelein JJP, Kim AM, Koenig W, Nissen S, et al. Lipid-reduction variability and antidrug-antibody formation with bococizumab. *N Engl J Med* 2017; 376:1517–26.
13. Wu S-J, Luo J, O'Neil KT, Kang J, Lacy ER, Canziani G, Baker A, Huang M, Tang QM, Raju TS, et al. Structure-based engineering of a monoclonal antibody for improved solubility. *Protein Eng Des Sel* 2010; 23:643–51.
14. Sun A, Benet LZ. Late-stage failures of monoclonal antibody drugs: a retrospective case study analysis. *PHA* 2020; 105:145–63.
15. *Pfizer Discontinues Global Development of Bococizumab, Its Investigational PCSK9* Inhibitor [Internet]. 2016; Available from: https://www.pfizer.com/news/press- release/press-release- detail/pfizer_discontinues_global_development_of_bococizumab_its_investigational_pc sk9_inhibitor
16. Ballantyne CM, Neutel J, Cropp A, Duggan W, Wang EQ, Plowchalk D, Sweeney K, Kaila N, Vincent J, Bays H. Results of bococizumab, a monoclonal antibody against proprotein convertase subtilisin/kexin type 9, from a randomized, placebo- controlled, dose-ranging study in statin-treated subjects with hypercholesterolemia. *Am J Cardiol* 2015; 115:1212–21.
17. Ridker PM, Revkin J, Amarenco P, Brunell R, Curto M, Civeira F, Flather M, Glynn RJ, Gregoire J, Jukema JW, et al. Cardiovascular efficacy and safety of bococizumab in high-risk patients. *N Engl J Med* 2017; 376:1527–39.
18. Liu CY, Ahonen CL, Brown ME, Zhou L, Welin M, Krauland EM, Pejchal R, Widboom PF, Battles MB. Structure-based engineering of a novel CD3ε-targeting antibody for reduced polyreactivity. *mAbs* 2023; 15:2189974.
19. Wagner KR, Fleckenstein JL, Amato AA, Barohn RJ, Bushby K, Escolar DM, Flanigan KM, Pestronk A, Tawil R, Wolfe GI, et al. A phase I/IItrial of MYO-029 in adult subjects with muscular dystrophy. *Ann Neurol* 2008; 63:561–71.
20. Singh P, Rong H, Gordi T, Bosley J, Bhattacharya I. Translational Pharmacokinetic/pharmacodynamic analysis of MYO-029 antibody for muscular dystrophy. *Clin Transl Sci* 2016; 9:302–10.
21. Nichols P, Li L, Kumar S, Buck PM, Singh SK, Goswami S, Balthazor B, Conley TR, Sek D, Allen MJ. Rational design of viscosity reducing mutants of a monoclonal antibody: hydrophobic versus electrostatic inter-molecular interactions. *mAbs* 2015; 7:212–30.
22. Bailly M, Mieczkowski C, Juan V, Metwally E, Tomazela D, Baker J, Uchida M, Kofman E, Raoufi F, Motlagh S, et al. Predicting antibody developability profiles through early stage discovery screening. *mAbs* 2020; 12:1743053.

23. Jain T, Sun T, Durand S, Hall A, Houston NR, Nett JH, Sharkey B, Bobrowicz B, Caffry I, Yu Y, et al. Biophysical properties of the clinical-stage antibody landscape. *Proc Natl Acad Sci U S A* 2017; 114:944–9.
24. Jarasch A, Koll H, Regula JT, Bader M, Papadimitriou A, Kettenberger H. Developability assessment during the selection of novel therapeutic antibodies. *J Pharm Sci* 2015; 104:1885–98.
25. Kingsbury JS, Saini A, Auclair SM, Fu L, Lantz MM, Halloran KT, Calero-Rubio C, Schwenger W, Airiau CY, Zhang J, et al. A single molecular descriptor to predict solution behavior of therapeutic antibodies. *Sci Adv* 2020; 6:eabb0372.
26. Kraft TE, Richter WF, Emrich T, Knaupp A, Schuster M, Wolfert A, Kettenberger H. Heparin chromatography as an in vitro predictor for antibody clearance rate through pinocytosis. *mAbs* 2020; 12:1683432.
27. Xu Y, Wang D, Mason B, Rossomando T, Li N, Liu D, Cheung JK, Xu W, Raghava S, Katiyar A, et al. Structure, heterogeneity and developability assessment of therapeutic antibodies. *mAbs* 2018; 11:239–64.
28. Yang X, Xu W, Dukleska S, Benchaar S, Mengisen S, Antochshuk V, Cheung J, Mann L, Babadjanova Z, Rowand J, et al. Developability studies before initiation of process development. *mAbs* 2013; 5:787–94.
29. Ahmed L, Gupta P, Martin KP, Scheer JM, Nixon AE, Kumar S. Intrinsic physicochemical profile of marketed antibody-based biotherapeutics. *Proc Natl Acad Sci* U S A 2021; 118:e2020577118.
30. Blanco MA. Computational models for studying physical instabilities in high concentration biotherapeutic formulations. *mAbs* 2022; 14:2044744.
31. Chennamsetty N, Voynov V, Kayser V, Helk B, Trout BL. Prediction of aggregation prone regions of therapeutic proteins. *J Phys Chem B* 2010; 114:6614–24.
32. Evers A, Malhotra S, Bolick W-G, Najafian A, Borisovska M, Warszawski S, Fomekong Nanfack Y, Kuhn D, Rippmann F, Crespo A, et al. SUMO: In Silico Sequence Assessment Using Multiple Optimization Parameters. In: Zielonka S, Krah S, editors. *Genotype Phenotype Coupling*. New York: Springer US; 2023 [cited *2023 Jul 13*]. page 383–98. Available from: https://link.springer.com/10.1007/978-1-0716-3279-6_22
33. Khetan R, Curtis R, Deane CM, Hadsund JT, Kar U, Krawczyk K, Kuroda D, Robinson SA, Sormanni P, Tsumoto K, et al. Current advances in biopharmaceutical informatics: guidelines, impact and challenges in the computational developability assessment of antibody therapeutics. *mAbs* 2022; 14:2020082.
34. Lauer TM, Agrawal NJ, Chennamsetty N, Egodage K, Helk B, Trout BL. Developability index: a rapid in silico tool for the screening of antibody aggregation propensity. *J Pharm Sci* 2012; 101:102–15.
35. Licari G, Martin KP, Crames M, Mozdzierz J, Marlow MS, Karow-Zwick AR, Kumar S, Bauer J. Embedding dynamics in intrinsic physicochemical profiles of market-stage antibody-based biotherapeutics. *Mol Pharm* 2023; 20:1096–111.
36. Negron C, Fang J, McPherson MJ, Stine WB, McCluskey AJ. Separating clinical antibodies from repertoire antibodies, a path to in silico developability assessment. *mAbs* 2022; 14:2080628.
37. Raybould MIJ, Marks C, Krawczyk K, Taddese B, Nowak J, Lewis AP, Bujotzek A, Shi J, Deane CM. Five computational developability guidelines for therapeutic antibody profiling. *Proc Natl Acad Sci U S A* 2019; 116:4025–30.
38. Sankar K, Krystek Jr SR, Carl SM, Day T, Maier JKX. AggScore: prediction of aggregation-prone regions in proteins based on the distribution of surface patches. *Proteins Struct Funct Bioinform* 2018; 86:1147–56.
39. Sormanni P, Aprile FA, Vendruscolo M. The CamSol method of rational design of protein mutants with enhanced solubility. *J Mol Biol* 2015; 427:478–90.

40. Fernández-Quintero ML, Ljungars A, Waibl F, Greiff V, Andersen JT, Gjølberg TT, Jenkins TP, Voldborg BG, Grav LM, Kumar S, et al. Assessing developability early in the discovery process for novel biologics. *mAbs* 2023; 15:2171248.
41. Mieczkowski C, Zhang X, Lee D, Nguyen K, Lv W, Wang Y, Zhang Y, Way J, Gries J- M. Blueprint for antibody biologics developability. *mAbs* 2023; 15:2185924.
42. Svilenov HL, Arosio P, Menzen T, Tessier P, Sormanni P. Approaches to expand the conventional toolbox for discovery and selection of antibodies with drug-like physicochemical properties. *mAbs* 2023; 15:2164459.
43. Zhang W, Wang H, Feng N, Li Y, Gu J, Wang Z. Developability assessment at early- stage discovery to enable development of antibody-derived therapeutics. *Antibody Therap* 2023; 6:13–29.
44. Hwang WYK, Foote J. Immunogenicity of engineered antibodies. *Methods* 2005; 36:3–10.
45. Avery LB, Wade J, Wang M, Tam A, King A, Piche-Nicholas N, Kavosi MS, Penn S, Cirelli D, Kurz JC, et al. Establishing in vitro in vivo correlations to screen monoclonal antibodies for physicochemical properties related to favorable human pharmacokinetics. *mAbs* 2018; 10:244–55.
46. Bumbaca Yadav D, Sharma VK, Boswell CA, Hotzel I, Tesar D, Shang Y, Ying Y, Fischer SK, Grogan JL, Chiang EY, et al. Evaluating the use of antibody variable region (Fv) charge as a risk assessment tool for predicting typical cynomolgus monkey pharmacokinetics. *J Biol Chem* 2015; 290:29732–41.
47. Grinshpun B, Thorsteinson N, Pereira JN, Rippmann F, Nannemann D, Sood VD, Fomekong Nanfack Y. Identifying biophysical assays and in silico properties that enrich for slow clearance in clinical-stage therapeutic antibodies. *mAbs* 2021; 13:1932230.
48. Gupta P, Makowski EK, Kumar S, Zhang Y, Scheer JM, Tessier PM. Antibodies with weakly basic isoelectric points minimize trade-offs between formulation and physiological colloidal properties. *Mol Pharm* 2022; 19:775–87.
49. Homer, Fitzgerald R. *The Iliad*. Oxford: Oxford University Press; 1998.
50. Kitchen DB, Decornez HY. Computational Techniques to Support Hit Triage. In: *Small Molecule Medicinal Chemistry*. John Wiley & Sons, Ltd; 2015 [cited 2022 Dec 27]. page 189–220. Available from: https://onlinelibrary.wiley.com/doi/abs/10.1002/9781118771723.ch7
51. Abanades B, Wong WK, Boyles F, Georges G, Bujotzek A, Deane CM. *ImmuneBuilder: Deep-Learning Models for Predicting the Structures of Immune Proteins*. Commun Biol 2023; 29: 575–582.
52. Chemical Computing Group (CCG). Research [Internet]. [cited 2022 Aug 25]; Available from: https://www.chemcomp.com/Research-Citing_MOE.htm
53. Schrödinger Release 2021–3. *BioLuminate*. New York: Schrödinger, LLC. 2021 [cited 2022 Aug 25]; Available from: https://www.schrodinger.com/products/bioluminate
54. Leaver-Fay A, Tyka M, Lewis SM, Lange OF, Thompson J, Jacak R, Kaufman K, Renfrew PD, Smith CA, Sheffler W, et al. ROSETTA3: an object-oriented software suite for the simulation and design of macromolecules. *Methods Enzymol* 2011; 487:545–74.
55. Leman JK, Weitzner BD, Lewis SM, Adolf-Bryfogle J, Alam N, Alford RF, Aprahamian M, Baker D, Barlow KA, Barth P, et al. Macromolecular modeling and design in Rosetta: recent methods and frameworks. *Nat Methods* 2020; 17:665–80.
56. Safdari Y, Farajnia S, Asgharzadeh M, Khalili M. Antibody humanization methods – a review and update. *Biotechnol Genetic Eng Rev* 2013; 29:175–86.
57. Vita R, Mahajan S, Overton JA, Dhanda SK, Martini S, Cantrell JR, Wheeler DK, Sette A, Peters B. The Immune Epitope Database (IEDB): 2018 update. *Nucleic Acids Res* 2019; 47:D339–43.

58. Datta-Mannan A, Estwick S, Zhou C, Choi H, Douglass NE, Witcher DR, Lu J, Beidler C, Millican R. Influence of physiochemical properties on the subcutaneous absorption and bioavailability of monoclonal antibodies. *mAbs* 2020; 12:1770028.
59. Grevys A, Frick R, Mester S, Flem-Karlsen K, Nilsen J, Foss S, Sand KMK, Emrich T, Fischer JAA, Greiff V, et al. Antibody variable sequences have a pronounced effect on cellular transport and plasma half-life. *iScience* 2022; 25:103746.
60. Schoch A, Kettenberger H, Mundigl O, Winter G, Engert J, Heinrich J, Emrich T. Charge-mediated influence of the antibody variable domain on FcRn-dependent pharmacokinetics. *Proc Natl Acad Sci U S A* 2015; 112:5997– 6002.
61. Hu D, Hu S, Wan W, Xu M, Du R, Zhao W, Gao X, Liu J, Liu H, Hong J. Effective optimization of antibody affinity by phage display integrated with high-throughput DNA synthesis and sequencing technologies. *PLoS One* 2015; 10:e0129125.
62. Larman HB, Jing Xu G, Pavlova NN, Elledge SJ. Construction of a rationally designed antibody platform for sequencing-assisted selection. *Proc Natl Acad Sci U S A* 2012; 109:18523–8.
63. Mathonet P, Ullman CG. The application of next generation sequencing to the understanding of antibody repertoires. *Front Immunol* 2013; 4:265.
64. Barreto K, Maruthachalam BV, Hill W, Hogan D, Sutherland AR, Kusalik A, Fonge H, DeCoteau JF, Geyer CR. Next-generation sequencing-guided identification and reconstruction of antibody CDR combinations from phage selection outputs. *Nucleic Acids Res* 2019; 47:e50.
65. Rajpal A, Beyaz N, Haber L, Cappuccilli G, Yee H, Bhatt RR, Takeuchi T, Lerner RA, Crea R. A general method for greatly improving the affinity of antibodies by using combinatorial libraries. *Proc Nat Acad Sci U S A* 2005; 102:8466–71.
66. Liu G, Zeng H, Mueller J, Carter B, Wang Z, Schilz J, Horny G, Birnbaum ME, Ewert S, Gifford DK. Antibody complementarity determining region design using high- capacity machine learning. *Bioinformatics* 2020; 36:2126–33.
67. Saka K, Kakuzaki T, Metsugi S, Kashiwagi D, Yoshida K, Wada M, Tsunoda H, Teramoto R. Antibody design using LSTM based deep generative model from phage display library for affinity maturation. *Sci Rep* 2021; 11:5852.
68. Mason DM, Friedensohn S, Weber CR, Jordi C, Wagner B, Meng SM, Ehling RA, Bonati L, Dahinden J, Gainza P, et al. Optimization of therapeutic antibodies by predicting antigen specificity from antibody sequence via deep learning. *Nat Biomed Eng* 2021; 5:600–12.
69. Makowski EK, Kinnunen PC, Huang J, Wu L, Smith MD, Wang T, Desai AA, Streu CN, Zhang Y, Zupancic JM, et al. Co-optimization of therapeutic antibody affinity and specificity using machine learning models that generalize to novel mutational space. *Nat Commun* 2022; 13:3788.
70. Hu R, Fu L, Chen Y, Chen J, Qiao Y, Si T. Protein engineering via Bayesian optimization-guided evolutionary algorithm and robotic experiments. *Brief Bioinform* 2023; 24:bbac570.
71. Parkinson J, Hard R, Wang W. The RESP AI model accelerates the identification of tight-binding antibodies. *Nat Commun* 2023; 14:454.
72. Tiller T, Schuster I, Deppe D, Siegers K, Strohner R, Herrmann T, Berenguer M, Poujol D, Stehle J, Stark Y, et al. A fully synthetic human Fab antibody library based on fixed VH/VL framework pairings with favorable biophysical properties. *mAbs* 2013; 5:445–70.
73. Azevedo Reis Teixeira A, Erasmus MF, D'Angelo S, Naranjo L, Ferrara F, Leal-Lopes C, Durrant O, Galmiche C, Morelli A, Scott-Tucker A, et al. Drug-like antibodies with high affinity, diversity and developability directly from next-generation antibody libraries. *mAbs* 13:1980942.

74. Arras P, Yoo HB, Pekar L, Schröter C, Clarke T, Krah S, Klewinghaus D, Siegmund V, Evers A, Zielonka S. A library approach for the de novo high-throughput isolation of humanized VHH domains with favorable developability properties following camelid immunization. *mAbs* 2023; 15:2261149.
75. Arras P, Yoo HB, Pekar L, Clarke T, Friedrich L, Schröter C, Schanz J, Tonillo J, Siegmund V, Doerner A, et al. AI/ML combined with next-generation sequencing of VHH immune repertoires enables the rapid identification of de novo humanized and sequence-optimized single domain antibodies: a prospective case study. *Front Mol Biosci* 2023; 10:1249247.
76. Amimeur T, Shaver JM, Ketchem RR, Taylor JA, Clark RH, Smith J, Citters DV, Siska CC, Smidt P, Sprague M, et al. *Designing Feature-Controlled Humanoid Antibody Discovery Libraries Using Generative Adversarial Networks* [Internet]. 2020 [cited 28 Sep 2022]; Available from: https://www.biorxiv.org/content/10.1101/2020.04.12.024844v2
77. Melnyk I, Das P, Chenthamarakshan V, Lozano A. *Benchmarking Deep Generative Models for Diverse Antibody Sequence Design* [Internet]. 2021 [cited 2022 Oct 20]; Available from: https://arxiv.org/abs/2111.06801
78. Shin J-E, Riesselman AJ, Kollasch AW, McMahon C, Simon E, Sander C, Manglik A, Kruse AC, Marks DS. Protein design and variant prediction using autoregressive generative models. *Nat Commun* 2021; 12:2403.
79. Shuai RW, Ruffolo JA, Gray JJ. IgLM: Infilling language modeling for antibody sequence design. *Cell Syst.* 2023; 14: 979–989.
80. Lim YW, Adler AS, Johnson DS. Predicting antibody binders and generating synthetic antibodies using deep learning. *mAbs* 2022; 14:2069075.
81. Chinery L, Wahome N, Moal I, Deane CM. Paragraph—antibody paratope prediction using graph neural networks with minimal feature vectors. *Bioinformatics* 2023; 39:btac732.
82. Wong WK, Robinson SA, Bujotzek A, Georges G, Lewis AP, Shi J, Snowden J, Taddese B, Deane CM. Ab-Ligity: identifying sequence-dissimilar antibodies that bind to the same epitope. *mAbs* 2021; 13:1873478.
83. Abanades B, Georges G, Bujotzek A, Deane CM. ABlooper: fast accurate antibody CDR loop structure prediction with accuracy estimation. *Bioinformatics* 2022; 38:1877– 80.
84. Leem J, Dunbar J, Georges G, Shi J, Deane CM. ABodyBuilder: automated antibody structure prediction with data-driven accuracy estimation. *mAbs* 2016; 8:1259–68.
85. Eguchi RR, Choe CA, Huang P-S. Ig-VAE: generative modeling of protein structure by direct 3D coordinate generation. *PLOS Comput Biol* 2022; 18:e1010271.
86. Riesselman AJ, Ingraham JB, Marks DS. Deep generative models of genetic variation capture the effects of mutations. *Nat Methods* 2018; 15:816–22.
87. Schneider C, Buchanan A, Taddese B, Deane CM. DLAB-Deep learning methods for structure-based virtual screening of antibodies. *Bioinformatics* 2021; 38:btab660.
88. Jumper J, Evans R, Pritzel A, Green T, Figurnov M, Ronneberger O, Tunyasuvunakool K, Bates R, Žídek A, Potapenko A, et al. Highly accurate protein structure prediction with AlphaFold. *Nature* 2021; 596:583–9.
89. Gligorijević V, Berenberg D, Ra S, Watkins A, Kelow S, Cho K, Bonneau R. Function-guided protein design by deep manifold sampling. *Bioinformatics* 2021. Available from: https://biorxiv.org/lookup/doi/10.1101/2021.12.22.473759
90. Dauparas J, Anishchenko I, Bennett N, Bai H, Ragotte RJ, Milles LF, Wicky BIM, Courbet A, de Haas RJ, Bethel N, et al. Robust deep learning–based protein sequence design using ProteinMPNN. *Science* 2022; 378:49–56.
91. Madani A, McCann B, Naik N, Keskar NS, Anand N, Eguchi RR, Huang P-S, Socher R. *ProGen: Language Modeling for Protein Generation* [Internet]. 2020 [cited 2022 Oct 26]; Available from: https://arxiv.org/abs/2004.03497

92. Lin Z, Akin H, Rao R, Hie B, Zhu Z, Lu W, Smetanin N, Verkuil R, Kabeli O, Shmueli Y, et al. *Evolutionary-Scale Prediction of Atomic Level Protein Structure with a Language Model* [Internet]. 2022 [cited 31 Dec 2022]; Available from: https://www.biorxiv.org/content/10.1101/2022.07.20.500902v3
93. Tsuchiya Y, Yamamori Y, Tomii K. Protein–protein interaction prediction methods: from docking-based to AI-based approaches. *Biophys Rev* 2022; 14:1341–8.
94. Bryant P, Pozzati G, Elofsson A. Improved prediction of protein-protein interactions using AlphaFold2. *Nat Commun* 2022; 13:1265.
95. Pantsar T, Poso A. Binding affinity via docking: fact and fiction. *Molecules* 2018; 23:1899.
96. Pierce BG, Hourai Y, Weng Z. Accelerating protein docking in ZDOCK using an advanced 3D convolution library. *PLoS One* 2011; 6:e24657.
97. Sodhi JK, Benet LZ. Successful and unsuccessful prediction of human hepatic clearance for lead optimization. *J Med Chem* 2021; 64:3546–59.

11 Use of Systems Biology Approaches toward Target Discovery, Validation, and Drug Development

Madhuresh Sumit and Venkata Gayatri Dhara

11.1 INTRODUCTION (INTRODUCTION TO SYSTEMS BIOLOGY AND ITS SCOPE IN DRUG DISCOVERY AND DEVELOPMENT)

Systems biology is a holistic approach comprised of computational and experimental tools employed to understand a complex biological system at multiple scales and dimensions (Kitano 2002). These tools are leveraged to capture the spatial and temporal complexity of a biological system. Dissecting such complexity through a systems approach can help understand non-linear and non-intuitive relationships among various components of a biological system as well as their emergent properties. It also provides insights into potential ways of modulating the system from within and from outside.

The systems biology approach allows for integrated model development, testing of multiple hypotheses, simulation of biological systems under various constraints,

DOI: 10.1201/9781003300311-11

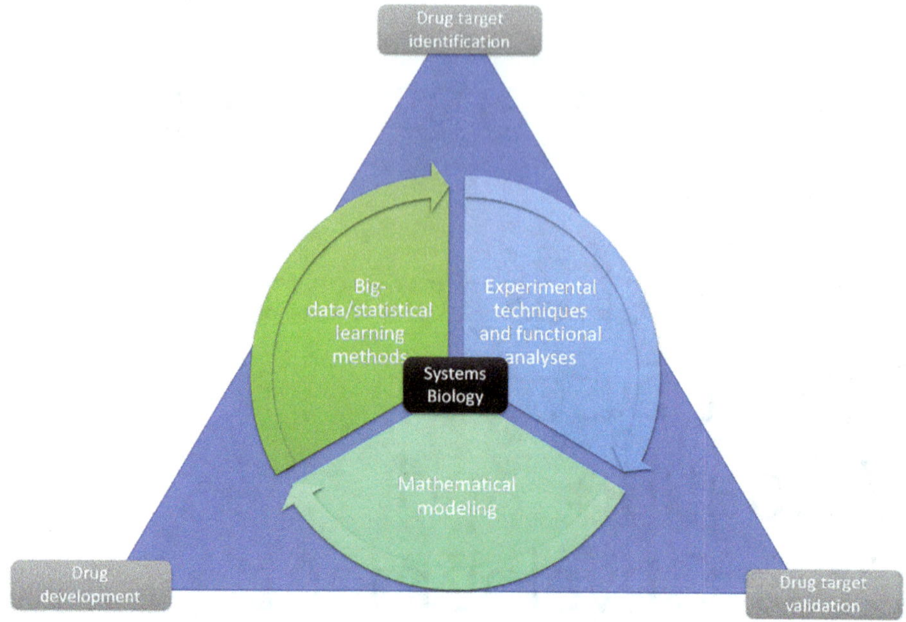

FIGURE 11.1 Overview of systems biology in biopharmaceutical R&D.

and mechanistic interpretation of the data at hand. Consequently, hypothesis testing is significantly informed and more predictive as compared to the hit-and-trial method and thus optimizes drug discovery and development processes (Butcher et al. 2004). Figure 11.1 shows a schematic of the components of the systems biology approach. The approach allows generating and *in silico* testing of biologically relevant hypotheses. This ability of systems biology has a two-pronged scope within biopharmaceutical informatics. On one hand, it enables the discovery of new biomarkers for diseases and identifies potential drug targets, by identifying cause-and-effect pathways potentially involved in a disease. On the other hand, it can provide insights into ways of improving productivity and product quality in biopharmaceutical development. Thus, it helps reduce development timelines and resource requirements with an enhanced probability of clinical success. Overall, the systems approach is utilized to potentially help reduce the cost of drug discovery and development, bringing novel biotherapeutics to market faster, and increasing their affordability and accessibility at the same time.

11.2 EXPERIMENTAL METHODS FOR SYSTEMS APPROACH IN BIOPHARMACEUTICAL DRUG DISCOVERY AND DEVELOPMENT

The systems biology approach can be classified broadly into experimental methods and computational methods (Figure 11.2). The experimental approach is generally based on

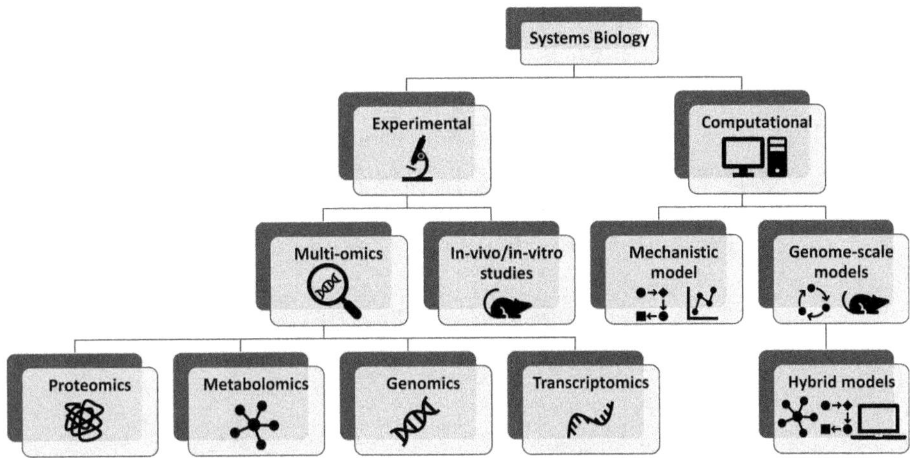

FIGURE 11.2 Systems biology tools utilized toward biopharmaceutical drug discovery and development.

creating a map of the cell with the genes, metabolites, proteins, and their glycoforms. The key components of cellular metabolism include metabolic pathways composed of enzymes and other proteins, and metabolites such as substrates, intermediates as well as end products. Deciphering the cellular metabolism involves identification and characterization of these components within the cell. This can be accomplished through a holistic analysis of its metabolites (metabolome) and the proteins/enzymes involved in those interactions (proteome), respectively. A comprehensive analysis of all the genes encoded (genome) and expressed (transcriptome) provides additional information about whether the cell contains the genetic code to manufacture the enzyme and the maximum level at which the quantity of the enzyme can be produced by the cells. Thus, experimental methods of systems biology rely on investigating the biological system at hand through a holistic analysis of its genome, transcriptome, proteome, glycome, lipidome, and/or metabolome (Piña et al. 2018). Table 11.1 summarizes a few of the major 'omics' approaches, their scope at the molecular level, and corresponding predictive capacity.

The 'omics' methods are often utilized in biopharmaceutical drug discovery and development *in tandem* to gain a multi-dimensional insight into complex biological systems (Hasin et al. 2017). For example, the integration of multi-dimensional omics information from diseased primary tissues can provide insights into mechanisms that underlie the disease development (Kreitmaier et al. 2023). Kreitmaier et al. summarized the insights from multi-omics analysis for several diseases including type 2 diabetes, osteoarthritis, and Alzheimer's disease. Similarly, drug discovery requires screening of several potential targets to narrow down the top candidates that can be most causally related to a disease condition. As an example, Li et al. (2020) developed experimental and bioinformatics pipelines incorporating multi-omics analyses to identify potential targets for ovarian cancer. Such pipelines may include microarray studies, disease model studies, and multi-omics studies among others. The overall goal is to understand how cells work and respond to their environment when external triggers, such as a

TABLE 11.1 Omics tools and techniques employed in biopharmaceutical drug discovery and development

SYSTEM/ OMICS METHODS	*MOLECULE*	*PREDICTIVE CAPACITY*	*METHODOLOGY*	*MOLECULAR SPECIES ANNOTATION*	*TYPICAL OUTPUT*
Transcriptome	mRNA	Actual processes (what a tissue/cell is set to do in its current state)	Microarray, RNA-seq[a]	NA, BLAST	20,000–34,000 transcripts
Proteome	Proteins	Physiological activities (what a tissue/cell is actually doing)	Protein digestion/ MALDI-TOF	Peptide libraries	100–1,000 peptides (counts)
Metabolome	Metabolites	Metabolic status (what the tissue/cell is actually consuming and producing)	GC-MS, LC-MS, NMR	HMDB, YMDB	30–500 metabolites (peak areas)
Lipidome[b]	Lipids	Complements transcriptome, proteome, and metabolome through interconnectedness	GC-MS, LC-MS	Lipid Database, Lipid Maps	100–1,000 lipids (areas)
Glycome	Glycans	Microheterogeneity of glycosylation linked to effector functions/biophysical properties of biotherapeutics	HILIC, LC-MS	SNFG	~300–3,000 glycan species

Source: Adapted from Piña et al. (2018).

[a] Relative fluorescence values in case of microarray, counts in case of RNA-seq.

[b] Lipidome is sometimes considered a subset of metabolome.

diseased state, are introduced to modify their regular function. A cellular-level scrutiny to decipher the functions and response of different pathways can be achieved using these omics technologies developed over the past few decades.

Once the omics data is generated, the data can be utilized to perform statistical and functional analyses in the context of the cell and disease model under investigation. These analyses could be performed at a 'molecular level' or at a broader 'functional level.' Molecular-level analyses involve comparing transcripts, proteins, or metabolites side by side for the control and disease conditions. For example, differential gene expression analysis methods such as DESeq, edgeR, limma, etc., are utilized to compare the transcriptome across the control and disease conditions (Rapaport et al. 2013). The molecules, however, are generally part of a broader biological function that they represent. Hence, analysis of transcriptome across various functions is desirable as well. These functions, or *a priori* defined set of genes/proteins or metabolites, could provide statistically stronger inferences while comparing different biological states. One such example of functional analyses is the gene set enrichment analysis (GSEA) (Subramanian et al. 2005). The *a priori* definition of gene set could be based on functional groups such as gene ontology, canonical pathways of metabolism, or specific signatures for a cell type or disease type. Additional complexity for the analyses at molecular and functional levels could be added by temporal variations in the omics state (Sumit et al. 2019a). Cellular pathway functional analyses identify the affected pathways due to a disease condition. Gene ontology functional analyses focus on the higher-level cellular functionalities that are impacted between the control and disease conditions. Overall, the omics and functional analyses tools available for biopharmaceutical drug discovery and development can potentially help in many ways. Prominent areas of impact of such analyses include a reduction in the percentage of false positives that emerge from the potential targets in drug discovery, identification of multiple informed choices for drug target validation, and enhancement of productivity and product quality attributes during the development of biologics or biotherapeutics.

11.3 COMPUTATIONAL METHODS FOR SYSTEMS APPROACH IN BIOPHARMACEUTICAL DRUG DISCOVERY AND DEVELOPMENT

Systems biology integrates experimental techniques such as omics and functional analyses (as described in Section 11.2) with mathematical modeling and simulation. The construction of quantitative/mathematical models for a systems-level biological network requires the application of computational methods to understand and dissect the behavior of the system. Several types of mathematical models have been developed to study biopharmaceutical drug discovery and development. These models and modeling approaches could be broadly classified into mechanistic or reaction kinetics modeling approach and structural constraints-based approach (Bruggeman and Westerhoff 2007). Additionally, with advancements in experimental data generation and computational

TABLE 11.2 Mathematical modeling approaches utilized to develop quantitative models for disease and development

MODELING APPROACH	*INFORMATION UTILIZED*	*TEMPORAL INFORMATION*	*DATA REQUIREMENT*	*TYPICAL OUTCOME*
Constraint-based approach	Reaction network and stoichiometry, metabolic fluxes	Not incorporated	+	Phenotypic range and systems limitations
Mechanistic approach	Stoichiometry and kinetic parameters, initial concentrations	Possible	++	Systems dynamics and non-linearity
Statistical learning-based approach	Dependent and independent variables; extensive amount of experimental data	Possible	+++	Prediction for systems wherein mechanism not fully understood

capabilities, a third approach *viz.* statistical and machine learning methods is being utilized in various cases of drug discovery and development. These approaches vary in the amount of information and data requirements as well as expected outcomes. Table 11.2 summarizes these three approaches with additional details.

Mechanistic models are based on fundamental reaction engineering principles and consist of all the reaction kinetics within a system under study (Kremling 2013). Generally, the system is described by a series of differential equations depicting enzymatic/non-enzymatic reactions along with stoichiometric balance. Information about interacting network motifs could also be added in the form of negative and positive feedback and feedforward loops (Milo et al. 2002). Additionally, mathematical models may incorporate biochemical variability and/or stochasticity to better explain the noise in the biological systems (Sumit et al. 2019b). In cases where kinetic parameter and rate constant details may not be available, or there is an experimentally known rate limiting step, a few linked equations could also be clubbed together to modulate the granularity of the model. Initial concentrations of the components and kinetic rate constants are estimated or inferred from experimental data or from literature. The set of differential equations is then solved for temporal or non-temporal inputs that could either depict a disease condition or a potential modulator of the disease condition. An example of utilizing mathematical models to depict the disease condition is modeling the tumor microenvironment to understand cancer metabolism from the perspective of the Warburg effect (Shamsi et al. 2018). In contrast, an example of utilizing mathematical models to understand the potential modulation or treatment of a disease condition is the mechanistic modeling of tumor progression and how tumors would respond to cytotoxic drug administration as a treatment strategy (Vavourakis et al. 2018). On the other hand, the constraint-based approach has become more popular in systems biology recently, especially with the increasing genome sequencing capabilities. For example, one can develop a metabolic model consisting of information about all the genes a species contains and can be expressed in the form of transcripts and proteins. These transcripts and proteins maybe involved in a vast complex network of metabolic reactions, and such models are generally termed genome-scale models. While it is challenging to develop a reaction kinetics/mechanistic model for a species owing to the lack of information on kinetics parameters, it is relatively easier to constrain stoichiometric fluxes for the entire network based on the transcriptomic state of the system and experimentally measurable fluxes. This is a great example where experimental systems biology and computational systems biology approaches are applied in tandem to develop systems-level understanding of fluxes in a metabolic network (Orth et al. 2010). Genome-scale metabolic flux balance models have been utilized in drug target discovery as well as during drug development for biopharmaceuticals. For example, Jerby and Ruppin (2012) utilized a genome-scale modeling approach to predict drug targets of cancer. Another example from drug development for biopharmaceuticals is to develop feeding strategies for improving productivity using the genome-scale modeling approach (Huang et al. 2020). Finally, the third approach, *viz.* the statistical learning approach is becoming more popular with the availability of more computational capabilities as well as a vast amount of experimental systems data such as transcriptomics, proteomics, or glycomics (Guerra and Glassey 2018; Rathore et al. 2022). For systems that may be constrained by limited mechanistic understanding, but have a large set of experimental data available, statistical learning approaches can be employed to perform supervised and unsupervised learning and make predictions about conditions not fully understood.

While mathematical modeling using one of these approaches could be a standalone exercise to develop a quantitative understanding of a biological system, when implemented together with experimental systems biology studies such as genomics, transcriptomics, proteomics, metabolomics, and/or glycomics, it can provide invaluable insights into the mechanism, controllability, and predictability of the biological system. Further, to incorporate the complexity of a biological system across multiple scales, multi-scale models such as agent-based models (ABMs) are also developed to gain insights into the system (Bonabeau 2002). In addition, in some cases, statistical learning approaches are incorporated into mechanistic models to generate hybrid models that leverage mechanisms and big data to gain deeper insights. Thus, such an integrated approach has the potential for the identification and validation of novel drug targets as well as precision control of the drug development processes (Figure 11.3). In the next sections, we explore such examples further.

11.4 APPLICATIONS OF SYSTEMS APPROACHES TO TARGET IDENTIFICATION AND VALIDATION IN DRUG DISCOVERY

Drug target identification and validation is the process of identifying therapeutic targets and subsequently developing therapeutic agents that could potentially produce effective modulation of the target to bring about a positive change in a disease condition in a safe and efficacious manner. It involves two major steps: (i) selection of a therapeutic

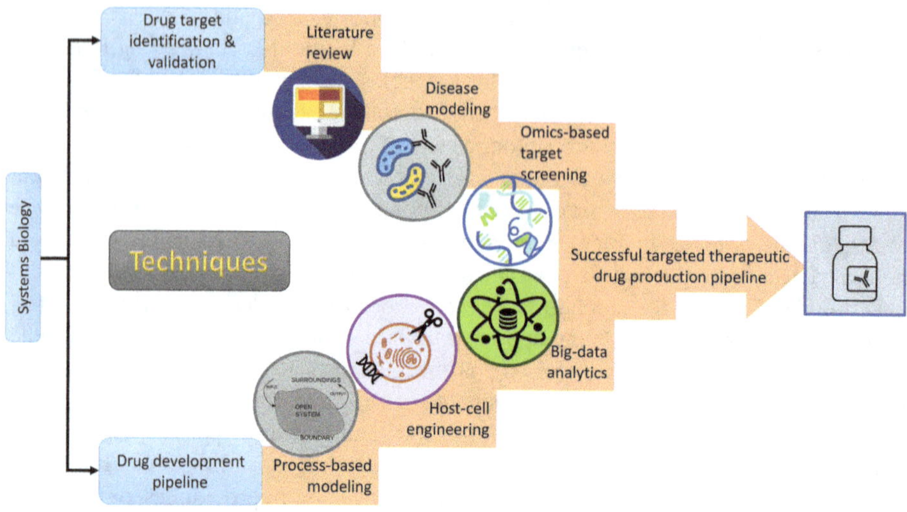

FIGURE 11.3 Systems approach toward drug target identification and validation.

target, and (ii) preclinical assessment of target modulation through potential interventions for safety and efficacy. The assessment is based on the effect of target modulation on the clinical endpoint, which is generally a direct or indirect marker of the diseases. The overall goal of drug discovery is to assess whether, and to what extent, this clinical endpoint is affected by the modulation of potential therapeutic targets (Young and Michelson 2011).

Identification of drug targets that have a high probability of success toward meeting the end goal is of prime importance to minimize failures during clinical trials and reduce the overall cost of drug discovery. Conventional methods of drug target identification are primarily based on literature surveys and/or phenotypic screening. However, such endpoint focused approach poses a big limitation as there is a limited possibility of reassessment for the choice of the target in case of a failure (Swinney and Anthony 2011). Swinney and Anthony performed the data analysis of success of a target-centric approach for first-in-class drugs including biologics approved by the US Food and Drug Administration between 1999 and 2008. In their analysis, they concluded that a lack of proper understanding of molecular mechanism of action (MMOA) results in an increasing number of failures and lower productivity in pharmaceutical research and development. As the approach is overall trial-and-error based, failure implies that one must start again from the beginning to identify new targets. While this target-based approach has been very successful in the past few decades (Swinney and Anthony 2011; Vincent et al. 2022), it is increasingly becoming inefficient and expensive due to lower odds of success in phase I and phase II trials (Seyhan 2019; Hay et al. 2014). Therefore, we need a more holistic approach that can significantly improve the odds of success in clinical trials. This is where systems biology tools such as multi-omics, computational models, statistical learning, and big data analytics are being utilized to make a more informed decision about the drug target identification and potential ways for its modulation (Ekins et al. 2019; Vincent et al. 2022; Paananen and Fortino 2020).

There are several advantages to the systems biology approach in drug target identification and validation. First, the approach is mechanistically driven and therefore, reduces the odds of failure during clinical trials. Second, it utilizes the top-down disease-centric scope, which is holistic and results in multiple drug target identification. Therefore, one always has backup options with the next-best target. Third, as an informed approach, it helps accelerate the drug development timeline and decreases the resource burden from discovery to development to commercialization. In these contexts, the systems biology approach can be implemented toward multi-omics functional analyses to dissect disease pathophysiology and develop mathematical models that describe such pathophysiology, and/or their interaction with potential biotherapeutics.

Several experimental tools within systems biology such as genomics, transcriptomics, proteomics, and metabolomics have been leveraged to study disease models for drug target identification. Table 11.3 shows a few recent examples of studies that incorporate omics analyses toward various disease models for drug target identification. In most cases, multiple omics dimensions are integrated to gain a deeper insight into

TABLE 11.3 Recent examples of studies performed in drug target discovery and target validation of biopharmaceuticals utilizing systems approach (omics tools and techniques)

OMICS TOOLS/ TECHNIQUES	*DISEASE MODEL*	*STUDY AND MAJOR OUTCOME*	*REFERENCE*
Genomics	Breast cancer	Putative target genes for treating breast cancer were identified using genomics and bioinformatic tools	Baxter et al. (2018)
Genomics	Parkinsons	Identified candidate drugs for Parkinsons disease by a method using GWAS data and *in silico* databases to screen the target loci known for the disease	Uenaka et al. (2018)
Transcriptomics	Breast cancer	Clonal transcriptomics helped discover pathway of action for drug resistance providing direction to a combinatorial therapy for triple negative breast cancer in mouse model	Wild et al. (2022)
Transcriptomics	Kidney disease	Leveraged single cell transcriptomics data and functional analysis to reveal that collecting duct cell plasticity, driven by Notch signaling, results in abnormal cell populations in chronic kidney disease	Park et al. (2018)
Transcriptomics	Ovarian cancer	Implemented multiple functional analysis tools to find significantly higher expression of CREB1 in normal ovarian tissues; also identified 25 CREB-related proteins as potential targets	Li et al. (2020)
Proteomics	Lymphoma	Described a pharmaco-proteomics approach to map the interactome of a tumor-enriched isoform of HSP90 (teHSP90)	Goldstein et al. (2015)

(*Continued*)

TABLE 11.3 (*Continued*) Recent examples of studies performed in drug target discovery and target validation of biopharmaceuticals utilizing systems approach (omics tools and techniques)

OMICS TOOLS/ TECHNIQUES	*DISEASE MODEL*	*STUDY AND MAJOR OUTCOME*	*REFERENCE*
Metabolomics	Melanoma	The study identified 21 plasma metabolites including amino acids, propionyl carnitine, phosphatidylcholines, and sphingomyelins as significantly altered in two B-RAF-mutant melanoma xenografts; identified selective MEK inhibitor as a potent drug	Ang et al. (2017)
Metabolomics	Tumor	Study identified that plasma concentrations of 26 metabolites, including amino acids, acylcarnitines, and phosphatidylcholines decreased in mice bearing PTEN-deficient tumors compared with non-tumor-bearing controls; identified class I PI3K inhibitor pictilisib as a potent drug	Ang et al. (2016)
Multi-omics	Type 2 diabetes	A review that characterizes multi-omics disease model for type 2 diabetes mellitus to identify overall change in the intestinal flora and metabolic disturbances	Wang et al. (2021)
Multi-omics	Pneumonia	Genomics, transcriptomics, and metabolic information were integrated in order to prioritize candidate targets in the purview of multi-drug resistance	Ramos et al. (2018)
Omics+ML	Tuberculosis	A whole-blood RNA therapy-end model was developed to identify hypothetical individual end-of-treatment time points with 22 RNAs	Singhania et al. (2018)
Multi-omics	COVID-19	Curated a highly informed integrated drug shortlist by combining structural diversity filtering along with experts' curation and drug target mapping on the depicted molecular pathways	Tomazou et al. (2021)
Multi-omics	Silicosis	Study revealed that arachidonic acid (AA) pathway metabolites, prostaglandin D2 (PGD2), and thromboxane A2 (TXA2) were significantly upregulated in silicosis lungs	Pang et al. (2021)
Multi-omics	Breast cancer	Identified three highly connected modules, EED, DHX9, and AURKA as potential candidate molecular targets for triple negative breast cancer	Turanli et al. (2019)

the disease and its modulation. For example, Turanli et al. utilized transcriptomic and proteomic analyses to identify three highly connected modules as potential drug targets for breast cancer (Turanli et al. 2019). Similarly, Ramos et al. integrated genomics, transcriptomics, and metabolomics information that helped prioritize candidate targets for an infectious disease in the purview of multi-drug resistance (Ramos et al. 2018).

Apart from omics tools and techniques, several types of mathematical models are also leveraged for drug target identification and validation. These include disease models, biotherapeutic models, dynamic pathway models, and data-driven statistical learning models. In the disease model, a systems-level mathematical or statistical framework is developed to describe a given disease *in silico*. Conventionally, kinetic parameters and rate constants from experiments, literature, and/or informed guess are leveraged to model the disease pathophysiology with ordinary differential equations (ODEs), also known as mechanistic models. Data from multi-omics such as transcriptomics, proteomics, and metabolomics are also being leveraged to improve the predictability of disease models. Another dimension of complexity in disease models is the interaction of components at multiple scales. This is generally captured by multi-scale hybrid models such as ABMs (Ji et al. 2017; Menezes et al. 2020). In some cases where data availability is limited to develop quantitative models, a systems-level qualitative model may be developed that is based on molecular cause-and-effect relationships. Table 11.4 captures a few of the recent examples of disease models/metabolic models. For example, Ji et al. developed a hybrid agent-based model to capture cell-cell interactions in bone marrow under multiple myeloma conditions (Ji et al. 2017). The model was utilized to predict the treatment effects of three key therapeutic drugs. The study found that the combination of these three drugs

TABLE 11.4 Recent examples of studies performed in drug target discovery and target validation of biopharmaceuticals utilizing computational systems biology approach

MODELING APPROACH	*DISEASE MODEL*	*STUDY AND MAJOR OUTCOME*	*REFERENCE*
Mechanistic	Metabolic disease	Study performed simulation of adiponectin exocytosis in in response to the reduction of β3ARs observed in adipocytes from animals with obesity-induced diabetes.	Lövfors et al. (2021)
Multi-scale mechanistic	Kidney disease	Developed a complete nephron model and tested inhibition of Na^+-glucose cotransporter 2 (SGLT2) along the proximal convoluted tubule. Results predicted that the segment's Na^+ reabsorption decreased significantly, resulting in natriuresis and osmotic diuresis.	Layton and Layton (2019)
Hybrid ODE and ABM	Cancer (multiple myeloma)	Developed and applied myeloma growth model to predict the treatment effects of three key therapeutic drugs. Study found that the combination of these three drugs can potentially suppress the growth of myeloma cells and reactivate the immune response.	Ji et al. (2017)
Multi-scale and biophysical	Cancer (Glioma)	The model predicts that cell migration depends on the relative balance between random motility and strength of chemo-attractants.	Kim et al. (2015)
Hybrid ODE and ABM	Tuberculosis	Captures various aspects of TB disease and predict that biomarkers in the blood may only faithfully represent events in the lung at early time points after infection.	Joslyn et al. (2022)

(*Continued*)

TABLE 11.4 (*Continued*) Recent examples of studies performed in drug target discovery and target validation of biopharmaceuticals utilizing computational systems biology approach

MODELING APPROACH	*DISEASE MODEL*	*STUDY AND MAJOR OUTCOME*	*REFERENCE*
ABM host-pathogen model	Pneumonia	Identified key contributors to alveolar infection in a mechanistic fashion. Results suggest that S. pneumonia interactions with alveolar epithelial cells contribute to overall infection dynamics as compared to interaction with macrophages.	Santos et al. (2018)
Mechanistic	Diabetic kidney disease	The study addresses the challenge of heterogeneity for clinical trials of novel anti-inflammatory therapies, by modeling how chronic inflammation affects kidney function in five compartments. Such models could be utilized for drug target validation.	Hofherr et al. (2022)
Mechanistic	COVID-19	Model tests the hypothesis that SARS-CoV-2 infects immune cells and, for this reason, induces high-level productions of inflammatory cytokines.	Reis et al. (2021)
Machine learning	Lung cancer	Developed machine learning model to explore novel antibody sequence space and to accelerate the development of highly potent, drug-like antibodies.	Makowski et al. (2022)

can potentially suppress the growth of myeloma cells and reactivate the immune response. Similarly, Layton and Layton (2019) developed a multi-scale mechanistic model for nephrons. The model was utilized to test the inhibition of Na^+-glucose cotransporter 2 (SGLT2) along the proximal convoluted tubule. The model predicted that the segment's Na^+ reabsorption decreased significantly, resulting in natriuresis and osmotic diuresis, thus providing insights into potential drug targets for kidney disease. In summary, multi-omics analyses and mathematical models have been utilized to gain insights into disease models and identify potential drug targets using a systems biology approach.

11.5 APPLICATION OF SYSTEMS BIOLOGY IN BIOPHARMACEUTICAL DEVELOPMENT – OPTIMIZING GROWTH AND PRODUCTIVITY

Once the therapeutic target has been identified and validated, the next step in the biopharmaceutical pipeline is to produce the therapeutic biomolecule or protein of interest. This means to develop a scalable process to generate material for various stages of studies, including regulatory toxicology studies and phase 1 and phase 2 clinical trials prior to approaching the Food and Drug Administration (FDA) for evaluation (Hu 2020). This set of steps, in the broader sense, can be termed biopharmaceutical development. It encompasses both the discovery and development of the drug target. A typical biopharmaceutical process development paradigm includes developing a cell line that can stably produce the biotherapeutic molecule (cell line development), generating the biomolecule in cell culture (upstream process development), purification of cell culture material to remove impurities (downstream process development), and the final formulation of the therapeutic into drug product (drug product development).

The upstream process development of biotherapeutics such as monoclonal antibodies is typically accomplished using recombinant DNA (rDNA) technology to engineer producer cell lines such as Chinese Hamster Ovary (CHO) cells, mouse myeloma (NS0) cells, among others (Dhara et al. 2018). As such, there are two major goals for biopharmaceutical product development: (i) enhance the productivity of the biologics (directly linked to the reduction in the cost of manufacturing) and (ii) enhance and maintain the target range for product quality attributes (directly linked to safety and efficacy of the drug product) (Figure 11.4). On both these fronts, systems biology techniques such as multi-omics analyses and process modeling are implemented to gain process/cellular enhancements.

The productivity of biotherapeutics in cell culture processes directly depends on the amount of cell biomass available and its capability to produce the protein of interest. Cell concentration is generally depicted by viable cell density (*VCD*). Specific productivity, the rate of protein production per cell per unit time, is depicted by q_P. Together, these two key factors determine overall volumetric productivity (V_P) as V_P is an integral of $q_{P*}VCD$ over the batch duration (time) and vessel volume. Conventionally, bioprocess parameters such as pH, temperature, dissolved oxygen, rate of agitation, etc., are utilized as levers to

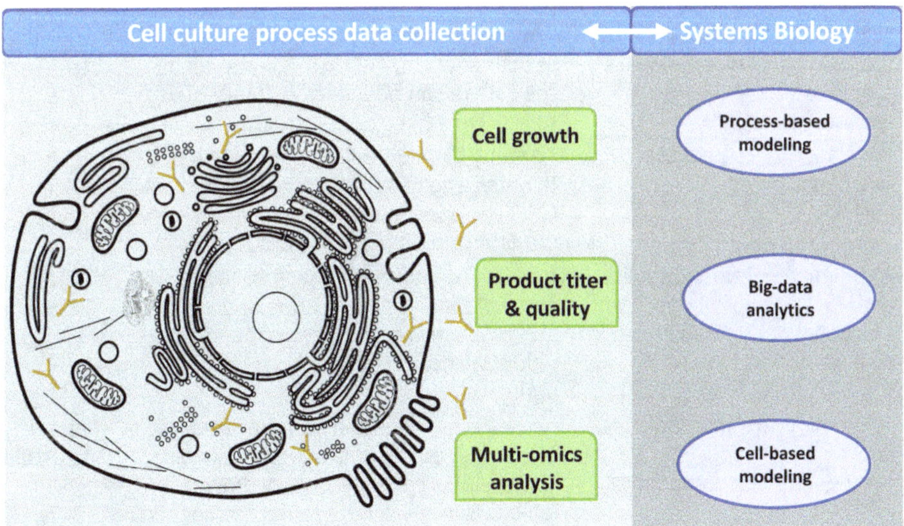

FIGURE 11.4 Systems approach toward biopharmaceutical development.

increase growth and productivity. In addition, perturbations in cell culture medium composition, feeding strategy, and perfusion are also leveraged to enhance cell growth and productivity (Ritacco et al. 2018; Hiller et al. 2017). However, these conventional levers have limitations as they cannot be utilized to perturb the cellular physiology from within to push the biological limit for growth and productivity (Hoang et al. 2022). Moreover, these methods require a significant number of experiments to determine the statistically optimal condition, as they are mostly trial-and-error based. The tools and techniques in systems biology can be leveraged to push the limits of growth and productivity by identifying physiological bottlenecks and developing mitigation strategies (Kildegaard et al. 2013). Moreover, as the design of experiments is based on mechanistic insights, it also helps in reducing the number of experiments needed for achieving optimal growth and productivity. Table 11.5 summarizes a few recent examples where a systems biology approach has been implemented to achieve improved cell growth and protein productivity. For instance, it is well established that high concentrations of lactate and ammonia result

TABLE 11.5 Recent examples of studies performed to enhance cell growth and productivity of biopharmaceuticals using systems approach

SYSTEMS APPROACH	*TOOLS/ TECHNIQUES EMPLOYED*	*STUDY AND MAJOR OUTCOME*	*REFERENCE*
Multi-omics and functional analyses	Transcriptomics, proteomics, metabolomics, and glycomics	Performed multi-omics profiling to investigate effects of pH on growth and productivity. Study concluded that pH set points differentially regulated various intracellular pathways including vesicular trafficking, cell cycle, and apoptosis, thereby impacting growth and productivity.	Lee et al. (2021)
Multi-omics and functional analyses	Proteomics and metabolomics	Study demonstrated that modifications to cellular environment by modifying feed in cell culture can be guided by omics studies to achieve enhanced cell growth.	Blondeel et al. (2016)
Omics and functional analyses	Proteomics	The study highlighted key pathways for targeted engineering to generate desirable CHO cell phenotypes with enhanced growth for biotherapeutic production.	Bryan et al. (2021)
Multi-omics and functional analyses	Transcriptomics, proteomics, and metabolomics	The study identifies key metabolic enzymes and metabolites as indicators of growth and productivity. Also demonstrates that redox balance is key to cellular health during cell culture processes.	Ali et al. (2019)
Multi-omics and functional analyses	Transcriptomics, and metabolomics	The study identifies and demonstrates intermediates and byproducts of amino acid metabolism as secondary growth inhibitors in cell culture for biotherapeutics production.	Chandra Mulukutla et al. (2017)

(*Continued*)

TABLE 11.5 (*Continued*) Recent examples of studies performed to enhance cell growth and productivity of biopharmaceuticals using systems approach

SYSTEMS APPROACH	*TOOLS/ TECHNIQUES EMPLOYED*	*STUDY AND MAJOR OUTCOME*	*REFERENCE*
Mathematical modeling	Genome-scale modeling	The study evaluates and compares parameters for the two phases of cell culture, i.e., growth phase and production phase.	Schinn et al. (2021)
Mathematical modeling	Genome-scale modeling	Developed a metabolic network-based modeling approach utilizing genome-scale model, and implemented it to develop feeding strategies to potential increase protein productivity.	Fouladiha et al. (2020)
Mathematical modeling	Constraint-based modeling	The study identifies unconventional objective functions to minimize non-essential nutrient uptake rate toward cell growth and productivity improvements in various CHO derived cell lines.	Chen et al. (2019)
Mathematical modeling	Metabolic flux analysis	The study suggests that productivity is related to the oxidative state of metabolism whereas cell growth can be characterized by glycolytic metabolic state.	Templeton et al. (2013)
Mathematical modeling	Genome-scale metabolic flux modeling	Implemented genome-scale model to identify feed supplements to enhance biotherapeutic productivity.	Huang et al. (2020)

in growth inhibition in cell cultures, thereby reducing the overall productivity (Pereira et al. 2018). Several process-related and cell engineering strategies have been devised to reduce the amount of these two growth inhibitors in cell culture. However, even with low concentrations of lactate and ammonia, cells in culture tend to plateau out in terms of growth (Chandra Mulukutla et al. 2017). Mulukutla *et al* employed omics techniques to identify and quantify several byproducts or intermediates of amino acid metabolism that accumulate in fed-batch cell culture and impact cell growth. Further, transcriptomics and metabolomics analyses resulted in the identification of cell engineering targets that could reduce the biosynthesis of these growth inhibitors. Metabolic engineering of CHO cells by knocking out such a gene target resulted in significant improvement in cell growth (Mulukutla et al. 2019). Similarly, mathematical modeling tools within systems biology have also been utilized to enhance protein productivity in producer cells. For instance, Huang *et al* implemented a genome-scale model along with transcriptomic and metabolomic analyses to systematically evaluate CHO cell culture and gain metabolic insights for bioprocess development (Huang et al. 2020). The genome-scale model was leveraged to identify an experimentally optimal condition for improved productivity. These are a couple of examples that demonstrate how systems biology tools and techniques, including multi-omics analyses and mathematical modeling, can be leveraged to enhance cell growth and productivity of biotherapeutics in cell culture.

11.6 APPLICATION OF SYSTEMS BIOLOGY IN BIOPHARMACEUTICAL DEVELOPMENT – CONTROLLING PRODUCT QUALITY ATTRIBUTES

Product quality attributes of biotherapeutics refer to the way the produced molecule exists in the drug substance apart from the amino acid backbone structure. Attributes such as post-translational modifications (glycosylation, deamidation, glycation, etc.), charge variants, fragmentation, and aggregation may be directly linked to the safety and efficacy of the biopharmaceutical drug product. While optimizing cell growth and productivity are the major drivers for developing a cost-effective biotherapeutics production process, controlling the aspects of the product quality attributes is equally important for developing safe and efficacious biotherapeutics. In cases where the product quality attributes could have a potential impact on drug product safety and/or efficacy, they are termed 'critical quality attributes' (CQAs). A well-defined control strategy is required to maintain a prescribed or clinically tested range of these CQAs during the manufacturing process (Rathore and Winkle 2009).

Conventionally, the biopharmaceutical development process involves the design of experiment (DoE) studies to identify process parameter ranges that are expected to result in products with CQAs within the prescribed range. However, given the complexity and heterogeneity within the biological systems, control of CQAs has been a challenging task. This is where systems biology tools can be implemented to gain detailed mechanistic insights into the influence of cell physiology and extracellular environment on quality attributes, and how it is impacted by cell culture process parameters (Kildegaard et al. 2013; Sha et al. 2016). As an example, this section will focus on N-linked glycosylation (N-glycosylation) to understand how systems biology can be implemented to gain better control and modulation of various glycosylated species during cell culture process development.

N-glycosylation is a post-translational modification that impacts various aspects of therapeutic proteins, including effector functions, pharmacokinetic clearance, safety, immunogenicity, stability, and shelf life. Various studies have been performed to understand the effect of cell culture process parameters and media components on N-glycosylation (Sha et al. 2016). However, to precisely control and modulate the glycosylated species, a deeper mechanistic insight is required. To this end, both multi-omics studies as well as mathematical modeling have been utilized to further the understanding of N-glycosylation control and modulation. Table 11.6 summarizes a few recent examples where systems biology has been implemented to achieve similar goals of controlling and modulating post-translational modifications such as N-glycosylation. For example, Krambeck *et al* developed a detailed reaction kinetics model for protein glycosylation to predict the various glycosylated species that can arise in cell culture when enzymes linked to N-glycosylation are perturbed (Krambeck et al. 2009). By leveraging mathematical modeling and integrating glycomics data from proteins produced

TABLE 11.6 Studies performed to enhance and control product quality attributes of biopharmaceuticals using systems approach

SYSTEMS APPROACH	*TOOLS/ TECHNIQUES EMPLOYED*	*STUDY AND MAJOR OUTCOME*	*REFERENCE*
Omics and functional analyses	Transcriptomics and metabolomics	The study reveals the effect of oxidative stress on protein sialylation using multi-omics approach.	Lewis et al. (2016)
Omics and functional analyses	Transcriptomics and metabolomics	The study identifies the root cause for temporal heterogeneity of glycan species in therapeutic proteins produced in cell culture and suggest mitigation strategies based on transcriptomic and metabolomics analyses.	Sumit et al. (2019a)
Omics and functional analyses	Proteomics	Differential activation of oxidative phosphorylation may result in differences in post-translational modification as well as charge variants of therapeutic proteins produced in CHO cell culture.	Strasser et al. (2021)
Omics and functional analyses	Transcriptomics, proteomics, and metabolomics	The study identifies key genes and pathways that are perturbed by pH variation in cell culture, leading to an impact on N-glycosylation, protein aggregation, and charge variant profiles apart from mAb productivity.	Lee et al. (2021)
Omics and functional analyses	Transcriptomics	The study implements omics analysis to understand biological causes of aggregation and also identifies potential strategies to control this product quality attribute.	Barzadd et al. (2022)
Mathematical modeling	Genome-scale modeling	Modeling framework to investigate the effect of ammonium of sialylation of monoclonal antibodies.	Savizi et al. (2021)
Mathematical modeling	Mechanistic, reaction kinetics model	Modeled structure-specific turnover rates of N-glycans, and prediction of N-glycan heterogeneity for glycosylation enzyme inhibitors.	Arigoni-Affolter et al. (2019)
Mathematical modeling	Mechanistic, reaction kinetics model	Extension of GLYMMER model to ten different CHO cell lines with varied activities of enzymes involved in N-glycosylation and NSD transport.	Krambeck et al. (2017)
Mathematical modeling	Mechanistic, reaction kinetics model	Utilized kinetic model to design and build 23 different transgenic pools to enhance galactosylation in protein therapeutics.	Stach et al. (2019)

(*Continued*)

TABLE 11.6 (*Continued*) Studies performed to enhance and control product quality attributes of biopharmaceuticals using systems approach

SYSTEMS APPROACH	*TOOLS/ TECHNIQUES EMPLOYED*	*STUDY AND MAJOR OUTCOME*	*REFERENCE*
Mathematical modeling	Machine learning and AI tools, glycomics	This study reviews the implementation of machine learning and AI tools on large set of glycomics data and proposed that such models can be analyzed to gain mechanistic insights into glycosylation machinery and how the machinery shapes glycans under different scenario.	Li et al. (2022)

by a variety of cell lines, they were able to further improve the understanding of glycan heterogeneity in therapeutic proteins (Krambeck et al. 2017). Such models have significantly helped improve the predictability and controllability of N-glycosylation. Similarly, Sumit et al. (2019) utilized multi-omics functional analyses to identify temporal bottlenecks in the N-glycosylation pathway during fed-batch production of therapeutic proteins. The study helped develop mitigation strategies to minimize temporal heterogeneity of glycan species in fed-batch processes toward better controllability of the system. Mathematical modeling and multi-omics analyses can also be utilized to identify gene targets that can modulated to precisely control N-glycosylation in cell culture (Chang et al. 2019). These examples demonstrate that systems biology tools and techniques, including multi-omics analyses and mathematical modeling, can be leveraged toward precision control of product quality attributes such as N-glycosylation for developing safe and efficacious biotherapeutics.

11.7 BIG DATA APPROACH IN BIOLOGICS DRUG DEVELOPMENT

During biopharmaceutical drug development, in particular cell culture process development, a vast amount of time series and discrete datasets are generated. This includes but is not limited to, process data from bioreactors, bioanalytical data such as nutrient, metabolite and byproduct concentrations, cell growth and viability, product titer and productivity, as well as data for product quality attributes. With limited success in phase 1 and phase 2 stages, it is imperative to utilize all the data generated from various programs toward a better process understanding and control (Hay et al. 2014). To this end, digital twins for upstream process development have been conceived and developed (Hoang et al. 2022). Such digital twins utilize hybrid models that incorporate mechanistic understanding from the systems biology approach and statistical learning approach for data with limited mechanistic insights (Macdonald 2022; Park et al. 2021). An overview of

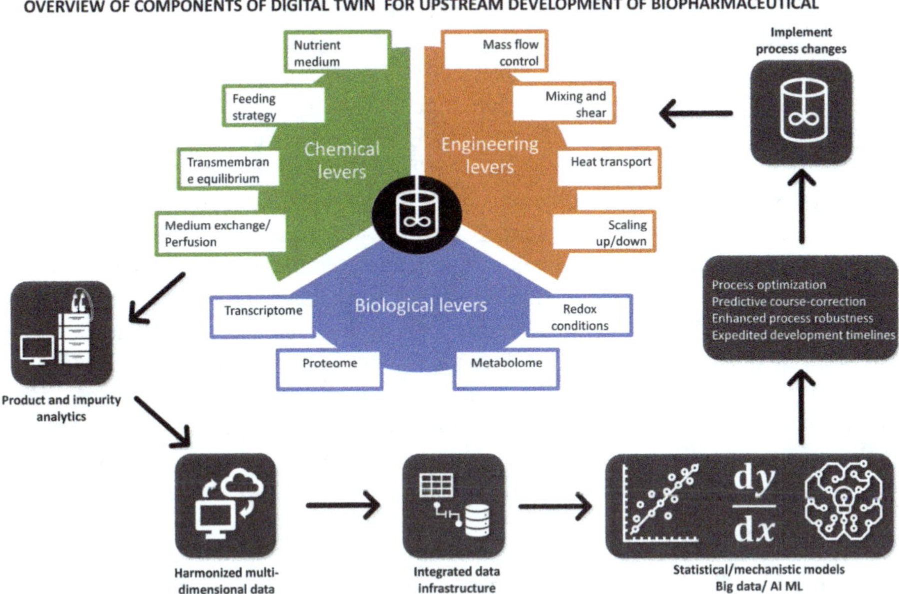

FIGURE 11.5 Digital twins and big data systems approach in biopharmaceutical drug discovery and development.

upstream digital twins is shown in Figure 11.5. Briefly, we can think of three major levers that can control process performance, improve productivity, and modulate product quality. The first lever is biological, and it includes the transcriptome, metabolome, proteome, and redox state of the cells, among many others. The second lever is chemical, and it includes medium composition, nutrient feed composition and feeding strategy, pH, transmembrane equilibrium of macro and micronutrients, etc. The third lever is a physical lever, also known as process control, and it includes agitation, power distribution per unit volume, heat transfer, scaling considerations, etc. As all three levers can influence productivity and product quality, efforts have been made to develop an integrated hybrid model that accounts for all these three levers. Thus, an integral part of the upstream digital twin is the omics state of the cells and its dynamics integrated into a mechanistic or genome-scale model that is part of a bigger hybrid model incorporating other sets of data. Such digital twins can help in process optimization, predicting course corrections, enhancing process robustness, and expediting development timelines (Park et al. 2021).

11.8 CONCLUSIONS AND FUTURE DIRECTIONS

Biopharmaceutical drug discovery and development is a challenging process owing to complexity the biological systems entail. The systems biology approach offers experimental and computational tools such as multi-omics analyses and quantitative mathematical modeling to dissect such complexities. These tools allow for systems-level analyses, integrated model development, multiple hypotheses testing, and simulation under various constraints, resulting in increased probability of success during drug discovery and development. The systems biology approach has been applied to drug target identification and validation by analyzing and comparing omics states of control and disease conditions both at the molecular level as well as functional level. In addition, systems-level quantitative disease models have been utilized to dissect underlying mechanisms of regulation. These models also provide a platform for *in silico* testing of hypotheses for drug target validation. Applications of the systems biology approach in drug development include enhancement of cell growth and productivity of producer cell lines engineered to produce recombinant biopharmaceuticals. This is achieved by utilizing functional analyses and mathematical modeling to identify metabolic bottlenecks and developing mitigation strategies either via the cell engineering approach or through media and process changes. Similar approaches have also been successfully implemented to dissect regulations underlying variations in product quality attributes of the biopharmaceuticals and have significantly aided in the control of these attributes for enhanced safety and efficacy.

With recent advancements in computational capabilities and automation, biopharmaceutical drug discovery and development are steering toward the next phase of acceleration to reduce development timelines and bring drugs faster to the patients. To this end, the systems approach (omics and computational modeling) is getting integrated with process analytical technologies (PATs) and advanced sensors and actuators to develop cyber-physical systems (CPS) that enable big data analysis and real-time control of process (Rathore et al. 2022). The use of hybrid models and statistical learning tools along with CPS will enable the industry to potentially control and modulate the system at a more precise level with predictive course correction. In summary, recent advances in the big data approach along with the systems biology approach appear to have a very positive outlook and tremendous scope in contributing significantly to biopharmaceutical drug discovery and development. Together with advances in data analytics, systems biology approach can significantly reduce the cost of drug discovery and development and bring novel biotherapeutics faster to the patients.

REFERENCES

Ali, Amr S., Ravali Raju, Rashmi Kshirsagar, Alexander R. Ivanov, Alan Gilbert, Li Zang, and Barry L. Karger. 2019. "Multi-Omics Study on the Impact of Cysteine Feed Level on Cell Viability and MAb Production in a CHO Bioprocess." *Biotechnology Journal* 14 (4). doi:10.1002/biot.201800352.

Ang, Joo Ern, Akos Pal, Yasmin J. Asad, Alan T. Henley, Melanie Valenti, Gary Box, Alexis De haven Brandon, et al. 2017. "Modulation of Plasma Metabolite Biomarkers of the MAPK Pathway with MEK Inhibitor RO4987655: Pharmacodynamic and Predictive Potential in Metastatic Melanoma." *Molecular Cancer Therapeutics* 16 (10): 2315–23. doi:10.1158/1535-7163.MCT-16-0881.

Ang, Joo Ern, Rupinder Pandher, Joo Chew Ang, Yasmin J. Asad, Alan T. Henley, Melanie Valenti, Gary Box, et al. 2016. "Plasma Metabolomic Changes Following PI3K Inhibition as Pharmacodynamic Biomarkers: Preclinical Discovery to Phase I Trial Evaluation." *Molecular Cancer Therapeutics* 15 (6): 1412–24. doi:10.1158/1535–7163.MCT-15–0815.

Arigoni-Affolter, Ilaria, Ernesto Scibona, Chia-Wei Lin, David Brühlmann, Jonathan Souquet, Hervé Broly, and Markus Aebi. 2019. "Mechanistic Reconstruction of Glycoprotein Secretion through Monitoring of Intracellular N-Glycan Processing." *Science Advances* 5 (11). doi:10.1126/sciadv.aax8930.

Barzadd, Mona Moradi, Magnus Lundqvist, Claire Harris, Magdalena Malm, Anna Luisa Volk, Niklas Thalén, Veronique Chotteau, et al. 2022. "Autophagy and Intracellular Product Degradation Genes Identified by Systems Biology Analysis Reduce Aggregation of Bispecific Antibody in CHO Cells." *New Biotechnology* 68 (May): 68–76. doi:10.1016/j.nbt.2022.01.010.

Baxter, Joseph S., Olivia C. Leavy, Nicola H. Dryden, Sarah Maguire, Nichola Johnson, Vita Fedele, Nikiana Simigdala, et al. 2018. "Capture Hi-C Identifies Putative Target Genes at 33 Breast Cancer Risk Loci." *Nature Communications* 9 (1). doi:10.1038/s41467-018-03411-9.

Blondeel, Eric J.M., Raymond Ho, Steffen Schulze, Stanislav Sokolenko, Simon R. Guillemette, Igor Slivac, Yves Durocher, et al. 2016. "An Omics Approach to Rational Feed: Enhancing Growth in CHO Cultures with NMR Metabolomics and 2D-DIGE Proteomics." *Journal of Biotechnology* 234: 127–38. doi:10.1016/j.jbiotec.2016.07.027.

Bonabeau, Eric. 2002. "Agent-Based Modeling: Methods and Techniques for Simulating Human Systems." *Proceedings of the National Academy of Sciences of the United States of America* 99 (SUPPL. 3): 7280–87. doi:10.1073/pnas.082080899.

Bruggeman, Frank J., and Hans V. Westerhoff. 2007. "The Nature of Systems Biology." *Trends in Microbiology* 15 (1): 45–50. doi:10.1016/j.tim.2006.11.003.

Bryan, Laura, Michael Henry, Ronan M. Kelly, Christopher C. Frye, Matthew D. Osborne, Martin Clynes, and Paula Meleady. 2021. "Mapping the Molecular Basis for Growth Related Phenotypes in Industrial Producer CHO Cell Lines Using Differential Proteomic Analysis." *BMC Biotechnology* 21 (1): 1–21. doi:10.1186/s12896-021-00704-8.

Butcher, Eugene C., Ellen L. Berg, and Eric J. Kunkel. 2004. "Systems Biology in Drug Discovery." *Nature Biotechnology*. doi:10.1038/nbt1017.

Chandra Mulukutla, Bhanu, Jaitashree Kale, Taylor Kalomeris, Michaela Jacobs, and Gregory W. Hiller. 2017. "Identification and Control of Novel Growth Inhibitors in Fed-Batch Cultures of Chinese Hamster Ovary Cells." *Biotechnol. Bioeng* 114: 1779–90. https://doi.org/10.1002/bit.26313

Chang, Michelle M., Leonid Gaidukov, Giyoung Jung, Wen Allen Tseng, John J. Scarcelli, Richard Cornell, Jeffrey K. Marshall, et al. 2019. "Small-Molecule Control of Antibody N-Glycosylation in Engineered Mammalian Cells." *Nature Chemical Biology* 15 (7): 730–36. doi:10.1038/s41589-019-0288-4.

Chen, Yiqun, Brian O. McConnell, Venkata Gayatri Dhara, Harnish Mukesh Naik, Chien Ting Li, Maciek R. Antoniewicz, and Michael J. Betenbaugh. 2019. "An Unconventional Uptake Rate Objective Function Approach Enhances Applicability of Genome-Scale Models for Mammalian Cells." *Npj Systems Biology and Applications* 5 (1). doi:10.1038/s41540-019-0103-6.

Dhara, Venkata Gayatri, Harnish Mukesh Naik, Natalia I. Majewska, and Michael J. Betenbaugh. 2018. "Recombinant Antibody Production in CHO and NS0 Cells: Differences and Similarities." *BioDrugs* 32 (6): 571–84. doi:10.1007/s40259-018-0319-9.

Ekins, Sean, Ana C. Puhl, Kimberley M. Zorn, Thomas R. Lane, Daniel P. Russo, Jennifer J. Klein, Anthony J. Hickey, and Alex M. Clark. 2019. "Exploiting Machine Learning for End-to-End Drug Discovery and Development." *Nature Materials*. doi:10.1038/s41563-019-0338-z.

Fouladiha, Hamideh, Sayed Amir Marashi, Fatemeh Torkashvand, Fereidoun Mahboudi, Nathan E. Lewis, and Behrouz Vaziri. 2020. "A Metabolic Network-Based Approach for Developing Feeding Strategies for CHO Cells to Increase Monoclonal Antibody Production." *Bioprocess and Biosystems Engineering* 43 (8): 1381–89. doi:10.1007/s00449-020-02332-6.

Goldstein, Rebecca L., Shao Ning Yang, Tony Taldone, Betty Chang, John Gerecitano, Kojo Elenitoba-Johnson, Rita Shaknovich, et al. 2015. "Pharmacoproteomics Identifies Combinatorial Therapy Targets for Diffuse Large B Cell Lymphoma." *Journal of Clinical Investigation* 125 (12): 4559–71. doi:10.1172/JCI80714.

Guerra, André C., and Jarka Glassey. 2018. "Machine Learning in Biopharmaceutical Manufacturing." *European Pharmaceutical Review* 23 (4): 62–65.

Hasin, Yehudit, Marcus Seldin, and Aldons Lusis. 2017. "Multi-Omics Approaches to Disease." *Genome Biology*. doi:10.1186/s13059-017-1215-1.

Hay, Michael, David W Thomas, John L Craighead, Celia Economides, and Jesse Rosenthal. 2014. "Clinical Development Success Rates for Investigational Drugs." *Nature Biotechnology* 32 (1): 40–51. doi:10.1038/nbt.2786.

Hiller, Gregory W, Ana Maria Ovalle, Matthew P Gagnon, Meredith L Curran, and Wenge Wang. 2017. "Cell-Controlled Hybrid Perfusion Fed-Batch CHO Cell Process Provides Significant Productivity Improvement Over Conventional Fed-Batch Cultures." *Biotechnology and Bioengineering* 114: 1438–47. https://doi.org/10.1002/bit.26259

Hoang, Duc, Bingyu Kuang, George Liang, Zhao Wang, and Seongkyu Yoon. 2022. "Modulation of Nutrient Precursors for Controlling Metabolic Inhibitors by Genome-Scale Flux Balance Analysis." *Biotechnology Progress*. doi:10.1002/btpr.3313.

Hofherr, Alexis, Julie Williams, Li Ming Gan, Magnus Söderberg, Pernille B.L. Hansen, and Kevin J. Woollard. 2022. "Targeting Inflammation for the Treatment of Diabetic Kidney Disease: A Five-Compartment Mechanistic Model." *BMC Nephrology* 23 (1): 1–17. doi:10.1186/s12882-022-02794-8.

Hu, Wei-Shou. 2020. Cell Culture Bioprocess Engineering, Second edition, edited by Wei-Shou Hu. Boca Raton : CRC Press. doi:10.1201/9780429162770.

Huang, Zhuangrong, Jianlin Xu, Andrew Yongky, Caitlin S. Morris, Ashli L. Polanco, Michael Reily, Michael C. Borys, Zheng Jian Li, and Seongkyu Yoon. 2020. "CHO Cell Productivity Improvement by Genome-Scale Modeling and Pathway Analysis: Application to Feed Supplements." *Biochemical Engineering Journal* 160 (April). doi:10.1016/j.bej.2020.107638.

Jerby, Livnat, and Eytan Ruppin. 2012. "Predicting Drug Targets and Biomarkers of Cancer via Genome-Scale Metabolic Modeling." *Clinical Cancer Research*. doi:10.1158/1078–0432.CCR-12–1856.

Ji, Zhiwei, Jing Su, Dan Wu, Huiming Peng, Weiling Zhao, Brian Nlong Zhao, and Xiaobo Zhou. 2017. "Predicting the Impact of Combined Therapies on Myeloma Cell Growth Using a Hybrid Multi-Scale Agent-Based Model." *Oncotarget* 8 (5): 7647–65. doi:10.18632/oncotarget.13831.

Joslyn, Louis R., Jennifer J. Linderman, and Denise E. Kirschner. 2022. "A Virtual Host Model of Mycobacterium Tuberculosis Infection Identifies Early Immune Events as Predictive of Infection Outcomes." *Journal of Theoretical Biology* 539: 111042. doi:10.1016/j.jtbi.2022.111042.

Kildegaard, Helene Faustrup, Deniz Baycin-Hizal, Nathan E. Lewis, and Michael J. Betenbaugh. 2013. "The Emerging CHO Systems Biology Era: Harnessing the 'Omics Revolution for Biotechnology." *Current Opinion in Biotechnology* 24 (6): 1102–7. doi:10.1016/j.copbio.2013.02.007.

Kim, Yangjin, Gibin Powathil, Hyunji Kang, Dumitru Trucu, Hyeongi Kim, Sean Lawler, and Mark Chaplain. 2015. "Strategies of Eradicating Glioma Cells: A Multi-Scale Mathematical Model with MiR-451-AMPK-MTOR Control." *PLoS One* 10 (1): 1–30. doi:10.1371/journal.pone.0114370.

Kitano, Hiroaki. 2002. "Systems Biology: A Brief Overview." *Science* 295 (5560): 1662–64. doi:10.1126/science.1069492.

Krambeck, Frederick J., Sandra V. Bennun, Mikael R. Andersen, and Michael J. Betenbaugh. 2017. "Model-Based Analysis of N-Glycosylation in Chinese Hamster Ovary Cells." *PLoS One* 12 (5): 1–30. doi:10.1371/journal.pone.0175376.

Krambeck, Frederick J., Sandra V. Bennun, Someet Narang, Sean Choi, Kevin J. Yarema, and Michael J. Betenbaugh. 2009. "A Mathematical Model to Derive N-Glycan Structures and Cellular Enzyme Activities from Mass Spectrometric Data." *Glycobiology* 19 (11): 1163–75. doi:10.1093/glycob/cwp081.

Kreitmaier, Peter, Georgia Katsoula, and Eleftheria Zeggini. 2023. "Insights from Multi-Omics Integration in Complex Disease Primary Tissues." *Trends in Genetics*. doi:10.1016/j.tig.2022.08.005.

Kremling, Andreas. 2013. Systems Biology: Mathematical Modeling and Model Analysis. CRC Press. https://books.google.com/books?hl=en&lr=&id=FXf6AQAAQBAJ&oi=fnd&pg=PP1&dq=systems+biology+mathematical+modeling+and+model+analysis&ots=PJmEm3GFRC&sig=7TPiMDS7x0XIiABc9l-Sewbr2gk#v=onepage&q&f=false.

Layton, Anita T., and Harold E. Layton. 2019. "A Computational Model of Epithelial Solute and Water Transport along a Human Nephron." *PLoS Computational Biology* 15 (2): 1–23. doi:10.1371/journal.pcbi.1006108.

Lee, Alison P., Yee Jiun Kok, Meiyappan Lakshmanan, Dawn Leong, Lu Zheng, Hsueh Lee Lim, Shuwen Chen, et al. 2021. "Multi-Omics Profiling of a CHO Cell Culture System Unravels the Effect of Culture PH on Cell Growth, Antibody Titer, and Product Quality." *Biotechnology and Bioengineering* 118 (11): 4305–16. doi:10.1002/bit.27899.

Lewis, Amanda M., William D. Croughan, Nelly Aranibar, Alison G. Lee, Bethanne Warrack, Nicholas R. Abu-Absi, Rutva Patel, et al. 2016. "Understanding and Controlling Sialylation in a CHO Fc-Fusion Process." *PLoS One* 11 (6). doi:10.1371/journal.pone.0157111.

Li, Haining, Austin W.T. Chiang, and Nathan E. Lewis. 2022. "Artificial Intelligence in the Analysis of Glycosylation Data." *Biotechnology Advances*. doi:10.1016/j.biotechadv.2022.108008.

Li, Ju Yueh, Chia Jung Li, Li Te Lin, and Kuan Hao Tsui. 2020. "Multi-Omics Analysis Identifying Key Biomarkers in Ovarian Cancer." *Cancer Control* 27 (1): 1–10. doi:10.1177/1073274820976671.

Lövfors, William, Christian Simonsson, Ali M. Komai, Elin Nyman, Charlotta S. Olofsson, and Gunnar Cedersund. 2021. "A Systems Biology Analysis of Adrenergically Stimulated Adiponectin Exocytosis in White Adipocytes." *Journal of Biological Chemistry* 297 (5): 101221. doi:10.1016/j.jbc.2021.101221.

Macdonald, Gareth John. 2022. "Digital Twins and AI Reshape Biopharmaceutical Manufacturing." *Genetic Engineering and Biotechnology News* 42 (8): 44–46. doi:10.1089/gen.42.08.13.

Makowski, Emily K., Patrick C. Kinnunen, Jie Huang, Lina Wu, Matthew D. Smith, Tiexin Wang, Alec A. Desai, et al. 2022. "Co-Optimization of Therapeutic Antibody Affinity and Specificity Using Machine Learning Models That Generalize to Novel Mutational Space." *Nature Communications* 13 (1). doi:10.1038/s41467-022-31457-3.

Menezes, Bruna, Cornelius Cilliers, Timothy Wessler, Greg M. Thurber, and Jennifer J. Linderman. 2020. "An Agent-Based Systems Pharmacology Model of the Antibody-Drug Conjugate Kadcyla to Predict Efficacy of Different Dosing Regimens." *AAPS Journal* 22 (2): 1–13. doi:10.1208/s12248-019-0391-1.

Milo, R., S. Shen-Orr, S. Itzkovitz, N. Kashtan, D. Chklovskii, and U. Alon. 2002. "Network Motifs: Simple Building Blocks of Complex Networks." *Science* 298 (5594): 824–27. doi:10.1126/science.298.5594.824.

Mulukutla, Bhanu Chandra, Jeffrey Mitchell, Pauline Geoffroy, Cameron Harrington, Manisha Krishnan, Taylor Kalomeris, Caitlin Morris, Lin Zhang, Pamela Pegman, and Gregory W. Hiller. 2019. "Metabolic Engineering of Chinese Hamster Ovary Cells towards Reduced Biosynthesis and Accumulation of Novel Growth Inhibitors in Fed-Batch Cultures." Metabolic Engineering 54 (July): 54–68. doi:10.1016/j.ymben.2019.03.001.

Orth, Jeffrey D., Ines Thiele, and Bernhard O. Palsson. 2010. "What Is Flux Balance Analysis?" *Nature Biotechnology* 28 (3): 245–48. doi:10.1038/nbt.1614.

Paananen, Jussi, and Vittorio Fortino. 2020. "An Omics Perspective on Drug Target Discovery Platforms." *Briefings in Bioinformatics* 21 (6): 1937–53. doi:10.1093/bib/bbz122.

Pang, Junling, Xianmei Qi, Ya Luo, Xiaona Li, Ting Shu, Baicun Li, Meiyue Song, et al. 2021. "Multi-Omics Study of Silicosis Reveals the Potential Therapeutic Targets PGD2 and TXA2." *Theranostics* 11 (5): 2381–94. doi:10.7150/thno.47627.

Park, Jihwan, Rojesh Shrestha, Chengxiang Qiu, Ayano Kondo, Shizheng Huang, Max Werth, Mingyao Li, Jonathan Barasch, and Katalin Suszták. 2018. "Single-Cell Transcriptomics of the Mouse Kidney Reveals Potential Cellular Targets of Kidney Disease." *Science* 360 (6390): 758–63. doi:10.1126/science.aar2131.

Park, Seo Young, Cheol Hwan Park, Dong Hyuk Choi, Jong Kwang Hong, and Dong Yup Lee. 2021. "Bioprocess Digital Twins of Mammalian Cell Culture for Advanced Biomanufacturing." *Current Opinion in Chemical Engineering* 33: 100702. doi:10.1016/j.coche.2021.100702.

Pereira, Sara, Helene Faustrup Kildegaard, and Mikael Rørdam Andersen. 2018. "Impact of CHO Metabolism on Cell Growth and Protein Production: An Overview of Toxic and Inhibiting Metabolites and Nutrients." *Biotechnology Journal*. doi:10.1002/biot.201700499.

Piña, Benjamin, Demetrio Raldúa, Carlos Barata, José Portugal, Laia Navarro-Martín, Rubén Martínez, Inmaculada Fuertes, and Marta Casado. 2018. "Functional Data Analysis: Omics for Environmental Risk Assessment." *Comprehensive Analytical Chemistry* 82: 583–611. doi:10.1016/bs.coac.2018.07.007.

Ramos, Pablo Ivan Pereira, Darío Fernández Do Porto, Esteban Lanzarotti, Ezequiel J. Sosa, Germán Burguener, Agustín M. Pardo, Cecilia C. Klein, et al. 2018. "An Integrative, Multi-Omics Approach towards the Prioritization of Klebsiella Pneumoniae Drug Targets." *Scientific Reports* 8 (1): 1–19. doi:10.1038/s41598-018-28916-7.

Rapaport, Franck, Raya Khanin, Yupu Liang, Azra Krek, Paul Zumbo, Christopher E Mason, Nicholas D Socci, and Doron Betel. 2013. "Comprehensive Evaluation of Differential Gene Expression Analysis Methods for RNA-Seq Data." *Genome Biology* 14: 1–14. https://doi.org/10.1186/gb-2013-14-9-r95.

Rathore, Anurag S., Saxena Nikita, Garima Thakur, and Somesh Mishra. 2022. "Artificial Intelligence and Machine Learning Applications in Biopharmaceutical Manufacturing." *Trends in Biotechnology* 41(4): 1–14. doi:10.1016/j.tibtech.2022.08.007.

Rathore, Anurag S., and Helen Winkle. 2009. "Quality by Design for Biopharmaceuticals." *Nature Biotechnology* 27 (1): 26–34. doi:10.1038/nbt0109–26.

Reis, Ruy Freitas, Alexandre Bittencourt Pigozzo, Carla Rezende Barbosa Bonin, Barbara de Melo Quintela, Lara Turetta Pompei, Ana Carolina Vieira, Larissa de Lima e. Silva, Maicom Peters Xavier, Rodrigo Weber dos Santos, and Marcelo Lobosco. 2021. "A Validated Mathematical Model of the Cytokine Release Syndrome in Severe COVID-19." *Frontiers in Molecular Biosciences* 8 (July): 1–13. doi:10.3389/fmolb.2021.639423.

Ritacco, Frank V., Yongqi Wu, and Anurag Khetan. 2018. "Cell Culture Media for Recombinant Protein Expression in Chinese Hamster Ovary (CHO) Cells: History, Key Components, and Optimization Strategies." *Biotechnology Progress*. doi:10.1002/btpr.2706.

Santos, Guido, Xin Lai, Martin Eberhardt, and Julio Vera. 2018. "Bacterial Adherence and Dwelling Probability: Two Drivers of Early Alveolar Infection by Streptococcus Pneumoniae Identified in Multi-Level Mathematical Modeling." *Frontiers in Cellular and Infection Microbiology* 8 (MAY): 1–19. doi:10.3389/fcimb.2018.00159.

Savizi, Iman Shahidi Pour, Ehsan Motamedian, Nathan E. Lewis, Ioscani Jimenez del Val, and Seyed Abbas Shojaosadati. 2021. "An Integrated Modular Framework for Modeling the Effect of Ammonium on the Sialylation Process of Monoclonal Antibodies Produced by CHO Cells." *Biotechnology Journal* 16 (8). doi:10.1002/biot.202100019.

Schinn, Song Min, Carly Morrison, Wei Wei, Lin Zhang, and Nathan E. Lewis. 2021. "Systematic Evaluation of Parameters for Genome-Scale Metabolic Models of Cultured Mammalian Cells." *Metabolic Engineering* 66 (July): 21–30. doi:10.1016/j.ymben.2021.03.013.

Seyhan, Attila A. 2019. "Lost in Translation: The Valley of Death across Preclinical and Clinical Divide – Identification of Problems and Overcoming Obstacles." *Translational Medicine Communications* 4 (1). doi:10.1186/s41231-019-0050–7.

Sha, Sha, Cyrus Agarabi, Kurt Brorson, Dong Yup Lee, and Seongkyu Yoon. 2016. "N-Glycosylation Design and Control of Therapeutic Monoclonal Antibodies." *Trends in Biotechnology* 34 (10): 835–46. doi:10.1016/j.tibtech.2016.02.013.

Shamsi, Milad, Mohsen Saghafian, Morteza Dejam, and Amir Sanati-Nezhad. 2018. "Mathematical Modeling of the Function of Warburg Effect in Tumor Microenvironment." *Scientific Reports* 8 (1). doi:10.1038/s41598-018-27303-6.

Singhania, Akul, Robert J. Wilkinson, Marc Rodrigue, Pranabashis Haldar, and Anne O'Garra. 2018. "The Value of Transcriptomics in Advancing Knowledge of the Immune Response and Diagnosis in Tuberculosis." *Nature Immunology* 19 (11): 1159–68. doi:10.1038/s41590-018-0225–9.

Stach, Christopher S., Meghan G. McCann, Conor M. O'Brien, Tung S. Le, Nikunj Somia, Xinning Chen, Kyoungho Lee, et al. 2019. "Model-Driven Engineering of N-Linked Glycosylation in Chinese Hamster Ovary Cells." *ACS Synthetic Biology* 8 (11): 2524–35. doi:10.1021/acssynbio.9b00215.

Strasser, Lisa, Amy Farrell, Jenny T.C. Ho, Kai Scheffler, Ken Cook, Patrick Pankert, Peter Mowlds, Rosa Viner, Barry L. Karger, and Jonathan Bones. 2021. "Proteomic Profiling of IgG1 Producing CHO Cells Using LC/LC-SPS-MS3: The Effects of Bioprocessing Conditions on Productivity and Product Quality." *Frontiers in Bioengineering and Biotechnology* 9 (April). doi:10.3389/fbioe.2021.569045.

Subramanian, Aravind, Pablo Tamayo, Vamsi K Mootha, Sayan Mukherjee, Benjamin L Ebert, Michael A Gillette, Amanda Paulovich, et al. 2005. "Gene Set Enrichment Analysis: A Knowledge-Based Approach for Interpreting Genome-Wide Expression Profiles." doi:10.1073/pnas.0506580102.

Sumit, Madhuresh, Sepideh Dolatshahi, An Hsiang Adam Chu, Kaffa Cote, John J. Scarcelli, Jeffrey K. Marshall, Richard J. Cornell, et al. 2019a. "Dissecting N-Glycosylation Dynamics in Chinese Hamster Ovary Cells Fed-Batch Cultures Using Time Course Omics Analyses." *IScience* 12: 102–20. doi:10.1016/j.isci.2019.01.006.

Sumit, Madhuresh, Andreja Jovic, Richard R. Neubig, Shuichi Takayama, and Jennifer J. Linderman. 2019b. "A Two-Pulse Cellular Stimulation Test Elucidates Variability and Mechanisms in Signaling Pathways." *Biophysical Journal* 116 (5): 962–73. doi:10.1016/j.bpj.2019.01.022.

Swinney, David C., and Jason Anthony. 2011. "How Were New Medicines Discovered?" *Nature Reviews Drug Discovery* 10 (7): 507–19. doi:10.1038/nrd3480.

Templeton, Neil, Jason Dean, Pranhitha Reddy, and Jamey D. Young. 2013. "Peak Antibody Production Is Associated with Increased Oxidative Metabolism in an Industrially Relevant Fed-Batch CHO Cell Culture." *Biotechnology and Bioengineering* 110 (7): 2013–24. doi:10.1002/bit.24858.

Tomazou, Marios, Marilena M. Bourdakou, George Minadakis, Margarita Zachariou, Anastasis Oulas, Evangelos Karatzas, Eleni M. Loizidou, et al. 2021. "Multi-Omics Data Integration and Network-Based Analysis Drives a Multiplex Drug Repurposing Approach to a Shortlist of Candidate Drugs against COVID-19." *Briefings in Bioinformatics* 22 (6): 1–24. doi:10.1093/bib/bbab114.

Turanli, Beste, Kubra Karagoz, Gholamreza Bidkhori, Raghu Sinha, Michael L. Gatza, Mathias Uhlen, Adil Mardinoglu, and Kazim Yalcin Arga. 2019. "Multi-Omic Data Interpretation to Repurpose Subtype Specific Drug Candidates for Breast Cancer." *Frontiers in Genetics* 10 (MAY): 1–12. doi:10.3389/fgene.2019.00420.

Uenaka, Takeshi, Wataru Satake, Pei Chieng Cha, Hideki Hayakawa, Kousuke Baba, Shiying Jiang, Kazuhiro Kobayashi, et al. 2018. "In Silico Drug Screening by Using Genome-Wide Association Study Data Repurposed Dabrafenib, an Anti-Melanoma Drug, for Parkinson's Disease." *Human Molecular Genetics* 27 (22): 3974–85. doi:10.1093/hmg/ddy279.

Vavourakis, Vasileios, Triantafyllos Stylianopoulos, and Peter A. Wijeratne. 2018. "In-Silico Dynamic Analysis of Cytotoxic Drug Administration to Solid Tumours: Effect of Binding Affinity and Vessel Permeability." *PLoS Computational Biology* 14 (10). doi:10.1371/journal.pcbi.1006460.

Vincent, Fabien, Arsenio Nueda, Jonathan Lee, Monica Schenone, Marco Prunotto, and Mark Mercola. 2022. "Phenotypic Drug Discovery: Recent Successes, Lessons Learned and New Directions." *Nature Reviews Drug Discovery*. doi:10.1038/s41573-022-00472-w.

Wang, Shuai, Hui Yong, and Xiao-Dong He. 2021. "Multi-Omics: Opportunities for Research on Mechanism of Type 2 Diabetes Mellitus." *World Journal of Diabetes* 12 (7): 1070–80. doi:10.4239/wjd.v12.i7.1070.

Wild, Sophia A., Ian G. Cannell, Ashley Nicholls, Katarzyna Kania, Dario Bressan, Gregory J. Hannon, and Kirsty Sawicka. 2022. "Clonal Transcriptomics Identifies Mechanisms of Chemoresistance and Empowers Rational Design of Combination Therapies." *ELife* 11: 1–36. doi:10.7554/eLife.80981.

Young, Daniel L., and Seth Michelson. 2011. *Systems Biology in Drug Discovery and Development*. Wiley.

12 Recent Advances in PK/PD and Quantitative Systems Pharmacology (QSP) Models for Biopharmaceuticals

Hardik Mody, Venkata Krishna Kowthavarapu, and Alison Betts

12.1 INTRODUCTION TO PK/PD AND QSP MODELING

Pharmacokinetic/pharmacodynamic (PK/PD) modeling and quantitative systems pharmacology (QSP) modeling are mathematical techniques which can be used to answer quantitative questions and enable decision-making across the drug discovery and development continuum. These approaches have sufficient flexibility and tractability to answer critical questions arising from early drug discovery to clinical development. As such, they are useful tools to increase efficiency and effectiveness in research and development (R&D), which can play an important role in offsetting the high cost and attrition of drug research. A recent publication by the Food and Drug

DOI: 10.1201/9781003300311-12

Administration (FDA) states that quantitative systems pharmacology (QSP) modeling and simulation are seen as critical tools for accelerating drug development and assisting in regulatory decisions [1].

The type of modeling approach required depends upon the granularity of the question asked, the data available, and the time required for decision-making (Table 12.1). For example, PK/PD models are useful for 'top-down' fitting of data, more empirically grounded, quicker and easier to develop and use, and good at extrapolating within a limited field of vision across different doses and subpopulations. QSP models use 'bottom-up' approaches to describe the dynamic interactions between drugs and complex biological systems. They are more mechanistic in nature, combining data and knowledge from various sources to construct a mathematical framework for the entire system. As such, they are useful for more complex hypothesis-driven questions. PK/PD modeling and QSP modeling are compared in Table 12.1 and discussed in more depth below.

12.1.1 PK/PD Modeling

The basic principles of pharmacokinetics (PK), pharmacodynamics (PD), and physiology form the foundation of PK/PD modeling. PK encompasses the factors affecting the time course of drug concentrations in relevant biological fluids and tissues after various routes of administration and represents the driving force for pharmacological and most toxicological effects [2]. PD is the study of a drug's molecular, biochemical, and physiological effects or actions. Through years of implementation in drug development, PK/PD modeling has demonstrated tremendous value in elucidating the relationship between the PK of a therapeutic intervention and the resulting PD effect [3].

In PK/PD analysis, relatively simple models can be used to understand the time-concentration-effect relationship through fitting of data. This 'top-down' approach enables

TABLE 12.1 Comparison of PK/PD and QSP models

	PK/PD	*QSP*
Level	Organism	Scale of interest, e.g., cellular, tissue/organ
Granularity	Low granularity Low in assumptions Fitting of experimental data with estimation of parameters Data driven	Higher granularity – contains more mechanistic information. Assumption rich – needs experimental data to calibrate Data integrative
Time scale/ impact	Faster questions/drug level	Slower questions/modality level
Use	Understanding of the time-concentration-effect relationship Dose predictions and TI Optimizing the design and interpretation of in vivo studies	Understanding target pharmacology Focus on setting project targets Enabler for biomarker selection and translational strategy Mechanistically informed PK, dose, and regimen predictions
Examples	Emax, indirect response, transduction models, TGI, etc.	PBPK, DILIsym, and disease-scale platform models

estimation of parameters describing potency (e.g., EC_{50}) and capacity or efficacy (e.g., E_{max}), which define the relationship between drug effect and concentration in plasma or in a specific tissue. PK/PD modeling can account for delays in drug response due to biodistribution to distal sites of action or transduction of biological signals. PK/PD can also be used to characterize physiological turnover and homeostasis, and stimulation or inhibition of these processes by therapeutic intervention [2]. In addition, PK/PD analysis routinely incorporates population variability and uncertainty, which can be useful in understanding variability in responses. This is called non-linear mixed-effects (NLME) modeling as it incorporates both fixed and random effects in order to understand inter- and intra-individual variabilities. PK/PD modeling is a helpful tool for optimizing the design and interpretation of in vivo studies. This may include predicting the outcome of a dose yet to be tested or determining the optimal time points for PK or response measures. In the translational space, the PK/PD parameters derived from relevant preclinical studies can be adjusted for the clinical scenario and used to provide prospective simulations to guide clinical dose level and frequency.

Potential drawbacks of PK/PD modeling are that it tends to focus on specific PD endpoints and so does not capture the behavior of the underlying system. As such, it may miss interactions between bio-signals or only capture behavior under a particular set of conditions. As such, PK/PD models may have limited capacity to extrapolate beyond collected datasets [4]. A major advantage of PK/PD models is that they don't require a lot of resources, including data or computational power to implement. Therefore, the results from PK/PD modeling can be realized in a short time frame, which makes them particularly suitable for influencing decision-making in the early stages of research. Over the years, simpler PK/PD modeling approaches have evolved to incorporate more mechanistic components, in so-called mechanism-based PK/PD models, to facilitate translation across species and/or between different patient populations. As such, the difference between PK/PD models and QSP is more of a continuum of increased complexity depending on the question to be answered.

12.1.2 QSP Modeling

With advances in computational power, access to greater biological knowledge, increased data availability, and a simultaneous explosion in the complexity of therapeutic modalities, more mechanistic questions are being asked within the drug discovery and development process. These questions require a more holistic quantitative description of the mechanism of action, and a different type of modeling approach is required. QSP models combine computational modeling and experimental data to examine the relationships between a drug, the biological system, and the disease process [5,6]. They are designed to investigate the effects of drug action on emergent behaviors of the underlying system, such as pathway, cell, tissue, organ, or multi-organ/whole-body process. To do so, QSP models integrate datasets from diverse studies, contexts, and spatio-temporal scales into a mathematical framework that reflects our knowledge of the system [7].

A key feature of these models is their explicit distinction between 'drug' and 'system' parameters. System-specific parameters typically include organ/tissue blood flow rates, receptor expression, internalization rates and turnover rates, cell lifespans, and homeostatic feedback mechanisms. Ideally, these parameters should be available from the literature or from prior experiments. Drug-specific parameters typically include PK parameters, such as clearance and volume of distribution, and pharmacological parameters, such as in vivo target affinity and intrinsic efficacy of compounds, and are usually estimated from PK/PD data gathered for the drug [8]. The strength of QSP models is that they only need to be as complex as the question they need to answer [9]. For example, QSP models can be used early in the drug discovery process to evaluate potential targets and to inform drug design for optimal efficacy and therapeutic index (TI). At this stage, system parameters can be extracted from the literature and hypothetical drug parameters can be used to perform exploratory simulations.

For more complex questions, disease-scale platform models are a type of QSP model which describes the interplay of multiple drug targets, pathways, and tissues in disease. These models will often characterize the untreated disease state and a broad range of disease phenotypes. They will enable comparisons, and hence differentiation, of a range of drugs and support multiple applications, including evaluation of the impact on efficacy and safety of monotherapies and combinations of drugs at different dosing regimens. These models give the added value that they can be reused, adapted, and repurposed for new treatments, questions, and indications [7].

Physiologically based pharmacokinetic (PBPK) modeling may be considered as a type of QSP modeling for characterizing and predicting the disposition of drugs. Traditionally, the plasma PK of drugs has been used to infer its tissue concentrations and interpret its PD or toxicodynamic effects. While this may be relevant for small-molecule drugs, it is not suitable for other modalities, such as biotherapeutics, where the plasma concentration may not accurately reflect the concentration at the site of action. As such, PBPK models have gained traction as a more mechanistic and realistic modeling approach to describe drug disposition [10]. These models are highly complex in nature and integrate drug-specific parameters, like intrinsic clearance and tissue partition coefficients, with a drug-independent structural model consisting of anatomical compartments (e.g., organs and tissues) connected via physiological processes, e.g., blood flow and lymph flow. The physiological nature of these models makes them relatively easy to scale between species, and the mechanistic nature makes them flexible enough to adapt to changes in pathological conditions.

Due to the mechanistic detail and larger scale, QSP model development requires more time and biological information compared with PK/PD model development. However, they offer a significant return on investment with respect to the fidelity of questions answered.

12.1.3 Why Are PK/PD Modeling and QSP Modeling Important for Biotherapeutics?

PK/PD and QSP models are useful tools to answer quantitative questions for all drug modalities. However, they are specifically useful for biotherapeutics as these drugs have some unique challenges compared to small-molecule drugs. First, the PK and PD of biotherapeutics are intrinsically linked. For most small molecules, the concentrations at which they are administered generally greatly exceed the concentration of the receptors that they are binding to. This is not true for biotherapeutic modalities, and as such, target-mediated drug disposition (TMDD) can be a major clearance mechanism [11]. Binding of biotherapeutics to soluble targets can also act as a significant sink for drugs. Also, the development of anti-drug antibodies to biotherapeutics can result in accelerated clearance of these drugs. TMDD, binding to soluble target sinks, and immunogenicity can restrict the concentration of drug available to exert its pharmacological effects [11,12]. In turn, heterogeneity of receptor expression on cells and the number of cells expressing the target can lead to variability in both PK and PD [12]. As such, it is of great importance to understand the PK/PD relationship of biotherapeutic drugs.

In addition, biotherapeutics display complex biodistribution, driven by their size [13,14]. Small molecules tend to have relatively rapid distribution driven by diffusion across membranes, which enables plasma concentrations to be used as a surrogate of tissue concentrations. Antibodies, cell therapies, and gene therapy vectors are much larger in size, and their diffusion across membranes is restricted. Their distribution is driven by processes, including extravasation and non-specific or receptor-mediated endocytosis. As a result, tissue concentrations are not in rapid equilibrium with plasma concentrations. In addition, many biotherapeutics have distal sites of action. As a result, there is often a need to predict tissue concentrations and relate these to pharmacological effects.

Many biotherapeutic modalities, including gene therapies, have intracellular mechanisms of action, providing a greater level of complexity. Often, data from disparate sources need to be integrated to understand the impact of multiple downstream processes. These can have multiple non-linearities leading to non-intuitive results. QSP modeling is ideal to tackle these and can be used to deconvolve complex mechanisms of action. In the next section, different biotherapeutic modalities will be introduced and their specific PK and PD considerations and challenges will be discussed.

12.2 PK/PD CHARACTERISTICS AND CONSIDERATIONS FOR BIOTHERAPEUTICS

12.2.1 Monoclonal Antibodies (mAbs)

Ever since the first FDA approval of a therapeutic monoclonal antibody (mAb) targeting CD3 in the 1980s, there has been a tremendous increase, especially in the last decade or two, in the discovery and development of biotherapeutics, including mAbs, and more recently novel modalities such as cell and gene therapies. With significant clinical successes and over 100 regulatory approvals for a variety of indications, including cancer and auto-immune diseases, biotherapeutics such as mAbs are changing treatment paradigms and are increasingly dominating R&D portfolio programs of large pharmaceutical and biotechnological companies for various therapeutic areas [15]. In addition, advances in antibody engineering technologies have enabled the successful development of novel format antibodies like ADCs, bispecific or multispecific Abs, antibody fragments, recycling Abs, and sweeping Abs. Such therapeutic modalities have several advantages as compared to conventional small molecules, including high potency, better target selectivity, less non-specific activity, and better half-lives [16–20].

Most therapeutic mAbs are of IgG (immunoglobulin G) format, which usually consists of two Fab fragments involved in antigen binding and target recognition and one Fc region involved in binding to various cell receptors such as FcRn (receptors mainly expressed on vascular endothelium and hematopoietic cells) and FcγR (receptors expressed on immune cells). With a much higher molecular weight (~150 kDa) and size (~14 nm) as compared to that of small molecules (usually less than 0.9 kDa and 1 nm in molecular weight and size, respectively), PK processes, including absorption, distribution, metabolism, and excretion (ADME), are quite different and unique for mAbs as compared to traditional small molecules [16–20]. For instance, glomerular filtration, tissue distribution, and cellular penetration are generally limited for large molecules such as mAbs. Rather, their PK and distribution are mainly governed by mechanisms such as pinocytosis for non-specific cellular uptake, extravasation, intracellular catabolism as a major pathway of elimination, target-mediated clearance, and salvage via FcRn recycling, which contribute to their longer half-life in the systemic circulation and perhaps limited tissue distribution [16,17].

The Fc region of the Ab binds to the FcRn receptors only at acidic pH with no or minimal binding at neutral pH in the blood. After pinocytosis, mAbs enter the early endosome where they bind to FcRn at pH 6. The Fc-FcRn complex protects the Abs from lysosomal degradation and is eventually recycled back from the endosomes to the cell membrane. Once at the cell surface, the Fc-FcRn complex dissociates due to weak binding at neutral pH, and eventually, Abs are released back into the systemic circulation, which contributes to their longer half-life [21–23]. Significant efforts have been carried out to further modulate the half-life of mAbs with the help of specific amino acid mutations in the Fc region of the Ab. Specific mutations can either increase or decrease the binding to the FcRn and thereby increase or decrease the half-life. Also, reducing the charge or isoelectric point (pI) of the mAbs can also increase their half-life given that the negative cell surface charge can cause repulsion with a negatively charged Ab and decrease pinocytosis [17,24]. Besides, the Fc region is also involved in engaging with the host immune system via FcγR expressed on various effector cells and mediating PD effects of mAbs. Most Abs mediate PD effects through one of the following mechanisms – neutralizing a target, suppressing a pathway, and either enhancing or suppressing immune effector function.

In addition to FcRn-mediated mechanisms, the PK of mAbs is also strongly influenced by its binding to the specific target, where it can undergo rapid clearance due to target-mediated endocytosis, the phenomenon commonly known as TMDD. Here, an Ab binds to the target, and the Ab-target complex rapidly internalizes and gets cleared from systemic circulation through protein catabolism. In contrast to FcRn recycling, which is non-specific and generally not saturable at usual doses of Ab, TMDD is capacity-limited and saturable, which is impacted by the dose of Ab, levels of the target, turnover of the target, and binding affinity of the Ab toward the target [25].

One of the MoAs for therapeutic mAbs is to neutralize the soluble target, which is pathogenic in nature. However, high dose levels and/or frequency of Abs and Ab-mediated target accumulation are major associated challenges. While research groups have explored increasing the exposure and decreasing the dosing frequency of Abs via enhanced FcRn binding affinity, the challenges associated with target accumulation remain. Other antibody formats such as recycling and sweeping antibodies have also been explored [26,27]. Recycling antibodies bind to the target in a pH-dependent manner with decreased binding at pH 5.5–6 as compared to neutral pH. As a result, once they enter the endosomes after pinocytosis, the target antigens are released from the complex for lysosomal degradation, while the Abs are salvaged via FcRn recycling. Sweeping antibodies also leverage improvements, including increased FcRn binding affinity and enhanced uptake of the Ab-target complex into the endosomes. Advancements in Ab engineering have led to the successful development of novel antibody formats, including bispecifics and multispecifics, which possess unique PK/PD attributes and considerations as compared to conventional Abs. For instance, a bispecific Ab has two binding domains, each specific to a different target and simultaneous engagement to form a trimolecular complex mediating PD effects. Such novel formats have the additional advantages of increased potency and reduced chances of resistance due to dual targeting. One such class of drugs is T-cell engagers for oncology with several molecules approved and many in early- and late-stage clinical development [12,18,28]. The bispecific T-cell engager (BiTE) molecules were the first-generation candidates with novel Ab formats without Fc and much smaller molecular weight and size as compared to full IgGs. Hence, their half-lives were much shorter as compared to conventional Abs. Next-generation molecules were half-life extension (HLE) with Fc-fusion proteins to improve half-life and reduce dosing frequency. Plenty of other Ab formats are leveraged for this class of drugs (e.g., bivalent or biparatopic) to further improve potency and specificity. Thus, each of these novel formats of Abs can have its own unique PK/PD characteristics and considerations.

Due to complex mechanisms and unique PK/PD characteristics, various PK/PD modeling and simulation (M&S) approaches are used in the discovery and development of therapeutic Abs [16,17]. Abs can typically exhibit either linear PK or non-linear PK. Linear PK properties are typically governed by non-specific mechanisms such as pinocytosis and FcRn recycling, which are usually not saturable at relevant doses of Abs. Such molecules display dose proportionality for exposure, and other PK attributes such as clearance and half-life are independent of the dose level. Linear PK profiles after IV administration of mAbs are typically biphasic (distribution and elimination) in nature where typical two-compartmental PK models can be used. Various preclinical species have been previously explored to characterize the PK of Abs and translate to humans. Among them, cynomolgus monkeys have been shown to translate best for the PK of mAbs in humans after using allometric scaling for the two species [29]. This can be attributed to the similar FcRn binding affinity for IgGs in the two species, which is known to be different for other species such as rodents. Other human FcRn transgenic mouse models such as Tg32 and Tg276, which express human FcRn, have also been explored as potential alternative models to characterize the PK of Abs, especially for rank-ordering candidates for PK, and to estimate linear PK parameters for non-cross-reactive molecules [30]. In addition to linear PK characteristics, Abs can also exhibit non-linear PK characteristics usually attributed to target binding and impacted by target levels, target turnover, and binding affinity. Such molecules do not display dose proportionality for exposures, and a clear impact of doses on clearances and half-lives is typically observed. In order to quantitatively characterize the non-linearity, empirical Michaelis-Menten (MM) or semi-mechanistic TMDD modeling approaches are utilized [31]. While MM approach uses a constant, Km, and the maximum rate of non-linear elimination, Vmax, as non-linear parameters, TMDD approach leverages target binding affinity (equilibrium dissociation constant, KD; association rate constant, Kon; and dissociation rate constant, koff), target concentrations, and target turnover (synthesis rate, ksyn; and elimination or internalization rate, kdeg/int) as related parameters. While the former approach is more empirical, knowledge of the interspecies differences for the different parameters can be included while characterizing non-linearity with the latter approach. Additional complex, mechanism-based PBPK models for mAbs have also been developed with the incorporation of measurable physiological (system-based) and drug-specific parameters [32]. More specifically, such models incorporate parameters related to plasma and lymph flow rates, rate of pinocytosis, lysosomal degradation, FcRn recycling, recirculation flow rate, lymphatic/vascular reflection coefficients, tissue volumes, and others. PBPK-based modeling approaches are extremely useful to characterize tissue distribution of mAbs (e.g., to brain or ocular compartments), triage drug candidates for lead selection, and leverage in vitro and in vivo data to apriori predict the distribution of Ab candidates to specific tissues and sites of action. As with PK, various modeling approaches are also used to characterize the PD of mAbs across different species and translate to humans. For instance, empirical, indirect response models have been used previously for oncology indications, where tumor static concentrations (TSCs) are estimated from preclinical mouse xenograft studies and used for efficacious dose projections in the clinic [33]. However, such empirical approaches typically do not account for interspecies differences on several parameters, including target expression or levels, binding affinity, and tumor dynamics. To address these gaps and given the complex MoA of Abs, more mechanistic QSP models have also been explored to characterize the PD of Abs and translate from preclinical species to humans [34].

12.2.2 Antibody-Drug Conjugates (ADCs)

Antibody-drug conjugates (ADCs) are a class of targeted therapies for cancer treatment that combine a specific antibody to a tumor antigen linked to a potent cytotoxic agent. The aim of this therapeutic is to target the cytotoxic drug (known as the payload) to tumor cells, thus maximizing efficacy while minimizing systemic toxicity. ADCs are clinically validated, with 14 ADCs currently FDA approved for various solid and hematological malignancies [35]. Following distribution into the tumor, ADCs bind to an over-expressed antigen on the surface of tumor cells. The ADC is internalized into the cell, and the payload is released in the endosomal or lysosomal compartment (via different mechanisms). The payload can then diffuse or be transported into the cytosol, where it can bind to its intracellular target, which triggers tumor cell killing. Alternatively, some ADCs (e.g., EDB-ADCs) rely on extracellular cleavage releasing membrane permeable payloads [36]. There are different types of payload classes, including microtubule inhibitors, DNA cross-linkers, and topoisomerase inhibitors, which are all potent cytotoxins. One potentially important aspect of the ADC mechanism is the 'bystander effect', whereby the cytotoxic drug released in the targeted cell can diffuse out of that cell and into other (non-target-expressing) tumor cells to exert its cytotoxic effect. This is important as solid tumors tend to be heterogeneous and not all cells in a tumor will express the targeted protein.

ADCs have been very successful, demonstrating transformative responses in the clinic, and consequently are one of the fastest growing classes of anticancer drugs. However, they have a complex mechanism of action, with many variables which need to be optimized to enable optimal delivery of the payload to express its cell-killing pharmacology. In addition, they are not truly targeted and can be taken up into non-malignant cells releasing their payload. As such, ADCs have been limited by several challenges in the clinic, including sub-optimal efficacy and dose-limiting toxicities. To understand the variables to be optimized, the complex disposition of ADCs must be delineated. Following administration of ADCs into the systemic circulation, they can deconjugate or be catabolized, which releases the payload. They can also bind to healthy cells expressing target or to soluble (shed receptors), both of which can be significant drug sinks. ADC clearance can therefore be a combination of first-order elimination and TMDD. The distribution of ADCs into tumors occurs mainly via paracellular passage across blood capillaries. Given the large size of ADCs, this is a slow process and often results in incomplete penetration of ADCs into the center of tumors. This can be exacerbated by the binding-site barrier effect, whereby binding of the ADC to the cells close to the blood supply of the tumor, and subsequent internalization, restricts the ADC from distributing deeper into the tumor [37]. In contrast, the payload has PK properties consistent with a small-molecule drug. As such, the released payload can diffuse out of cells into the extracellular space and into the systemic circulation where they tend to be rapidly cleared, limiting general tissue toxicity. In addition, payloads can be actively transported out of tumors by efflux transporters, which can limit the intracellular exposure to the cytotoxic drug. The combination of large (e.g. antibody) and small molecules (e.g. linker, payload) in an ADC contributes to their complex ADME properties.

The inherent complexity of ADCs lends itself well to the use of mathematical modeling and simulation, to map out the mechanism of action and to consider the impact of multiple variables. Several mathematical models have been published for ADCs over recent years, evolving from empirical and semi-mechanistic PK/PD models, toward more mechanism-based models [38]. PK/PD models have proven very useful in the preclinical and clinical development of ADCs to maximize information obtained from experimental data, while minimizing resource utilization. These models have been used to establish in vitro to in vivo correlation of ADC efficacy [33], quantify and translate from in vivo studies to the clinic [39], and differentiate between ADCs binding to the same target [40]. However, they are limited in their ability to predict efficacy across different targets and to inform design parameters. QSP models contain sufficient mechanistic details to enable an understanding of the processes critical to an ADC's performance and to perform multiscale predictions. These models describe cellular mechanisms, tumor penetration, preclinical to clinical translation, and clinical simulations [38]. A seminal paper by Shah and coworkers presented a bench-to-bedside translation of brentuximab vedotin using a multiscale QSP model [41]. This model provided translation from preclinical experiments to humans to successfully predict clinical outcomes for brentuximab vedotin. A similar model structure and translational strategy was also applied by others for the successful prediction of clinical outcomes for T-DM1 [42,43] and inotuzumab ozogamicin [44]. Further details are provided in the case studies (section 4). There are fewer publications on QSP models for ADC toxicity [45,46], which is a definite gap in the science.

12.2.3 Cell Therapies

Cellular immunotherapy or adoptive cell therapy (ACT) includes tumor-infiltrating lymphocytes (TILs), engineered T-cell receptor T cells (TCR-T), chimeric antigen receptor T cells (CAR-Ts), chimeric antigen receptor natural killer cells (CAR-NK), and other cell types. Here, patients are usually subjected to autologous or allogeneic cell therapies, where on autologous therapy, they are first subjected to leukapheresis, and immune cells are isolated, often transduced ex vivo, expanded to large numbers, and then infused back for treatment. In contrast, allogeneic cell therapies are *'off-the-shelf'* and derived from normal healthy donors or other cell sources. The expanded and genetically modified immune cells better interact with the target on the tumor cells, in an MHC-dependent or independent manner (depending on the specific type of cell therapy), which subsequently leads to the induction of key signaling events such as immune cell activation and proliferation, cytokine production, and eventually tumor cell lysis [47,48]. Currently, there are several approved autologous CAR-T therapies targeting CD19 in hematological malignancies and B-cell maturation antigen (BCMA) for multiple myeloma patients [49]. These therapies have resulted in unprecedented clinical outcomes for relapsed/refractory terminally ill patients; hence, there has been an exponential increase in research for the development of cell-based immunotherapies, especially for the treatment of cancer.

Cellular therapies like CAR-Ts exhibit unique pharmacokinetics (PK) that differ greatly from conventional therapeutics. The PK profiles of small or large molecules would typically capture the processes of ADME. However, some of these aspects are not directly applicable to novel modalities like CAR-Ts. In contrast to small or large molecules, CAR-Ts undergo rapid proliferation or expansion after infusion in patients; thus, both total cells infused and those expanded in vivo are quantitated while characterizing the PK. Hence, they are considered as 'living biologics' or 'replicating therapeutics' and their in vivo kinetic disposition is termed as 'cellular kinetics' (CK) [50].

As shown in Figure 12.1, a typical CK profile of CAR-T therapy is multiphasic and includes four distinct phases, including margination or distribution, expansion, contraction, and persistence [50]. **Margination or Distribution:** After administration in patients, CAR-Ts rapidly disappear from the bloodstream within a few hours and get extensively distributed in peripheral tissues (e.g., lung, spleen, lymph, and bone marrow), which corresponds to the margination or distribution phase. **Expansion:** Upon target antigen recognition and engagement, CAR-Ts undergo rapid proliferation and expansion along with the release of cytokines and subsequent killing and clearance of the tumor cells. The expansion phase is followed by biexponential decline of CAR-Ts

A

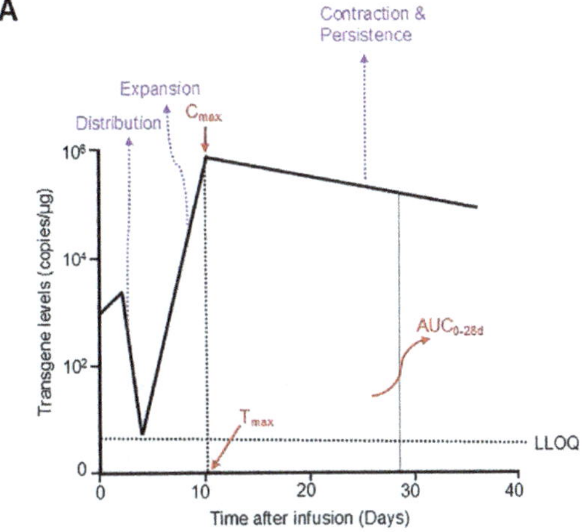

B

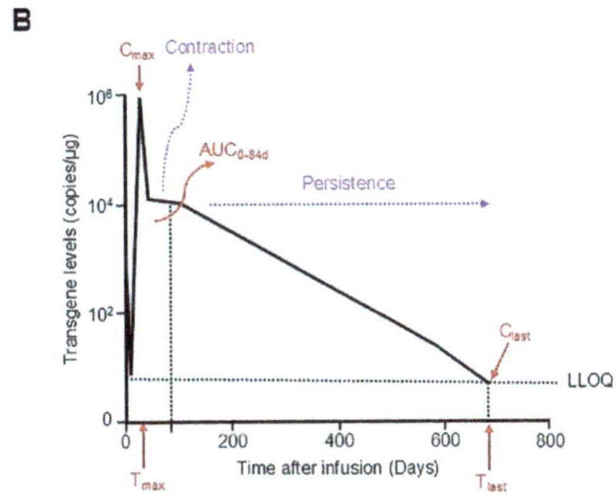

FIGURE 12.1 A typical cellular kinetics (CK) profile for CAR-Ts demonstrating multi-phasic nature and key CK parameters. Cmax, maximum observed concentration; Tmax, time of Cmax; AUC0-28d, area under the concentration-time curve from time zero to 28 days after dosing; AUC0-84d, area under the concentration-time curve from time zero to 84 days after dosing; Clast, last quantifiable concentration; Tlast, time of last quantifiable concentration. Figure adapted from [50].

that are termed as contraction and persistence phases. **Contraction:** The contraction is likely due to loss of antigen stimulation upon tumor clearance. Other likely mechanisms include T-cell exhaustion or activation-induced cell death (AICD). **Persistence:** The low levels of CAR-Ts in the persistence phase are thought to represent the memory-type cells which sustain for longer periods of time in the patient's body. Overall, the CK profile of CAR-Ts is described by two general sets of PK parameters to capture the 'expansion' and 'persistence' phases: Cmax (maximal/peak expansion), Tmax (time to reach Cmax), and AUC (area under the curve) over a shorter period (e.g., AUC_{0-28d}) are used to describe the expansion phase, while Clast (last measurable concentration), Tlast (time of last measurable concentration), and half-life are used to capture the 'persistence' phase of CAR-Ts [50].

The multiphasic CK profile of CAR-Ts is attributed to how the CAR-Ts get distributed in patients post-infusion and how they interact (MOA) with the tumors and patient's immune system. Monitoring CK in preclinical and clinical studies can help better understand CAR-T in vivo expansion and persistence and dose-exposure-response correlation. However, significant variabilities have been reported across different clinical trials and the dose-exposure-response relationship is often convoluted for CAR-T therapies. Typically, there is a lack of clear correlation between dose and exposure as well as dose and response (safety or efficacy) for CAR-Ts [51,52]. In contrast, CAR-T expansion (Cmax and AUC_{0-28d}) has been shown to correlate well with responses in clinical studies. Since cytokine release is anticipated with this MOA, expansion is also linked with adverse events or toxicity, including cytokine release syndrome (CRS). Besides, a variety of factors, including patient-related (e.g., tumor burden) or product-related (e.g., CD4:CD8 or immune phenotype), can potentially impact the CK profile and overall clinical performance (safety and efficacy) of CAR-Ts [51,52].

Typical PK compartmental models that have been traditionally utilized are not applicable for CK of CAR-Ts. Non-compartmental analysis (NCA) and/or model-based approaches (NLME model with piecewise function) along with covariate analysis are typically used to characterize CK, as well as evaluate the dose-exposure-response (safety and efficacy) relationship and impact of various covariates. Besides, various mechanistic, PBPK, and QSP modeling approaches have been developed for CAR-T therapy that have been extensively summarized elsewhere along with their potential applications [51–54].

12.2.4 Gene Therapies

Gene therapy is a form of treatment that involves the delivery of genetic material to cells, in order to modify the expression of a specific gene or set of genes. The goal of gene therapy is to treat or prevent diseases that are caused by genetic mutations or deficiencies [55]. There are several different types of gene therapy, including gene replacement therapy, gene editing, and gene silencing.

More than a thousand clinical experiments have been conducted with gene therapy since it was first developed more than 20 years ago [55]. Nonviral vectors have appealing characteristics for clinical use, but they are inefficient in vivo, prompting future advancements in these vectors. It is important for vector development and optimization to understand how nonviral vectors behave within cells. The model-based approach is an effective tool for comprehending and describing the various mechanisms that gene transfer systems should overcome inside the body. A model-based approach enables the known use of PK/PD modeling in conventional therapeutics [55].

PK considerations for gene therapies include how the therapeutic gene is delivered to the target cells, the efficiency of the delivery method, and the stability of the therapeutic gene within the target cells [56]. The most common delivery method for gene therapy is the use of viral vectors. These are viral particles that have been genetically modified to carry the therapeutic gene into the target cells. Some of the most commonly used viral vectors for gene therapy include adenoviruses, adeno-associated viruses (AAVs), and lentiviruses. Each of these viral vectors has its own set of advantages and disadvantages, and the choice of vector will depend on the specific disease being targeted and the characteristics of the therapeutic gene [56]. Once the therapeutic gene has been delivered to the target cells, it must be expressed in order for the therapy to be effective. The level of gene expression will depend on a number of factors, including the efficiency of the delivery method, the stability of the therapeutic gene within the target cells, and the activity of the therapeutic gene. Factors that can affect the stability of the therapeutic gene include the presence of epigenetic modifications, such as methylation or histone modification, and the activity of the cell's DNA repair machinery [56]. Additionally, the therapeutic gene may be subject to degradation by cellular enzymes, such as nucleases.

PD considerations include the level of expression of the therapeutic gene within the target cells, the duration of gene expression, and the safety and efficacy of the therapy in treating the target disease [57]. The safety and efficacy of the therapy will depend on the specific disease being targeted and the nature of the therapeutic gene. Gene replacement therapy involves the delivery of a functional copy of a gene that is missing or mutated in a patient with a genetic disorder [57]. This type of therapy is typically used to treat diseases that are caused by a single-gene mutation, such as cystic fibrosis or hemophilia. In contrast, gene editing involves the precise modification of a specific gene or set of genes. This type of therapy is typically used to treat diseases that are caused by a specific genetic mutation, such as sickle cell anemia or Tay-Sachs disease. Meanwhile, gene silencing involves the inhibition of the expression of a specific gene or set of genes. This type of therapy is typically used to treat diseases that are caused by the over-expression of a specific gene, such as cancer [51,57].

Other considerations for gene therapies include the potential for off-target effects, the risk of mutagenesis, and the potential for immune response to the viral vector or therapeutic gene. Additionally, the specific disease being targeted will also play a role in the design and development of a gene therapy. For example, the PK/PD considerations for a gene therapy used to treat a chronic disease like cancer will be different than for a gene therapy used to treat a genetic disorder like cystic fibrosis [57].

12.2.5 Vaccines

A vaccine is a biological preparation that provides active acquired immunity to a particular disease. The goal of a vaccine is to prevent infection by stimulating the immune system to recognize and fight the pathogen that causes the disease. There are several different types of vaccines, including inactivated, live attenuated, subunit, and DNA/RNA vaccines [58].

For vaccines, PK factors include the route of administration, the dosage, and the schedule of administration. The most common routes of administration for vaccines include injection (subcutaneous, intramuscular, and intradermal) and oral. The choice of route will depend on the specific vaccine, the population being vaccinated, and the disease being targeted [58,59]. Some vaccines, such as live attenuated vaccines, may only be administered via one route, while others, such as subunit vaccines, may be administered via different routes. The dosage and schedule of administration are also important PK considerations. The dosage will depend on the specific vaccine and the population being vaccinated. Infants, for example, may require a different dosage than adults. Some vaccines, such as the measles, mumps, and rubella (MMR) vaccine, may require two or three doses for full protection, while others, such as the tetanus vaccine, may only require a single dose [59].

Immunogenicity refers to the ability of the vaccine to stimulate an immune response. The immunogenicity of a vaccine will depend on the specific vaccine and the population being vaccinated. Some vaccines, such as live attenuated vaccines, may be highly immunogenic, while others, such as inactivated vaccines, may be less immunogenic. Safety of the vaccine is an important consideration, as the vaccine should not harm the person receiving it [58,59]. Adverse effects of the vaccine are expected but should be minimal and acceptable for the benefits the vaccine provides.

Efficacy refers to the ability of the vaccine to protect against the disease. The efficacy of a vaccine will depend on the specific vaccine and the population being vaccinated. Some vaccines, such as the human papillomavirus (HPV) vaccine, may be highly effective in preventing infection, while others, such as the influenza vaccine, may be less effective. The efficacy of a vaccine is also dependent on the coverage rate, i.e., the percentage of the population that receives the vaccine; the more the coverage, the more effective the vaccine is in controlling the spread of the disease [59]. Vaccines include the potential for off-target effects, the risk of mutagenesis, and the potential for immune response to the viral vectors or therapeutic genes [60]. Additionally, the specific disease being targeted will also play a role in the design and development of a vaccine. For example, a vaccine for a highly contagious disease like measles will have different PK/PD considerations than a vaccine for a less contagious disease like hepatitis B [58–60].

In conclusion, PK/PD characteristics and considerations for vaccines are complex and multifaceted and require careful study and optimization in order to develop safe and effective vaccines. The route of administration, dosage, schedule of administration, immunogenicity, safety, and efficacy are all important factors to consider when developing a vaccine [58]. Additionally, the specific disease being targeted, the population being vaccinated, and the potential for off-target effects, mutagenesis, and immune response must be considered.

12.2.6 mRNA/siRNA/Oligonucleotide Therapeutics

mRNA, siRNA, and oligonucleotide therapeutics are based on the use of nucleic acids to modulate gene expression and have the potential to treat a wide range of diseases, including genetic disorders, cancer, and viral infections [61,62].

mRNA therapeutics involve the delivery of synthetic mRNA molecules to cells, with the goal of inducing the production of a specific protein. siRNA therapeutics involve the delivery of small interfering RNAs (siRNAs) to cells, with the goal of inhibiting the expression of specific genes [63]. Oligonucleotide therapeutics involve the delivery of short, synthetic nucleic acid molecules to cells, with the goal of modulating gene expression. The mRNA/siRNA/oligonucleotide molecules are typically delivered to cells via a lipid nanoparticle (LNP) delivery system, which allows for the efficient and targeted delivery of the mRNA to cells. PK considerations for these therapeutics include the efficiency of the LNP delivery system, the stability of the mRNA/siRNA/oligonucleotide within the target cells, and the pharmacokinetics of the LNP delivery system. The LNP delivery system should be able to effectively target the cells where these therapeutics will be needed and should be stable enough to allow for efficient translation into the desired protein/inhibition of gene expression [61,63].

The level of gene expression will depend on a number of factors, including the efficiency of the LNP delivery system, the stability of the nucleic acid within the target cells, and the activity of the nucleic acid. Factors that can affect the stability of the nucleic acid include the presence of epigenetic modifications, such as methylation or histone modification, and the activity of the cell's DNA repair machinery [62]. Moreover, the nucleic acid may be subject to degradation by cellular enzymes, such as nucleases.

There is a potential for off-target effects, the risk of mutagenesis, and immune response to the LNP delivery system or nucleic acid. Additionally, the specific disease being targeted will also play a role in the design and development of the therapy. For example, a therapy for a chronic disease like cancer will have different PK/PD considerations than a therapy for a genetic disorder like cystic fibrosis [62,63].

12.3 USE OF M&S APPROACHES IN THE DISCOVERY AND DEVELOPMENT OF BIOTHERAPEUTICS AND NOVEL MODALITIES

Computational and mathematical modeling approaches can aid in a better understanding of the complexities associated with biotherapeutics and novel modalities. To that end, quantitative decision-making to guide the design of the molecule, final lead selection, preclinical to clinical translation, and designing clinical studies can substantially boost the overall R&D efficiency and productivity of biotherapeutic drugs.

Various model-based approaches (top-down and/or bottom-up) can be leveraged at various stages of the discovery and development of biotherapeutics and novel modalities. At early stages in discovery, it can be used in early target feasibility assessments to identify a promising and viable target for a particular disease or indication. Known biology of the target (e.g., whether it is a driver of the disease or upstream/downstream of specific signaling pathways, endogenous ligands, etc.) and learnings from competitive space (e.g., successes and failures) or clinical precedence with the target can be leveraged to inform the models and for GO/NO-GO decisions. A quantitative framework can also guide target molecule profile (e.g., ideal binding affinity range, desired mechanism of action, and molecular format) and modality selection for a particular target with a higher probability of clinical success during early stages of discovery. In addition, they can help with the designing of in vitro and in vivo (PK/PD, safety, and efficacy) studies (e.g., selection of doses and dosing regimen), as well as leveraging the preclinical data for triaging lead molecules and selecting the final lead candidate. Interspecies scaling and translation to humans for PK, safety, and efficacy can also be accomplished with allometric scaling and translational quantitative PK/PD approaches. Such approaches eventually support the design of clinical studies, including starting dose selection for the first-in-human (FIH) study, dose-escalation strategy based on efficacious dose and therapeutic window predictions, and RP2D selection. Finally, reverse translation learnings from successful and failed clinical programs to guide the development of next-generation molecules can also be accomplished.

Depending on the stage of the program and the availability of data, various modeling-based analyses can be carried out for biotherapeutics. A quantitative framework should typically integrate various types of data (in vitro, in vivo, and clinical) and data from various sources (literature and/or molecule-specific) based on the model structure and objective of the modeling work. On the one hand, classic, empirical, and compartmental PK or PK/PD models can be used to answer simple questions such as dose-exposure-response analysis, human dose and TI predictions, or designing of preclinical or clinical studies. On the other hand, semi-mechanistic or mechanistic models such as TMDD-, PBPK-, and QSP-based approaches can be leveraged to better understand the mechanism of action in a quantitative fashion where information on the molecule properties and target pharmacology is also included.

Over the past decade or two, the use of M&S approaches has exponentially grown for various purposes in the discovery and development of drugs. For instance, population PK/PD models for dose selection or PBPK models to characterize drug-drug interactions (especially for small molecules) are routinely used [64]. Moreover, regulatory agencies are now more open towards the use of model-informed drug development (MIDD) to support product development and regulatory filings than ever before. In addition, recent advances in analytical technologies have enabled the capturing of newer, diverse, and more comprehensive datasets, during the preclinical and clinical developmental stages, which were perhaps not available earlier. Similarly, of late, there has been a huge investment in advancing computational approaches, including the use of real-world data, big data, AI, and ML in the drug development industry [65]. Overall, there are increasing trends in developing and using mathematical modeling approaches for biotherapeutics. Fit-for-purpose mindset depending on factors such as scientific question, data availability, stage of the program, model assumptions, gaps/limitations, and relevance is crucial for the effective and impactful utilization of MIDD in biotherapeutics.

In the next section, we highlight case studies from literature where various M&S approaches (population PK, PK/PD, TMDD, PBPK/PD, and/or QSP) were leveraged for various applications at different stages of drug development across different modalities and therapeutic indications.

12.4 CASE STUDIES

12.4.1 Application of Mechanistic PK/PD Models to Inform Early Drug Discovery Decisions for Biotherapeutics

Key decisions at the early stages of biotherapeutic drug discovery include determining the feasibility of drugging a target, prioritizing between targets, and defining optimal drug design properties. To avoid the synthesis and screening of large numbers of tool molecules, these decisions can be informed initially using model-based approaches and data from the public domain. A model framework describing the intended pharmacology of the biotherapeutic can be constructed and parameterized with physiological or system properties available from the literature, including compartment volumes, cell numbers, receptor expression levels, soluble protein concentrations, and internalization rates. Drug properties, such as affinity of the individual binding arms and PK half-life, can be set to nominal values and tested for sensitivity. These models can be used to predict PK (including non-linearities due to TMDD), target engagement, and effective dose.

Marcantonio et al. [66] presented a workflow for the application of mechanistic PK/PD models, for early feasibility analysis (EFA) of biotherapeutics in early drug discovery stages. The approach was validated by the prediction of clinically efficacious dose, without fitting to PK or PD data, for nine approved biotherapeutics across a range of targets and indications. Separate mechanistic PK/PD models were developed for each target class analyzed, including soluble targets, membrane receptors, and bispecific targets. All were human models describing drug administration, PK, target binding, and target dynamics in one or more compartments. For each drug, a criterion for defining effective dose (e.g., 90% sustained target inhibition) was chosen. Models were then simulated to predict PK, target engagement, and target inhibition at different doses, to determine the dose required to achieve the criterion. This model-predicted effective dose was compared to clinically approved doses for each drug. For the nine biotherapeutics tested, the predicted efficacious dose from the modeling analyses was generally within ~ threefold of the clinically approved doses.

The PK/PD model-based approach was able to successfully predict the effective doses of two anti-TNF-α drugs: adalimumab and infliximab, for the treatment of rheumatoid arthritis (RA). Despite binding to the same target, these drugs have very different PK and binding properties, leading to different approved doses and regimens. The modeling analysis was able to provide mechanistic insight as to why the effective doses are significantly higher than the dose predicted from a more straightforward exposure versus potency comparison. The efficacious concentrations are much higher than the Kd (~ 1,000-fold), which is due to an increase of total TNF-α above baseline on drug binding to this soluble target. This has been shown to occur due to HLE effects where the short-lived soluble targets form long-lived complexes with the administered antibodies. These analyses demonstrate the advantage of applying a mechanistic PK/PD model for dose predictions. In addition, a sensitivity analysis showed that TNF-α half-life was a sensitive target parameter impacting predicted clinical dose, enabling prioritization of experiments during drug development.

The EFA approach was also able to accurately predict the effective dose for amivantamab, an anti-EGFR, anti-c-met bispecific antibody, approved for the treatment of patients with non-small cell lung cancer. Interestingly, PK data from a monospecific EGFR mAb (panitumumab) and an anti-c-met antibody (emibetuzumab) were used to benchmark target expression levels, as they both exhibit non-linear PK due to TMDD. These target expression levels were then applied to predict PK and the effective dose of amivantamab, showing the flexibility of this model-based approach to make use of data already available from the literature. These analyses show how model-based approaches can be used to inform key decisions for biotherapeutics early in drug discovery, before the collection of PK/PD data.

12.4.2 QSP Modeling of ADCs for Preclinical to Clinical Translation and Optimization of Doses for Different Oncology Indications

A QSP model was developed for inotuzumab ozogamicin, a CD22-targeting ADC for B-cell malignancies [44]. The model was used for preclinical to clinical translation and to optimize doses and regimens for a new indication being explored (acute lymphocytic leukemia [ALL]) versus the original indication (non-Hodgkin's lymphoma [NHL]), which had been terminated due to lack of superiority versus standard of care.

The model included a plasma PK model characterizing the disposition and clearance of inotuzumab ozogamicin and its released payload N-Ac-γ-calicheamicin DMH, a tumor disposition model describing ADC diffusion into the tumor extracellular environment, and a cellular model describing inotuzumab ozogamicin binding to CD22 on tumor cells, internalization, intracellular N-Ac-γ-calicheamicin DMH release, binding to DNA, or efflux from the tumor cell (Figure 12.2). Key preclinical data used to parameterize the model included receptor expression on cancer cell lines, binding affinities, internalization rates, and PK data. The model was used to predict intracellular payload concentrations, which were linked to a tumor growth inhibition (TGI) model and calibrated to TGI data from mouse xenograft studies. The preclinical model was translated to the clinic by incorporating human PK for inotuzumab ozogamicin and clinically relevant tumor volumes, tumor growth rates, and values for CD22 expression in the relevant patient populations. The resulting stochastic models predicted progression-free survival (PFS) rates for inotuzumab ozogamicin in patients comparable to the observed clinical results. The model suggested that a fractionated dosing regimen was superior to a conventional dosing regimen for ALL but not for NHL.

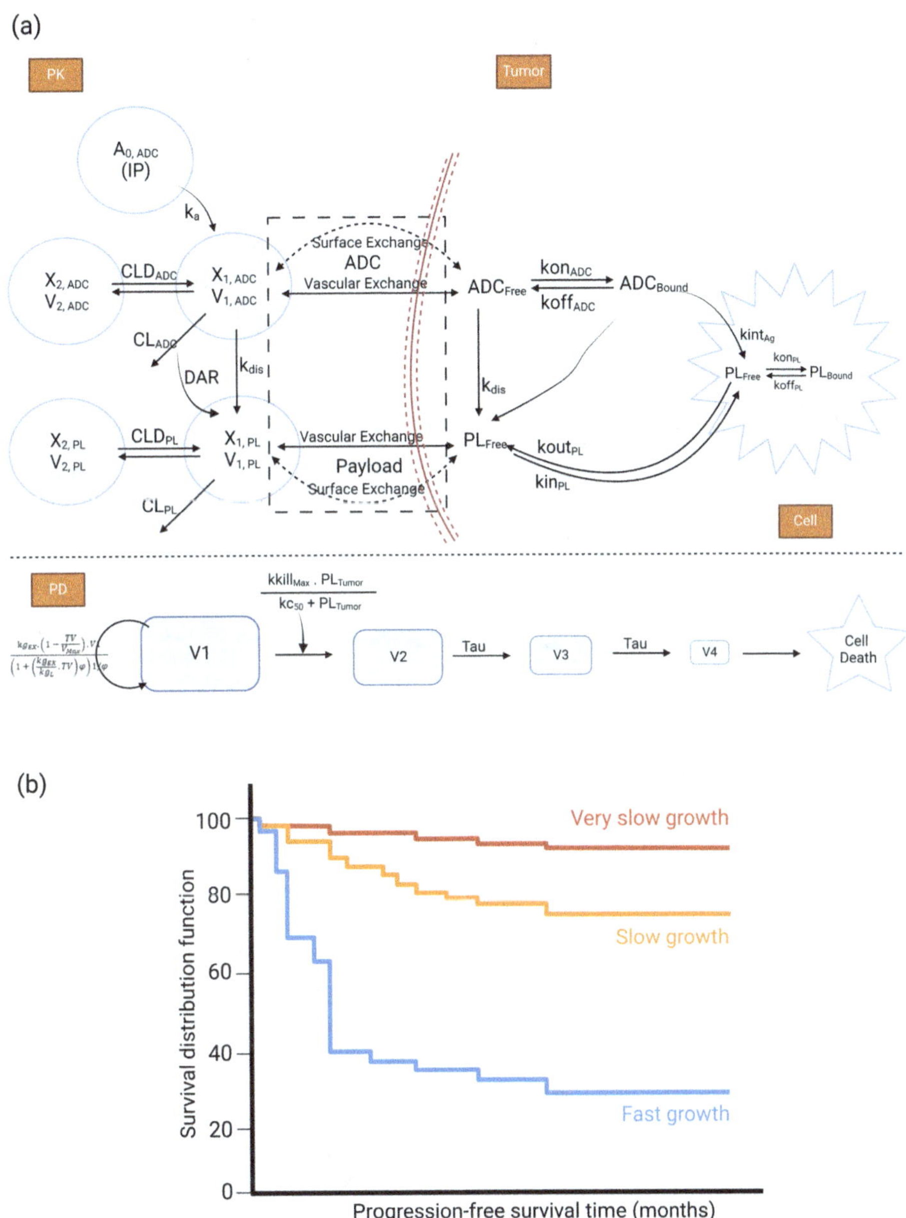

FIGURE 12.2 (a) A schematic representation of a QSP model with relevant parameters describing the PK, penetration into solid tumors (dotted box), and PD. Parameters related to the drug penetration into the solid tumors were excluded for liquid tumor indications. (b) Representative model-based PFS rate predictions in NHL patients (with different growth rates) for nimotuzumab ozogamicin at 1.8 mg/m^2 every 4 weeks. Figure adapted from [44].

A sensitivity analysis was performed to give insight into the parameters defining, or even limiting, efficacy of inotuzumab versus NHL. Tumor growth rate was found to be the most sensitive parameter and suggested that for the more aggressive NHL subtypes like diffuse large B-cell lymphoma (DLBCL) patients would require significantly higher doses for efficacy, compared with slower-growing NHL subtypes such as follicular lymphoma. Calicheamicin efflux from the tumor cell was a sensitive parameter, which is important as N-Ac-γ-calicheamicin DMH is known to be a substrate for MDR1, an efflux transporter that is upregulated on many tumor cell types. The least sensitive parameter was CD22 receptor expression, which indicated the suitability of this receptor as an ADC target due to its high expression across B cells and rapid internalization rate. These findings suggest that MDR1 status in patients would be a more useful diagnostic of efficacy than CD22 receptor expression. In summary, the QSP model developed for inotuzumab ozogamicin was able to give useful mechanistic insight into optimal dosing regimens and sensitive parameters impacting outcomes. This knowledge could be applied to optimize the design of ADCs in the discovery phase of research and/or for the selection of predictive diagnostics in the clinic.

12.4.3 A Translational Platform PBPK Model for Antibody Disposition in the Brain

For many monoclonal antibodies (mAbs) and other targeted drug modalities, the molecular target may be located within the tissues. As a result, their pharmacodynamic (PD) and exaggerated PD/toxic effects are a function of tissue concentrations. Unlike small molecules, the tissue pharmacokinetics (PK) of mAbs cannot be assumed to be in rapid equilibrium with plasma concentrations. One particularly challenging case is the delivery of mAbs to specific sites of action in the brain. Methodologies available to measure brain exposure of antibodies in the clinic are usually limited to cerebrospinal fluid (CSF) concentrations, which do not represent delivery of mAbs across the blood-brain barrier (BBB) endothelial cells. Preclinically, mAb concentrations in brain homogenate are routinely used to determine mAb exposure, which may not accurately represent concentration at the site of action.

An alternative option is to use a model-based approach such as PBPK modeling to predict the concentration of mAbs in different regions of the brain. These models are particularly attractive, as they incorporate anatomical and physiological factors which are very amenable to translation across species. PBPK models account for mechanism specifics such as FcRn or target binding and are flexible enough to adapt to changes in pathological conditions.

A platform PBPK model for the disposition of mAbs in the brains of mouse, rat, monkey, and human was developed by [67] Chang et al. The model accounts for known anatomy and physiology of the brain, including the presence of distinct BBB and blood-CSF barrier (BCSFB). Physiological processes responsible for the disposition of non-targeting mAbs in the brain, including CSF circulation, movement of macromolecules within the brain, and FcRn-mediated mAb disposition, are described in the model. The model builds upon a platform model previously developed to characterize the plasma and tissue disposition of mAbs in preclinical species and human [10]. The complete PBPK model includes 16 tissue compartments (blood, lung, heart, kidney, muscle, skin, liver, adipose, thymus, bone, small intestine, large intestine, spleen, pancreas, other/carcass, and brain) and a lymph node compartment, connected to each other in an anatomical manner using blood and lymph flow. Each tissue compartment, except the brain, is composed of vascular, endosomal, interstitial, and cellular sub-compartments (Figure 12.3). The brain compartment is divided into CSF circulation system and brain parenchyma and describes the limited

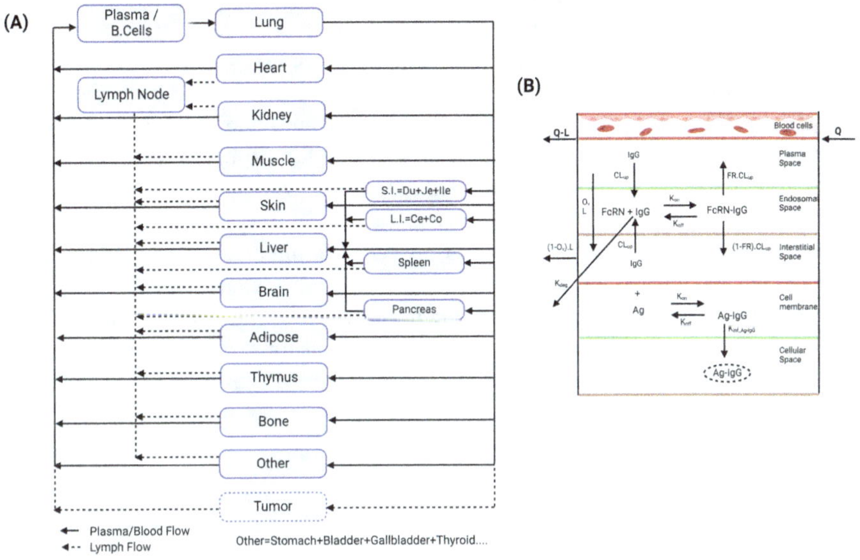

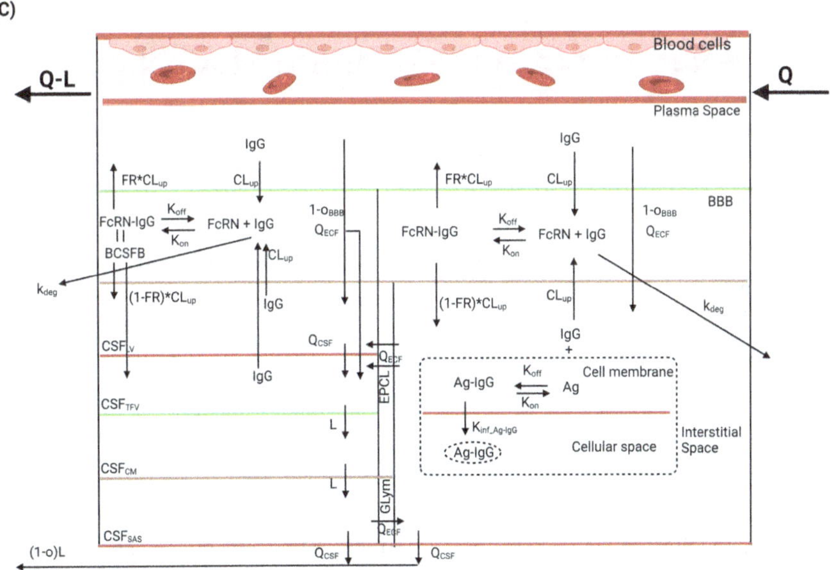

FIGURE 12.3 Representative scheme of the whole-body and brain-specific PBPK model for mAb disposition across species. (A) Whole-body PBPK model represented at the anatomical level, (B) Standard tissue compartment within the PBPK framework, and (C) Brain compartment divided into brain parenchyma and CSF sections (Figure adapted from [67]).

mAb entry into the brain via bulk flow through the BBB or BCSFB or via non-specific pinocytosis/transcytosis. FcRn-mediated efflux, recycling, and transcytosis of mAb across BBB and BCSFB were also incorporated into the brain capillary endosomal spaces.

First, the model was used to characterize PK data (from disparate sources) describing the disposition of mAbs in the rat brain, including data determined using microdialysis to quantify PK in different brain regions. Most of the model parameters were fixed based on literature-reported values, and only three parameters were estimated using rat data. The rat PBPK model was then translated to mouse, monkey, and human, simply by changing the values of physiological parameters corresponding to each species. The translated PBPK models were validated by *a priori* prediction of brain PK of mAbs in all three species, comparing predicted exposures with observed data. The platform PBPK model was able to predict all the validation PK profiles reasonably well (within threefold), without estimating any parameters.

The platform PBPK model described provides an unprecedented quantitative tool for the prediction of mAb PK at the site of action in the brain, and preclinical to clinical translation of mAbs is developed against central nervous system (CNS) disorders. The proposed model could be further expanded to account for target engagement, disease pathophysiology, and novel mechanisms, to support the discovery and development of novel CNS targeting mAbs.

12.4.4 Empirical Model-based Cellular Kinetic Analysis of CAR-Ts in Clinical Studies, Investigation of Dose-exposure-response Relationship, and Covariate Modeling

The first empirical, NLME population model for CAR-T therapy was developed by Stein et al. [68] to characterize the clinical CK of an anti-CD19 CAR-T therapy (tisa-cel) and to investigate the impact of various covariates on the CAR-T peak and expansion. Specifically, the study also focused on assessing the impact of CRS-treating therapies (tocilizumab and corticosteroids) on the kinetics of in vivo tisa-cel expansion, comparing the peak levels and the expansion rates in patients treated with tocilizumab or corticosteroid therapy against those who did not receive such therapies. Data from relapsed/refractory B-cell ALL patients on tisa-cel in two phase II studies were used for this modeling exercise. As shown in Figure 12.4, the model assumes two types of cell population (short-lived effector and long-lived memory phenotype) and uses piecewise function to capture three distinct phases of CAR-T CK profile – expansion, contraction, and persistence – each with distinct rate constants. Based on the modeling-based analysis, the doubling time for the expansion phase, the half-life for the initial decline/contraction phase, and the half-life for the terminal persistence phase were estimated. Besides, both NCA-based and model-based estimates for Cmax and AUC at day 28 were determined to have a good correlation. The model predicted that there was no effect of CRS-treating therapies

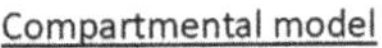

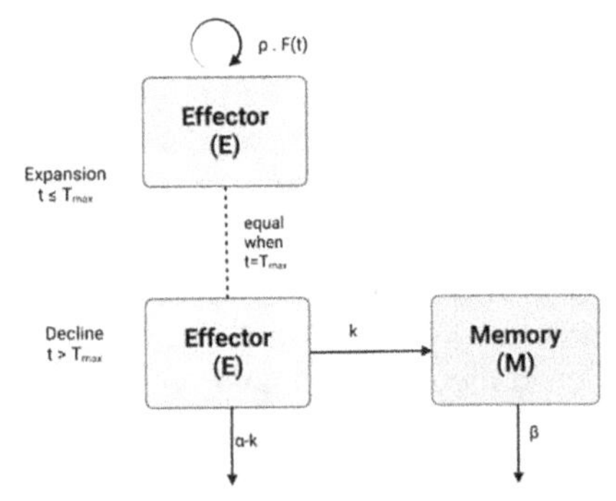

Definitions

$\rho = \log(fold_x)\,/T_{max}$

$k = F_B\,.(\alpha - \beta)$

Equations

$$\frac{dE}{dt} = \rho\,.\,F(t)\,.\,E$$

$$\frac{dE}{dt} = -\,\alpha\,.\,E$$

$$\frac{dM}{dt} = k\,.\,E - \beta\,.M$$

Initial Conditions

$$at\ t = 0: \quad E(0) = \frac{C_{max}}{fold_x}$$

$$at\ t = T_{max}: \quad E(T_{max}) = C_{max}$$

$$M(T_{max}) = 0$$

Comedication Effect: F(t)

$$F(t) = f_{toci}\,(t).f_{ster}\,(t)$$

$$f_{toci}\,(t) = \begin{cases} 1 & t \le T_{toci} \\ F_{toci} & t > T_{toci} \end{cases}$$

$$f_{ster}\,(t) = \begin{cases} 1 & t \le T_{ster} \\ F_{ster} & t > T_{ster} \end{cases}$$

FIGURE 12.4 A compartmental model to characterize T-cell kinetics. Key definitions, initial conditions, and equations for the model are shown in the figure. The model includes effector cells (E) with an exponential growth rate ρ up to the time to maximal expansion (T_{max}) and then they decline at the rate (α-k) or get converted to the memory cells at the rate k. The memory cells then undergo decline at a rate β. Additional details on the model can be found at [68]. Cmax, maximal concentration; foldx, fold expansion; Fster, effect of steroids; Ftoci, effect of tocilizumab; Tster, time to maximal expansion with steroid therapy; Ttoci, time to maximal expansion with tocilizumab therapy; FB, fraction of transgene copies present during the decline at the gradual rate β, starting from Tmax. Reproduced from [68].

(tocilizumab and corticosteroids) on the CAR-T expansion rate. The model estimated that the Cmax was twofold higher in patients who required tocilizumab, which is consistent with other reported findings that show higher exposure (Cmax) leads to higher CRS and therefore the need for medication. Finally, none of the evaluated covariates were determined to significantly correlate with the peak CAR-T levels. It should be noted that the pre-infusion tumor burden after lymphodepletion, which has previously been shown to affect CAR-T kinetics in several studies, was not measured and hence not included as a covariate in the analysis. In addition, some patients may have received multiple doses of CRS-treating medications, but the model evaluated the impact of only the first dose. In addition, no interaction term was included due to the sequential administration of corticosteroids after tocilizumab in most cases. Other co-medications (e.g., IL-6 and TNF-α antagonist) were utilized only in a few patients, hence not considered in the analysis. Overall, such a modeling framework provides a methodology to characterize clinical CK profiles of CAR-T therapy and to examine the impact of prophylactic CRS-treating medications on CK and efficacy for future clinical studies. The Stein model was recently adopted by Ogasawara et al. [69] and Wu et al. [70] to carry out a similar analysis for liso-cel (another anti-CD19 CAR-T therapy) in LBCL patients and cilta-cel (anti-BCMA CAR-T therapy) in MM patients.

Liu et al. [71] further developed a modified cellular kinetic model. In contrast to the Stein model, the piecewise model here was used to capture four (instead of three) distinct phases – rapid distribution, expansion, contraction, and persistence, of the multiphasic CAR-T CK profile. For this analysis, the authors utilized published, individual cellular kinetic data from different CAR-T therapies (anti-CD19, anti-BCMA, and anti-EGFR) across seven different clinical trials, including diverse patient populations, such as B cell Acute Lymphocytic Leukemia (ALL), Chronic Lymphocytic Leukemia (CLL), Diffuse Large-B Cell Lymphoma (DLBCL), Glioblastoma (GBM), Non-Small Cell Lung Cancer (NSCLC), and Multiple Myeloma (MM). The main goal of the meta-analysis was to characterize, compare, and contrast CK profiles across multiple tumor types, identify differences between responders and non-responders or hematological malignancies and solid tumors, and investigate the impact of different covariates on clinical outcomes. The model predicted similar multiphasic CK profiles for different CAR-T therapies across various tumor types. Across most trials, the model-based analysis determined a significantly higher proliferation rate constant and thereby higher proliferative capacity (Cmax) in responders against non-responders. In addition, longer duration of contraction and differentiation attributing to a more durable persistence phase was consistent in responders. The model also determined higher proliferative capacity (Cmax) and longer duration of proliferation in hematological malignancies, including ALL, CLL, and MM, as compared to lymphomas and solid tumors, which can be attributed to the limited antigen accessibility and distribution requirement of the CAR-Ts to the solid tumors. In contrast, analysis revealed higher contraction and memory differentiation rates in solid tumors. In addition, the doses (available for CLL, MM, GBM, and NSCLC) showed a weak correlation with responses and no correlation with any model-estimated parameters, thereby suggesting a steep dose-response curve consistent with previous reports. In contrast to previously published reports, baseline tumor burden of MM and CLL showed no correlation with CAR-T proliferation or expansion and responses. The authors indicated the use of circulating biomarkers to assess tumor burden as one of the factors contributing to the inconsistency. Finally, analysis determined the CD4:CD8 ratio of the CAR-T product to be close to one in responders while higher or diverse ratios in non-responders for MM and NSCLC trials. Overall, such a model-based meta-analysis approach can be used to systematically compare cellular kinetic profiles, determine critical drivers of clinical outcomes across different CAR-T products, clinical trials, and indications, and guide the future discovery and development of next-generation cell therapies.

12.4.5 Mechanistic, Multiscale PK/PD and PBPK Modeling for In Vitro to In Vivo Correlation (IVIVC) and Preclinical to Clinical Translation for CAR-T Therapy

Singh et al. built a mechanism-based, multiscale PK/PD modeling framework to better understand IVIVC for CAR-T efficacy and to potentially guide preclinical to clinical PK/PD translation for cell therapy [72]. Using a stepwise bottom-up approach, the study also aimed at identifying critical determinants for the PK/PD of CAR-Ts. First, an in vitro cell-level PD model was developed to quantitatively investigate the impact of drug-specific (CAR density and CAR-target affinity) and system-specific (target density and tumor burden) parameters on in vitro functional readouts such as target cell killing, CAR-T cell proliferation, and cytokine release. Previously published comprehensive in vitro datasets from different affinity variant CAR-Ts targeting EGFR or HER2 across different experimental conditions (different E:T ratios, cancer cell lines with varying target cell density, etc.) were used. As shown in Figure 12.5, the in vitro PD model assumes that the formation of CAR-target complexes was governed by affinity-driven target-mediated interactions between the two cell populations (CAR-T and tumor cells) and that the CAR-tumor complexes drive tumor cell depletion, CAR-T expansion, and cytokine release. The model was able to characterize comprehensive in vitro datasets simultaneously and estimate key parameters for tumor cell killing, CAR-T expansion, and cytokine release (e.g., IC50, which is defined as 'the number of CAR-target complexes/tumor cells' required to induce 50% of the maximum killing rate of CAR-Ts', K_{max} for killing).

Next, a PBPK model was developed to capture the in vivo CK and tissue biodistribution of CAR-Ts in xenograft mouse models. Previously published biodistribution studies with radiolabeled anti-EGFR and anti-CD19 CAR-Ts in mouse xenograft models were leveraged for this purpose. The model assumes blood and other major organs as representative compartments linked via blood and lymphatic flows with each tissue compartment further sub-compartmentalized as vascular space and extravascular space (Figure 12.5). First-order transmigration organ rates characterizing the migration or movement of CAR-Ts from vascular space to extravascular space while organ-specific lymphatic flow rates governing their circulation back into the blood were utilized in the model structure. The developed PBPK model was able to simultaneously characterize the distribution of multiple CAR-Ts to major tissues, including tumor, liver, lung, spleen, kidney, and lymph nodes. Finally, an integrated PBPK/PD model was established to understand the in vivo expansion of CAR-Ts and its relationship with TGI in mouse xenograft efficacy studies. Datasets from mouse xenograft and efficacy studies with CAR-Ts targeting EGFR, CD19, BCMA, or HER2 were used. In addition to the PBPK structure, the model now also included a PD component (similar to the in vitro PD model) that assumes affinity-driven target-mediated

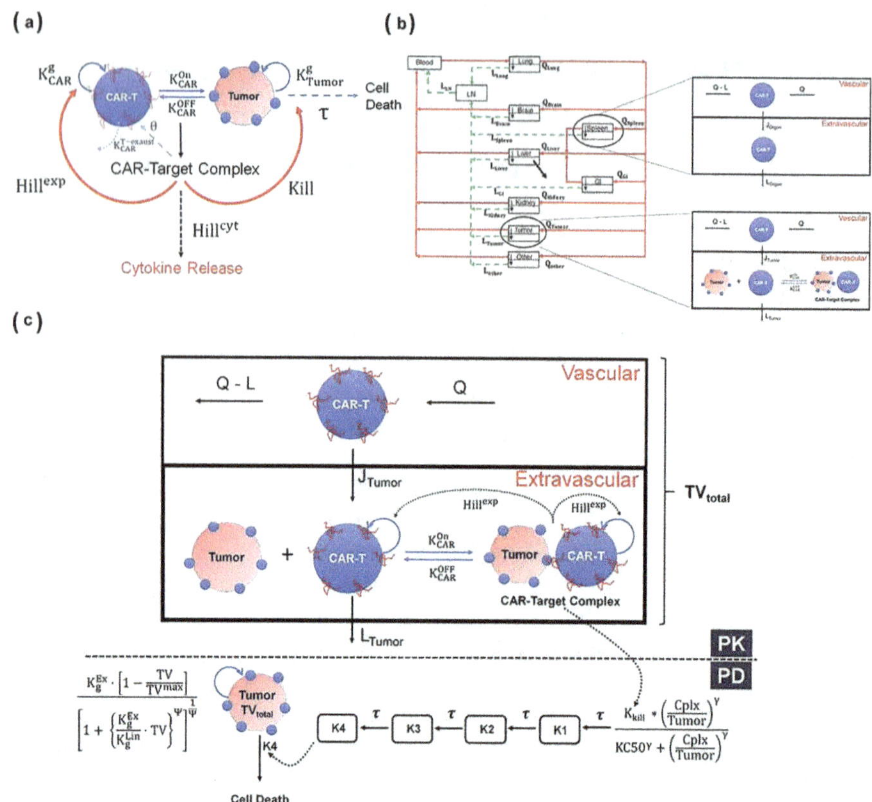

FIGURE 12.5 (a) A schematic diagram of a cell-level pharmacodynamic model for CAR-T cell activity: A dynamic population of CAR-T cells and tumor cells was assumed in an in vitro system. Upon target-mediated interaction among the two cell populations, there is the formation of CAR-target complexes, which simultaneously mediate the tumor cell depletion, expansion of CAR-T cells, and release of cytokines. (b) A schematic diagram of a physiologically based pharmacokinetic (PBPK) model to characterize the disposition of CAR-T cells: The model is compartmentalized into blood and relevant tissues, anatomically arranged via blood flows (red arrows) and lymphatic flows (green arrows). Each tissue is further subcompartmentalized into vascular and extravascular spaces. A first-order elimination of CAR-T cells (solid black arrow) is characterized from liver extravascular space. In a typical tissue, CAR-T cells extravasate from vascular space to extravascular space via first-order transmigration (JOrgan) rates, eventually circulating back to the bloodstream via organ-specific lymphatic flow. Within the tumor extravascular space, there is the formation of CAR-target complexes, whereas only unbound CAR-T cells can circulate back via lymphatic flow. (c) A schematic diagram of the PBPK-PD model to characterize CAR-T cell expansion and tumor growth inhibition: The diagram illustrates only the 'tumor compartment' of the full PBPK model structure, where upon the formation of CAR-target complexes in tumor extravascular space, there is expansion of total (unbound + tumor-bound) CAR-T cells and depletion of total tumor volume (TVtotal), which comprises vascular and extravascular spaces. Only the unbound CAR-T cells can leave the tumor tissue via lymphatic (LTumor) flows. The 'number of CAR-target complexes per tumor cells' undergoes a series of signal transduction steps (K1–K4), before they ultimately induce killing of inherently growing tumor cells (Kg) to induce TGI. Reproduced from [72].

interactions between CAR-T and tumor cells and that the CAR-tumor complexes drive TGI. The model was able to characterize the unique, multiphasic expansion kinetics of CAR-Ts observed in the mouse studies. Besides, the TGI profiles were simultaneously captured well and in vivo killing potency parameters, specific to each CAR-T, were estimated.

The IVIVC analysis showed that the in vitro potency parameters were consistently estimated to be ~10- to 20-fold higher as compared to the in vivo estimates, which could be attributed to the higher probability of CAR-Ts to come in direct contact and interact with tumor cells in the in vitro static conditions as compared to the in vivo situation. Additional model-based simulations suggested a 'threshold dose', below which there was no response, above which there was a steep dose-exposure-response for a narrow dose range, and after that the dose-exposure-response relationship was flat. In contrast, tumor burden was predicted to be a more sensitive parameter for CAR-T expansion kinetics as compared to the cell dose, which is consistent with several published studies. Similarly, sensitivity analysis demonstrated CAR affinity, CAR density, and antigen density to be positively correlated with CAR-T expansion. Overall, the developed multiscale PK/PD modeling framework can be used to determine the ideal properties of a CAR-T lead (e.g., CAR-target affinity and CAR density), compare and select lead candidate(s), better understand IVIVC, establish the PK/PD relationship in a preclinical setting, and finally translate the learnings to the clinical setting for CAR-T therapies.

12.4.6 QSP Model for Preclinical to Clinical Translation of CD3 Bispecific Antibodies

A translational QSP model was developed for a CD3 bispecific antibody (bsAb), P-cadherin/CD3 DART® (Pcad-LP-DART), targeting P-cadherin for solid tumor indications [34]. CD3 bsAbs can simultaneously engage with the CD3 receptors on the cytolytic T cells and target antigen on the tumor cells, which then leads to the formation of trimolecular complexes and subsequently T-cell activation and tumor cell lysis. The model was used to characterize the PK/PD relationship for Pcad-LP-DART in preclinical mouse efficacy studies and eventually translate to humans to predict efficacy.

As shown in Figure 12.6, the model incorporated the PK of Pcad-LP-DART, T-cell distribution in the central compartment (systemic circulation) and tumor, and binding of the drug to the T cells and soluble P-cadherin in the systemic circulation. In addition, the model assumed the binding of Pcad-LP-CART to CD3 receptors on T cells and target on the tumor cells to form the trimolecular complexes or immune synapses within the tumor, which are then linked to the tumor cell killing. Data, including PK of Pcad-LP-DART, T-cell distribution, TGI, and binding affinities of the drug to the target and T cells and target and CD3 levels on tumor and T cells, respectively, were used to build the model. Overall, the preclinical QSP model was able to characterize the PK/PD data, relevant parameters were estimated, and TSC (defined as minimum trimer concentration required for efficacy) was calculated across mouse models. Next, the preclinical

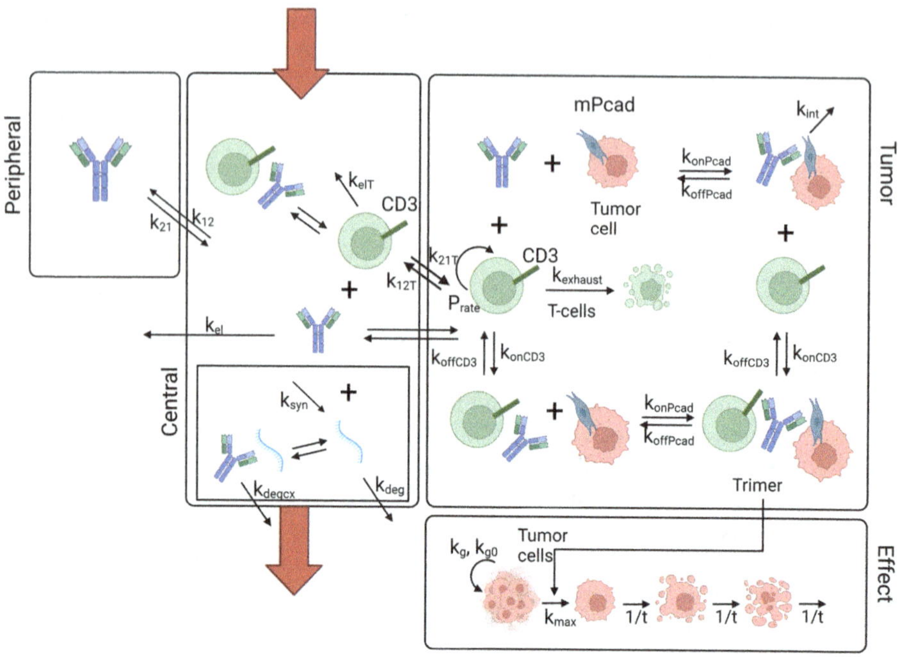

FIGURE 12.6 A diagram showing a translational QSP model for CD3 bispecific antibodies with key events including distribution of drug to different compartments, binding of the drug to target antigens on tumor cells and CD3 receptors on T-cells, and formation of trimers between drugs, T-cells, and tumor cells which is linked to tumor cell killing. Figure adapted from [34].

QSP model was translated to humans by integrating predicted human PK and their clinically relevant estimates for T-cell concentration at the tumor site, tumor volumes, levels of target, and CD3 receptors, as well as soluble P-cadherin in patients. Here, the human PK was predicted from a cynomolgus monkey using conventional two-compartmental modeling and allometric scaling approaches. The impact of soluble P-cadherin on the PK of Pcad-LP-DART was evaluated, and potential clinically efficacious doses were predicted. The model-based simulations identified target levels and T-cell infiltration at the tumor site as sensitive parameters that severely impacted the efficacious dose projections in humans.

Overall, the translational mechanistic model integrated in silico, in vitro, and in vivo datasets, characterized the PK/PD relationship, and can be used to potentially predict efficacy in the clinic. Such a modeling framework can be used at different stages of discovery and development of CD3 bispecifics, including to guide molecule design (e.g., binding affinity), triage different leads, and facilitate ideal lead selection, as well as support preclinical to clinical translation for dose projections and guide the FIH clinical study design.

12.4.7 QSP Model for mRNA Therapeutic Lipid Nanoparticle for Preclinical to Clinical Translation

Crigler-Najjar syndrome type 1 (CN1) is an autosomal recessive disease that is caused by a significant drop in the activity of the UGT1A1 enzyme. It is hypothesized that administering hUGT1A1-modRNA (a modified messenger RNA encoding for UGT1A1) as a LNP will restore hepatic production of UGT1A1, allowing normal glucuronidation and elimination of bilirubin in patients. To better understand the mechanisms of hUGT1A1-modRNA and to direct the design of the FIH clinical investigations, a QSP model was proposed that incorporates preclinical data from studies in Gunn rats with the known physiological distinctions between humans and Gunn rats. The inherent kinetics of the target plays a major role in the selection of an optimal dose and dosing regimen, making it difficult to directly apply allometric coefficients derived from large datasets comprising small compounds or biologics to the modRNA platform [73]. Drug properties and biological mechanisms (such as the intracellular half-life of mRNA and UGT1A1) govern the expected dose regimen, and these factors are not easily scaled allometrically based on body weights or surface areas. Because QSP models may leverage datasets from several contexts (e.g., in vitro and in vivo) and multiple species/indications (e.g., rats, healthy volunteers, and CN1 subjects), they are anticipated to provide more predictive power than the empirical PK/PD techniques (Figure 12.7) [73].

The model is able to provide projections on the consequences of these parameters, which in this instance are non-linear and are not expected to follow typical allometric scaling laws. In the end, a model with such a mechanistic basis can be utilized to facilitate more informed clinical decision-making.

12.4.8 QSP Model to Gather Mechanistic Insights for Gene Delivery in Sickle Cell Disease

To better understand the interplay between erythropoiesis, gene therapy, and autologous stem cell transplantation, authors built a multiscale mathematical model [74]. The model was used to simulate the effects of a novel exogenous globin that generates anti-sickling hemoglobin on patients with sickle cell disease (SCD). In most cases, SCD is caused by a deficiency in a single gene and cannot be treated at this time. Therefore, it is a promising candidate for gene therapy. Preclinical research shows that anti-sickling behavior is conferred by the insertion of an alternative globin gene. However, clinical data demonstrating the quantitative relationship between therapy parameters and post-treatment results are limited. Understanding these connections and developing a logical framework to create and improve gene therapies for SCD were the primary objectives of this case study [74].

Red blood cells (RBCs) with an anti-sickling hemoglobin should be produced by these stem cells indefinitely. There is a lack of patient data from early clinical studies, and the cost of this sophisticated, multi-step treatment is prohibitive. The goal was to

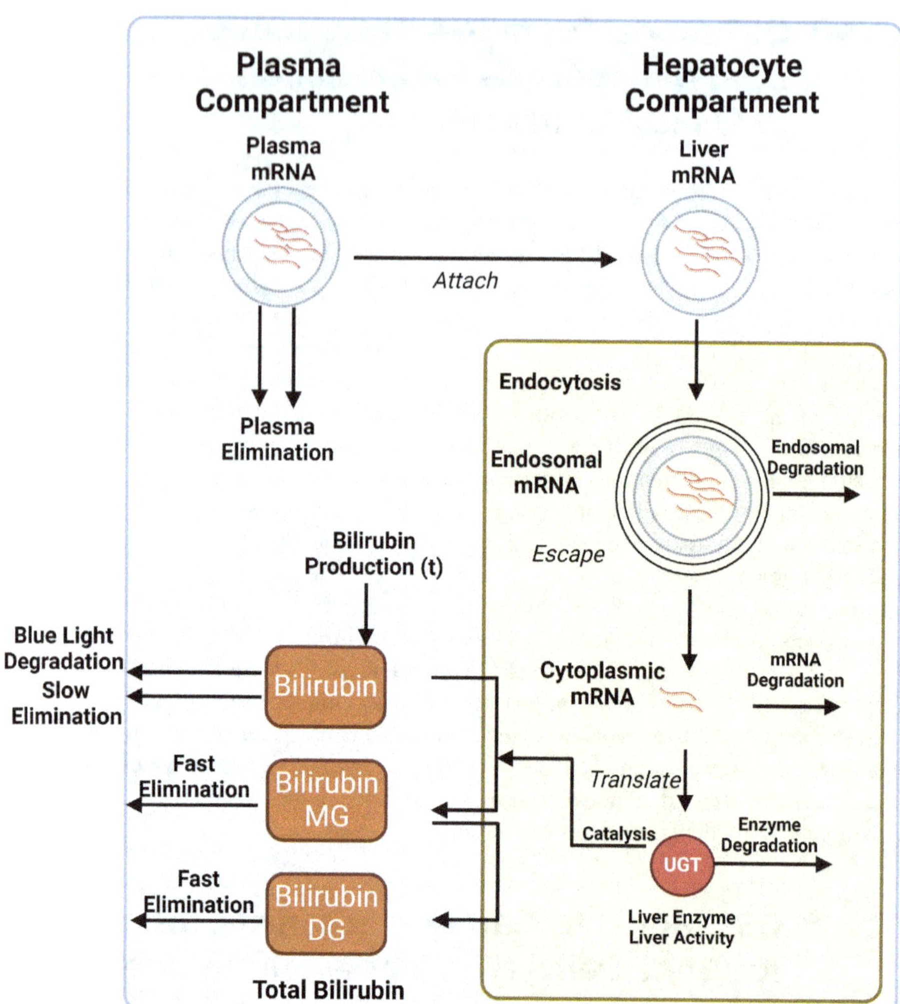

FIGURE 12.7 Schematic representation for the quantitative systems pharmacology (QSP) model of hUGT1A1-modRNA. The model outlines the clearance of lipid nanoparticles (LNP) from plasma due to liposomal instability, their uptake via endocytosis into liver hepatocytes, the release of mRNA from endosomes into the cytoplasm, subsequent translation of mRNA to synthesize uridine-diphosphate-glucuronosyltransferase (UGT1A1 protein), and the glucuronidation processes forming diglucuronides (DG) and monoglucuronides (MG) (Figure adapted from [73]).

assess the effect of treatment factors on engraftment success, peripheral RBC counts, and anti-sickling hemoglobin levels over time. These parameters include initial stem cell dose, lentiviral transduction effectiveness, and bone marrow preconditioning intensity. The model of RBC formation from progenitor cells in the bone marrow and of hemoglobin assembly from its constituent globin monomers is based on ordinary differential equations. RBC and hemoglobin levels seen in healthy and SCD phenotypes are reflected in the model. The kinetics of stem cell engraftment and RBC carrying the therapeutic gene product can be predicted using treatment simulations. Following therapy, there is an initial phase of reconstitution caused by short-lived stem cells, followed by a sustained RBC production from steady engraftment of long-lived stem cells [74]. Relationships between treatment parameters and efficacy were assessed using sensitivity analysis of the model. Predicted long-term success is highest when the initial dose of transduced stem cells and the intensity of myeloablative bone marrow preconditioning are maximized. Using a QSP method, this study demonstrated the efficacy of SCD gene treatments.

12.5 CONCLUSIONS AND FUTURE PERSPECTIVES

To overcome the challenges in drug discovery and development and to deconvolute the complexities of novel biotherapeutic modalities, innovative approaches are needed. Mathematical modeling is a key tool which has been shown to increase efficiency and effectiveness in drug discovery and development and can be used to facilitate molecule design, lead selection, and preclinical to clinical translation and to optimize clinical trials for biotherapeutics. In this chapter, two different types of mathematical modeling are reviewed: PK/PD modeling, which involves fitting of data in a 'top-down' manner, and QSP modeling, which integrates data from disparate sources in a 'bottom-up' approach, to examine the relationships between a drug, the biological system, and the disease process. These modeling tools are not mutually exclusive, and the appropriate method should be implemented to address the question in hand.

PK/PD modeling and QSP modeling are particularly useful tools for quantitative understanding of biotherapeutics as these drugs have some unique challenges. Biotherapeutics have evolved very rapidly over the last two decades and consist of a diverse set of therapies, including antibodies and their derivatives such as multispecific antibodies and ADCs, modified RNA molecules, cell and gene therapies, and vaccines. Each of these modalities has unique PK/PD behaviors, which are reviewed in this chapter. For example, the PK and PD of mAbs are often interdependent due to processes such as TMDD and immunogenicity. For bispecific antibodies such as T-cell retargeting drugs, which cross-link immune cells and tumor cells, the formation of a trimolecular complex between the mAb, the T cell, and the tumor cell drives the PD (and toxicodynamic) responses and not drug concentration alone. This leads to interesting mechanistic complexities, including bell-shaped concentration-response relationships. Cell therapies such as CAR-T cells are even more complex as they are 'living drugs' which proliferate, differentiate, actively traffic between tissues, and engage in two-way communication with the patient's immune system. The resultant pharmacology is different from that of small molecules or mAbs, as there is little relationship between administered dose and exposure. In the face of these complexities, PK/PD modeling and QSP modeling have become key quantitative tools to guide biotherapeutic design and development.

Consistent with the evolution in biotherapeutic modalities, PK/PD modeling and QSP modeling are now routinely used across all stages of drug discovery and development, from very early discovery programs to large-scale phase 3/4 patient studies. The principles and methods underpinning the quantitative approaches remain relatively constant across this continuum, but what changes are the questions that are asked. The scope and impact of the application of PK/PD and QSP modeling and simulation are showcased in the case studies that are presented in this chapter. For example, at early stages, the questions may include 'What is the best target to invest exploratory resources in?' and 'What are the optimal design properties for my biotherapeutic drug?'. This is exemplified in a case study applying mechanistic PK/PD models for early feasibility analyses to inform key decisions for biotherapeutics in early drug discovery before the collection of data [66]. Once a lead molecule has been identified and is being progressed toward clinical studies, the question may become 'How do I integrate in vitro and in vivo data collected in the preclinical phase to translate to humans and predict safe and effective clinical doses?'. Case studies presented herein showcase mechanistic modeling frameworks applied for preclinical to clinical translation of T-cell engager molecules [44], CAR-T cells [43], and a LNP delivering a modified mRNA [73]. In clinical development, the questions become 'What dose and patient population do we select for the phase 3 study?' or 'How can I use data from one indication to inform optimal dose and regimen for a second indication?'. This is exemplified in a case study where a mechanistic model is applied for an ADC to treat NHL and used to recommend a different regimen for ALL [44]. In addition to answering specific questions, perhaps a more nuanced impact of applying a modeling-based strategy in the discovery and development of biotherapeutics is the mechanistic understanding that comes from a rigorous examination of the system dynamics and the impact of the drug in perturbing this system.

To conclude, modeling and simulation approaches will continue to evolve as a critical component to tackling attrition in the drug discovery and development process.

REFERENCES

1. Zhu, H.; Huang, S.M.; Madabushi, R.; Strauss, D.G.; Wang, Y.; Zineh, I. Model-Informed Drug Development: A Regulatory Perspective on Progress. *Clin. Pharmacol. Ther.* **2019**, *106*, 91–93.
2. Mager, D.E.; Jusko, W.J. Development of Translational Pharmacokinetic–Pharmacodynamic Models. *Clin. Pharmacol. Ther.* **2008**, *83*, 909–912.
3. Bradshaw, E.L.; Spilker, M.E.; Zang, R.; Bansal, L.; He, H.; Jones, R.D.O.; Le, K.; Penney, M.; Schuck, E.; Topp, B.; et al. Applications of Quantitative Systems Pharmacology in Model-Informed Drug Discovery: Perspective on Impact and Opportunities. *CPT Pharmacomet. Syst. Pharmacol.* **2019**, *8*, 777–791.
4. der Graaf, P.H.; Benson, N. Systems Pharmacology: Bridging Systems Biology and Pharmacokinetics-Pharmacodynamics (PKPD) in Drug Discovery and Development. *Pharm. Res.* **2011**, *28*, 1460–1464.

5. Sorger, P.K.; Allerheiligen, S.R.B.; Abernethy, D.R.; Altman, R.B.; Brouwer, K.L.R.; Califano, A.; D'Argenio, D.Z.; Iyengar, R.; Jusko, W.J.; Lalonde, R.; et al. Quantitative and Systems Pharmacology in the Post-Genomic Era: New Approaches to Discovering Drugs and Understanding Therapeutic Mechanisms. In *Proceedings of the NIH White Paper by the QSP Workshop Group*; **2011**; Vol. 48, pp. 1–47. https://pharmaceutical.report/Resources/Whitepapers/fca8a762-30c7-4d49-a04c-3b97a266db3c_SystemsPharmaWPSorger2011.pdf
6. Helmlinger, G.; Sokolov, V.; Peskov, K.; Hallow, K.M.; Kosinsky, Y.; Voronova, V.; Chu, L.; Yakovleva, T.; Azarov, I.; Kaschek, D.; et al. Quantitative Systems Pharmacology: An Exemplar Model-Building Workflow with Applications in Cardiovascular, Metabolic, and Oncology Drug Development. *CPT Pharmacomet. Syst. Pharmacol.* **2019**, *8*, 380–395.
7. Musante, C.J.; Ramanujan, S.; Schmidt, B.J.; Ghobrial, O.G.; Lu, J.; Heatherington, A.C. Quantitative Systems Pharmacology: A Case for Disease Models. *Clin. Pharmacol. Ther.* **2017**, *101*, 24–27.
8. Agoram, B.M.; Martin, S.W.; van der Graaf, P.H. The Role of Mechanism-Based Pharmacokinetic–Pharmacodynamic (PK–PD) Modelling in Translational Research of Biologics. *Drug Discov. Today* **2007**, *12*, 1018–1024.
9. Agoram, B.M.; Demin, O. Integration Not Isolation: Arguing the Case for Quantitative and Systems Pharmacology in Drug Discovery and Development. *Drug Discov. Today* **2011**, *16*, 1031–1036.
10. Shah, D.K.; Betts, A.M. Towards a Platform PBPK Model to Characterize the Plasma and Tissue Disposition of Monoclonal Antibodies in Preclinical Species and Human. *J. Pharmacokinet. Pharmacodyn.* **2012**, *39*, 67–86.
11. Wang, W.; Wang, E.Q.; Balthasar, J.P. Monoclonal Antibody Pharmacokinetics and Pharmacodynamics. *Clin. Pharmacol. Ther.* **2008**, *84*, 548–558.
12. Betts, A.; van der Graaf, P.H. Mechanistic Quantitative Pharmacology Strategies for the Early Clinical Development of Bispecific Antibodies in Oncology. *Clin. Pharmacol. Ther.* **2020**, *108*, 528–541.
13. Shah, D.K.; Betts, A.M. Antibody Biodistribution Coefficients: Inferring Tissue Concentrations of Monoclonal Antibodies Based on the Plasma Concentrations in Several Preclinical Species and Human. *Proc. MAbs* **2013**, *5*, 297–305. https://www.ncbi.nlm.nih.gov/pmc/articles/PMC3893240/
14. Li, Z.; Krippendorff, B.-F.; Sharma, S.; Walz, A.C.; Lavé, T.; Shah, D.K. Influence of Molecular Size on Tissue Distribution of Antibody Fragments. *MAbs* **2016**, *8*, 113–119.
15. Senior, M. Fresh from the Biotech Pipeline: Fewer Approvals, but Biologics Gain Share. *Nat. Biotechnol.* **2023**, 1.
16. Tang, Y.; Cao, Y. Modeling Pharmacokinetics and Pharmacodynamics of Therapeutic Antibodies: Progress, Challenges, and Future Directions. *Pharmaceutics* **2021**, *13*, 422.
17. Haraya, K.; Tsutsui, H.; Komori, Y.; Tachibana, T. Recent Advances in Translational Pharmacokinetics and Pharmacodynamics Prediction of Therapeutic Antibodies Using Modeling and Simulation. *Pharmaceuticals* **2022**, *15*, 508.
18. Trivedi, A.; Stienen, S.; Zhu, M. al; Li, H.; Yuraszeck, T.; Gibbs, J.; Heath, T.; Loberg, R.; Kasichayanula, S. Clinical Pharmacology and Translational Aspects of Bispecific Antibodies. *Clin. Transl. Sci.* **2017**, *10*, 147.
19. Van Der Graaf, P.H.; Gabrielsson, J. Pharmacokinetic–Pharmacodynamic Reasoning in Drug Discovery and Early Development. *Future Med. Chem.* **2009**, *1*, 1371–1374.
20. Tibbitts, J.; Canter, D.; Graff, R.; Smith, A.; Khawli, L.A. Key Factors Influencing ADME Properties of Therapeutic Proteins: A Need for ADME Characterization in Drug Discovery and Development. *Proc. MAbs* **2016**, *8*, 229–245.
21. Garg, A.; Balthasar, J.P. Physiologically-Based Pharmacokinetic (PBPK) Model to Predict IgG Tissue Kinetics in Wild-Type and FcRn-Knockout Mice. *J. Pharmacokinet. Pharmacodyn.* **2007**, *34*, 687–709.

22. Qi, T.; Cao, Y. In Translation: FcRn across the Therapeutic Spectrum. *Int. J. Mol. Sci.* **2021**, *22*, 3048.
23. Keizer, R.J.; Huitema, A.D.R.; Schellens, J.H.M.; Beijnen, J.H. Clinical Pharmacokinetics of Therapeutic Monoclonal Antibodies. *Clin. Pharmacokinet.* **2010**, *49*, 493–507.
24. Igawa, T.; Tsunoda, H.; Tachibana, T.; Maeda, A.; Mimoto, F.; Moriyama, C.; Nanami, M.; Sekimori, Y.; Nabuchi, Y.; Aso, Y.; et al. Reduced Elimination of IgG Antibodies by Engineering the Variable Region. *Protein Eng. Des. Sel.* **2010**, *23*, 385–392.
25. Ryman, J.T.; Meibohm, B. Pharmacokinetics of Monoclonal Antibodies. *CPT Pharmacomet. Syst. Pharmacol.* **2017**, *6*, 576–588.
26. Klaus, T.; Deshmukh, S. PH-Responsive Antibodies for Therapeutic Applications. *J. Biomed. Sci.* **2021**, *28*, 1–14.
27. Igawa, T.; Haraya, K.; Hattori, K. Sweeping Antibody as a Novel Therapeutic Antibody Modality Capable of Eliminating Soluble Antigens from Circulation. *Immunol. Rev.* **2016**, *270*, 132–151.
28. Rathi, C.; Meibohm, B. Clinical Pharmacology of Bispecific Antibody Constructs. *J. Clin. Pharmacol.* **2015**, *55*, S21–S28.
29. Ménochet, K.; Yu, H.; Wang, B.; Tibbitts, J.; Hsu, C.-P.; Kamath, A. V; Richter, W.F.; Baumann, A. Non-Human Primates in the PKPD Evaluation of Biologics: Needs and Options to Reduce, Refine, and Replace. A BioSafe White Paper. *Mabs* **2022**, *14*, 2145997.
30. Betts, A.; Keunecke, A.; van Steeg, T.J.; van der Graaf, P.H.; Avery, L.B.; Jones, H.; Berkhout, J. Linear Pharmacokinetic Parameters for Monoclonal Antibodies Are Similar within a Species and across Different Pharmacological Targets: A Comparison between Human, Cynomolgus Monkey and HFcRn Tg32 Transgenic Mouse Using a Population-Modeling Approach. *MAbs* **2018**, *10*, 751–764. https://pubmed.ncbi.nlm.nih.gov/29634430/
31. Dua, P.; Hawkins, E.; Van Der Graaf, P.H. A Tutorial on Target-Mediated Drug Disposition (TMDD) Models. *CPT Pharmacomet. Syst. Pharmacol.* **2015**, *4*, 324–337.
32. Baxter, L.T.; Zhu, H.; Mackensen, D.G.; Butler, W.F.; Jain, R.K. Biodistribution of Monoclonal Antibodies: Scale-up from Mouse to Human Using a Physiologically Based Pharmacokinetic Model. *Cancer Res.* **1995**, *55*, 4611–4622.
33. Shah, D.K.; Loganzo, F.; Haddish-Berhane, N.; Musto, S.; Wald, H.S.; Barletta, F.; Lucas, J.; Clark, T.; Hansel, S.; Betts, A. Establishing in vitro–in vivo Correlation for Antibody Drug Conjugate Efficacy: A PK/PD Modeling Approach. *J. Pharmacokinet. Pharmacodyn.* **2018**, *45*, 339–349.
34. Betts, A.; Haddish-Berhane, N.; Shah, D.K.; van der Graaf, P.H.; Barletta, F.; King, L.; Clark, T.; Kamperschroer, C.; Root, A.; Hooper, A.; et al. A Translational Quantitative Systems Pharmacology Model for CD3 Bispecific Molecules: Application to Quantify T Cell-Mediated Tumor Cell Killing by P-Cadherin LP DART®. *AAPS J.* **2019**, *21*, 1–16.
35. Fu, Z.; Li, S.; Han, S.; Shi, C.; Zhang, Y. Antibody Drug Conjugate: The "Biological Missile" for Targeted Cancer Therapy. *Signal Transduct. Target. Ther.* **2022**, *7*, 93.
36. Hooper, A.T.; Marquette, K.; Chang, C.-P.B.; Golas, J.; Jain, S.; Lam, M.-H.; Guffroy, M.; Leal, M.; Falahatpisheh, H.; Mathur, D.; et al. Anti-Extra Domain B Splice Variant of Fibronectin Antibody–Drug Conjugate Eliminates Tumors with Enhanced Efficacy When Combined with Checkpoint Blockade. *Mol. Cancer Ther.* **2022**, *21*, 1462–1472.
37. Cilliers, C.; Guo, H.; Liao, J.; Christodolu, N.; Thurber, G.M. Multiscale Modeling of Antibody-Drug Conjugates: Connecting Tissue and Cellular Distribution to Whole Animal Pharmacokinetics and Potential Implications for Efficacy. *AAPS J.* **2016**, *18*, 1117–1130.
38. Lam, I.; Pilla Reddy, V.; Ball, K.; Arends, R.H.; Mac Gabhann, F. Development of and Insights from Systems Pharmacology Models of Antibody-Drug Conjugates. *CPT Pharmacomet. Syst. Pharmacol.* **2022**, *11*, 967–990.
39. Haddish-Berhane, N.; Shah, D.K.; Ma, D.; Leal, M.; Gerber, H.-P.; Sapra, P.; Barton, H.A.; Betts, A.M. On Translation of Antibody Drug Conjugates Efficacy from Mouse Experimental Tumors to the Clinic: A PK/PD Approach. *J. Pharmacokinet. Pharmacodyn.* **2013**, *40*, 557–571.

40. Betts, A.; Clark, T.; Jasper, P.; Tolsma, J.; van der Graaf, P.H.; Graziani, E.I.; Rosfjord, E.; Sung, M.; Ma, D.; Barletta, F. Use of Translational Modeling and Simulation for Quantitative Comparison of PF-06804103, a New Generation HER2 ADC, with Trastuzumab-DM1. *J. Pharmacokinet. Pharmacodyn.* **2020**, *47*, 513–526.
41. Shah, D.K.; Haddish-Berhane, N.; Betts, A. Bench to Bedside Translation of Antibody Drug Conjugates Using a Multiscale Mechanistic PK/PD Model: A Case Study with Brentuximab-Vedotin. *J. Pharmacokinet. Pharmacodyn.* **2012**, *39*, 643–659.
42. Singh, A.P.; Maass, K.F.; Betts, A.M.; Wittrup, K.D.; Kulkarni, C.; King, L.E.; Khot, A.; Shah, D.K. Evolution of Antibody-Drug Conjugate Tumor Disposition Model to Predict Preclinical Tumor Pharmacokinetics of Trastuzumab-Emtansine (T-DM1). *AAPS J.* **2016**, *18*, 861–875.
43. Singh, A.P.; Shah, D.K. Application of a PK-PD Modeling and Simulation-Based Strategy for Clinical Translation of Antibody-Drug Conjugates: A Case Study with Trastuzumab Emtansine (T-DM1). *AAPS J.* **2017**, *19*, 1054–1070.
44. Betts, A.M.; Haddish-Berhane, N.; Tolsma, J.; Jasper, P.; King, L.E.; Sun, Y.; Chakrapani, S.; Shor, B.; Boni, J.; Johnson, T.R. Preclinical to Clinical Translation of Antibody-Drug Conjugates Using PK/PD Modeling: A Retrospective Analysis of Inotuzumab Ozogamicin. *AAPS J.* **2016**, *18*, 1101–1116.
45. Bender, B.C.; Schaedeli-Stark, F.; Koch, R.; Joshi, A.; Chu, Y.-W.; Rugo, H.; Krop, I.E.; Girish, S.; Friberg, L.E.; Gupta, M. A Population Pharmacokinetic/Pharmacodynamic Model of Thrombocytopenia Characterizing the Effect of Trastuzumab Emtansine (T-DM1) on Platelet Counts in Patients with HER2-Positive Metastatic Breast Cancer. *Cancer Chemother. Pharmacol.* **2012**, *70*, 591–601.
46. Ait-Oudhia, S.; Zhang, W.; Mager, D.E. A Mechanism-Based PK/PD Model for Hematological Toxicities Induced by Antibody-Drug Conjugates. *AAPS J.* **2017**, *19*, 1436–1448.
47. June, C. H.; O'Connor, R.S.; Kawalekar, O.U.; Ghassemi, S.; Milone, M.C. CAR T Cell Immunotherapy for Human Cancer. *Science (New York, N.Y.)*, **2018**, *359*, 1361–1365.
48. June, C.H.; Sadelain, M. Chimeric Antigen Receptor Therapy. *N. Engl. J. Med.* **2018**, *379*, 64–73.
49. Boettcher, M.; Joechner, A.; Li, Z.; Yang, S.F.; Schlegel, P. Development of CAR T Cell Therapy in Children—A Comprehensive Overview. *J. Clin. Med.* **2022**, *11*, 2158.
50. Mueller, K.T.; Maude, S.L.; Porter, D.L.; Frey, N.; Wood, P.; Han, X.; Waldron, E.; Chakraborty, A.; Awasthi, R.; Levine, B.L.; et al. Cellular Kinetics of CTL019 in Relapsed/Refractory B-Cell Acute Lymphoblastic Leukemia and Chronic Lymphocytic Leukemia. *Blood* **2017**, *130*, 2317–2325.
51. Huang, W.; Li, J.; Liao, M.Z.; Liu, S.N.; Yu, J.; Jing, J.; Kotani, N.; Kamen, L.; Guelman, S.; Miles, D.R. Clinical Pharmacology Perspectives for Adoptive Cell Therapies in Oncology. *Clin. Pharmacol. Ther.* **2022**, *112*, 968–981.
52. Kast, J.; Nozohouri, S.; Zhou, D.; Yago, M.R.; Chen, P.-W.; Ahamadi, M.; Dutta, S.; Upreti, V. V Recent Advances and Clinical Pharmacology Aspects of Chimeric Antigen Receptor (CAR) T-Cellular Therapy Development. *Clin. Transl. Sci.* **2022**, *15*, 2057–2074.
53. Chaudhury, A.; Zhu, X.; Chu, L.; Goliaei, A.; June, C.H.; Kearns, J.D.; Stein, A.M. Chimeric Antigen Receptor T Cell Therapies: A Review of Cellular Kinetic-Pharmacodynamic Modeling Approaches. *J. Clin. Pharmacol.* **2020**, *60*, S147–S159.
54. Nukala, U.; Rodriguez Messan, M.; Yogurtcu, O.N.; Wang, X.; Yang, H. A Systematic Review of the Efforts and Hindrances of Modeling and Simulation of CAR T-Cell Therapy. *AAPS J.* **2021**, *23*, 1–20.
55. Parra-Guillén, Z.P.; González-Aseguinolaza, G.; Berraondo, P.; Trocóniz, I.F. Gene Therapy: A Pharmacokinetic/Pharmacodynamic Modelling Overview. *Pharm. Res.* **2010**, *27*, 1487–1497.
56. Sun, K.; Liao, M.Z. Clinical Pharmacology Considerations on Recombinant Adeno-Associated Virus–Based Gene Therapy. *J. Clin. Pharmacol.* **2022**, *62*, S79–S94.

57. McIntosh, A.; Sverdlov, O.; Yu, L.; Kaufmann, P. Clinical Design and Analysis Strategies for the Development of Gene Therapies: Considerations for Quantitative Drug Development in the Age of Genetic Medicine. *Clin. Pharmacol. Ther.* **2021**, *110*, 1207–1215.
58. Wang, W.; Zhou, H. Pharmacological Considerations for Predicting PK/PD at the Site of Action for Therapeutic Proteins. *Drug Discov. Today Technol.* **2016**, *21*, 35–39.
59. Wolfreys, A.; Kilgour, J.; Allen, A.D.; Dudal, S.; Freke, M.; Jones, D.; Karantabias, G.; Krantz, C.; Moore, S.; Mukaratirwa, S.; et al. Review of the Technical, Toxicological, and PKPD Considerations for Conducting Inhalation Toxicity Studies on Biologic Pharmaceuticals—the Outcome of a Cross-Industry Working Group Survey. *Toxicol. Pathol.* **2021**, *49*, 261–285.
60. Wolf, J.J. Special Considerations for the Nonclinical Safety Assessment of Vaccines. In *Nonclinical Development of Novel Biologics, Biosimilars, Vaccines and Specialty Biologics*; Elsevier, **2013**; pp. 243–255. https://www.sciencedirect.com/science/article/abs/pii/B9780123948106000101
61. Curreri, A.; Sankholkar, D.; Mitragotri, S.; Zhao, Z. RNA Therapeutics in the Clinic. *Bioeng. Transl. Med.* **2023**, *8*, e10374.
62. Feng, R.; Patil, S.; Zhao, X.; Miao, Z.; Qian, A. RNA Therapeutics-Research and Clinical Advancements. *Front. Mol. Biosci.* **2021**, *8*, 710738.
63. Nair, J.K.; Attarwala, H.; Sehgal, A.; Wang, Q.; Aluri, K.; Zhang, X.; Gao, M.; Liu, J.; Indrakanti, R.; Schofield, S.; et al. Impact of Enhanced Metabolic Stability on Pharmacokinetics and Pharmacodynamics of GalNAc–SiRNA Conjugates. *Nucleic Acids Res.* **2017**, *45* (19), 10969–10977. https://pubmed.ncbi.nlm.nih.gov/28981809/
64. Madabushi, R.; Seo, P.; Zhao, L.; Tegenge, M.; Zhu, H. Role of Model-Informed Drug Development Approaches in the Lifecycle of Drug Development and Regulatory Decision-Making. *Pharm. Res.* **2022**, *39*, 1669–1680.
65. Selvaraj, C.; Chandra, I.; Singh, S.K. Artificial Intelligence and Machine Learning Approaches for Drug Design: Challenges and Opportunities for the Pharmaceutical Industries. *Mol. Divers.* **2021**, 1–21.
66. Marcantonio, D.H.; Matteson, A.; Presler, M.; Burke, J.M.; Hagen, D.R.; Hua, F.; Apgar, J.F. Early Feasibility Assessment: A Method for Accurately Predicting Biotherapeutic Dosing to Inform Early Drug Discovery Decisions. *Front. Pharmacol.* **2022**, 2081.
67. Chang, H.-Y.; Wu, S.; Meno-Tetang, G.; Shah, D.K. A Translational Platform PBPK Model for Antibody Disposition in the Brain. *J. Pharmacokinet. Pharmacodyn.* **2019**, *46*, 319–338.
68. Stein, A.M.; Grupp, S.A.; Levine, J.E.; Laetsch, T.W.; Pulsipher, M.A.; Boyer, M.W.; August, K.J.; Levine, B.L.; Tomassian, L.; Shah, S.; et al. Tisagenlecleucel Model-Based Cellular Kinetic Analysis of Chimeric Antigen Receptor–T Cells. *CPT Pharmacometrics Syst. Pharmacol.* **2019**, *8*, 285–295.
69. Ogasawara, K.; Dodds, M.; Mack, T.; Lymp, J.; Dell'Aringa, J.; Smith, J. Population Cellular Kinetics of Lisocabtagene Maraleucel, an Autologous CD19-Directed Chimeric Antigen Receptor T-Cell Product, in Patients with Relapsed/Refractory Large B-Cell Lymphoma. *Clin. Pharmacokinet.* **2021**, *60*, 1621–1633.
70. Wu, L.; Su, Y.; Li, C.; Zhou, W.; Jackson, C.; ien Sun, Y.-N.; Zhou, H. Population-Based Cellular Kinetic Characterization of Ciltacabtagene Autoleucel in Subjects with Relapsed or Refractory Multiple Myeloma. *Clin. Transl. Sci.* **202**, *15* (12), 3000–3011. https://pubmed.ncbi.nlm.nih.gov/36204820/
71. Liu, C.; Ayyar, V.S.; Zheng, X.; Chen, W.; Zheng, S.; Mody, H.; Wang, W.; Heald, D.; Singh, A.P.; Cao, Y. Model-Based Cellular Kinetic Analysis of Chimeric Antigen Receptor-T Cells in Humans. *Clin. Pharmacol. Ther.* **2021**, *109*, 716–727.

72. Singh, A.P.; Zheng, X.; Lin-Schmidt, X.; Chen, W.; Carpenter, T.J.; Zong, A.; Wang, W.; Heald, D.L. Development of a Quantitative Relationship between CAR-Affinity, Antigen Abundance, Tumor Cell Depletion and CAR-T Cell Expansion Using a Multiscale Systems PK-PD Model. *Proc. MAbs* **2020**, *12*, 1688616.
73. Apgar, J.F.; Tang, J.-P.; Singh, P.; Balasubramanian, N.; Burke, J.; Hodges, M.R.; Lasaro, M.A.; Lin, L.; Miliard, B.L.; Moore, K.; et al. Quantitative Systems Pharmacology Model of HUGT1A1-ModRNA Encoding for the UGT1A1 Enzyme to Treat Crigler-Najjar Syndrome Type 1. *CPT Pharmacomet. Syst. Pharmacol.* **2018**, *7*, 404–412.
74. Zheng, B.; Wille, L.; Peppel, K.; Hagen, D.; Matteson, A.; Ahlers, J.; Schaff, J.; Hua, F.; Yuraszeck, T.; Cobbina, E.; et al. A Systems Pharmacology Model for Gene Therapy in Sickle Cell Disease. *CPT Pharmacomet. Syst. Pharmacol.* **2021**, *10*, 696–708.

13 The Artificial Intelligence Revolution

Transforming the Design and Optimization of Multispecific Antibodies

Per Jr. Greisen, Ziwei Pang, and Fernando Garces

13.1 INTRODUCTION

Multispecific antibodies (MsAbs) have emerged as a new class of biologics with added therapeutic benefits over the classical monoclonal antibodies (mAbs) [1]. This engineered mechanism recognizing two or more epitopes located on the same or distinct targets expands the functionality of conventional mAbs, allowing for diverse applications, such as recruiting immune cells to destroy tumor cells or crosslinking distinct cell surface proteins [2–4]. Moreover, MsAbs are new molecular entities (NMEs) that display significant variation in size, configuration, valencies, flexibility, and angle of approach of their binding modules, biodistribution, and pharmacokinetic attributes [2,3]. However, the development of MsAbs quickly exposed the limitations of current workflows, including the computational tools available and the overall approach to a

DOI: 10.1201/9781003300311-13

high complexity in protein engineering. The limited success rate with only nine MsAbs approved thus far in more than two decades of research highlights the need to consider new approaches [5]. While much of the low MsAb approval rate can be attributed to sub-optimal prediction of the biology, this is likely due to the complexity of these new and non-native quaternary structures and challenges in target identification. Furthermore, the developability of these NMEs largely remains a significant challenge [6]. Since the advent of AlphaFold2 in the 14th Critical Assessment of Protein Structure Prediction (CASP14) [7], machine learning (ML)-based methods have emerged as the next-generation of technologies with the potential to increase the success rate in developing MsAbs. Currently, an increasing number of new startups specializing in developing artificial intelligence (AI)/ML models to address the discovery and engineering of biologics are emerging, attracting strong interest from the pharmaceutical industry. The main difference between AI and physics-based approaches lies in their methodology. Physics-based models rely on fundamental principles of physics like Newton's equations of motion, while knowledge-based AI/ML models learn from patterns in data and incorporate expert insights. Large language models (LLMs) initially gained traction in protein modeling with the use of long short-term memory (LSTM) architectures, including specialized variations like the multiplicative LSTM (mLSTM) [8,9] (Figure 13.1). A key advance was the development of the UniRep representation, which allowed LSTMs to learn complex protein sequence patterns more effectively. Models built on top of UniRep achieved several breakthroughs, including predicting the stability of natural and designed proteins, the function of diverse mutants, and even accelerating protein engineering tasks with vastly improved efficiency [10]. One intriguing finding was the mLSTM's ability to enhance protein characteristics using zero-shot or few-shot learning techniques, even with limited data (low N) [11]. The advent of transformers revolutionized natural language processing by introducing parallelizable self-attention mechanisms, enabling them to process entire sequences simultaneously and capture long-range dependencies within text, vastly outperforming the sequential limitations of recurrent neural networks (RNNs) and LSTMs [12]. Building upon this, Meta's Evolutionary Scale Modeling (ESM) model, utilizing transformer-based models, further advanced LLM capabilities in this domain [13]. Further advancements led to the development of larger LLMs, such as ESM-1b with 650 million parameters and ESM-2 with a significant increase to 15 billion parameters. These larger models demonstrated improved

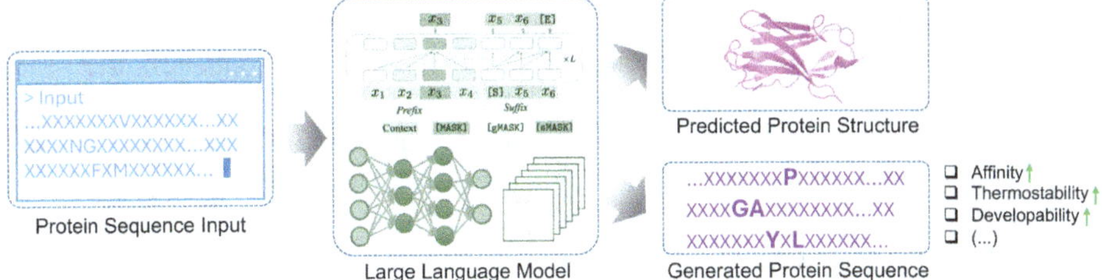

FIGURE 13.1 Protein design with large language models (LLMs). The LLM, trained on extensive protein data, generates a new protein sequence that meets design goals such as affinity, thermostability, among others.

performance compared to their smaller predecessors. Moreover, researchers found strong correlations between the language model's understanding of protein sequences (measured by perplexity) and the accuracy of structure prediction. The close link between language modeling and protein structure highlights the potential of LLMs with even greater scale in parameters, data, and computation for further breakthroughs [14]. Companies like BioMap are pioneering this field using large-scale AI models to transform protein design. Their xTrimoPGLM model, with over 100 billion parameters, exemplifies this potential with its ability to process and generate protein sequences tailored to specific inputs. These breakthroughs and the field's rapid advancement demonstrate the immense potential of LLMs to transform our understanding of proteins [15]. A significant development in LLMs for protein design is the use of labels (e.g., function, stability, enzyme activity, among others) to guide the generation of sequences with desired properties [16]. Expanding these models up to 6.5 billion protein sequences has demonstrated that larger models have an improved ability to predict and control these properties [17].

Here, we will describe the AI/ML technologies currently being applied to the development of mAbs and also offer insights about how the same or improved versions can be utilized to facilitate the design and engineering of MsAbs.

13.2 AI/ML: A GAME CHANGER FOR ANTIBODY DESIGN

The use of AI techniques is accelerating the pace and quality of mAb discovery and optimization, and for biologics, in general. These techniques are ushering in a new era of precision and efficiency, and they are changing the way that mAbs are designed and developed. AI methodologies have been pivotal in enhancing various aspects of mAb design, including affinity, developability, pharmacokinetics (PK), pharmacodynamics (PD), immunogenicity, and more, demonstrating a profound impact on the therapeutic efficacy and safety of these critical biologics [1].

Computational design of libraries has enabled the optimization of protein function, laying the groundwork for more targeted and effective generation of libraries for protein optimization compared with rational or random library generation [18]. One of the parameters important for therapeutic mAbs is getting the right affinity, which is often not present in the initial parental mAb coming from discovery platforms. AI techniques optimize affinity and simplify approaches by inserting beneficial mutations in the mAb's complementarity-determining regions (CDRs), while preserving and even improving specificity [19].

For this, computational methods, including structural modeling tools like Rosetta [20] or MOE [21], have been instrumental in enabling researchers to finetune affinity through precise modifications. For instances where no structural information is available, homology models can now be generated using advanced ML-based tools like AlphaFold 2 [7], AbBuilder [22], or xTrimoABFold [23], offering unprecedented insights into mAb structure in the absence of experimentally determined structures (crystallographic, CryoEM, and nuclear magnetic resonance (NMR) data). However, despite these advances, predicting the impact of sequence manipulation on affinity remains a significant challenge, plagued with large efforts of trial and error, underscoring the complexity of mAb optimization.

In addition, the use of AI has already proven beneficial to improve several attributes related to the mAb developability ranging from removing chemical liabilities to improving protein expression levels. Chemical liabilities like deamidation, isomerization, tryptophan/methionine oxidations, but not limited to, are not always possible to remediate nor are always predicted accurately, leading to unnecessary and expensive engineering efforts to generate sequence variants that retain the initial properties, including binding affinity. AI-based models combining the prediction of attributes like protein folding, thermostability, expression levels, and protein aggregation have already proven to be helpful in guiding protein engineers to first, prioritize which chemical liabilities present on the therapeutics candidates (e.g., mAbs) must be removed from those ones that not, and second, when required, generating sequences with high success rate once tested in the laboratory [24]. AI has also been applied to address other development concerns like mAb viscosity and aggregation propensity. Traditional methods often struggle to reliably predict these properties, highlighting the value of AI-based approaches [25,26]. Therefore, incorporating this approach into the optimization process saves time and resources, building a workflow that can be controlled and whose success can also be predicted. Another parameter that should be taken into the optimization process early is deimmunization. Here, the goal is to mitigate immunogenic concerns that could arise during clinical trials [27]. Deimmunization of CDRs (particularly the foreign H3) and potentially framework regions (FWRs) is crucial for optimization. Neural networks can predict potential T-cell epitopes (e.g., using NetMHC-II [28]), helping reduce the risk of immunogenic responses through protein engineering [29]. AI has also transformed the speed with which one can humanize murine mAbs through LLMs like BioPhi [30], which previously required data-driven approaches with feedback from experiments. AI tools not only speed up mAb optimization but also inform the decision-making, leading to faster workflows with higher success rate in identifying mAbs with the best overall drug-like properties.

Altogether, data-driven strategies have been applied to multiobjective optimization to generate mAbs with suitable properties as efficiently as possible. However, the continued evolution of AI/ML tools and methodologies promises to further refine our understanding and capabilities in this space, paving the way for the development of more effective and safer therapeutic mAbs.

13.3 MULTISPECIFIC ANTIBODY DESIGN

The transition from mAbs to the more complex arena of MsAbs (Figure 13.2) calls for a significant evolution in the application of AI/ML strategies. While AI/ML has proven revolutionary in optimizing mAb design across various facets: affinity, developability,

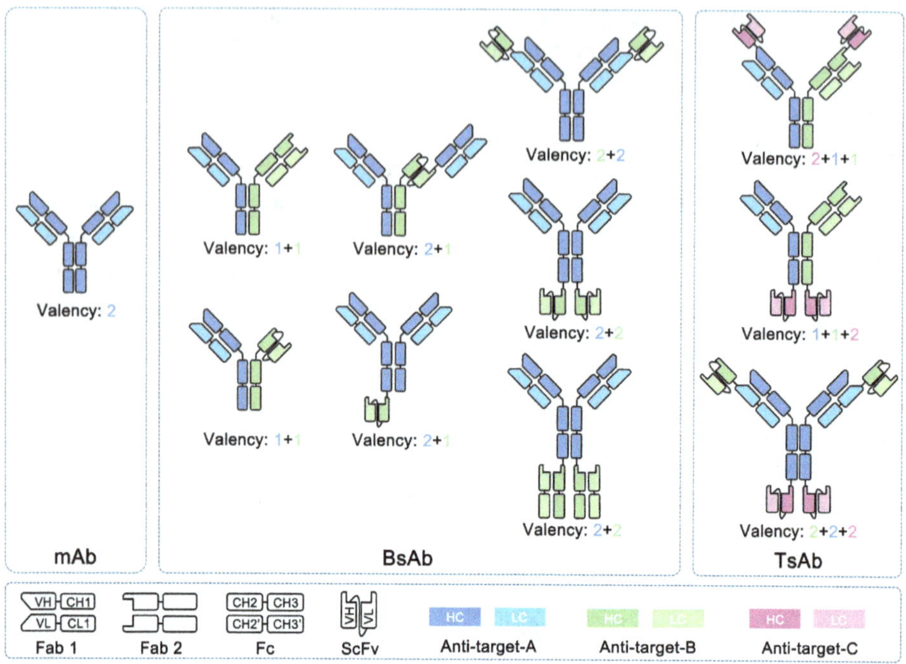

FIGURE 13.2 Schematic representation highlighting the binding specificity (shown in different colors) of monoclonal (mAb) and multispecific antibodies (MsAb), specifically bispecific antibodies (BsAbs) and trispecific antibodies (TsAb). Moreover, the number of repeats in building blocks (e.g., Fabs and ScFvs) directly correlates with the binding valency that these molecules can exhibit.

pharmacokinetics, and pharmacodynamics, to name a few, the leap to MsAbs introduces a new layer of complexity that challenges existing computational frameworks. MsAbs, designed to engage multiple targets simultaneously (Figure 13.2), demand a refined approach to address not only the challenges inherent in mAb optimization but also those unique to their multifaceted often new quaternary structure.

One of the primary challenges in MsAb design is managing the intricate relationship between the angle of approach that the warheads are required to adapt and the location of those same corresponding epitopes in the case of membrane-bound targets (Figure 13.3a). Moreover, prior understanding of the target copy number and its distribution on the surface of the cell will inform the design of molecules with the suitable valency that matches the surfaceoma (all extracellular domains of proteins anchored to the cell membrane) of the targeting cell (Figure 13.3a). This complexity significantly affects how these NMEs are assembled into a suitable quaternary structure with the right balance of steric hindrance, valency, and degrees of freedom. Important considerations that will predetermine developability and their therapeutic efficacy. Companies like Chugai [31], Amgen [32], and Genentech [33] have been at the forefront, exploring the nuances of epitope compatibility to ensure that the designed MsAbs can achieve the desired clinical outcomes without unintended cross-reactivity [31].

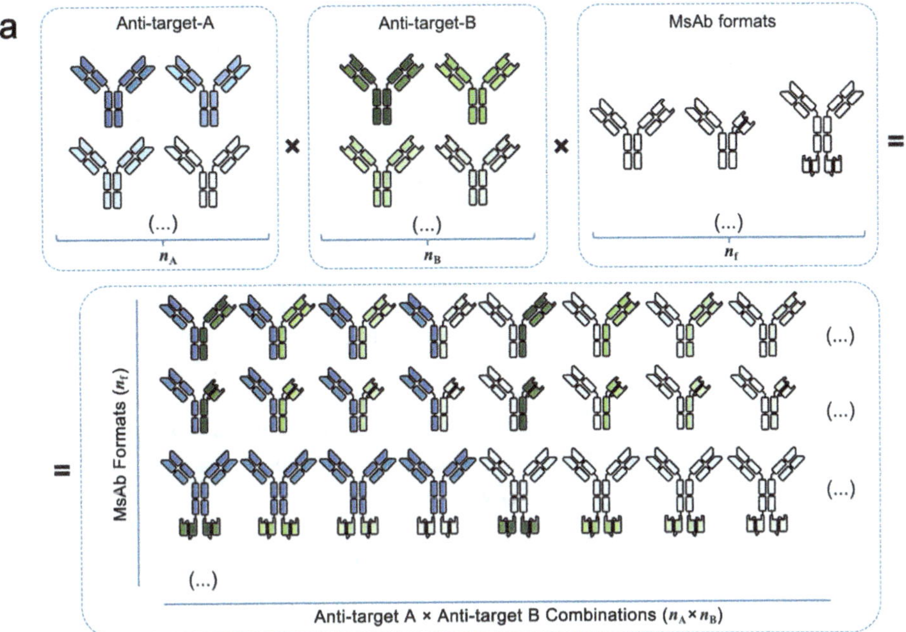

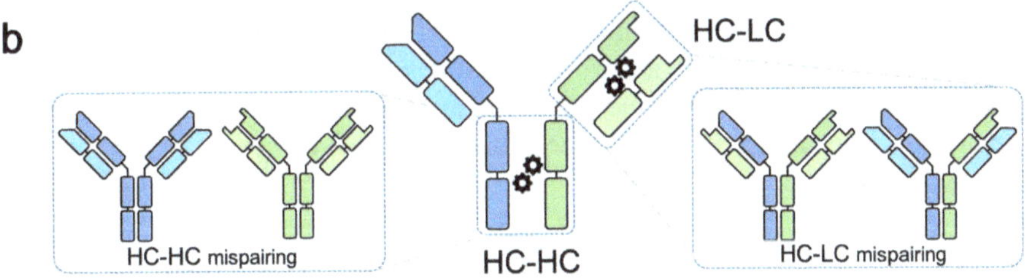

FIGURE 13.3 Schematic representation of the molecule assembly for MsAbs. (a) The combinatorial design approach for bispecific antibodies targeting antigens A and B. Multiple format variations (e.g., IgG-like, Fab, and scFv-based) are explored to optimize binding interactions and potential biological effects based on the location of target epitopes. (b) Assembly of MsAbs often requires engineering approaches to drive effective cognate pairing of the heavy (HC) and light (LC) chains when multiple chains are expressed in a single-cell host. Techniques like knob-in-hole (KiH) and charged pairing mutations (CPMs) (black wheel) have been developed to minimize the chain mispairing.

Moreover, the kinetics of binding presents another layer of complexity. The efficacy of an MsAb is not just a function of its ability to bind to multiple targets but also how these binding interactions influence each other in the context of their kinetics. The effects of binding kinetics (k_{ON} and k_{OFF}) can dramatically affect the clinical efficacy of the therapeutic molecule [34], necessitating sophisticated AI/ML models to predict and optimize these kinetic profiles. While this complexity has previously been addressed using a combination of surface plasmon resonance (SPR) measurements and ordinary differential equations (ODE) to understand bivalent interactions [35], coarse-grained simulations [36], or sophisticated analytic solutions [37], accurately estimating mutational effects on kinetics has remained challenging. Nonetheless, the critical importance of faster or slower binding rates on clinical efficacy cannot be overstated. Therefore, there is an urgent need for sophisticated AI/ML models to precisely predict and optimize these kinetic profiles, ensuring the development of MsAbs with the desired therapeutic properties.

The combinatorial explosion in screening potential building blocks (BB) combinations for MsAbs further underscores the need for AI/ML adaptation (Figure 13.3a). Indeed, in addition to the large number of BBs often generated for each target, we also need to consider the type of BBs (which can include fragment antigen-binding (Fab), single-chain variable fragments (scFvs), VHH, cytokines, and de novo binders) and its defining physicochemical properties [1]. Traditional approaches to mAb development, which already involve screening vast libraries of candidates, are exponentially complicated when each candidate can comprise multiple, distinct binding entities (Figure 13.4a). Indeed, the assembly of MsAbs comprising multiple polypeptide chains requires sophisticated engineering solutions to drive correct chain pairing, including the heavy chain-heavy chain and/or heavy chain-light chain (Figure 13.3b). Although several solutions have been reported, the chain pairing results in a more native-like quaternary arrangement, which translates into a better stability profile. Examples of such technologies are Genentech's knob-into-hole (KiH) [33] and Amgen's chain-pairing mutations (CPMs), already clinically validated and that are broadly applied in the MsAbs' design [38,39]. Moreover, the deployment of a common light chain [40] is also an elegant approach to the generation of MsAbs molecules in a single-cell expression system. Highly dependent on the success of the cognate chain pairing is the purification, where the appearance of sub-products (mispairing) often adds complexity while decreasing the total recovery of the targeted product. Nevertheless, it is paramount before functional characterization (Figure 13.4b) [41,42].

AI/ML can manage this complexity by efficiently navigating the combinatorial space, predicting optimal combinations that would be infeasible to identify through empirical methods alone.

Additionally, modeling interactions between the different entities within an MsAb molecule extends beyond the capabilities of many current computational tools designed for mAb. The ability to accurately predict how different parts within an MsAb will interact with each other and with multiple targets requires the development of new algorithms and models that can handle this increased complexity. For example, two given BBs, each with acceptable charge and hydrophobic patch distribution on their surface, could form super patches (e.g., hydrophobic areas exceeding 300 $Å^2$) once plugged in together, generating unexpected liabilities like aggregation [43]. In summary, the adaptation and refinement of AI/ML tools for MsAb design is a new frontier in therapeutic development. Addressing the unique challenges of MsAbs—ranging from epitope compatibility and kinetic profiling to the combinatorial explosion in screening and modeling inter-entity interactions—requires innovative computational strategies. As the field progresses, the evolution of these computational tools will undoubtedly play a pivotal role in unlocking the potential of MsAb, paving the way for next-generation therapeutics.

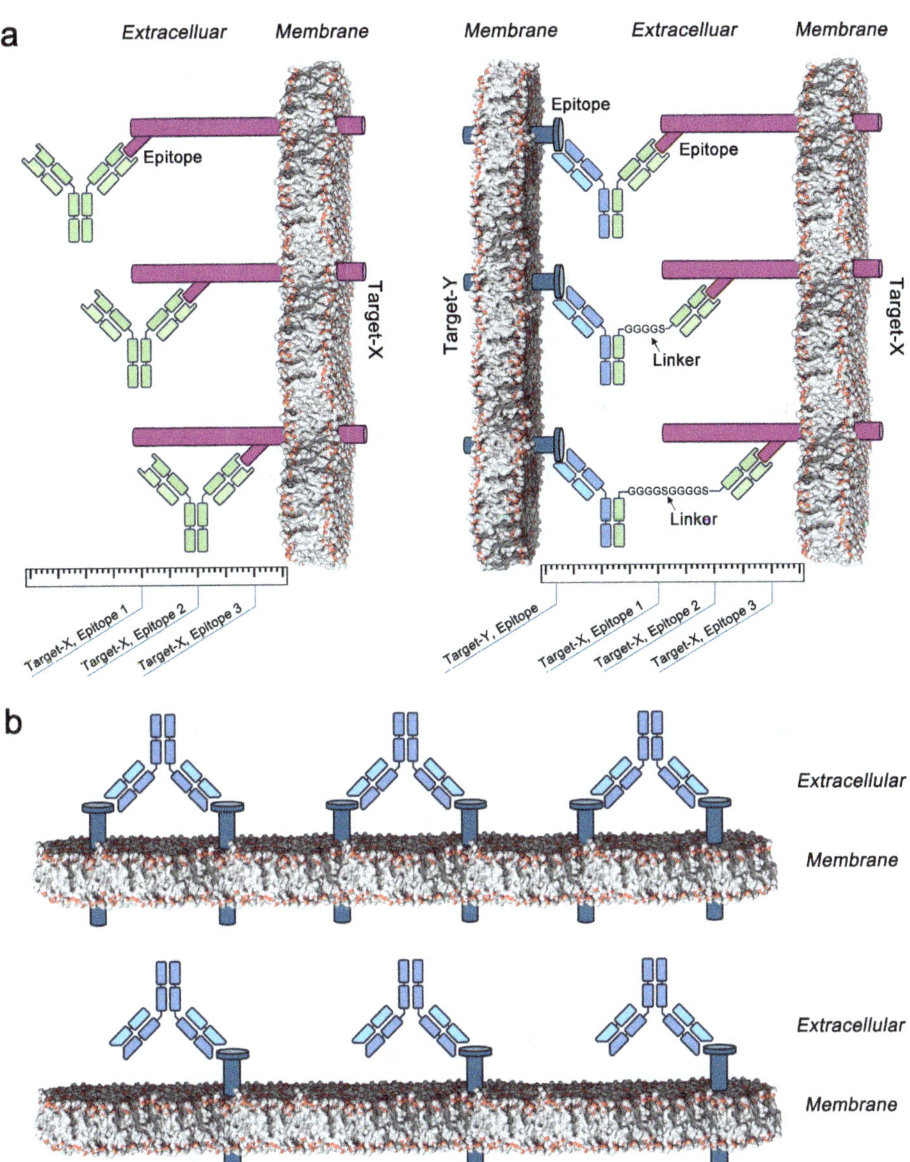

FIGURE 13.4 MsAb quaternary structure dictates binding properties, which directly influences biological activity. (a) Schematic representation highlighting the molecular basis of the relationship between the format rearrangements of MsAbs (e.g., linker length) and optimal biology. For example, in the case of T-cell engagers (and others), the location of the targeting epitope on the TAA (tumor-associated antigen) (membrane proximal vs. distal) can be critical for tumor cell killing. (b) A representation of the relationship between MsAbs' valency and the target copy number and distribution on the surface of the cell.

13.4 ADAPTING AI TO THE DESIGN OF MULTISPECIFIC ANTIBODIES

13.4.1 Structure Prediction and Modeling

The advent of AI/ML technologies will bring significant advancements to the design of MsAbs, enabling precise predictions of optimal formats, binding affinities, and stability. This will allow tailoring MsAb for specific indications and for each route of administration (ROA). These tools can model the complementarity of epitopes, aiding in the strategic placement of different antigen-binding sites to enhance therapeutic efficacy while minimizing the need for extensive screening. Despite increased complexity, AI-driven linker design promises optimal binding domain functionality and separation in mAbs, as demonstrated in small molecules [44]. The effect of the linker and its composition has been demonstrated for bispecific diabodies [45] and seen in single-chain Fvs [46] but has mostly been done using rational design of the linkers. This is crucial for ensuring that each binding arm of an MsAb can engage its target without interference from adjacent arms.

13.4.2 Developability Prediction and Optimization

Often, the requirements in terms of stability and cost of goods for MsAbs are similar to those of mAbs, which can be a tall order for these NMEs. Predicting and optimizing the developability of MsAbs presents unique challenges, given their complexity. A more suitable approach can be for the AI/ML models to first predict the critical attributes of each BB, including solubility, aggregation potential, thermostability, and yield, among others, and then later to the MsAb entity as a whole. Despite the scarcity of data on the combined properties that MsAbs can generate, AI-driven approaches can leverage existing knowledge from mAb development to inform predictions and optimizations. In addition, LLMs for biology like PGLM [15] can offer an exciting opportunity to fill in the gaps in the absence of prior knowledge and could have the potential to become foundation models over time when more data is integrated into them. This holistic view of developability ensures that the individual BBs and overall MsAb design meet the necessary criteria for successful development and production. By understanding and manipulating these molecular characteristics, AI-driven methods streamline the design process, reducing the complexity and resource requirements traditionally associated with developing MsAbs.

13.4.3 Virtual Screening and Lead Identification

AI/ML technologies can revolutionize virtual screening processes, enabling the efficient evaluation of vast libraries of antibody fragments. These tools can identify promising candidates by screening multiple structure complexes and optimizing the linkers that connect different binding entities. Through structural analysis, AI algorithms determine the best balance between flexibility and rigidity of these linkers, ensuring optimal orientation and functionality of the binding domains. This accelerates the discovery phase, allowing researchers to focus on the most promising MsAbs candidates for further development.

Data-driven design coupled with automation have revolutionized biologics development. By integrating the 'Design-Build-Test-Learn' cycle with rigorous design of experiments (DoE), researchers can leverage data to intelligently guide the generation of novel and rapidly improved therapeutic candidates [47–49]. Data-driven approaches are rapidly becoming more sophisticated, enabling faster design cycles and the integration of more parameters from both in silico and experimental assays with the convergence of experiments and algorithmic development. The integration of active learning algorithms like Bayesian optimization accelerates biologics optimization [50], paving the way for fully automatic protein engineering laboratories [51]. Indeed, elaborated automation efforts are rapidly accelerating this process, generating vast datasets in real time [52]. Robust data infrastructure, molecule registration for data curation, proper DoE-controlling repeats, references, and low-signal to noise and algorithms capable of identifying subtle patterns are crucial for translating data into actionable insights for ML [53]. Design-Build-Test-Learn cycles can be run in just under 2 weeks [41], enabling fast improvements in the molecule design (Figure 13.5).

13.4.4 In Silico Modeling and Simulation

AI/ML simulations are essential tools in MsAb design, offering insights into potential off-targets and efficacy. While the absence of off-target effects observed in mAbs is an essential pre-condition, it does not guarantee the same for multispecific constructs; therefore, careful modeling and real-world validation remain crucial. Meticulous screening of the human proteome with ML algorithms can predict off-target interactions, minimizing unwanted side effects for small molecules [54]. Additionally, electrostatic calculations can be used to predict specificity of mAbs, improving their precision [55]. Furthermore, AI models facilitate the exploration of optimal geometries and affinities between binding arms as has been explored for mAbs using MD simulations [56], ensuring that the MsAb can effectively engage multiple targets with high specificity and efficacy.

The complexity of determining optimal epitopes varies depending on the targets and their localization, whether they are membrane-bound or soluble where the complexity can increase as shown by a cytokine receptor using MD simulations [57]. These in silico approaches offer a powerful means to predict and refine the therapeutic potential of MsAbs before advancing to costly and time-consuming experimental stages.

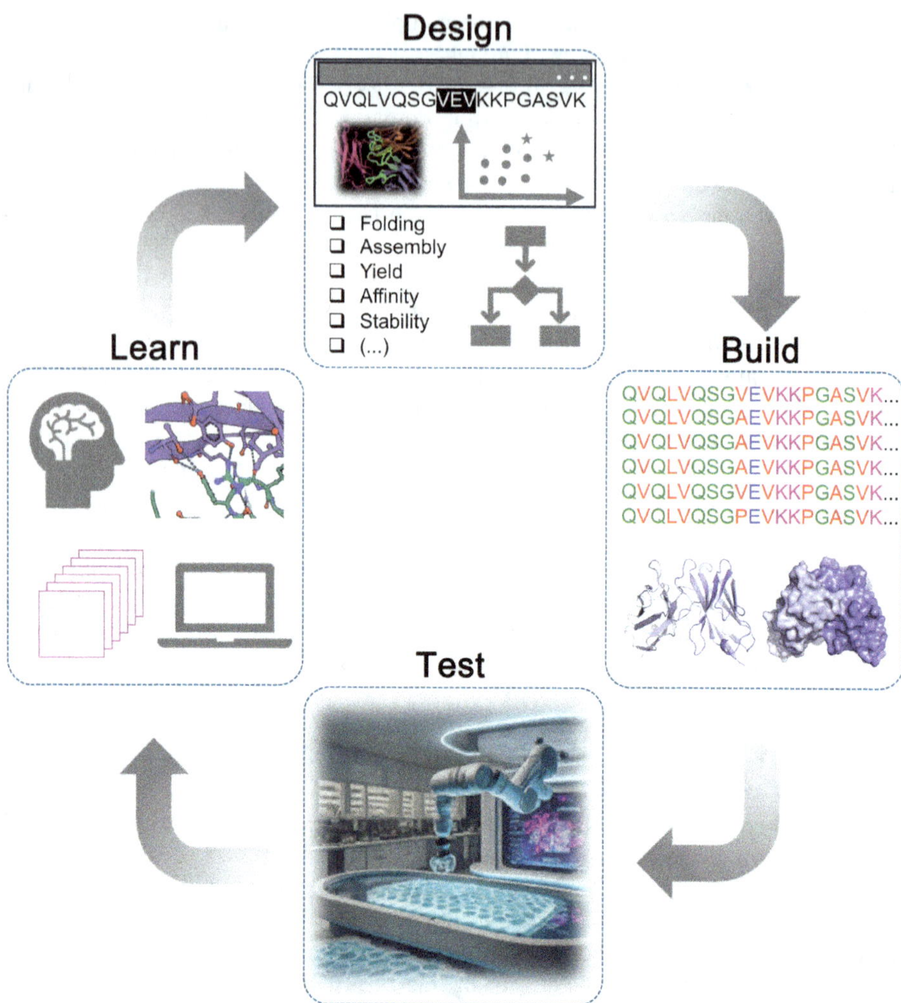

FIGURE 13.5 Optimizing biologics with the design-build-test-learn cycle. This iterative cycle is employed to optimize biologics such as mAbs using a data-driven approach. Scientists begin by designing the initial sequence, carefully considering features like affinity, yield, and stability that require improvement. The protein's properties are then analyzed to predict potential interactions and effectiveness. After this analysis, selected sequences are built and experimentally tested for these crucial features. The resulting data is fed back into an active learning loop, refining the design process in subsequent iterations. This cycle enables the development of highly specific and effective biologics for use in disease treatment, diagnostics, and therapeutic delivery.

13.5 THE FUTURE: BEYOND OPTIMIZATION

13.5.1 Market Trends and Commercialization

The advent of AI-driven MsAb development marks a revolutionary shift in the biopharmaceutical industry, offering new therapeutic strategies for a range of diseases previously deemed untreatable. Innovations in therapeutics for cancer, autoimmune disorders, and rare diseases, exemplified by breakthrough treatments like blinatumomab [32], Hemlibra [58], and Mim8 [48], underscore the transformative potential of MsAbs. These therapies have expanded the horizon of treatable conditions, leveraging the unique capability of MsAbs to address complex pathological mechanisms that single-targeted therapies cannot.

AI/ML technologies stand at the forefront of this transformation, poised to significantly increase the number of MsAbs by streamlining the development process. Through in silico screening, these technologies can sift through vast libraries of candidate molecules, a task that is impractical with traditional in vitro and in vivo methods due to the sheer number of potential combinations (Figure 13.3a). This computational approach not only accelerates the identification of viable candidates but also enhances the quality of leads, promising a future where MsAbs are developed faster, more cost effective, and more importantly, a future where more MsAbs reach the clinic and help patients in need.

13.5.2 Logic Gates, Biosensors, and De Novo Design

The field of logic gate-based antibody design and de novo mAb design represents an exciting frontier in antibody engineering. MsAbs introduce the possibility of embedding computational intelligence within therapeutic molecules. For example, logic gates such as AND, OR, NOR, NOT, XOR, and NAND can be engineered into MsAbs to refine their targeting capabilities and reduce off-target effects, thereby minimizing drug toxicity and increasing efficacy in the clinic [59] (Figure 13.6).

Similarly, other technologies also known as molecular switches responsive to alterations in pH and adenosine triphosphate (ATP) levels, among other biophysical properties, can also be engineered into MsAbs, which has been shown for mAbs [60,61]. This sophisticated level of control enables the design of MsAbs that are activated only in specific tissues or microenvironments, enhancing therapeutic precision and safety.

De novo design of antibodies using AI/ML offers huge potential. It allows for the generation of mAbs with highly specific epitopes and precise cognate chain pairing, which is ideal for the integration into MsAb formats that require chain pairing. However, other properties, including affinity and pharmacokinetics (PK)/pharmacodynamics (PD) characteristics, may still require further validation within the context of MsAbs. These properties are often influenced by quaternary structure [62]. Altogether, AI algorithms are crucial in this endeavor, enabling the prediction of affinity, optimizing molecule design for efficacy and safety, and generating sequences with enhanced manufacturability and stability.

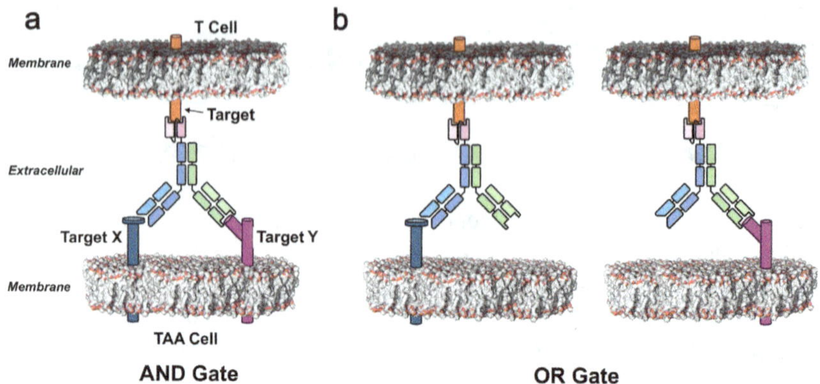

FIGURE 13.6 Illustration of MsAb formats designed to deliver on AND and OR gate mechanisms of conditional activation. The T-cell engager MsAb molecule binds to two antigens (Target X and Target Y) on the target cell (TAA), triggering a signaling pathway within the T cell. In the AND gate format (a), T-cell activation requires binding to both Target X and Target Y simultaneously. Alternatively, in the OR gate format (b), T-cell activation occurs if the MsAb binds to either Target X or Target Y.

13.5.3 Challenges and Opportunities

Despite the promising advancements, the integration of AI/ML into the design and engineering of MsAbs still presents several challenges. The optimal pairing of different BBs to achieve the desired therapeutic effect remains a complex task. However, the capabilities of AI/ML in de novo design and developability prediction should offer a pathway to overcome these obstacles. By refining sequences, e.g., removing liabilities and precisely predicting interactions, these technologies expedite the development of new therapies, significantly reducing time to clinic.

However, the prospect of accelerating the development of MsAbs hinges on the ability of AI/ML to refine and streamline the design process. As these technologies continue to evolve, they will enable the creation of more effective, safer, and more targeted therapies, opening new avenues for treating a wide range of diseases and delivering on the promise of personalized medicine.

13.6 CONCLUSION

MsAbs hold the promise to revolutionize treatment options, addressing limitations faced by mAbs. However, their complex structures demand advanced computational approaches like AI/ML, which have the potential to revolutionize MsAb design. When AI/ML reaches its full potential, it will dramatically accelerate the design and optimization of MsAbs. Embracing this technology shift is crucial for unlocking the full clinical potential of MsAbs.

ACKNOWLEDGMENTS

The authors are thankful to Dr. Sandeep Kumar's invitation to contribute to this book and for his additional guidance and review. Also, they extend their thanks to Leonard Wossnig for pre-reading and providing helpful suggestions.

REFERENCES

[1] D. Keri, M. Walker, I. Singh, K. Nishikawa, and F. Garces, "Next generation of multispecific antibody engineering," *Antib. Ther.*, vol. 7, no. 1, pp. 37–52, 2023, doi: 10.1093/abt/tbad027.

[2] A. F. Labrijn, M. L. Janmaat, J. M. Reichert, and P. W. H. I. Parren, "Bispecific antibodies: a mechanistic review of the pipeline," *Nat. Rev. Drug Discov.*, vol. 18, no. 8, pp. 585–608, 2019, doi: 10.1038/s41573-019-0028-1.

[3] U. Brinkmann and R. E. Kontermann, "Bispecific antibodies," *Science*, vol. 372, no. 6545, pp. 916–917, 2021, doi: 10.1126/science.abg1209.

[4] G. Fan, Z. Wang, M. Hao, and J. Li, "Bispecific antibodies and their applications," *J. Hematol.* Oncol., vol. 8, no. 1, p. 130, 2015, doi: 10.1186/s13045-015-0227-0.

[5] X. Lyu, Zhao, Q., Hui, J., Wang, T., Lin, M., Wang, K., et al., "The global landscape of approved antibody therapies," Antib. Ther., vol. 5, no. 4, pp. 233–257, 2022, doi: 10.1093/abt/tbac021.

[6] D. Sun, W. Gao, H. Hu, and S. Zhou, "Why 90% of clinical drug development fails and how to improve it?," *Acta Pharm. Sin.* B, vol. 12, no. 7, pp. 3049–3062, 2022, doi: 10.1016/j.apsb.2022.02.002.

[7] J. Jumper, Evans, R., Pritzel, A., Green, T., Figurnov, M., Ronneberger, O., et al., "Highly accurate protein structure prediction with AlphaFold," *Nature*, pp. 1–11, 2021, doi: 10.1038/s41586-021-03819-2.

[8] A. Radford, R. Jozefowicz, and I. Sutskever, "Learning to generate reviews and discovering sentiment," arXiv, 2017, doi: 10.48550/arxiv.1704.01444.

[9] S. Hochreiter and J. Schmidhuber, "Long short-term memory," *Neural Comput.*, vol. 9, no. 8, pp. 1735–1780, 1997, doi: 10.1162/neco.1997.9.8.1735.

[10] E. C. Alley, G. Khimulya, S. Biswas, M. AlQuraishi, and G. M. Church, "Unified rational protein engineering with sequence-based deep representation learning," *Nat. Methods*, vol. 16, no. 12, pp. 1315–1322, 2019, doi: 10.1038/s41592-019-0598-1.

[11] S. Biswas, G. Khimulya, E. C. Alley, K. M. Esvelt, and G. M. Church, "Low-N protein engineering with data-efficient deep learning," *Nat. Methods*, vol. 18, no. 4, pp. 389–396, 2021, doi: 10.1038/s41592-021-01100-y.

[12] A. Vaswani, Shazeer, N., Parmar, N., Uszkoreit, J., Jones, L., Gomez, A. N., et al., "Attention is all you need," *arXiv*, 2017, doi: 10.48550/arxiv.1706.03762.

[13] A. Rives, Meier, J., Sercu, T., Goyal, S., Lin, Z., Liu, J., et al., "Biological structure and function emerge from scaling unsupervised learning to 250 million protein sequences," *Proc. Natl. Acad. Sci. U. S. A*, vol. 118, no. 15, p. e2016239118, 2021, doi: 10.1073/pnas.2016239118.

[14] Z. Lin, Akin, H., Rao, R., Hie, B., Zhu, Z., Lu, W., et al., "Evolutionary-scale prediction of atomic-level protein structure with a language model," *Science*, vol. 379, no. 6637, pp. 1123–1130, 2023, doi: 10.1126/science.ade2574.

[15] B. Chen, Cheng, X., Li, P., Geng, Y.A., Gong, J., Li, S., et al., "xTrimoPGLM: unified 100B-scale pre-trained transformer for deciphering the language of protein," *arXiv*, 2024, doi: 10.48550/arxiv.2401.06199.

[16] A. Madani, Krause, B., Greene, E. R., Subramanian, S., Mohr, B. P., Holton, J. M., et al., "Large language models generate functional protein sequences across diverse families," *Nat. Biotechnol.*, vol. 41, no. 8, pp. 1099–1106, 2023, doi: 10.1038/s41587-022-01618-2.

[17] E. Nijkamp, J. A. Ruffolo, E. N. Weinstein, N. Naik, and A. Madani, "ProGen2: exploring the boundaries of protein language models," *Cell Syst.*, vol. 14, no. 11, pp. 968–978.e3, 2023, doi: 10.1016/j.cels.2023.10.002.

[18] I. Cherny, P. Greisen, Y. Ashani, S. D. Khare, G. Oberdorfer, H. Leader, et al., "Engineering V-type nerve agents detoxifying enzymes using computationally focused libraries," ACS Chem. Biol., vol. 8, no. 11, pp. 2394–2403, 2013.

[19] E. K. Makowski, P. C. Kinnunen, J. Huang, L. Wu, M. D. Smith, T. Wang, et al., "Co-optimization of therapeutic antibody affinity and specificity using machine learning models that generalize to novel mutational space," *Nat. Commun.*, vol. 13, no. 1, p. 3788, 2022, doi: 10.1038/s41467-022-31457-3.

[20] J. K. Leman, B. D. Weitzner, S. M. Lewis, J. Adolf-Bryfogle, N. Alam, et al., "Macromolecular modeling and design in Rosetta: recent methods and frameworks," *Nat. Methods*, vol. 17, no. 7, pp. 665–680, 2020, doi: 10.1038/s41592-020-0848-2.

[21] Molecular Operating Environment (MOE), 2024.0601 Chemical Computing Group ULC, 910-1010 Sherbrooke St. W., Montreal, QC H3A 2R7, 2024.

[22] J. Leem, J. Dunbar, G. Georges, J. Shi, and C. M. Deane, "ABodyBuilder: automated antibody structure prediction with data–driven accuracy estimation," *mAbs*, vol. 8, no. 7, pp. 1259–1268, 2016, doi: 10.1080/19420862.2016.1205773.

[23] Y. Wang, X. Gong, S. Li, B. Yang, Y. Sun, C. Shi, et al., "xTrimoABFold: de novo antibody structure prediction without MSA," *arXiv*, 2022, doi: 10.48550/arxiv.2212.00735.

[24] J. F. Sydow, F. Lipsmeier, V. Larraillet, M. Hilger, B., Mølhøj, M., et al., "Structure-based prediction of asparagine and aspartate degradation sites in antibody variable regions," *PLoS One*, vol. 9, no. 6, p. e100736, 2014, doi: 10.1371/journal.pone.0100736.

[25] P.-K. Lai, A. Gallegos, N. Mody, H. A. Sathish, and B. L. Trout, "Machine learning prediction of antibody aggregation and viscosity for high concentration formulation development of protein therapeutics," *mAbs*, vol. 14, no. 1, p. 2026208, 2022, doi: 10.1080/19420862.2022.2026208.

[26] E. K. Makowski, H. T. Chen, T. Wang, L. Wu, J. Huang, M. Mock, et al., "Reduction of monoclonal antibody viscosity using interpretable machine learning," *mAbs*, vol. 16, no. 1, p. 2303781, 2024, doi: 10.1080/19420862.2024.2303781.

[27] N. Kovalova, J. Boyles, Y. Wen, D. R. Witcher, P. L., Brown-Augsburger et al., "Validation of a de-immunization strategy for monoclonal antibodies using cynomolgus macaque as a surrogate for human," *Biopharm. Drug Dispos.*, vol. 41, no. 3, pp. 111–125, 2020, doi: 10.1002/bdd.2222.

[28] K. K. Jensen, M. Andreatta, P. Marcatili, S. Buus, J. A. Greenbaum, Z. Yan, et al., "Improved methods for predicting peptide binding affinity to MHC class II molecules," *Immunology*, vol. 154, no. 3, pp. 394–406, 2018, doi: 10.1111/imm.12889.

[29] T. D. Jones, L. J. Crompton, F. J. Carr, and M. P. Baker, "Therapeutic antibodies, methods and protocols," *Methods Mol.* Biol., vol. 525, pp. 405–423, 2008, doi: 10.1007/978-1-59745-554-1_21.

[30] D. Prihoda, J. Maamary, A. Waight, V., Fayadat-Dilman, L., Svozil, D., et al., "BioPhi: a platform for antibody design, humanization, and humanness evaluation based on natural antibody repertoires and deep learning," *mAbs*, vol. 14, no. 1, p. 2020203, 2022, doi: 10.1080/19420862.2021.2020203.

[31] T. Kitazawa, T. Igawa, Z. Sampei, A. Muto, T. Kojima, T. Soeda, et al., "A bispecific antibody to factors IXa and X restores factor VIII hemostatic activity in a hemophilia A model," *Nat. Med.*, vol. 18, no. 10, pp. 1570–1574, 2012, doi: 10.1038/nm.2942.

[32] K. Hagop, A. Stein, N. Gökbuget, A.K. Fielding, A.C. Schuh, J.M. Ribera, et al., "Blinatumomab versus chemotherapy for advanced acute lymphoblastic leukemia," *N. Engl. J. Med.*, vol. 376, no. 9, pp. 836–847, 2017, doi: 10.1056/nejmoa1609783.

[33] J. B. B. Ridgway, L. G. Presta, and P. Carter, "'Knobs-into-holes' engineering of antibody CH3 domains for heavy chain heterodimerization," *Protein Eng., Des. Sel.*, vol. 9, no. 7, pp. 617–621, 1996, doi: https://doi.org/10.1093/protein/9.7.617.

[34] G. Vauquelin, "Effects of target binding kinetics on in vivo drug efficacy: koff, kon and rebinding," *Br. J. Pharmacol.*, vol. 173, no. 15, pp. 2319–2334, 2016, doi: 10.1111/bph.13504.

[35] K. Nguyen, K. Li, K. Flores, G. D. Tomaras, S. M. Dennison, and J. M. McCarthy, "Parameter estimation and identifiability analysis for a bivalent analyte model of monoclonal antibody-antigen binding," *Anal.* Biochem., vol. 679, p. 115263, 2023, doi: 10.1016/j.ab.2023.115263.

[36] C. D. Michele, P. D. L. Rios, G. Foffi, and F. Piazza, "Simulation and theory of antibody binding to crowded antigen-covered surfaces," *PLoS Comput. Biol.*, vol. 12, no. 3, p. e1004752, 2016, doi: 10.1371/journal.pcbi.1004752.

[37] E. F. Douglass, C. J. Miller, G. Sparer, H. Shapiro, and D. A. Spiegel, "A comprehensive mathematical model for three-body binding equilibria," *J Am Chem Soc*, vol. 135, no. 16, pp. 6092–6099, 2013, doi: 10.1021/ja311795d.

[38] B. Estes, A. Sudom, D. Gong, D. A. Whittington, V. Li, C. Mohr, et al., "Next generation Fc scaffold for multispecific antibodies," iScience, vol. 24, no. 12, p. 103447, 2021, doi: 10.1016/j.isci.2021.103447.

[39] K. Gunasekaran, M. Pentony, M. Shen, L. Garrett, C. Forte, A. Woodward, et al., "Enhancing antibody fc heterodimer formation through electrostatic steering effects applications to bispecific molecules and monovalent IgG," *J. Biol. Chem.*, vol. 285, no. 25, pp. 19637–19646, 2010, doi: 10.1074/jbc.m110.117382.

[40] A. M. Merchant, Z. Zhu, J. Q. Yuan, A. Goddard, C. W. Adams, L. G. Presta, et al., "An efficient route to human bispecific IgG," *Nat. Biotechnol.*, vol. 16, no. 7, pp. 677–681, 1998, doi: 10.1038/nbt0798-677.

[41] D. Li, A. C. Partin, L. Zhao, I. Chen, M. L. Michaels, Z. Wang, et al., "Protocol for high-throughput cloning, expression, purification, and evaluation of bispecific antibodies," STAR Protoc., vol. 3, no. 2, p. 101428, 2022, doi: 10.1016/j.xpro.2022.101428.

[42] D. Gong, T. P. Riley, K. P. Bzymek, A. R. Correia, D. Li, C. Spahr, et al., "Rational selection of building blocks for the assembly of bispecific antibodies," *mAbs*, vol. 13, no. 1, p. 1870058, 2021, doi: 10.1080/19420862.2020.1870058.

[43] F. Waibl, M. L. Fernández-Quintero, A. S. Kamenik, J. Kraml, F. Hofer, H. Kettenberger, et al., "Conformational ensembles of antibodies determine their hydrophobicity," *Biophys. J.*, vol. 120, no. 1, pp. 143–157, 2021, doi: 10.1016/j.bpj.2020.11.010.

[44] J. Guo, F. Knuth, C. Margreitter, J. P. Janet, K. Papadopoulos, O. Engkvist, et al., "Link-INVENT: generative linker design with reinforcement learning," *Digit. Discov.*, vol. 2, no. 2, pp. 392–408, 2023, doi: 10.1039/d2dd00115b.

[45] F. L. Gall, U. Reusch, M. Little, and S. M. Kipriyanov, "Effect of linker sequences between the antibody variable domains on the formation, stability and biological activity of a bispecific tandem diabody," *Protein Eng. Des. Sel.*, vol. 17, no. 4, pp. 357–366, 2004, doi: 10.1093/protein/gzh039.

[46] J. M. Schanzer, K. Wartha, R. Croasdale, S. Moser, K. P. Künkele, C. Ries, et al., "A novel glycoengineered bispecific antibody format for targeted inhibition of epidermal growth factor receptor (EGFR) and insulin-like growth factor receptor type I (IGF-1R) demonstrating unique molecular properties," *J. Biol. Chem.*, vol. 289, no. 27, pp. 18693–18706, 2014, doi: 10.1074/jbc.m113.528109.

[47] S. Govindarajan, B. Mannervik, J. A. Silverman, K. Wright, D. Regitsky, U. Hegazy, et al., "Mapping of amino acid substitutions conferring herbicide resistance in wheat glutathione transferase," *ACS Synth. Biol.*, vol. 4, no. 3, pp. 221–227, 2015, doi: 10.1021/sb500242x.

[48] H. Østergaard, J. Lund, P. J. Greisen, S. Kjellev, A. Henriksen, N. Lorenzen, et al., "FVIIIa-mimetic bispecific antibody (Mim8) ameliorates bleeding upon severe vascular challenge in hemophilia A mice," *Blood*, 2021, doi: 10.1182/blood.2020010331.

[49] P. A. Romero, A. Krause, and F. H. Arnold, "Navigating the protein fitness landscape with Gaussian processes," *Proc. Natl. Acad. Sci. U. S. A*, vol. 110, no. 3, pp. E193–E201, 2013, doi: 10.1073/pnas.1215251110.

[50] Y. Zainchkovskyy, J. Ferkinghoff-Borg, A. Bennett, T. Egebjerg, N. Lorenzen, P. Greisen Jr, et al., "Probabilistic thermal stability prediction through sparsity promoting transformer representation," arXiv, 2022, doi: 10.48550/arxiv.2211.05698.

[51] J. T. Rapp, B. J. Bremer, and P. A. Romero, "Self-driving laboratories to autonomously navigate the protein fitness landscape," *Nat. Chem. Eng.*, vol. 1, no. 1, pp. 97–107, 2024, doi: 10.1038/s44286-023-00002-4.

[52] S. Kitano, C. Lin, J. L. Foo, and M. W. Chang, "Synthetic biology: learning the way toward high-precision biological design," *PLoS Biol.*, vol. 21, no. 4, p. e3002116, 2023, doi: 10.1371/journal.pbio.3002116.

[53] L. Wossnig, N. Furtmann, A. Buchanan, S. Kumar, and V. Greiff, "Best practices for machine learning in antibody discovery and development," *arXiv*, 2023, doi: 10.48550/arxiv.2312.08470.

[54] H. A. Guvenilir and T. Doğan, "How to approach machine learning-based prediction of drug/compound–target interactions," *J. Cheminform.*, vol. 15, no. 1, p. 16, 2023, doi: 10.1186/s13321-023-00689-w.

[55] N. Sinha, S. Mohan, C. A. Lipschultz, and S. J. Smith-Gill, "Differences in electrostatic properties at antibody–antigen binding sites: implications for specificity and cross-reactivity," *Biophys. J.*, vol. 83, no. 6, pp. 2946–2968, 2002, doi: 10.1016/s0006–3495(02)75302-2.

[56] A. Tiwari, A. K. Abraham, J. M. Harrold, A. Zutshi, and P. Singh, "Optimal affinity of a monoclonal antibody: guiding principles using mechanistic modeling," *AAPS J.*, vol. 19, no. 2, pp. 510–519, 2017, doi: 10.1208/s12248-016-0004-1.

[57] C.-Y. Yang, "Comparative analyses of the conformational dynamics between the soluble and membrane-bound cytokine receptors," Sci. Rep., vol. 10, no. 1, p. 7399, 2020, doi: 10.1038/s41598-020-64034-z.

[58] J. Mahlangu, J. Oldenburg, I. Paz-Priel, C. Negrier, M. Niggli, M. E. Mancuso, et al., "Emicizumab prophylaxis in patients who have hemophilia A without inhibitors," *N. Engl. J. Med.*, vol. 379, no. 9, pp. 811–822, 2018, doi: 10.1056/nejmoa1803550.

[59] R. C. Abbott, H. E. Hughes-Parry, and M. R. Jenkins, "To go or not to go? Biological logic gating engineered T cells," *J. Immunother. Cancer*, vol. 10, no. 4, p. e004185, 2022, doi: 10.1136/jitc-2021–004185.

[60] F. Mimoto, K. Tatsumi, S. Shimizu, S. Kadono, K. Haraya, M. Nagayasu, et al., "Exploitation of elevated extracellular ATP to specifically direct antibody to tumor microenvironment," Cell Rep., vol. 33, no. 12, p. 108542, 2020, doi: 10.1016/j.celrep.2020.108542.

[61] H. Muramatsu, T. Kuramochi, H. Katada, A. Ueyama, Y. Ruike, K. Ohmine, et al., "Novel myostatin-specific antibody enhances muscle strength in muscle disease models," *Sci. Rep.*, vol. 11, no. 1, p. 2160, 2021, doi: 10.1038/s41598-021-81669-8.

[62] Q. Liu, P. Acharya, M. A. Dolan, P. Zhang, C. Guzzo, J. Lu, et al., "Quaternary contact in the initial interaction of CD4 with the HIV-1 envelope trimer," *Nat. Struct. Mol.* Biol., vol. 24, no. 4, pp. 370–378, 2017, doi: 10.1038/nsmb.3382.

Index

Note: **Bold** page numbers refer to tables and *italic* page numbers refer to figures.

For Product Safety Concerns and Information please contact our EU representative GPSR@taylorandfrancis.com
Taylor & Francis Verlag GmbH, Kaufingerstraße 24, 80331 München, Germany

www.ingramcontent.com/pod-product-compliance
Lightning Source LLC
LaVergne TN
LVHW020603110826
845149LV00002B/366

* 9 7 8 1 0 3 2 2 9 1 6 8 0 *